SYNTHETIC MODULATED STRUCTURES

This is a volume in the
Materials Science and Technology series.
Editors: *A. S. Nowick and G. G. Libowitz*

A complete list of the books in the series appears at the end of the volume.

SYNTHETIC MODULATED STRUCTURES

Edited by

LEROY L. CHANG

IBM Thomas J. Watson Research Center
Yorktown Heights, New York

B. C. GIESSEN

Department of Chemistry
Northeastern University
Boston, Massachusetts

1985

ACADEMIC PRESS, INC.

(Harcourt Brace Jovanovich, Publishers)

Orlando San Diego New York London

Toronto Montreal Sydney Tokyo

ACADEMIC PRESS, INC.
Orlando, Florida 32887

United Kingdom Edition published by
ACADEMIC PRESS INC. (LONDON) LTD.
24–28 Oval Road, London NW1 7DX

Library of Congress Cataloging in Publication Data
Main entry under title:

Synthetic modulated structures.

Includes index.
1. Layer structure (Solids) 2. Superlattices as materials. I. Chang, Leroy L. II. Giessen, B. C.,
Date.
QD921.S944 1985 530.4'1 84-18397
ISBN 0-12-170470-X (alk. paper)

PRINTED IN THE UNITED STATES OF AMERICA

85 86 87 88 9 8 7 6 5 4 3 2 1

CONTENTS

LIST OF CONTRIBUTORS

Numbers in parentheses indicate the pages on which the authors' contributions begin.

TROY W. BARBEE, JR. (313) Department of Materials Science and Engineering, Stanford University, Stanford, California 94305

M. R. BEASLEY (365) Department of Applied Physics, Stanford University, Stanford, California 94305

LEROY L. CHANG (113) IBM Thomas J. Watson Research Center, Yorktown Heights, New York 10598

GOTTFRIED H. DÖHLER (163) Max-Planck-Institut für Festkörperforschung, 7000 Stuttgart 80, Federal Republic of Germany

L. ESAKI (3) IBM Thomas J. Watson Research Center, Yorktown Heights, New York 10598

CHARLES M. FALCO (339) Departments of Physics and of Optical Sciences and Arizona Research Laboratories, The University of Arizona, Tucson, Arizona 85721

A. C. GOSSARD (215) AT&T Bell Laboratories, Murray Hill, New Jersey 07974

A. L. GREER (419) Division of Applied Sciences, Harvard University, Cambridge, Massachusetts 02138

K. HESS (257) Electrical Engineering Research Laboratory and Coordinated Science Laboratory, University of Illinois at Urbana-Champaign, Urbana, Illinois 61801

N. HOLONYAK, JR. (257) Electrical Engineering Research Laboratory and Materials Research Laboratory, University of Illinois at Urbana-Champaign, Urbana, Illinois 61801

T. C. MCGILL* (77) Thomas J. Watson, Sr., Laboratory of Applied Physics, California Institute of Technology, Pasadena, California 91125

D. B. MCWHAN (43) AT&T Bell Laboratories, Murray Hill, New Jersey 07974

E. E. MENDEZ (113) IBM Thomas J. Watson Research Center, Yorktown Heights, New York 10598

A. PINCZUK (215) AT&T Bell Laboratories, Murray Hill, New Jersey 07974

KLAUS PLOOG (163) Max-Planck-Institut für Festkörperforschung, 7000 Stuttgart 80, Federal Republic of Germany

S. T. RUGGIERO† (365) Section of Applied Physics, Yale University, New Haven, Connecticut 06520

* Present address: Department of Applied Physics, California Institute of Technology, Pasadena, California 91125.

† Present address: Department of Physics, University of Notre Dame, Notre Dame, Indiana 46556.

IVAN K. SCHULLER (339) Materials Science and Technology Division, Argonne National Laboratory, Argonne, Illinois 60439

J. N. SCHULMAN* (77) Department of Physics and Astronomy, University of Hawaii, Honolulu, Hawaii 96822

F. SPAEPEN (419) Division of Applied Sciences, Harvard University, Cambridge, Massachusetts 02138

* Present address: Hughes Research Laboratories, Malibu, California 90265.

PREFACE

At the suggestion of Professor A. S. Nowick, we undertook to put together the present book on synthetic modulated structures as a volume in the Materials Science and Technology Series of Academic Press. The past decade has been marked by rapid progress in thin-film deposition techniques. This development has made it possible to prepare modulated structures with a high degree of material quality and interfacial smoothness. As a class of materials these structures exhibit properties not present in homogeneous solids, thereby providing an opportunity for the investigation of physical phenomena in a medium that can be tailored to prescription. Vast progress has been made in both semiconductors and metals. Activities in these two areas are among the most rapidly expanding fields of solid-state science.

The book is organized into three parts. Part I consists of two chapters. Chapter 1 serves as an introduction by providing a perspective of developments in the semiconductor area that have also affected the metal area. Chapter 2 deals with structural characterization with a coverage of both of these areas. Parts II and III are concerned with semiconductors and metals, respectively. The philosophies of organization in these two parts are different, a reflection of different interest in the two areas. In Part II the emphasis is on electronic properties in different configurations, starting with Chapter 3 on theoretical aspects. Chapters 4, 5, and 6 treat, in sequence, structures with compositional modulation, doping modulation, and a combination of the two; the preparation technique is described individually in each chapter. This part ends with Chapter 7 on laser properties. Part III begins with Chapter 8 on preparation methods. This is followed by three consecutive chapters focusing on distinctive fields of interest in metals: transport and magnetic properties in Chapter 9, superconductivity in Chapter 10, and diffusion in Chapter 11; this last chapter also includes a brief section on diffusion in semiconductors.

This volume is intended primarily as a reference book for workers in the general area of modulated structures; however, it contains sufficient tutorial material to serve as an introduction as well. The chapters have been written with the intent of covering the major points with plentiful references, so that interested readers can have access to more extensive or specific treatments. It is our hope that current workers in this area will find this book useful in offering new perspectives in their own field of specialization and that others will be stimulated to enter this exciting frontier area of research.

PART I

GENERAL CONSIDERATIONS

1 HISTORY AND PERSPECTIVE OF SEMICONDUCTOR SUPERLATTICES

L. Esaki

IBM Thomas J. Watson Research Center
Yorktown Heights, New York

I. INTRODUCTION

When the term *synthetic modulated structures* is broadly used to describe any periodically perturbed materials with a repetition greater than the basic unit cell dimension [e.g., see Cowley *et al.* (1979)], it may encompass a variety of multilayer structures including interference optical filters, x-ray diffraction gratings, and neutron monochromators that were frequently attempted in the past. One of the main interests in this chapter, however, is the investigation of electronic properties or quantum effects associated with such modulated structures of semiconductors and metals. Recent advances in thin-film techniques have demonstrated the feasibility of growing modulated structures that are sufficiently well defined to exhibit expected or unexpected quantum phenomena.

Research on a synthetic modulated semiconductor structure, referred to as a semiconductor superlattice, was initiated with a proposal in 1969–1970 by Esaki and Tsu (1969, 1970): a one-dimensional periodic structure consisting of alternating ultrathin layers, with its period less than the electron mean free path. In the insert of Fig. 1, such a superlattice structure is schematically shown. The

ISBN 0-12-170470-X

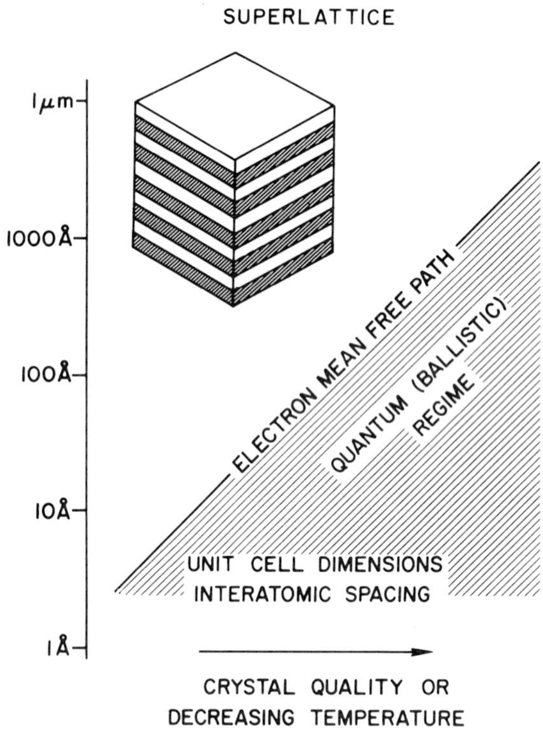

Fig. 1. Schematic illustration of a quantum regime (hatched) with a superlattice in the insert.

electron mean free path, an important parameter for the observation of quantum effects, depends heavily on crystal quality and also on temperatures, the effective mass, and so on. If characteristic dimensions such as a superlattice period and layer thicknesses (widths of potential wells or barriers) in semiconductor nanostructures are reduced to less than the electron mean free path, the entire electron system enters into a quantum regime with the assumption of the presence of ideal interfaces, as illustrated in Fig. 1. Our effort for the semiconductor superlattice is viewed as a search for novel phenomena in such a regime.

Although a similar idea can also be applied to modulated metal structures, novel effects to be researched for such structures are unlike those in the semiconductor case because of the existence of a large number of free electrons. In addition, metal structures appear to suffer more from interdiffusion between layers than do semiconductor structures. Therefore, this effect itself becomes an important subject for studies. In the case of superconductors, the critical parameter is obviously the coherence length in the superconducting state instead of the electron mean free path. Thus, the material quality is not as important as in the normal state.

The idea of the semiconductor superlattice occurred to us when examining the possible observation of resonant tunneling through double and multiple barriers. Such resonant tunneling arises from the interaction of electron waves with potential barriers (Bohm, 1951). The deBroglie wavelength λ is calculated to be 50 and 100 Å for electrons with an effective mass m^* of $0.1m_0$ and energies E of 0.32 and 0.08 eV, respectively, from the equation $\lambda = h(2m^*E)^{-1/2}$, where h is Planck's constant. In Fig. 2, the transmission coefficient T^*T is plotted as a function of electron energy for a double barrier. As shown in the insert of the figure, if the energy of incident electrons coincides with the resonant bound states within the potential well, the electrons tunnel through both barriers without attenuation. Such unity transmissivity arises because, for the electron waves of the resonant energies E_1 and E_2, the reflected waves from inside interfere destructively with those from outside, so that only transmitted waves remain (Bohm, 1951).

In this article, we first describe highlights during the early period (1969–1974) when most of the superlattice studies, including the structure synthesis by molecular beam epitaxy (MBE), were being carried out in our group (Esaki 1974a,b). This pioneering work is believed to have provided the foundation for subsequent progress, which proceeded rather rapidly in scope as well as in depth, stimulating explorations on quantum wells and other heterostructures.

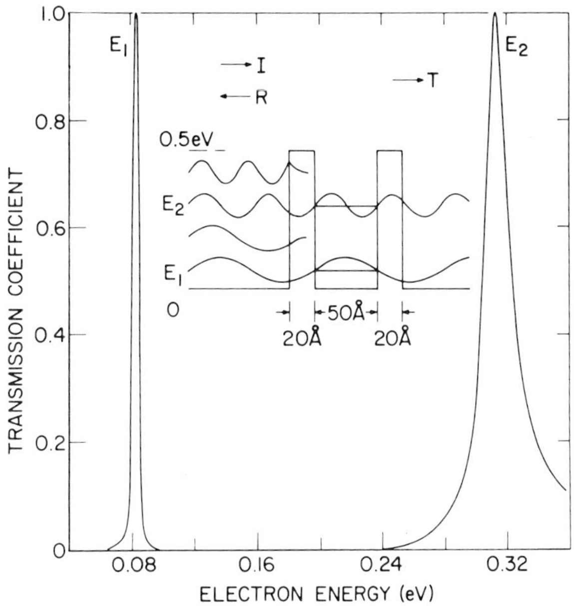

Fig. 2. Transmission coefficient T^*T versus electron energy for a double barrier shown in the insert, where, if the energy of incident electrons coincides with those of the bound states E_1 and E_2, the electrons tunnel through both barriers without attenuation.

Our early efforts focused on transport properties in the direction of the one-dimensional periodic potential. Later, however, it was noticed that optical studies such as absorption, luminescence, lasers, magnetoabsorption and inelastic light scattering, and magnetoquantum transport of the two-dimensional electron gas system in the layer plane prevailed. Such expanded activities resulted in new discoveries and inventions, leading to further proliferation of heterostructures in general. Some of the most significant developments in recent years will be itemized in Section III.

II. EARLY PERIOD

A. Proposal of a Semiconductor Superlattice

Esaki and Tsu (1970) envisioned two types of synthesized superlattices, doping and compositional, as shown in Fig. 3a,b, respectively, where, in either

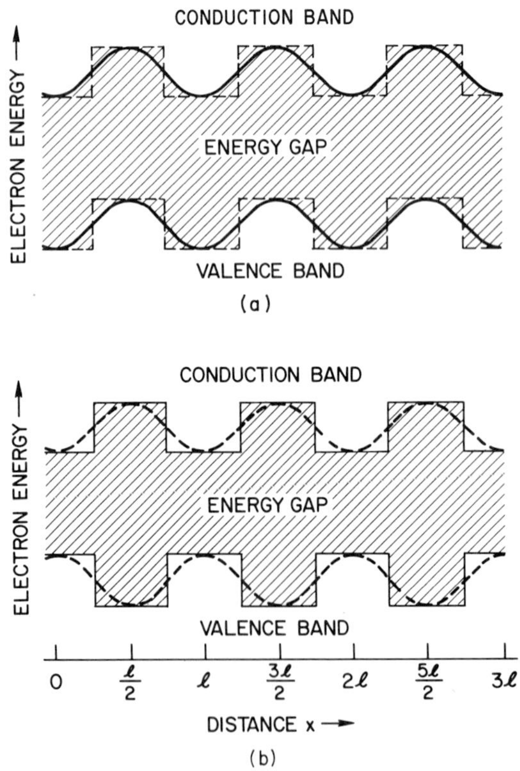

Fig. 3. Spatial variation of the conduction and valence band edges in two types of superlattices: (a) a doping superlattice of alternating *n*- and *p*-type layers; (b) a compositional superlattice with alternation of crystal composition.

case, a superlattice potential was introduced by a periodic variation of impurities or composition during epitaxial growth. The techniques of thin films, then, were rapidly advancing. It was theoretically shown that such a synthesized structure possesses unusual electronic properties (not seen in the host semiconductors) arising from predetermined quantum states of two-dimensional character.

The introduction of the superlattice potential clearly perturbs the band structure of the host materials: The degree of such perturbation depends on its amplitude and periodicity. Since the superlattice period is usually much greater than the original lattice constant, the Brillouin zone is divided into a series of minizones giving rise to narrow allowed subbands E_1 and E_2 separated by forbidden regions in the conduction band of the host crystal, as shown in Fig. 4a. Thus, this results in a highly perturbed energy–wave vector relationship for conduction electrons, as schematically illustrated in Fig. 4b. Figure 5 shows the density of states $\rho(E)$ for electrons in a superlattice in the energy range including the first three subbands: E_1 between (a) and (b), E_2 between (c) and (d), and E_3

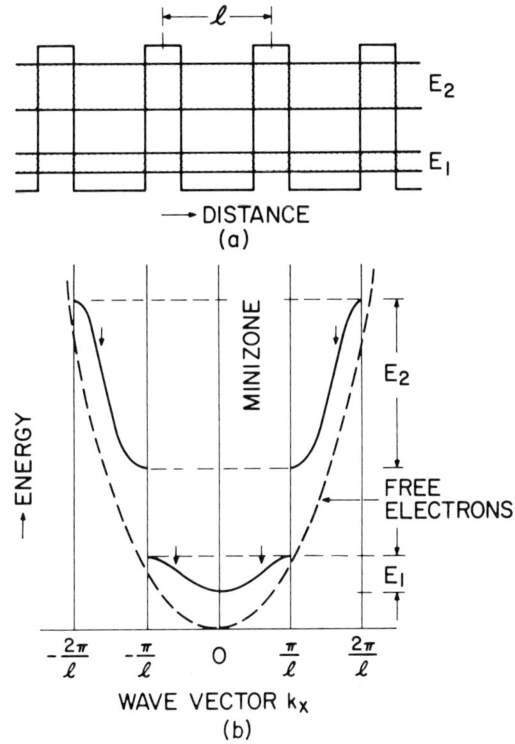

Fig. 4. (a) Potential profile of a superlattice and (b) its energy–wave vector relationship in the minizones.

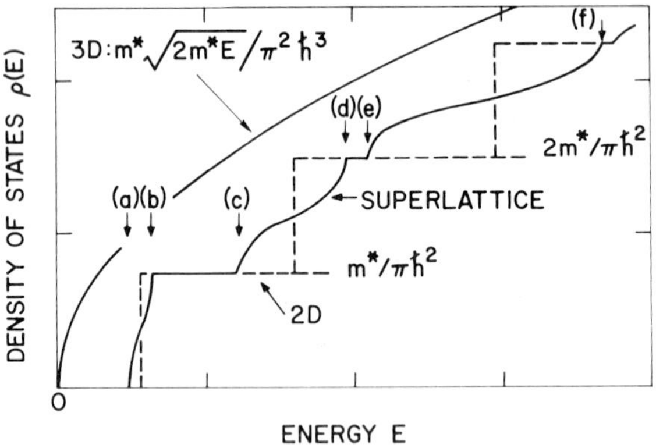

Fig. 5. Comparison of the density of states for the three-dimensional and two-dimensional electron systems with that of a superlattice.

between (e) and (f) (indicated by arrows in the figure), in comparison with the parabolic curve for the three-dimensional electron system and the staircaselike density of states for the two-dimensional system.

Although the situation is analogous to that for the Kronig–Penney band model (Kronig and Penney, 1930), it is seen here that the familiar properties are observed in a new domain of physical scale. In an extreme case, where quantum wells are sufficiently apart from each other, allowed bands become discrete states, and then electrons are completely two-dimensional, in which case the density of states is illustrated by the dashed line in Fig. 5.

The allowed subbands in a superlattice can be calculated from the following expression, assuming a one-dimensional, periodic square-well potential (Gol'dman and Krivchenokov, 1961),

$$-1 \le \cos\left(\frac{a(2mE)^{1/2}}{\hbar}\right)\cosh\left(\frac{b[2m(V-E)]^{1/2}}{\hbar}\right) + \left(\frac{V}{2E}-1\right)\left(\frac{V}{E}-1\right)^{-1/2}\sin\left(\frac{a(2mE)^{1/2}}{\hbar}\right)\sinh\left(\frac{B[2m(V-E)]^{1/2}}{\hbar}\right) \le 1,$$

where E is the electron energy in the superlattice direction, V the barrier height (0.4 V), a the well width, b the barrier width, and m the effective mass ($0.1m_0$). Assuming equal width ($a = b$), the allowed bands E_1, E_2, E_3, and E_4 calculated as a function of the width are shown with the hatched regions in Fig. 6, including discrete energy levels for a single square well with the width a.

The dynamics of conduction electrons in such narrow bands was analyzed: A simplified path integration method (Chambers, 1952) was used to obtain a

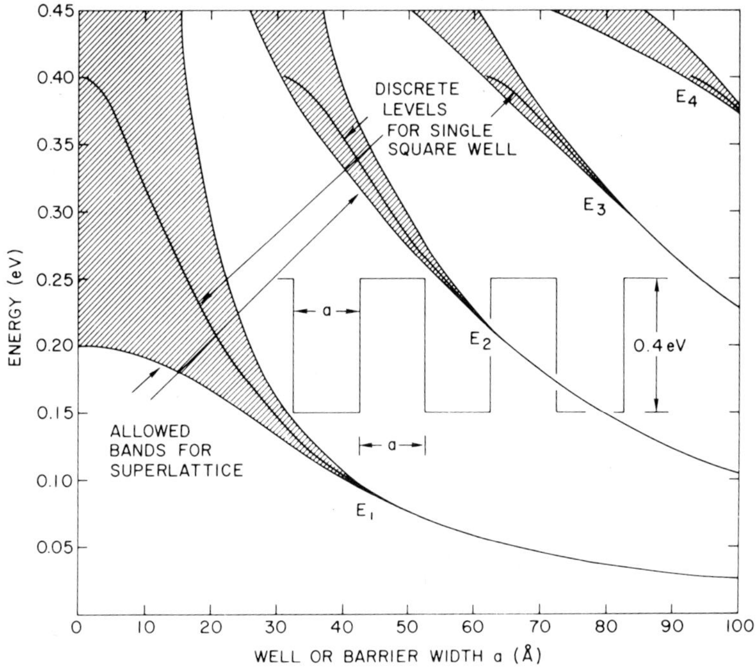

Fig. 6. Allowed energy bands E_1, E_2, E_3, and E_4 (hatched) calculated as a function of well or barrier width a, including the potential profile of a superlattice. Discrete energy levels for a single square well are also shown.

relation between the applied field F and the average drift velocity v_d. The equations of motion are

$$\hbar\, \partial k_x/\partial t = eF \qquad \text{and} \qquad v_x = \hbar^{-1}\, \partial E_x/\partial k_x;$$

the velocity increment in a time interval dt is

$$dv_x = eF\hbar^{-2}(\partial^2 E_x/\partial k_x^2)\, dt.$$

The average drift velocity, taking into account the scattering time τ, is written as

$$\begin{aligned} v_d &= \int_0^\infty \exp(-t/\tau)\, dv_x \\ &= eF\hbar^{-2} \int_0^\infty (\partial^2 E_x/\partial k_x^2) \exp(-t/\tau)\, dt. \end{aligned}$$

If a sinusoidal approximation is used for the E_x–k_x relationship,

$$v_d = \frac{eF\tau}{m^*(0)}\left[1 + \left(\frac{eF\tau l}{\hbar}\right)^2\right]^{-1};$$

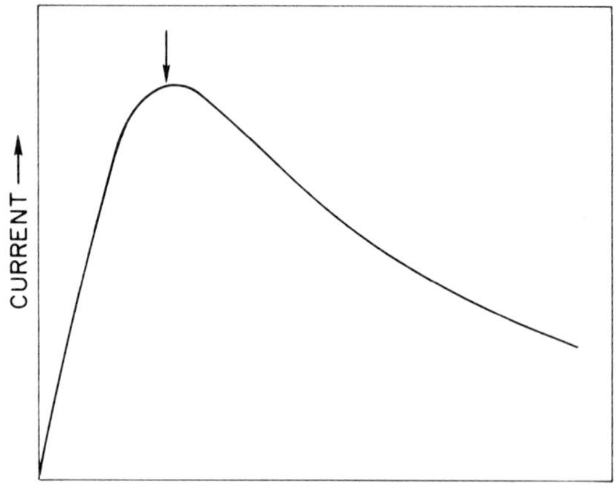

Fig. 7. A theoretical current–field (voltage) characteristic.

thus the current–field curve is given by

$$J = env_{\mathrm{d}},$$

where $m^*(0)$ is determined by the curvature of $E_x(k_x)$ at $k_x = 0$ and n the electron concentration. The current J, plotted as a function of F in Fig. 7, has the maximum at $eF\tau l/\hbar = 1$ (indicated by an arrow) and thereafter decreases, corresponding to a decreasing average drift velocity and giving rise to a differential negative resistance. This result indicates that if in applying modestly high electric fields conduction electrons gain enough energies to reach beyond inflection points (indicated by arrows in Fig. 4), they will be decelerated rather than accelerated by such electric fields.

It is worthwhile mentioning that Gnutzmann and Clauseker (1974) pointed out an interesting possibility: namely, the occurrence of a direct-gap superlattice made of indirect-gap host materials because of Brillouin-zone folding as a result of the introduction of the new superlattice periodicity, which was later re-examined by Madhukar (1982). The idea may suggest the synthesis of new optical materials.

B. Experimental Efforts

Esaki *et al.* (1970) reported the first experimental result on a superlattice synthesized by Blakeslee and Aliotta (1970) with the chemical vapor deposition (CVD) technique. The structure was obtained by a periodic variation of the phosphorus content x in the $GaAs_{1-x}P_x$ system. The period was made as thin as 200 Å with x varying from 0–0.1 to 0.5. Although the superlattice formation was clearly demonstrated by electron microscopy, x-ray diffraction, and cathode luminescence, transport measurements failed to show any predicted quantum effect. In this system, a relatively large lattice mismatch, 1.8%, between GaAs and $GaAs_{0.5}P_{0.5}$ inevitably generates a strain at the interfaces.

It was recognized at the beginning that, while the structure was undoubtedly of considerable theoretical interest, the formation of such a refined layer structure would be a formidable task. Nevertheless, the proposal of a semiconductor superlattice inspired a numer of materials scientists. In addition to the above-mentioned effort, Cho (1971) and Woodall (1972) demonstrated that it was feasible to produce alternating layers of lattice-matched GaAs and $Ga_{1-x}Al_xAs$ with the MBE and LPE techniques, respectively, while it is now well known that the MBE technique prevails. In 1972, we established a MBE system especially designed for the superlattice structure, which led to the successful growth of a GaAs–GaAlAs superlattice (Chang *et al.*, 1973a,b). Esaki *et al.* (1972) found that this superlattice exhibited a negative resistance in its transport properties, which was, for the first time, interpreted on the basis of the predicted quantum effect. Figure 8 shows the current–voltage characteristic at room temperature for such a superlattice having 100 periods, with each period consisting of a GaAs well 60 Å thick and a $Ga_{0.5}Al_{0.5}As$ barrier 10 Å thick. The structure was grown between two GaAs layers doped to an electron concentration of 5×10^{17} cm^{-3} on a similarly doped substrate. The electron concentration in the superlattice region is estimated to be approximately 10^{16} cm^{-3}. The curve trace of Fig. 8 (compare to Fig. 7) was obtaind from a 100-ns pulse measurement with a repetition rate of 10 kHz. The samples were evaluated with various techniques including HEED, He-ion backscattering, and Raman scattering. The

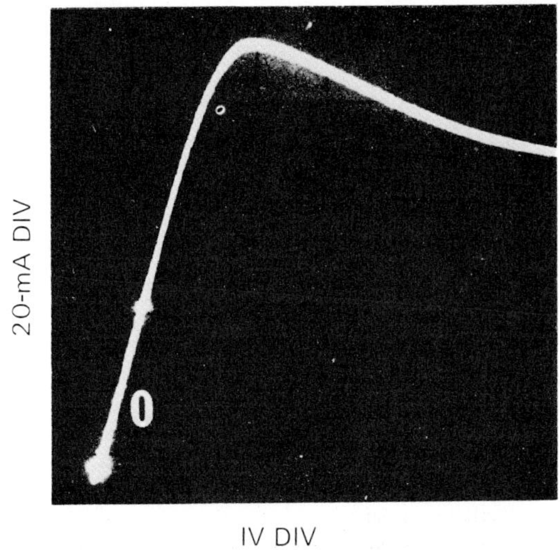

Fig. 8. Current–voltage characteristic for a 70-Å period superlattice.

compositional profile of such a structure was directly verified by the simultaneous use of ion sputter etching of the sample surface and Auger electron spectroscopy (Ludeke *et al.,* 1974).

C. Multibarrier Tunneling

In 1973, Tsu and Esaki computed the transport properties from the tunneling point of view, leading to the derivation of $I-V$ curves. The superlattice band model previously presented assumes an infinite periodic structure. In reality, however, not only a finite number of periods is prepared with alternating epitaxy, but also the electron mean free path is limited. In addition, two terminal electrodes for transport measurements, which create interfaces between the superlattice and such electrodes, are unavoidable. The potential profile of the realistic system is schematically illustrated in Fig. 9, in which (a) and (b) are an n-period superlattice of a finite length l used for calculations of the reflection R and transmission T amplitudes and the tunnel current J as a function of applied voltage V, respectively. Assuming the periodic structure made of GaAs and GaAlAs layers, the barrier height and width and the well width are taken to be 0.5 eV, 20 Å, and 50 Å, respectively, for calculations. The electrodes are made of n-GaAs layers, of which the Fermi energy is indicated as E_f in the figure. Incidentally, one will notice in this system that one-dimensional localization results from the fact that the two terminal electrodes serve as reservoirs that randomize the phase of electron waves.

The three-dimensional Schrödinger equation for the one-dimensional pe-

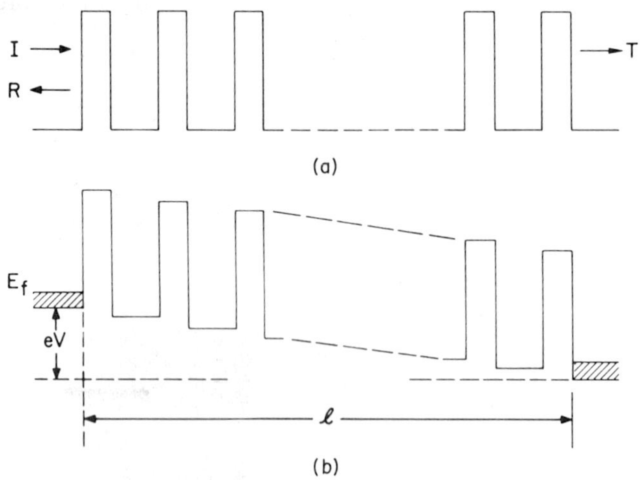

Fig. 9. (a) Finite superlattice of length l, where I, R, and T are the incidence, reflection, and transmission amplitudes; (b) potential profile with the application of a voltage V.

riod potential $U(x)$, where x is the direction of the multibarrier, can be separated into transverse and longitudinal parts; namely, the total energy E is written by the sum of the longitudinal and transverse energies

$$E = E_l(U) + \hbar^2 k_t^2/2m^*,$$

and the wave function is expressed by the product

$$\psi = \psi_l \cdot \psi_t.$$

The electron wave functions in the left and right contacts are

$$\psi_1 = \psi_l[\exp(ik_1 x) + R\exp(-ik_1 x)] \qquad \text{and} \qquad \psi_n = \psi_t[T\exp(ik_n x)],$$

respectively. By matching wave functions and their first derivatives at each interface, one can derive the reflection and transmission amplitudes, R and T. In Fig. 10, ln T^*T is plotted as a function of electron energy for a double, a

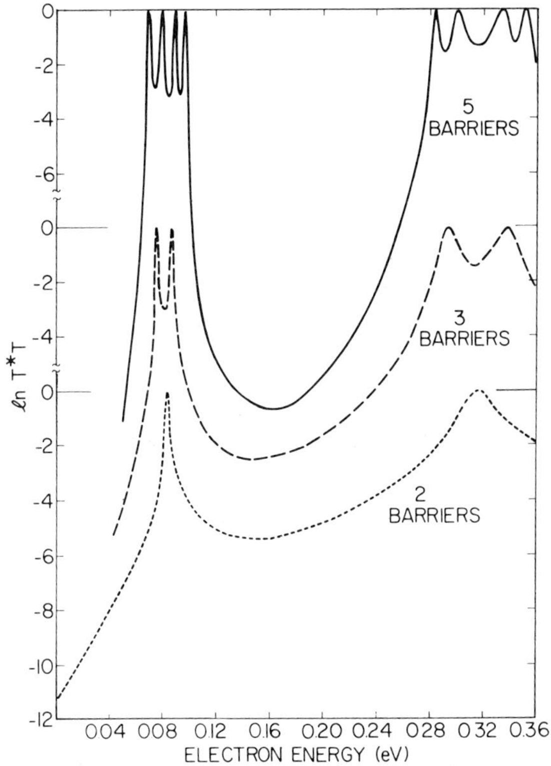

Fig. 10. Plot of ln T^*T (transmission coefficient) versus electron energy showing peaks at the energies of the bound states in the potential wells. The curves labeled 2 barriers, 3 barriers, and 5 barriers correspond to one, two, and four wells, respectively. [From Tsu and Esaki (1973).]

triple, and a quintuple barrier structure. Note that the resonant energies for the triple-barrier (double-well) case are split into doublets and that those for the quintuple barrier (quadruple-well) case are split into quadruplets. In the double-well case, each single-well quantum state is split into a symmetric and an antisymmetric combination. This is extended to n wells. In the large n limit, a bandlike condition is ultimately achieved, which corresponds to the superlattice band model.

Because of a separation of variable, the transmission coefficient is only a function of electron energy in the longitudinal direction. The tunnel current J, integrated over the transverse direction, is given by

$$J = \frac{em^*kT}{2\pi^2\hbar^3} \int_0^\infty T^*T \ln\left(\frac{1 + \exp[(E_f - E_l)/kT]}{1 + \exp[(E_f - E_l - eV)/kT]}\right) dE_t,$$

where E_l is the electron energy in the longitudinal direction. In the low-temperature limit, the above expression becomes

$$J = \frac{em^*}{2\pi^2\hbar^3} \int_0^{E_f} (E_f - E_t)T^*T \, dE_l, \qquad V \geq E_f,$$

and

$$J = \frac{em^*}{2\pi^2\hbar^3}\left[V \int_0^{E_f - V} T^*T \, dE_l + \int_{E_f - V}^{E_f} (E_f - E_t)T^*T \, dE_l\right], \qquad V < E_f.$$

We believe that this multibarrier tunneling model provided a useful insight into the transport mechanism and laid the foundation for the following experimental investigations.

D. Search for Quantum States

In early 1974, Chang *et al.* (1974a) observed resonant tunneling in double barriers, and subsequently Esaki and Chang (1974) measured quantum transport properties for a superlattice having a tight-binding potential.

The $I-V$ and $dI/dV - V$ characteristics are shown in Fig. 11 for a double barrier with a well of 50 Å and two barriers of 80 Å made of $Ga_{0.3}Al_{0.7}As$. The structure is completed by having two outside GaAs elctrodes doped to 10^{18} cm^{-3}: the substrate on one side and a top overgrowth on the other. The schematic energy diagram is shown in the inset, in which the first two quantized levels are indicated.

The resonance, in this case, is achieved under applied voltages by aligning the Fermi level of the electrode with the bound quantum states. The current should show peaks at voltages equal to twice the energies of the bound quantum states referred to the original conduction band edge in the well. The situation is shown as a and c in the energy diagrams and is also marked correspondingly in

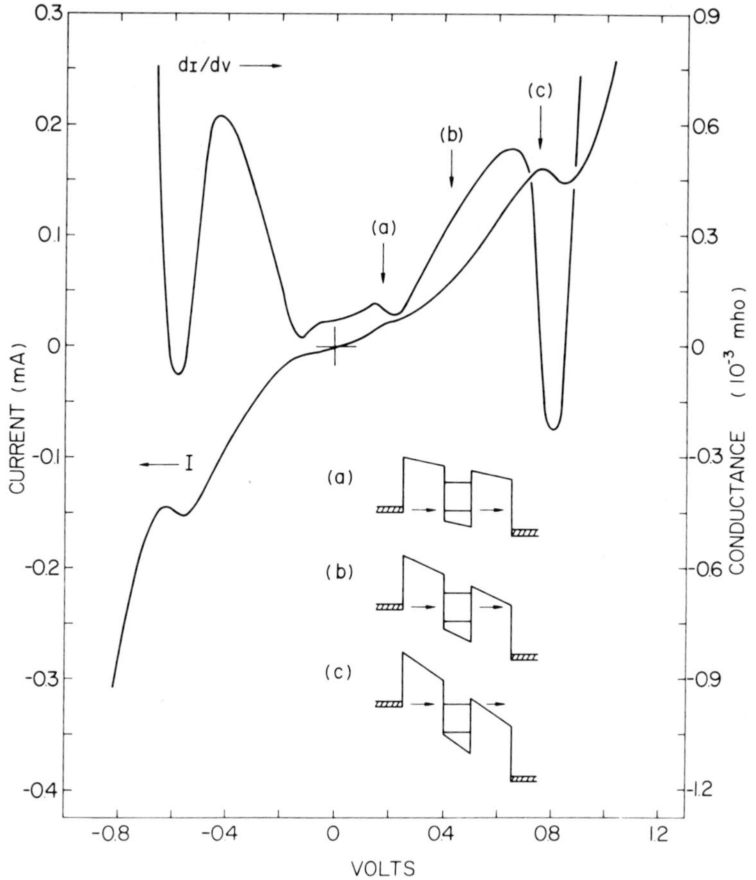

Fig. 11. Current–voltage and conductance–voltage characteristics of a double-barrier structure at 77 K. Conditions at resonance (a) and (c) and off-resonance (b) are indicated.

the experimental curves in Fig. 11, where two sets of singularities are clearly visible. The current at the lower level shows only a slight hump, but that at the higher level shows a pronounced peak and thus a negative resistance. The situation at the opposite voltage polarity is similar except that the singularities occur at somewhat lower values; the asymmetry is believed to be due to nonidentical barriers or an asymmetric barrier shape. The broadness of the singularities comes mainly from the fluctuations in both thicknesses and potentials that may exist in the formation of the double-barrier structure. A large number of double barriers were made with three different well widths, 40, 50, and 65 Å. Although there is some understandable spread, measured values for the quantized levels have been verified to coincide with the calculated ones. This agreement undoubtedly has confirmed the observation of resonant tunneling and, consequently, the energies of the bound quantum states.

Transport properties were investigated on a superlattice that comprises 50 periods, with each period consisting of a GaAs well 45 Å thick and an AlAs well 40 Å thick. The electron concentrations in the superlattice region and contacting GaAs regions are approximately 10^{17} and 10^{18} cm^{-3}, respectively. Because of a tight-binding potential, the conduction bandwidth is estimated to be as narrow as 5 meV in comparison with 40 meV for the previous superlattice (Esaki *et al.*, 1972) presented in Fig. 8. The $I-V$ and $dI/dV-V$ characteristics of this superlattice are shown in Fig. 12. The current at low voltages will be carried by a marginal band-conduction mechanism. As the voltage is increased to a value such that the drop across each period is comparable with the bandwidth, the band conduction cannot be maintained. This decrease of overlapping of the wave functions reduces the current, resulting in a negative differential resistance. With further increase in applied voltage, electrons will proceed by tunneling across a newly created, localized high field domain. Slight nonuniformity in

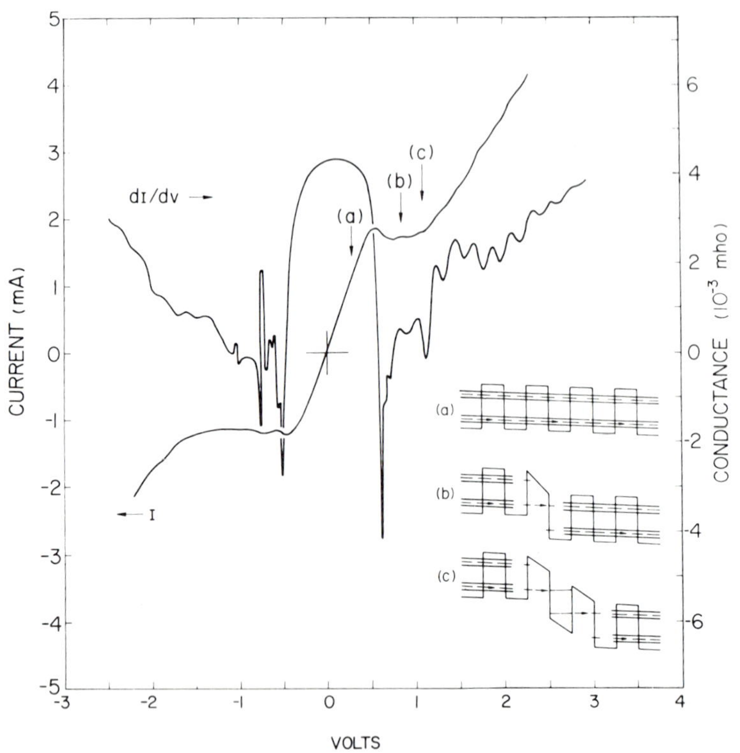

Fig. 12. Current–voltage and conductance–voltage characteristics of a superlattice at 77 K. Band-type conduction (a) and localized tunneling conduction (b) and (c) are indicated.

the shape of barriers or wells will lead to domain formation, which is indeed inherent in media having a voltage-controlled negative resistance. Once the first domain is formed, the tunneling process across it limits the current and keeps band-type conduction alive in other regions. This situation is maintained until the first band and the second band across this domain become aligned. A current maximum then occurs and the domain expands. The cases marked by b and c in Fig. 12 illustrate the events. The conductance in the negative resistance region is complicated by the association of instabilities. At higher applied voltages, however, a well-behaved oscillatory phenomenon is seen with a period of 0.24 V. As expected, this value agrees with the energy difference between the first and second allowed bands, calculated to be 0.24 eV for the present superlattice.

These experiments probably constitute the first clear observation of quantum states in both single and multiple potential wells. The impact of the achievement of superlattices or multibarrier structures is manyfold. The elegance of one-dimensional quantum physics, which had long remained a textbook exercise, could now, for the first time, be practiced in a laboratory (do-it-yourself quantum mechanics!). It should be emphasized that such an achievement would not have been accomplished without the key contribution of material preparation techniques.

During the past 10 years, the MBE technique was further perfected, producing smooth layers and abrupt changes in ultrathin dimensions. It is not surprising that with such an improvement in technique Sollner *et al.* (1983) demonstrated a dramatic improvement in the current singularity and negative resistance in resonant double-barrier tunneling through a GaAs quantum well and carried out detecting and mixing at frequencies as high as 2.5 THz.

E. X-Ray Analysis

For analyses of GaAs–$GaAs_{1-x}P_x$ (Esaki *et al.*, 1970) and GaAs–AlAs (Chang *et al.*, 1974b) superlattice structures, the high-angle as well as the low-angle x-ray scattering data provided relevant information (Segmüller *et al.*, 1977) including the accurate determination of the superlattice period. Figure 13 shows the high-angle diffraction patterns for an 88-Å-period GaAs–AlAs superlattice, obtained in the vicinity of (a) the (002), (b) the (004), and (c) the (006) reflections. The substrate peak is the highest one in the center. To its left in (b) and (c), the zero-order reflection of the superlattice is visible. The weak superlattice reflections are designated as +1, +2, +3, and −1, −2, −3 satellites. Figure 14 shows low-angle x-ray interferences for a 90.5-Å-period superlattice, in which several interference maxima are observed corresponding to the superlattice period. From these data, the numbers of Ga and Al layers in the superlattice unit cell can be accurately determined. On the other hand, the relative intensities of the high-angle reflections (Fig. 13) are a sensitive measure

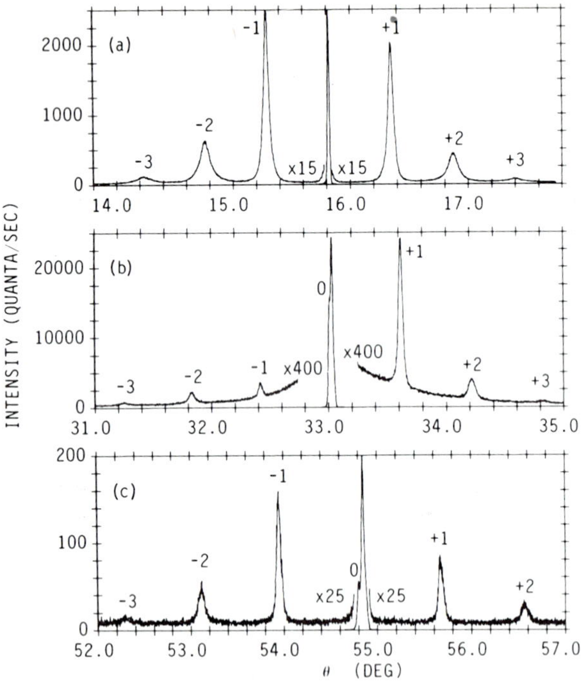

Fig. 13. High-angle diffraction patterns for a GaAs–AlAs superlattice with a period of 88 Å prepared on the (001) face, obtained in the vicinity of (a) the (002), (b) the (004), and (c) the (006) reflections. Scale factors of the diffraction curves on both sides of the substrate peak are indicated.

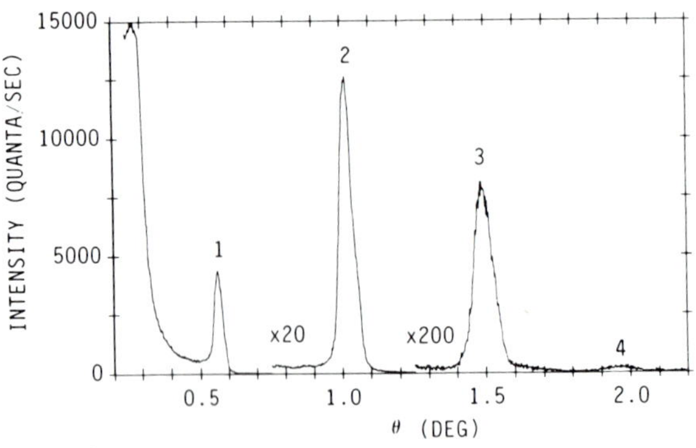

Fig. 14. Low-angle scattering pattern for a GaAs–AlAs superlattice with a period of 90.5 Å prepared on the (001) face of the GaAs substrate.

of the elastic strain present in the lattice. The measured elastic strain agreed with the value computed theoretically on the assumption that the strain is not relieved by misfit dislocations at the GaAs–AlAs interfaces.

In any case, some deviations from the ideal structure are evident from the fact that the half-widths of the superlattice reflections vary and the fit between the observed and calculated satellite intensities is not exact. At the GaAs–AlAs interface there will be mixed layers even if there is no diffusion because at the instant one beam is shut off and the other beam switched on, the surface will, in general, have a partial layer with an atomic step in it. This causes a rounding of the edges of the composition profile at the interface and produces incommensurate superlattice periods. Recently, Weisbuch *et al.* (1981) reported some estimates of interface roughness for GaAs–GaAlAs multiquantum well structures, analyzed from broadening of luminescence linewidths, which suggested the presence of the islandlike structure with a one-monolayer height and a lateral size of 300 Å or more.

F. Comments on Interfaces

One should realize that the superlattice proposal by Esaki and Tsu (1970) was made on a rather optimistic presupposition of a "clean" and yet atomically smooth interface that provides only an abrupt potential step with few undesirable localized states. In reality, however, the presence of possible misfit dislocations and other defect complexes, as well as roughness or disorder at the interface, may not be completely avoidable. Indeed, the early experimental results described in this chapter suffered considerably from such departures from an idealized model on which simple theories are based. Recently, advanced epitaxy techniques such as MBE have provided dramatically improved possibilities for achievement of nearly ideal interfaces in some favorable systems.

Considerable efforts have been made to understand the electronic structure at interfaces [e.g., see Cohen (1980)] or heterojunctions (Harrison, 1977; Frensley and Kroemer, 1977). Even in an ideal situation, the discontinuity at the interface provides formidable tasks in theoretical handling: Propagating and evanescent Bloch waves should be matched across the interface, satisfying continuity conditions on the envelope wave functions (Bastard, 1981a,b; White and Sham, 1981). The interface chemical bonds differing from the bulk may give rise to localized states in addition to bond relaxation in some circumstances. The extrinsic effects due to localized states around defect sites near the interface should be distinguished from the intrinsic properties arising from the discontinuities of the wave functions. In favorable systems such as GaAs–GaAlAs, the intrinsic localized states are virtually nonexistent in the energy range in which we are interested.

III. IMPORTANT DEVELOPMENTS

A. Remarks

Following the pioneering period, a series of interesting developments have been witnessed on this subject. Expanding activities resulted in observations of new phenomena and exploration into novel structures, leading to ramification of the superlattice field. Such structures include quantum wells and nearly ideal heterojunctions.

In this section, some of the important developments will be itemized in chronological order, as much as possible, by categorizing them according to host materials used, GaAs–$Ga_{1-x}Al_xAs$ and InAs–GaSb(–AlSb), followed by all other systems. Such categorization, however, often encounters difficulties because similar studies extend over a variety of materials; the same experimental techniques and theoretical methods can be easily applied to different systems. Nevertheless, most significant studies from scientific as well as technological aspects have been carried out with GaAs–$Ga_{1-x}Al_xAs$ and InAs–GaSb systems. In Fig. 15, three typical examples for the relationship of bandedge energies at heterojunctions (left) and their respective superlattices or multiquantum wells (right) are shown together with corresponding band bending

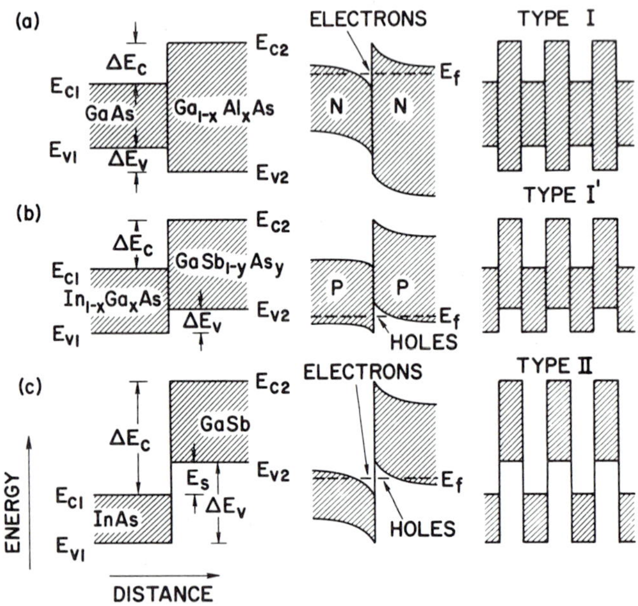

Fig. 15. Relationship of band-edge energies at the heterojunction interfaces for (a) GaAs–$Ga_{1-x}Al_xAs$, (b) $In_{1-x}Ga_xAs$–$GaSb_{1-y}As_y$, and (c) InAs–GaSb (left) and the energy-band diagrams of their respective superlattices or multiquantum wells (right), with band bending and carrier confinement at the heterojunction interfaces (middle).

and carrier confinement at the interface (middle). The conduction band discontinuity ΔE_c is equal to the difference in the electron affinities of the two semiconductors. In case (c), the band bending and electron and hole confinements occur, as indicated, regardless of the polarity of employed semiconductors, whereas in cases (a) and (b), those depend on the semiconductor polarity (*n* or *p*).

As seen in Fig. 15, in the GaAs–$Ga_{1-x}As_xAs$ system, the total difference in the energy gaps is shared by potential steps ΔE_c and ΔE_v, while the InAs–GaSb system exhibits a rather unusual band-edge relationship; namely, the conduction band edge of InAs E_{c1} is lower than the valence band edge of GaSb E_{v2} by E_s. The superlattices made of these three types of semiconductor pairs are identified as type I, type I′, and type II, as illustrated from the top to the bottom on the right side. Although a variety of superlattices have been synthesized with many semiconductor pairs, all of them will fall into these three categories.

B. GaAs–GaAlAs

1. Optical Absorption

Dingle *et al.* (1974, 1975) observed pronounced structure in the optical absorption spectrum representing bound states in isolated and coupled quantum wells. For the former, GaAs well widths in the range between 70 and 500 Å were prepared. The GaAs wells were separated by $Ga_{1-x}Al_xAs$ barriers that were normally thicker than 250 Å. In low-temperature measurements for such structures, several exciton peaks, associated with different bound-electron and bound-hole states, were resolved. For the latter study, a series of structures, with GaAs well widths in the range between 50 and 200 Å and $Ga_{1-x}Al_xAs$ ($0.19 < x < 0.27$) barrier widths between 12 and 18 Å, were grown by MBE on GaAs substrates. The spectra at low temperatures clearly indicated the evolution of resonantly split discrete states into the lowest band of a superlattice. In all experiments, in order to enhance the total GaAs absorption, as many as 50 (or 80) GaAs layers were grown in a single structure. These observations demonstrated the great precision of MBE in fabricating thin and uniform layers.

2. Lasers

Van der Ziel *et al.* (1975) observed optically pumped laser oscillation from the above-mentioned quantum-well structures at 15 K. The laser oscillation occurred at energies that are slightly below the exciton associated with the lowest energy ($n = 1$) bound state in GaAs quantum wells (50–500 Å thick).

In 1978, Dupuis and Dapkus and Holonyak *et al.* succeeded in room-temperature laser operation of quantum-well $Ga_{1-x}Al_xAs$–GaAs laser diodes, where the well width was around 200 Å. They observed lasing transitions all the way to the ($n = 5$) bound state in the quantum well corresponding to 1.8 eV

(0.69 μm), which is 0.29 eV higher than that of bulk GaAs. It should be added that these devices were prepared by metalorganic chemical vapor deposition (MOCVD). In view of the recent progress of the MOCVD, this technique appears to be a serious contender to the MBE technique for the growth of quantum wells or superlattices.

More recently, Tsang (1981) succeeded in attaining a threshold current density J_{th} as low as 250 A/cm^2 in MBE-grown $Ga_{1-x}Al_xAs$–GaAs lasers by incorporating a multiquantum-well structure in the active region. This was achieved as a result of utilizing the beneficial effects arising from the two-dimensional density of states of the confined carriers (Fig. 5). It is generally observed that, in multiquantum-well lasers, the beam width in the direction perpendicular to the junction plane and the temperature dependence of J_{th} are significantly reduced in comparison with the regular double-heterostructure (DH) lasers.

3. Photocurrent and Luminescence

Tsu *et al.* (1975) made photocurrent measurements that shed light on the interrelationship between quantum states and anomalous transport properties in a superlattice. Superlattice structures of different configurations were grown on *n*-type 10^{18}-cm^{-3} GaAs substrates. The total thickness in all cases is of the order of 1 μm, which is comparable with the absorption length for the photons involved. The superlattice region is undoped and an ohmic contact is made on the bottom of the substrate. A semitransparent Au film about 100 Å thick serves as the top electrode.

The experimental results indicated that, as the period becomes narrower, the photocurrent increases and the peak positions shift to higher energies with fewer numbers being observed. The measurements are in good agreement with the calculated energies and bandwidths for each transition.

More recently, Mendez *et al.* (1982; 1983a), with a somewhat similar experimental arrangement, observed an electric field-induced quenching of photoluminescence from quantum wells. Six identical GaAs quantum wells (width = 35 Å) in GaAlAs medium were formed in the space charge region of the Schottky barrier. With an increasing applied field perpendicular to the quantum wells, the luminescence intensity decreases and becomes completely quenched at an average field of a few tens of kilovolts per centimeter (applied voltage = −0.4 V). This is accompanied by a shift to lower energies of the peak positions. The observation is interpreted as caused by the field, which induces a separation of electrons and holes in the quantum well with the modification of the quantum states in the wells, resulting in luminescence quenching.

4. Raman Scattering

Manuel *et al.* (1976) reported the observation of enhancement in the Raman cross section for photon energies near electronic resonance in GaAs–

$Ga_{1-x}Al_xAs$ superlattices of a variety of configurations. Both the energy positions and the general shape of the resonant curves agree with those derived theoretically on the basis of the two-dimensionality of the quantum states in such superlattices. Polarization studies indicate a major contribution to the scattering from forbidden processes.

Among recent developments in the field have been the introduction of inelastic light scattering as a spectroscopic tool in the investigation of the two-dimensional electron system. The significance of resonant inelastic light scattering was first pointed out by Burstein *et al.* (1979), who claimed that the method yields separate spectra of single-particle and collective excitations that will lead to the determination of electronic energy levels in quantum wells as well as Coulomb interactions. Subsequently, Abstreiter and Ploog (1979) and Pinczuk *et al.* (1979) observed light scattering by intersubband single-particle excitations, between discrete energy levels, of two-dimensional electrons in $GaAs-Ga_{1-x}Al_xAs$ heterojunctions and quantum wells.

Meanwhile, Colvard *et al.* (1980) reported the observation of Raman scattering from folded acoustic longitudinal phonons in a GaAs (13.6-Å)–AlAs (11.4-Å) superlattice. The superlattice periodicity is expected to result in Brillouin-zone folding (as previously mentioned) and the appearance of gaps in the phonon spectrum for wave vectors satisfying the Bragg condition. An explanation of the data was given with a simple theory that involves the Kronig–Penney electron model and the phonons in the elastic continuum limit.

5. Magnetoquantum Effects

In 1977, Chang *et al.* reported the first observation of the oscillatory magnetoresistance (Shubnikov–de Haas effect) in $GaAs-Ga_{1-x}Al_xAs$ superlattices with the current flowing in the plane of the layers. The superlattice provides a unique opportunity to create made-to-order Fermi surfaces by controlling the energy and bandwidth of the subbands, which are determined by the barrier and well thicknesses as well as the barrier height. The observed oscillations manifest the electronic subband structure, which becomes increasingly two-dimensional in character as the bandwidth is narrowed.

In 1980, Tsui and Englert reported the first observation of magnetophonon resonances in the two-dimensional electron system in GaAs–GaAlAs heterojunctions and superlattices. The resonances, detected in the magnetic field dependence of the samples resistance at temperatures between 100 and 180 K, result from inelastic scattering of electrons between Landau levels with resonant absorption of GaAs LO phonons under the resonance condition of $N\hbar\omega_c = \hbar\omega_0$, where $\hbar\omega_c$ is the Landau-level spacing, ω_0 the phonon frequency, and N an integer. The result showed that, in this system, the electron–phonon interaction with the polar mode of GaAs is dominant and its strength is consistent with that expected from polar coupling of two-dimensional electrons to the LO phonons of bulk GaAs. Englert and Tsui (1982) made more detailed studies on this effect in higher magnetic fields so that the fundamental resonance ($N = 0$) can be observed.

Though the two-dimensional electron systems, including the Si surface inversion layer, have been intensively studied under high magnetic fields for almost two decades, the experimental methods customarily employed are magnetoresistance and Hall effect. In both cases, however, the step from the experimental data to the density of states is difficult because electronic transport in a combination of the density of states and electron scattering. Störmer *et al.* (1983b) observed the oscillatory magnetic moment as a function of magnetic field (de Haas–van Alphen effect) at 1.5 K for two-dimensional electrons in a modulation-doped GaAs–GaAlAs heterostructure. Although the strength of the signal was enhanced by stacking 4000 layers, the amplitude of the oscillations was about a factor of 30 weaker than expected, which was attributed to sample inhomogeneities. Nevertheless, the measurements of the low-temperature magnetic moment and susceptibility provide a useful tool for the investigation of the density of states per se.

6. Modulation Doping and High-Speed Devices

It is usually the case that free carriers (electrons or holes) in semiconductors are created by doping impurities (donors or acceptors). Thus, carriers inevitably suffer from impurity scattering. There are a few exceptions. Insulated-gate-effect devices (MOSFET) are one such example in which electrons or holes are induced by applied gate voltages; InAs–GaSb superlattices may be another example in which electrons and holes are produced solely by the unique band-edge relationship, as will be described later.

Now, in superlattices, it is possible to spatially separate carriers and their parent impurity atoms by doping impurities in the regions of the potential hills, as shown in Fig. 16. In the original article [Esaki and Tsu (1969), p. 9], this concept was expressed in the general terms as

> the scattering time is an important factor in the described effect . . ., if the superlattice structure is formed in such a manner that most scattering centers such as foreign atoms, imperfections, etc., are concentrated in the neighborhood of the potential hills, one can show that electrons would suffer less from such scattering center. . . .

In 1978, Dingle *et al.* successfully implemented such a concept in modulation-doped GaAs–GaAlAs superlattices, as illustrated in the top of Fig. 16, thus achieving electron mobilities that exceed the Brooks–Herring predictions. Modulation doping was performed by synchronizing the silicon (*n*-dopant) and aluminum fluxes in the MBE so that the dopant was distributed only in the GaAlAs layers and was absent from the GaAs layer.

In 1980, Mimura *et al.* and Delagebeaudeuf *et al.* applied modulation-doped GaAs–GaAlAs heterojunctions to fabricate a new high-speed FET called

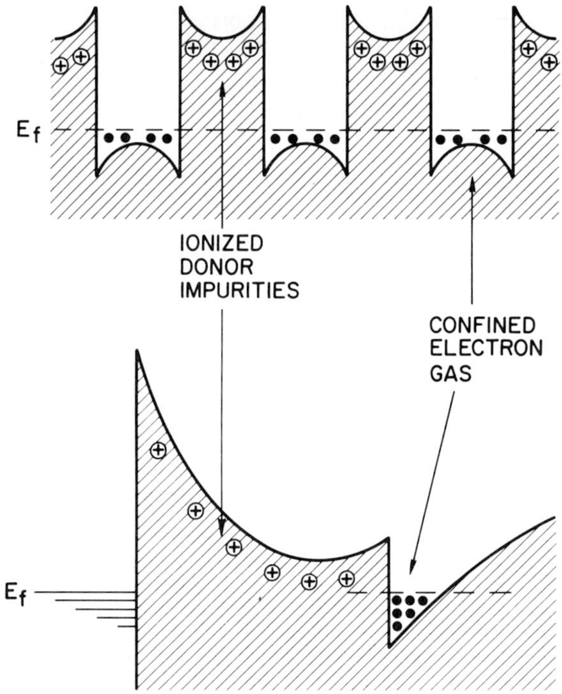

Fig. 16. Modulation doping for a superlattice (top) and a heterostructure with an attached Schottky junction (bottom).

HEMT (high electron mobility transistor) or TEGFET (two-dimensional electron gas field-effect transistor), of which the energy diagram is shown in the bottom of Fig. 16. The device, if operated at 77 K, apparently exhibits a high-speed performance three times superior to that of the conventional GaAs MESFET.

7. Quantized Hall Effect

In 1980, Klitzing *et al.* demonstrated an interesting proposition that quantized Hall resistance can be used for precision determination of the fine structure constant α by using two-dimensional electrons in the inversion layer of a Si MOSFET (metal–oxide–semiconductor field-effect transistor). Subsequently, Tsui and Gossard (1981) found that, with the modulation-doped GaAs–GaAlAs heterostructures, the pronounced characteristic of the quantized Hall effect can be easily observed, primarily because of their high electron mobilities, which led to the determination of α with a great accuracy such as $\alpha^{-1} = 137.035965(12)$ (0.089ppm) (Tsui *et al.*, 1982a).

The quantized Hall effect in the two-dimensional electron system is observable at sufficiently high magnetic fields and low temperatures when there is little overlap in the density of state of neighboring Landau levels; in such a range of magnetic fields as to locate the Fermi level between the Landau levels, the parallel component of resistance ρ_{xx} vanishes and the Hall resistance ρ_{xy} goes through plateaus. This surprising result can be understood by the argument that the localized states do not take part in quantum transport (Ando and Uemura, 1974). At the plateaus, the Hall resistance is given by $\rho_{xy} = h/e^2 i = \mu_0 c/2\alpha i \approx$ 25,813 Ω/i, where i is the number of filled Landau levels, h Planck's constant, e the electronic charge, μ_0 the vacuum permeability, and c the speed of light in vacuum.

Tsui *et al.* (1982b) discovered a striking phenomenon: the existence of an anomalous quantized Hall effect, a Hall plateau in ρ_{xy} and a dip in ρ_{xx}, at a fractional filling factor of $\frac{1}{3}$, in the extremum quantum limit at temperatures lower than 4.2 K. Laughlin (1983) as an explanation of such a fractional filling, presented variational ground-state and excited-state wave functions that describe the condensation of a two-dimensional electron gas into a new state of matter, an incompressible quantum fluid. The elementary excitations of this quantum fluid are fractionally charged, and this elegant theory predicts a series of ground states characterized by the variational parameter m ($m = 3, 5, \ldots$), decreasing in density and terminating in a Wigner crystal. Störmer *et al.* (1983a), in addition to previously reported structures at filling factors $\nu = \frac{1}{3}$ and $\frac{2}{3}$, observed new structures at $\nu = \frac{4}{3}, \frac{5}{3}, \frac{2}{5}, \frac{3}{5}, \frac{4}{5}$, and $\frac{2}{7}$. The results suggested that fractional quantization of the Hall effect exists in multiple series, each based on the inverse of an odd integer. There exists no full explanation for all of these experimental results at this moment.

8. Impurity States and Excitons in Quantum Wells

Bastard (1981b, 1982) performed a variational calculation of the hydrogenic impurity ground states in a quantum well, obtaining the binding energy E as a function of well thickness and impurity position. Figure 17a shows the binding energies in units of R_0^*: the three-dimensional effective Rydberg, at the edge and the center, indicated by crosses and dots, respectively, as a function of the normalized well thickness L/a_0^*, where a_0^* is the three-dimensional effective Bohr radius. The electron confinement in a potential well lifts the usual ground-bound state degeneracy with respect to the impurity position. The dependence of the binding energy on the impurity position leads to the formation of an impurity band. The density of states manifests itself in acceptor → conduction or valence → donor absorption process as two peaks located at the band extrema, as shown in Fig. 17b, in which the dimensionless frequency χ is given by $\hbar\omega = E_g - E_i^{max} + \omega E_i^{max}$, where E_i refers to donor or acceptor binding energy. Photoluminescence from GaAs quantum wells measured by Miller *et al.* (1982a,b) apparently confirmed Bastard's prediction. Bastard *et al.* (1982a) derived the exciton binding energy in quantum wells with variation calculations. Similar calculations were reported by Greene and Bajaj (1983a,b).

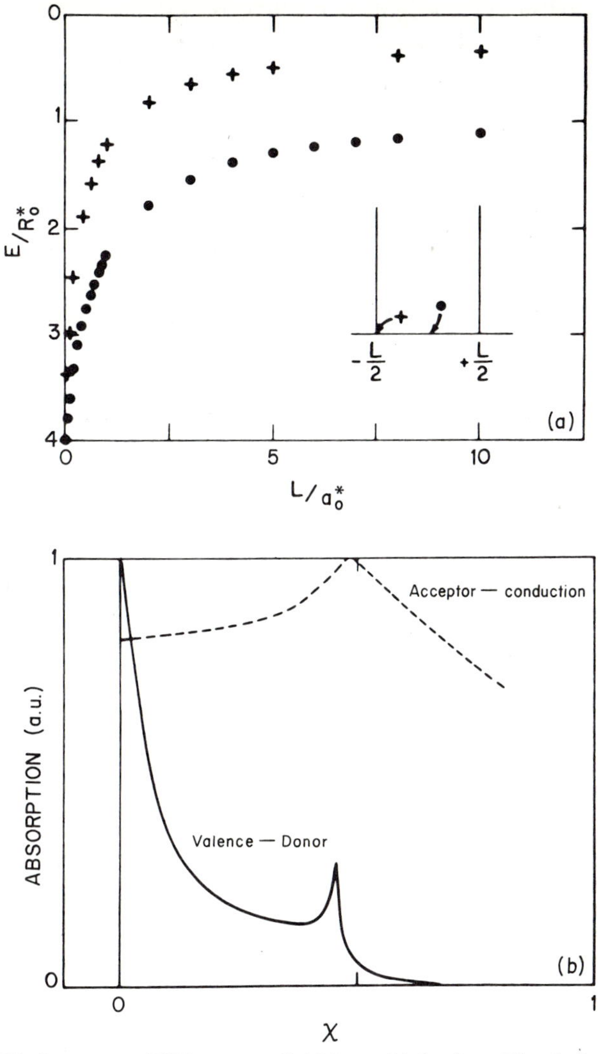

Fig. 17. (a) Binding energy E/R_0^* versus well thickness L/a_0^*, where R_0^* and a_0^* are the three-dimensional effective Rydberg and Bohr radius, respectively; (b) absorption coefficient versus the dimensionless frequency χ ($m_v/m_c = 6$, $L/a_0^* = 1$). [From Bastard (1982).]

9. Excitonic Optical Bistability

In 1982, Gibbs *et al.* reported excitonic optical bistability observed with a GaAs–GaAlAs superlattice etalon. In this case, laser light at frequencies just below the exciton absorption energy is slightly absorbed, producing carriers that change the absorptivity and the polarizability associated with the exciton resonance, thus giving rise to an intensity-dependent refractive index. The two-dimensionality of electrons confined in the quantum wells increases the binding energy of the free excitons in GaAs and permits room-temperature bistable operation of such a superlattice etalon. This may suggest the possibility of a new optical bistable device.

10. Avalanche Photodiodes with Heterostructures

Superlattices or heterostructures have been exploited to obtain improved avalanche photodiodes. As is known, a large difference in the ionization rates for electrons and holes, α and β, respectively, is desirable for a low noise performance. Unfortunately, most III–V semiconductors have α nearly equal to β. Therefore, Chin *et al.* (1980) proposed increasing α/β by using a GaAs–GaAlAs superlattice structure, as shown in Fig. 18. In this structure, since the conduction band-edge discontinuity ΔE_c is considerably larger than the valence band-edge discontinuity ΔE_v, the ionization rate for electrons will be enhanced, but not for holes, thus resulting in an artificial increase of α/β. Capasso *et al.* (1982), indeed, obtained an increased α/β of 10 with a $p-i-n$ structure consisting of 50 alternating 450 Å GaAs and 550 Å $Ga_{0.55}Al_{0.45}As$ layers. In addition, with the application of band-gap engineering such as graded-gap or periodic structures, a number of new photodiodes or photomultipliers have been explored (Capasso, 1983) including channeling avalanche detectors (Tanoue and Sakaki, 1982), which spatially separate electrons and holes.

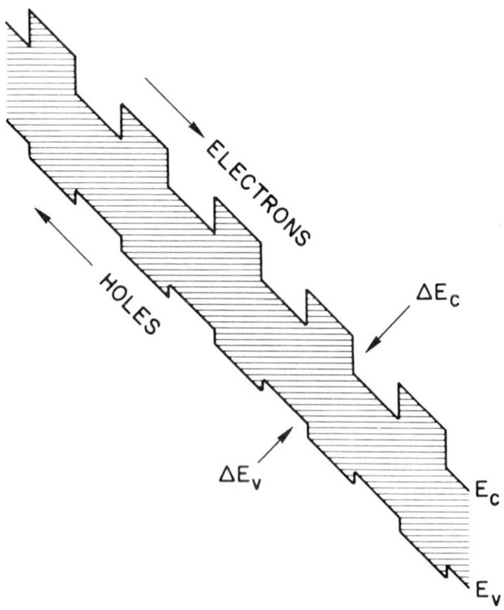

Fig. 18. Energy-band diagram of a superlattice avalanche photodiode. ΔE_c and ΔE_v are the band-edge discontinuities of the conduction and valence bands, respectively.

11. Attempts for One-Dimensionality

The introduction of a superlattice potential facilitates the reduction of the dimensionality of carriers from three to two at the lower limit in the case of a series of quantum wells. On the other hand, the electrons on the surface inversion layers, as seen in Si MOSFETs, always constitute a two-dimensional gas. Thus, the electron system on the surface inversion layer of a superlattice crystal, as schematically illustrated in Fig. 19, will exhibit the dimensionality between one and two, depending on the superlattice potential profile. It is expected here that, by using a superlattice being *p*-type or having no carriers, surface electrons are generated either by an external field or by modulation doping from a cladding GaAlAs layer (not shown in the figure). If the GaAlAs barriers in the superlattice are sufficiently thick so that wave functions in the electron pockets on the surface have no overlap, then a confined one-dimensional electron gas will be obtained. (This structure is an outgrowth of discussions with L. L. Chang and P. J. Stiles.)

Sakaki (1980) proposed a V-grooved MOSFET of a one-dimensional electron gas with a single quantum well, virtually similar to that shown in Figure 19, suggesting extremely high electron mobilities on the basis of the scattering

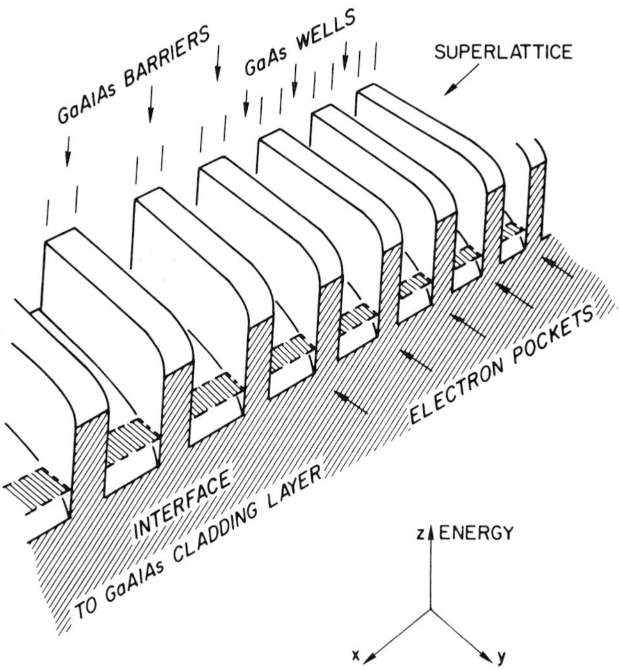

Fig. 19. Quasi-one-dimensional electron system on the surface inversion layer of a superlattice crystal.

probability from Coulomb potential. Petroff *et al.* (1982) attempted the realization of a one-dimensional carrier-confining structure with MBE-growth GaAs and GaAlAs layers.

C. InAs–GaSb(–AlSb)

1. Theoretical Treatments

In 1976, in search for an alternative to the GaAs–AlAs system, in which the introduction of the superlattice periodicity will give a greater modification to the electronic band structure of the host materials, the InAs–GaSb system appeared as a good candidate because of its unique band-edge relationship at the interface, as shown in Fig. 15. At that time, of course, such alignment of the band gaps of the two hosts was not well known. The band-edge relationship was estimated from the values of the electron affinities determined by Gobeli and Allen (1966) which might have had experimental uncertainties. Later, however, Sakaki *et al.* (1977) confirmed that this was indeed the case by observing an unusual nonrectifying characteristic in the $p-n$ junctions made of n InAs and p GaSb, which was the direct consequence of "interpenetration" between the GaSb valence band and the InAs conduction band. At the interface, shown in the center at the bottom of Fig. 15, electrons that "flood" from the GaAs valence band to the InAs conduction band, leaving holes behind, produce a strong dipole layer consisting of the two-dimensional electron and hole gases. An approximate classical solution for the band bending (Sai-Halasz *et al.*, 1978) can be obtained rather simply by using a self-consistent Fermi–Thomas approximation.

First, Sai-Halasz *et al.* (1977) made a one-dimensional calculation for the InAs–GaSb superlattices labeled type II at the bottom right-hand side of Fig. 15 and treated each host material in Kane's two-band framework. The GaAs–AlAs superlattices, on the other hand, can be adequately treated in one-band approaches. In the InAs–GaSb superlattices, although potential wells for electrons and holes are located in the different semiconductors, quantized levels in the electron wells will be very close in energy to those in the hole wells. Thus, in such cases, there exists a strong interaction between them.

Following an analytical treatment, one obtains

$$\cos(kd_0) = \cos(k_1 d_1)\cos(k_2 d_2) - F\sin(k_1 d_1)\sin(k_2 d_2), \tag{1}$$

$$F = \frac{1}{2}\left(\frac{ik_1 + u_1'/u_1}{ik_2 + u_2'/u_2} + \frac{ik_2 + u_2'/u_2}{ik_1 + u_1'/u_2}\right), \tag{2}$$

where the superlattice period d_0 is the sum of the thickness d_1 and d_2 of alternating layers and $u_i = u_i(k_i, x)$, $u_i' = du_i(k_i, x)/dx|_{x=0}$. The allowed bands in the superlattice correspond to the energy range in which Eq. (1) has a solution for real k. For such allowed bands, Eq. (1) provides the $E-k$ relationship. In this calculation, we found that the energy gap is strongly dependent on the layer thickness or the period, decreasing from ~0.6 eV to 0. It should be mentioned that the present result is reduced to the well-known Kronig–Penney solution if the logarithmic derivatives u_i'/u_i in Eq. (2) are set to 0.

Next a band calculation was performed with the linear-combination-atomic-orbitals (LCAO) method (Sai-Halasz *et al.*, 1978a) handling a large size of the primitive cell and ignoring charge redistribution at the interface. The calculated subband structure is strongly dependent on the period. Figure 20 shows calculated subband energies and bandwidths for electrons and light and heavy holes as a function of period, together with the energy gaps of GaSb and InAs in the left, assuming $d_1 = d_2$, where the semiconducting energy gap is determined by the energy difference between the electron and heavy hole ground subbands. This gap decreases with an increase in the period, becoming 0 at 170 Å, as seen in the figure, corresponding to a semiconductor-to-semimetal transition.

This indicates that, by a choice of the period, it is possible to synthesize a tailor-made narrow-gap semiconductor or a semimetal with an electronic band structure that bears little resemblance to that of the host semiconductors. Once the system becomes semimetallic with increase in the period, the charge redistribution at the interface will give rise to severe band bending.

2. Optical Absorption and Luminescence

Optical absorption measurements (Sai-Halasz *et al.*, 1978b) were made on InAs–GaSb superlattices with periods ranging between 30 and 60 Å, which are

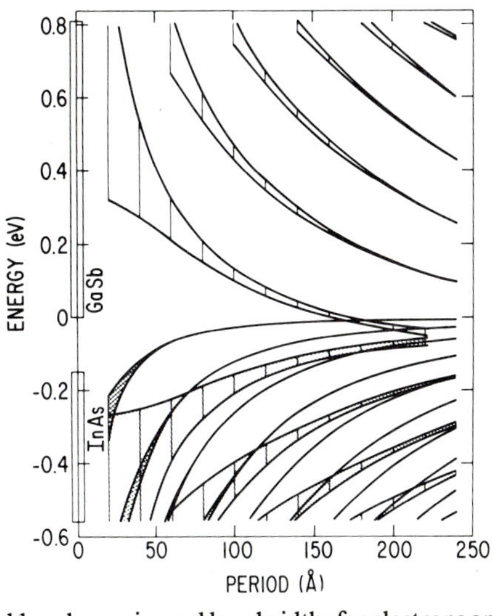

Fig. 20. Calculated subband energies and bandwidths for electrons and light and heavy holes as a function of period, assuming $d_1 = d_2$.

in the semiconductor regime as indicated in Fig. 20. Samples consist of several hundred layers, typically 2.5–3 μm thick, grown on GaSb substrates. The energy gaps derived from the measured absorption edges as a function of periods, gave an estimate for $E_s \sim 0.15$ eV, where E_s is the energy difference between the top of the GaSb valence band and the bottom of the InAs conduction band, as shown in Fig. 15c.

Recently, Voisin *et al.* (1981) observed luminescence from the similar semiconducting InAs–GaSb superlattices. It was found that, in addition to radiative recombination between the electron and hole ground subbands, the luminescence spectra exhibit a low-energy tail that is believed to arise from impurities and interface defects.

3. Semiconductor–Semimetal Transitions and Shubnikov–de Haas Oscillations

In order to verify the critical layer thickness (Chang *et al.,* 1979) for the predicted semiconductor–semimetal transition, a number of InAs–GaSb superlattices with a variety of periods were grown on (100)GaSb substrates whose carriers largely freeze out at low temperatures. The total thickness of the superlattice region is typically ~ 2 μm. No intentional doping was introduced, but residual impurities usually provide a background electron concentration of the order of 10^{16} cm^{-3}. The approximate electron concentration and mobility parallel to the plane were derived from Hall measurements at 4.2 K in low magnetic fields because Hall voltages are primarily induced by electrons that have higher mobilities than holes. The measured electron concentration exhibited a sudden increase of an order of magnitude in the neighborhood of 100 Å. This increase indicates the onset of electron transfer from GaSb to InAs: the transition from the semiconducting state to the semimetallic state as the ground conduction subband E_1 moves below the ground valence subband H_1. The observation is in good agreement with the theoretical prediction, presented in Fig. 20, giving a critical layer thickness for this transition of 85 Å when the energy difference E_s is 0.15 eV at the interface.

It should be mentioned that, since this increase in the electron concentration is not due to doping of impurities, mobilities beyond 10^4 cm/V sec at 4.2 K can be achieved, levels much higher than bulk InAs with comparable electron concentrations, somehow similar to the modulation doping.

L. L. Chang *et al.* (1981) made extensive Shubnikov–de Haas experiments on superlattice samples covering the entire semimetallic regime from the onset of the semiconductor–semimetal transition to the heterojunction limit.

4. Far-Infrared Magnetoabsorption and Cyclotron Resonance

Far-infrared magnetoabsorption experiments (Guldner *et al.,* 1980, 1981; Maan *et al.,* 1981) were performed at 1.6 K for semimetallic 120–80, 200–100, and 1000–1000-Å superlattices with radiation near normal incidence to the layers. Infrared sources are H_2O ($\lambda = 118\ \mu m$), HCN ($\lambda = 337\ \mu m$) and DCN ($\lambda = 198\ \mu m$) molecular lasers and also carcinotrons ($\lambda = 1-2\ \mu m$). The transmission signal for each wavelength exhibits oscillations with increase in magnetic field. The data for the 120–80-Å superlattice indicate that the energies at which absorption maxima occur are directly proportional to the field and all lines converge to -38 ± 2 meV at zero field. We interpret such absorption as being due to interband transitions from H_1 to E_1 Landau levels. If these transitions are assumed to occur at a selection rule, $\Delta N = 0$, the converged value should correspond to the negative energy gap at the semimetallic superlattice, $E_1 - H_1$. A similar experiment for the 200–1000-Å sample yielded $E_1 - H_1 = 61 \pm 4$ meV. These values are in good agreement with the calculated ones. The observation of two cyclotron masses for the 1000–1000-Å sample (which can be considered as a series of isolated heterojunctions) appears to manifest the multiple subband structure, in which each subband has its own effective mass (Takaoka *et al.*, 1983).

5. Polytype Superlattices

The superlattice studies in the past were limited to systems involving two host semiconductors or their pseudobinary alloys. We are now considering the introduction of a third constituent such as AlSb in the present InAs–GaSb system (Esaki *et al.*, 1981), which provides an additional degree of freedom.

The lattice constant of AlSb (6.136 Å) is compatible with those of GaSb (6.095 Å) and InAs (6.058 Å) for heteroepitaxy. These three semiconductors, closely matching in lattice constant and yet significantly differing in band parameters, represent a rather unique combination among III–V compound semiconductors.

This triple-constituent system (type III) leads to a new concept of man-made polytype superlattices (ABCABC, ABAC, ACBC, etc.), which can never be achieved with the dual-constituent system, as shown in Fig. 21, where A is AlSb,

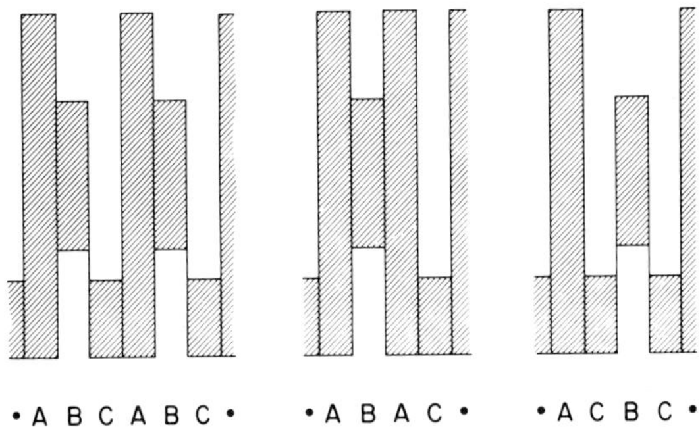

Fig. 21. Potential-energy profiles for polytype (type III) superlattices.

B GaSb, and C InAs. Since the AlSb layers are expected to be potential barriers, the number of electrons and holes in the structure will be changed by applied electric fields. Thus, this system appears to offer an electrically controllable medium.

Takaoka *et al.* (1983) studied transport properties of GaSb–AlSb–InAs multiheterojunctions, one of the basic elements in the proposed superlattice. Such structures were prepared by successive MBE growths on *p*-type GaSb substrates with (100) surface orientation at temperatures of 500°C for GaSb, 450–500°C for AlSb, and 450°C for InAs (Chang *et al.*, 1982). Heteroepitaxy between GaSb and AlSb with a lattice mismatch of 0.66% appears to proceed smoothly, whereas that between InAs and AlSb with a lattice mismatch of 1.26% indicates a sign of some departure from the ideal case in the observation of streaking RHEED patterns. It is known that the small lattice mismatch is likely accommodated by relaxation involving interfacial atoms, as seen in the case of the InAs–GaSb superlattice (Saris *et al.*, 1980).

The formation of a GaSb–AlSb superlattice was confirmed metallurgically from x-ray diffraction and optically from electroreflectance and photoluminescence measurements (Mendez *et al.*, 1983b). The luminescence spectra exhibited emission peaks associated with interband transitions between quantum states as well as acceptor impurities. Voisin *et al.* (1983) measured optical transmission near the band gap in a series of GaSb–AlSb superlattices, of which the spectra showed a staircaselike structure characteristic of the two-dimensional electron system. The absorption steps were attributed to transitions between the valence and conduction subbands in the GaSb quantum wells. The energy of the absorption edge, however, was found to be smaller than expected, which was interpreted as a result of the strain effect. Incidentally, photoluminescence from GaSb–AlSb superlattices at 300 K was previously studied by Naganuma *et al.* (1981).

6. InAs Quantum Wells

Indium arsenide quantum wells have been investigated with MBE-grown heterostructures GaSb–InAs–GaSb and AlSb–InAs–AlSb. Bastard *et al.* (1982b) performed self-consistent calculations for the electronic properties of GaSb–InAs–GaSb heterostructures (including the effect of high magnetic fields) and predicted the existence of a semiconductor-to-semimetal transition when the InAs thickness exceeds a threshold, as a result of electron transfer from GaSb, similar to the situation of the InAs–GaSb superlattice. This was experimentally verified by Chang *et al.* (1984).

Mendez *et al.* (1984) observed the quantized Hall effect for two-dimensional carriers confined in GaSb–InAs–GaSb heterostructures. In contrast to previously investigated systems such as GaAs–GaAlAs, the present heterostructures offer a new medium in which a two-dimensional electron gas formed in

InAs coexists with a two-dimensional hole gas, of the same density, formed in the GaSb side: The intrinsic nature of the band-edge relationship determined the carrier density.

D. Other Semiconductor Superlattice Structures

1. Other Semiconductor Systems

A variety of semiconductor pairs have been exploited to synthesize superlattices and heterostructures (Chang, 1983). Compared to GaAs–GaAlAs and InAs–GaSb, however, other superlattice studies are still in their infancy; in many cases, experimental results do not show as high a degree of sophistication as we have seen in the GaAs–GaAlAs system, partly because of their inferior structural quality, even though a number of intriguing proposals were made on the theoretical basis.

Nevertheless, high-quality interfaces can be obtained with InP-based lattice-matched systems. Rezek *et al.* (1977) succeeded with a double heterojunction (DH) laser using InP–InGaPAs multiple thin layers. Brummell *et al.* (1983) and Voos (1983) reported magnetotransport as well as other important experiments in MOCVD-grown InP–InGaAs superlattices and heterojunctions.

Osbourn (1982) proposed that strained-layer superlattices from a variety of material systems including the alloys of GaAs–GaP and GaAs–InAs on a premise that lattice-mismatched heterostructures can be grown with essentially no misfit defect generation if the layers are sufficiently thin because the mismatch is accommodated by uniform lattice strain (van der Merwe, 1963). The GaAs–GaAsP system, which was theoretically analyzed in this proposal (Osbourn, 1982), actually was the first superlattice experimentally investigated (Esaki *et al.*, 1970). Ludowise *et al.* (1983) observed stimulated emission in strained-layer GaAs–$GaAs_{0.75}P_{0.25}$ superlattices with photoexcitation. Osbourn *et al.* (1982) and Gourley and Biefeld (1982) claimed the growth of a direct-gap (2.03-eV), strained-layer superlattice with indirect-gap hosts GaP and $GaAs_{0.4}P_{0.6}$ due to Brillouin-zone folding. Camras *et al.* (1983) presented data showing that Zn diffused into a strained-layer GaP–$GaAs_{0.4}P_{0.6}$ superlattice enhanced the interdiffusion of As and P at the interfaces, resulting in a disordered $GaAs_{0.2}P_{0.8}$ bulk crystal. Fritz *et al.* (1983) reported doping and transport properties of a GaAs–$Ga_{0.8}In_{0.2}As$ strained-layer superlattice. Kim and Madhukar (1982) theoretically treated GaP–AlP superlattices with varying thickness and showed that the superlattices exhibit a direct-gap behavior. Greene (1976) and Eltoukhy *et al.* (1977) reported the formation of GaSb–InSb periodic structures (polycrystalline) by multitarget rf sputtering, which were used for metallurgical studies including interdiffusion.

The Ge–GaAs system is perhaps the oldest example of heteroepitaxy (Anderson, 1960; Esaki *et al.*, 1964) with nearly perfect lattice matching. Recently, the electronic structure at this interface, which involves both polar and nonpolar faces, has drawn considerable attention for theoretical considerations (Baraff *et al.*, 1977; Pickett *et al.*, 1977; Harrison *et al.*, 1978; Pollmann and Pantelides, 1980) (including Ge–ZnSe(100) interfaces). The heteroepitaxial growth of a compound semiconductor or an elemental semiconductor, however, turns out to present unsurmountable technical problems such as formation of two possible sublattices, charge neutrality, and diffusion. Such problems apparently prevent the achievement of the structural perfection. Nevertheless, Petroff *et al.* (1979) and C.-A. Chang *et al.* (1981) reported the MBE growth of Ge–$Ga_{1-x}Al_xAs$ (ultrathin layer) and Ge–GaAs superlattices, respectively, together with the results of their metallurgical analyses. Bauer and Mikkelsen (1982) studied surface processes controlling Ge–GaAs(100) heterojunction formation. Madhukar and Delgado (1981) calculated the electronic structure of Si–GaP(100) interface and superlattice.

Kasper *et al.* (1975) reported the growth of a Si–$S_{0.85}Ge_{0.15}$ superlattice by UHV epitaxy with its metallurgical characterization including dislocations. Manasevit *et al.* (1982) observed enhanced Hall mobilities in (100)-oriented, *n*-type Si–$Si_{1-x}Ge_x$ multilayer films.

Schulman and McGill (1979) theoretically treated a CdTe–HgTe superlattice and showed that its band gap is adjustable from 0 to 1.6 eV, depending on the thickness of the CdTe and HeTe layers. The usefulness of this superlattice for infrared optoelectric devices was suggested. Faurie *et al.* (1982) reported the MBE growth of CdTe–HgTe superlattices that were characterized by Auger electron spectroscopy and ion microprobe profiling measurements. Guldner (1983) *et al.* performed far-infrared magnetoabsorption experiments for such superlattices. The results indicated that the superlattice is a quasi-zero-energy-gap semiconductor and provided, for the first time, the offset energy between the HgTe and CdTe valence bands.

Kinoshita and Fujiyasu (1980) successfully prepared PbTe–$Pb_{0.8}Sn_{0.2}Te$ superlattices by the hot-wall technique on BaF_2 substrates, which were profiled by sputtering-Auger electron spectroscopy in conjunction with the investigation of the interdiffusion of Pb and Sn. Shubnikov–de Haas oscillations (Kinoshita *et al.*, 1982) were observed for two-dimensional electrons with a mobility of 1.9×10^5 cm^2/V sec in modulation Bi-doped superlattices. Partin (1982) reported the MBE growth of PbTe–$Pb_{0.97}Ge_{0.03}Te$ superlattices with their metallurgical characterization.

2. Ultrathin Layer Superlattices

Gossard *et al.* (1976) reported the achievement of an ultrathin layer superlattice by alternate monolayer depositions of GaAs and AlAs. Transmission electron microscopy showed such a MBE-grown structure to be perfectly epitaxial with layered composition modulation of the expected periodicity, although dark-field transmission electron micrographs indicated the presence of disordered regions along with ordered monolayer domains. The electronic properties were studied by optical absorption and luminescence. Petroff *et al.* (1979) studied the MBE growth of $Ge-Ga_{1-x}Al_xAs$ ultrathin-layer superlattices, as previously mentioned. They also attempted alternate monolayer depositions with two components with a sizable lattice mismatch such as a pair of GaAs and InAs.

3. Doping superlattices *(n–i–p–i)*

Döhler (1972) and Ploog *et al.* (1981) pursued a doping superlattice that was included in the original proposal (Esaki and Tsu, 1970) as shown in the top of Fig. 3. The periodic rise and fall of the conduction and valence bands are caused by a periodic variation of impurity concentration, where electrons are attracted to minima in the conduction band while holes are attracted to maxima in the valence band. If this superlattice is illuminated, extra electrons and holes are created, which are to be spatially separated because of the already-mentioned reason. Therefore, they can have anomalously long lifetimes. An interesting consequence of this fact is that the amplitude of the periodic potential is reduced by the extra carrier, leading to a crystal that has a variable energy gap. Döhler *et al.* (1981) indeed observed that the photon energies in luminescence, which represent the band gap, were varied from 1.3 eV to 1.53 eV by the laser excitation intensity. In addition, Raman experiments provided an evidence of electronic subbands in purely space-charge-induced quantum wells. The doping superlattice used here was grown by MBE, which consists of 20 *n*- and *p*-doped layers with equal doping concentration of 10^{18} cm^{-3}. The individual layer thickness is 400 Å.

IV. SUMMARY

It has been more than a decade since Esaki and Tsu (1970) in their original proposal stated that "the study of superlattices and observation of quantum mechanical effects on a new physical scale may provide a valuable area of investigation in the field of semiconductor physics." In the meantime, the evolution of MBE as a technique for the growth of ultrathin layers of high-quality semiconductors allowed access to such a new quantum regime. Indeed, in recent years, considerable attention has been given to the engineering of artificial structures such as superlattices, quantum wells, and other multiheterojunctions. Obviously, the intriguing physics, particularly regarding the electron gas system of the reduced dimensionality involved in such structures, has provided fuel for this advancement. The described InAs–GaSb research is one example that elucidates the salient features involved in synthesized superlattices. In that case, important electronic parameters such as energy gaps (including negative gaps) can be engineered by introducing made-to-order superlattice potentials: Some of its unique properties may not even exist in any "natural" crystal.

We believe that efforts in this direction apparently have opened up a new area of interdisciplinary investigations in the fields of materials science and device physics. A variety of materials, including III–V, II–VI, and IV–VI compounds, as well as elemental semiconductors, have been exploited for the synthesis of superlattices. In fact, the progress in the semiconductor superlattice has inspired investigations on metallic superlattices and even amorphous multilayered structures (Abeles and Tiedje, 1983). In the area of semiconductor devices, ideas originating from the superlattice have found their applications in heterojunction field-effect transistors (HEMT or TEGFET), quantum-well lasers, and novel avalanche photodiodes. Superlattice is now one of the important topics appearing at number of conferences and symposia, from which new results are flowing with enthusiasm. We hope that this article, which cannot possibly cover every landmark, provides a little flavor to this excitement.

ACKNOWLEDGMENT

Our investigation on superlattices was sponsored in part by the U.S. Army Research Office.

REFERENCES

Abeles, B., and Tiedje, T. (1983). *Phys. Rev. Lett.* **51,** 2003.
Abstreiter, G., and Ploog, K. (1979). *Phys. Rev. Lett.* **42,** 1308.
Anderson, R. L. (1960). *IBM J. Res. Dev.* **4,** 283.
Ando, T., and Uemura, Y. (1974). *J. Phys. Soc. Jpn.* **36,** 959.
Baraff, G. A., Appelbaum, A., and Hamann, D. R. (1977). *J. Vac. Sci. Technol.* **14,** 999.
Bastard, G. (1981a). *Phys. Rev. B* **24,** 5693.
Bastard, G. (1981b). *Phys. Rev. B* **24,** 4714.
Bastard, G. (1982). *Surf. Sci.* **113,** 165.
Bastard, G., Mendez, E. E., Chang, L. L., and Esaki, L. (1982a). *Phys. Rev. B* **26,** 1974.
Bastard, G., Mendez, E. E., Chang, L. L., and Esaki, L. (1982b). *J. Vac. Sci. Technol.* **21,** 531.
Bauer, S., and Mikkelsen, J. C., Jr. (1982). *J. Vac. Sci. Technol.* **21,** 491.
Blakeslee, A. E., and Aliotta, C. F. (1970). *IBM J. Res. Dev.* **14,** 686.
Bohm, D. (1951). "Quantum Theory." Prentice-Hall, Englewood Cliffs, New Jersey.
Brummell, M. A., Nicholas, R. J., Portal, J. C., Razeghi, M., and Poisson, M. A. (1983). *Physica* **117B/118B,** 753.
Burstein, E., Pinczuk, A., and Buchner, S. (1979). *In* "Physics of Semiconductors 1978" (B. L. H. Wilson, ed.), p. 1231, Institute of Physics Conf. Ser. 43, London.
Camras, M. D., Holonyak, N., Hess, K., Jr., Ludowise, M. J., Dietze, W. T., and Lewis, C. R. (1983). *Appl. Phys. Lett.* **42,** 185.
Capasso, F. (1983). *J. Vac. Sci. Technol. B* **1,** 457.
Capasso, F., Tsang, W. T., Hutchinson, A. L., and Williams, G. F. (1982). *Appl. Phys. Lett.* **40,** 38.
Chambers, R. G. (1952). *Proc. Phys. Soc. London A* **65,** 458.
Chang, C.-A., Segmüller, A., Chang, L. L., and Esaki, L. (1981). *Appl. Phys. Lett.* **38,** 912.
Chang, C.-A., Takaoka, H., Chang, L. L., and Esaki, L. (1982). *Appl. Phys. Lett.* **40,** 983.
Chang, C.-A., Mendez, E. E., Chang, L. L., and Esaki, L. To appear in *Surf. Sci.*

Chang, L. L. (1983). *J. Vac. Sci. Technol. B* **1,** 120.
Chang, L. L., Esaki, L., Howard, W. E., and Ludeke, R. (1973a). *J. Vac. Sci. Technol.* **10,** 11.
Chang, L. L., Esaki, L., Howard, W. E., Ludeke, R., and Schul, G. (1973b). *J. Vac. Sci. Technol.* **10,** 655.
Chang, L. L., Esaki, L., and Tsu, R. (1974a). *Appl. Phys. Lett.* **24,** 593.
Chang, L. L., Esaki, L., Segmüller, A., and Tsu, R. (1974b). *Proc. Int. Conf. Phys. Semicond. 12th, Stuttgart, Germany,* p. 688. B. G. Teubner, Stuttgart.
Chang, L. L., Sakaki, H., Chang, C.-A., and Esaki, L. (1977). *Phys. Rev. Lett.* **38,** 1489.
Chang, L. L., Kawai, N. J., Sai-Halasz, G. A., Ludeke, R., and Esaki, L. (1979). *Appl. Phys. Lett.* **35,** 939.
Chang, L. L., Kawai, N. J., Mendez, E. E., Chang, C.-A., and Esaki, L. (1981). *Appl. Phys. Lett.* **38,** 30.
Chin, R., Holonyak, N., Jr., and Stillman, G. E. (1980). *Electron. Lett.* **16,** 467.
Cho, A. Y. (1971). *Appl. Phys. Lett.* **19,** 467.
Cohen, M. L. (1980). "Advances in Electronics and Electron Physics," Vol. 51, p. 1. Academic Press, New York.
Colvard, C., Merlin, R., Klein, M. V., and Gossard, A. C. (1980). *Phys. Rev. Lett.* **45,** 298.
Cowley, J. M., Cohen, J. B., Salamon, M. B., and Wuensch, B. J. (eds.) (1979). "Modulated Structures." AIP Conf. Proc. 53. Amer. Inst. Phys., New York.
Delagebeaudeuf, D., Delescluse, P., Etienne, P., Laviron, M., Chaplart, J., and Linh, N. T. (1980). *Electron. Lett.* **16,** 667.
Dingle, R., Wiegmann, W., and Henry, C. H. (1974). *Phys. Rev. Lett.* **33,** 827.
Dingle, R., Gossard, A. C., and Wiegmann, W. (1975). *Phys. Rev. Lett.* **34,** 1327.
Dingle, R., Stormer, H. L., Gossard, A. C., and Wiegmann, W. (1978). *Appl. Phys. Lett.* **33,** 665.
Döhler, G. H. (1972). *Phys. Status Solidi B* **52,** 79, 533.
Döhler, G. H., Künzel, H., Olego, D., Ploog, K., Ruden, P., Stolz, H. J., and Abstreiter, G. (1981). *Phys. Rev. Lett.* **47,** 864.
Dupuis, R. D., and Dapkus, P. D. (1978). *Appl. Phys. Lett.* **32,** 295.
Eltoukhy, A. H., Ziiko, J. L., Wickersham, C. E., and Greene, J. E. (1977). *Appl. Phys. Lett.* **31,** 156.
Englert, T., and Tsui, D. C. (1982). *Solid State Commun.* **44,** 1301.
Esaki, L. (1974a). "Les Prix Nobel en 1973." Imprimerie Royale P. A. Norstedt, Stockholm.
Esaki, L. (1974b). *Science* **183,** 1149.
Esaki, L., and Chang, L. L. (1974). *Phys. Rev. Lett.* **33,** 495.
Esaki, L., and Tsu, R. (1969). *IBM Res. Note* **RC-2418.**
Esaki, L., and Tsu, R. (1970). *IBM J. Res. Develop.* **14,** 61.
Esaki, L., Howard, W. E., and Heer, J. (1964). *Appl. Phys. Lett.* **4,** 3.
Esaki, L., Chang, L. L., and Tsu, R. (1970). *Proc. Int. Conf. Low Temp. Phys, 12th, Kyoto, Japan,* p. 551. Keigaku, Tokyo.
Esaki, L., Chang, L. L., Howard, W. E., and Rideout, V. L. (1972). *Proc. Int. Conf. Phys. Semicond., 11th, Warsaw, Poland,* p. 431. Polish Scientific Publishers, Warsaw.
Esaki, L., Chang, L. L., and Mendez, E. E. (1981). *Jpn. J. Appl. Phys.* **20,** L529.
Faurie, J. P., Million, A., and Piaguet, J., (1982). *Appl. Phys. Lett.* **41,** 713.
Frensley, W. R., and Kroemer, H. (1977). *Phys. Rev. B* **16,** 2642.
Fritz, I. J., Dawson, L. R., and Zipperian, T. E. (1983). *J. Vac. Sci. Technol. B* **1,** 387.
Gibbs, H. M., Tarng, S. S., Jewell, J. L., Weinberger, D. A., Tai, K., Gossard, A. C., McCall, S. L., Passner, A., and Wiegmann, W. (1982). *Appl. Phys. Lett.* **41,** 221.
Gnutzmann, U., and Clauseker, K. (1974). *Appl. Phys.* **3,** 9.
Gobeli, G. W., and Allen, F. G. (1966). *In* "Semiconductors and Semimetals" (R. K. Willardson and A. C. Beer, eds.) Vol. 2, p. 263. Academic Press, New York.
Gol'dman, I. I., and Krivchenokov, V. (1961). "Problems in Quantum Mechanics," Addison-Wesley, Reading, Massachusetts.

Gossard, A. C., Petroff, P. M., Weigmann, W., Dingle, R., and Savage, A. (1976). *Appl. Phys. Lett.* **29,** 323.
Gourley, P. L., and Biefeld, R. M. (1982). *J. Vac. Sci. Technol.* **21,** 473.
Greene, J. E., Wickersham, C. E., and Zilko, J. L. (1976). *J. Appl. Phys.* **47,** 2289.
Greene, R. L., and Bajaj, K. K. (1983a). *Solid State Commun.* **45,** 825.
Greene, R. L., and Bajaj, K. K. (1983b). *Solid State Commun.* **45,** 831.
Guldner, Y., Vieren, J. P., Voisin, P., Voos, M., Chang, L. L., and Esaki, L. (1980). *Phys. Rev. Lett.* **45,** 1719.
Guldner, Y., Vieren, J. P., Voisin, P., Voos, M., Maan, J. C., Chang, L. L., and Esaki, L. (1982). *Solid State Commun.* **41,** 755.
Guldner, Y., Bastard, G., Vieren, J. P., Voos, M., Faurie, J. P., and Million, A., (1983). *Phys. Rev. Lett.* **51,** 907.
Harrison, W. A. (1977). *J. Vac. Sci. Technol.* **14,** 1016.
Harrison, W. A., Kraut, E. A., Waldrop, J. R., and Grant, R. W. (1978). *Phys. Rev. B* **18,** 4402.
Holonyak, N., Jr., Kolbas, R. M., Rezek, E. A., and Chin, R. (1978). *J. Appl. Phys.* **49,** 5392.
Kasper, E., Herzog, H. J., and Kibbel, H. (1975). *Appl. Phys.* **8,** 199.
Kim, J. Y., and Madhukar, A. (1982). *J. Vac. Sci. Technol.* **21,** 526.
Klitzing, K. V., Dorda, G., and Pepper, M. (1980). *Phys. Rev. Lett.* **45,** 494.
Kinoshita, H., and Fujiyasu, H. (1980). *J. Appl. Phys.* **51,** 5845.
Kinoshita, H., Takaoka, S., Murase, K., and Fujiyasu, H. (1982). *Proc. Int. Symp. Molecular Beam Epitaxy Related Clean Surface Tech., 2nd, Tokyo, Japan,* p. 61.
Kronig, R. de L., and Penney, W. J. (1930). *Proc. Roy. Soc. London A* **130,** 499.
Laughlin, R. B. (1983). *Phys. Rev. Lett.* **50,** 1395.
Ludeke, R., Esaki, L., and Chang, L. L. (1974). *Appl. Phys. Lett.* **24,** 417.
Ludowise, M. J., Dietze, W. T., Lewis, C. R., Holonyak, N., Jr., Hess, K., Camras, M. D., and Nixon, M. A. (1983). *Appl. Phys. Lett.* **42,** 257.
Maan, J. C., Guldner, Y., Vieren, J. P., Voisin, P., Voos, M., Chang, L. L., and Esaki, L. (1981). *Solid State Commun.* **39,** 683.
Madhukar, A. (1982). *J. Vac. Sci. Technol.* **20,** 149.
Madhukar, A., and Delgado, J. (1981). *Solid State Commun.* **37,** 199.
Manasevit, H. M., Gergis, I. S., and Jones, A. B. (1982). *Appl. Phys. Lett.* **41,** 464.
Manuel, P., Sai-Halasz, G. A., Chang, L. L., Chang, C.-A., and Esaki, L. (1976). *Phys. Rev. Lett.* **25,** 1701.
Mendez, E. E., Bastard, G., Chang, L. L., Esaki, L., Morkoc, H., and Fischer, R. (1982). *Phy. Rev. B* **26,** 7101.
Mendez, E. E., Bastard, G., Chang, L. L., Esaki, L., Morkoc, H., and Fischer, R. (1983a). *Physica* **117B/118B,** 711.
Mendez, E. E., Chang, C.-A., Takaoka, H., Chang, L. L., and Esaki, L. (1983b). *J. Vac. Sci. Technol.* **B1,** 152.
Mendez, E. E., Chang, L. L., Chang, C-A., Alexander, L. F., and Esaki, L. (1984). To appear in *Surf. Sci.*
Miller, R. C., Gossard, A. C., Tsang, W. T., and Munteanu, O. (1982a). *Solid State Commun.* **43,** 519.
Miller, R. C., Gossard, A. C., Tsang, W. T., and Munteanu, O. (1982b). *Phys. Rev. B* **25,** 3871.
Mimura, T., Hiyamizu, S., Fujii, T., and Nanbu, K. (1980). *Jpn. J. Appl. Phys.* **19,** L225.
Naganuma, M., Suzuki, Y., and Okamoto, H. (1981). *Proc. Int. Symp. GaAs Related Comp.* (T. Sugano, ed.), p. 125. Institute of Physics, University of Reading, Berkshire.
Osbourn, G. C. (1982). *J. Appl. Phys.* **53,** 1586.
Osbourn, G. C., Biefeld, R. M., and Gourley, P. L. (1982). *Appl. Phys. Lett.* **40,** 173.
Partin, D. L. (1982). *J. Vac. Sci. Technol.* **21,** 1.
Petroff, P. M., Gossard, A. C., Savage, A., and Wiegmann, W. (1979). *J. Cryst. Growth* **46,** 172.

Petroff, P. M., Gossard, A. C., Logan, R. A., and Wiegmann, W. (1982). *Appl. Phys. Lett.* **41,** 635.
Pickett, W. E., Louis, S. G., and Cohen, M. L. (1977). *Phys. Rev. Lett.* **39,** 109.
Pinczuk, A., Stormer, H. L., Dingle, R., Worlock, J. M., Wiegmann, W., and Gossard, A. C. (1979). *Solid State Commun.* **32,** 1001.
Ploog, K., Fischer, A., Döhler, G. H., and Künzel, H. (1980). *In* "Gallium Arsenide and Related Compounds 1980" (H. W. Thim, ed.), p. 721. Institute of Physics Conf. Ser. 56, London.
Pollmann, J., and Pantelides, S. T. (1980). *Phys. Rev. B* **21,** 709.
Rezek, E. A., Holonyak, N., Jr., Vojak, B. A., Stillman, G. E., Rossi, J. A., Keune, D. L., and Fairing, J. D. (1977). *Appl. Phys. Lett.* **31,** 288.
Sai-Halasz, G. A., Tsu, R., and Esaki, L. (1977). *Appl. Phys. Lett.* **30,** 651.
Sai-Halasz, G. A., Esaki, L., and Harrison, W. A. (1978a). *Phys. Rev. B* **18,** 2812.
Sai-Halasz, G. A., Chang, L. L., Welter, J. M., Chang, C.-A., and Esaki, L. (1978b). *Solid State Commun.* **25,** 935.
Sakaki, H. (1980). *Jpn. J. Appl. Phys.* **19,** L735.
Sakaki, H., Chang, L. L., Ludeke, R., Chang, C.-A., Sai-Halasz, G. A., and Esaki, L. (1977). *Appl. Phys. Lett.* **31,** 211.
Saris, F. W., Chu, W. K., Chang, C.-A., Ludeke, R., and Esaki, L. (1980). *Appl. Phys. Lett* **37,** 931.
Schulman, J. N., and McGill, T. C. (1979). *Appl. Phys. Lett.* **34,** 883.
Segmüller, A., Krishna, P., and Esaki, L. (1977). *J. Appl. Cryst.* **10,** 1.
Sollner, T. C. L. G., Goodhue, W. E., Tannenwald, P. E., Parker, C. D., and Peck, D. D. (1983). *Appl. Phys. Lett.* **43,** 588.
Störmer, H. L., Chang, A., and Tsui, D. C. (1983a). *Phys. Rev. Lett.* **50,** 1953.
Störmer, H. L., Haavasoja, T., Narayanamurti, V., Gossard, A. C., and Wiegmann, W. (1983b). *J. Vac. Sci. Technol.* **B1,** 423.
Takaoka, H., Chang, C.-A., Mendez, E. E., Chang, L. L., and Esaki, L. (1983). *Physica* **117B/118B,** 741.
Tanoue, T., and Sakaki, H. (1982). *Appl. Phys. Lett.* **41,** 67.
Tsang, W. T. (1981). *Appl. Phys. Lett.* **39,** 786.
Tsu, R., and Esaki, L. (1973). *Appl. Phys. Lett.* **22,** 562.
Tsu, R., Chang, L. L., Sai-Halasz, G. A., and Esaki, L. (1975). *Phys. Rev. Lett.* **34,** 1509.
Tsui, D. C., and Englert, T. (1980). *Phys. Rev. Lett.* **44,** 341.
Tsui, D. C., and Gossard, A. C. (1981). *Appl. Phys. Lett.* **38,** 550.
Tsui, D. C., Gossard, A. C., Field, B. F., Cage, M. E., and Dziuba, R. F. (1982a). *Phys. Rev. Lett.* **48,** 3.
Tsui, D. C., Störmer, H. L., and Gossard, A. C. (1982b). *Phys. Rev. Lett.* **48,** 1559.
van der Merwe, J. H. (1963). *J. Appl. Phys.* **34,** 117.
van der Ziel, J. P., Dingle, R., Miller, R. C., Wiegmann, W., and Nordland, W. A., Jr. (1975). *Appl. Phys. Lett.* **26,** 463.
Voisin, P., Bastard, G., Goncalves da Silva, C. E. T., Voos, M., Chang, L. L., and Esaki, L. (1981). *Solid State Commun.* **39,** 79.
Voisin, P., Bastard, G., and Voos, M. (1983). *J. Vac. Sci. Technol.* **B1,** 409.
Voos, M. (1983). *J. Vac. Sci. Technol.* **B1,** 404.
Weisbuch, C., Dingle, R., Gossard, A. C., and Wiegmann, W. (1981). *Solid State Commun.* **38,** 709.
White, S. R., and Sham, L. J. (1981). *Phys. Rev. Lett.* **47,** 879.
Woodall, J. M. (1972). *J. Cryst. Growth* **12,** 32.

2 STRUCTURE OF CHEMICALLY MODULATED FILMS

D. B. McWhan

AT&T Bell Laboratories
Murray Hill, New Jersey

I. INTRODUCTION

The synthesis of single crystals with controlled modulation of the chemical composition is an outstanding example of the progress in material science over the past decade. Crystals can be grown with modulation wavelengths that span the crossover region from three- to two-dimensional behavior in electronic, superconducting, and magnetic properties, and many technologically important devices can be made. The idea of growing chemically modulated films dates back to the work of DuMond and Youtz (1935, 1940), but the present activity in this field stems from the demonstration of the feasibility of growing modulated semiconducting films by vapor-phase (Esaki *et al.*, 1970) molecular-beam (Cho, 1970), and liquid-phase epitaxi (Woodall, 1972). This activity has spread to the synthesis of metallic multilayer films, building on the early diffusion studies of Hilliard and his co-workers (1979). The understanding of the growth processes is still in an early stage, and the quality of the films varies markedly from system to system. In this chapter the evolution of both chemical and crystalline order in modulated films is discussed. The emphasis is on the use of x-ray diffraction to trace this evolution, but other probes such as transmission electron microscopy, ion backscattering, and Auger profiling are included. First, a qualitative description of the development of crystalline order is presented. In Section III idealized models for the diffraction from multilayers of varying degrees of order are developed with reference to specific examples of

ISBN 0-12-170470-X

multilayer systems. This section is subdivided into discussions of one-dimensional and three-dimensional order and crystal perfection. In Section IV a survey of multilayer systems that have been reported through 1982 is given.

II. QUALITATIVE DESCRIPTION OF MULTILAYERS

This book deals with synthetic, chemically modulated films that are a small subset of modulated structures [see examples in "Modulated Structures" (1979)]. There are many minerals and alloys that have long-wavelength chemical modulations, and these are usually accompanied by a modulation of the interplanar spacing as a result of differences in atomic size. In many cases the modulation wavelength is incommensurate in that the wavelength is not a small integral multiple of the interplanar spacing in the direction of the modulation.

The classification of modulated structures also includes materials that undergo displacive transitions as a function of temperature or pressure from an undistorted phase to a phase with an incommensurate or commensurate distortion wave. Often there are further transitions from an incommensurate to a commensurate phase as in $NaCO_3$, K_2SeO_4 (Iizumi *et al.*, 1977), and 2H–$TaSe_2$ (Fleming *et al.*, 1980a). As shown by Sinha (1980), phase transitions in lower-dimensional materials are an area of strong theoretical and experimental interest, and many of the theoretical concepts such as the Frank and van der Merwe model (1949) for a commensurate–incommensurate transition were developed within the context of epitaxial growth of one material or another.

The structure of chemically modulated films is dependent on the properties of the interfaces. How the two materials match at each succeeding interface influences the continued growth of the film. There is an enormous amount of literature on epitaxial growth of semiconductors and metals (Matthews, 1975), but except for the semiconductor systems, most attempts to grow multilayers have been based on an empirical approach rather than on controlling composition to obtain good lattice matching. The quality of a multilayer is dependent on variables such as substrate symmetry, temperature, and cleanliness, and systematic studies of these variables have been reported for several semiconductor systems. In favorable cases not only is it possible to obtain smooth epitaxial growth, but also it is possible to fix the crystallographic axis along the growth direction by a judicious choice of substrate symmetry and temperature. Nb_nTa_m multilayers are one of the first systems in which this fine tuning has been successful (Durbin *et al.*, 1982).

In discussing chemically modulated films it is necessary to distinguish between chemical order and structural order. It is possible to envision extremes in which one type of order might be almost perfect while the other might be nonexistent. For example, a chemically modulated film could be made under conditions that would lead to amorphous blocks of each material. Assuming that the deposition rates and the shutters that controlled the thickness of each block were monitored precisely, the resulting modulated film would have sharp, smooth interfaces and a high degree of chemical order. The film would be structurally disordered because each material was amorphous. Conversely, a film composed of mutually soluble materials could be grown as a perfect single crystal, but because of interdiffusion the chemical modulation could be a sinusoidal wave whose amplitude was less than 100%. The film would be ordered in the crystallographic sense that an average lattice is defined, but the chemical order would be fairly low.

A second equally important distinction must be made between the concepts of order and coherence in chemically modulated films. Order is a measure of the amplitude and shape of the composition wave, while coherence is a measure of the regularity of the composition wave throughout the modulated film (both laterally and along the growth direction). For example, a modulated film could have a sinusoidal composition wave in which the composition of a component varies only from 49 to 51% in a regular way throughout the film. The film would have a low degree of order, but it would be completely coherent. In the other extreme, the composition wave might be a square wave with the composition varying abruptly from 100% of one component to 100% of the other, but the modulation wavelength might vary as a result of fluctuations during the deposition. The resulting film would be completely ordered but would have limited coherence.

To study physical properties, high-quality crystals are desirable, and as the preparative techniques are improved, more and more high-quality multilayers will be synthesized. At the present time the systems that have been reported in the literature span almost the whole range in terms of structural order. In contrast to this, the average chemical order is remarkably good. For example, tungsten–carbon multilayers have been made for use as x-ray mirrors (Underwood *et al.*, 1979). In these films the tungsten is polycrystalline with a strong [110] texture and the carbon is amorphous. The films have high chemical order but low structural order. On the other hand, $(GaAs)_n(AlAs)_m$ (Dernier *et al.*, 1977; Fleming *et al.*, 1980b) and Nb_nTa_m (Durbin *et al.*, 1982; Hertel *et al.*, 1982) multilayers have high structural order, but the interfaces are on average two atomic layers thick so the chemical order is good but not perfect.

Assuming that the chemical order is high, it seems reasonable to classify multilayers principally on the basis of the degree of structural order. As shown schematically in Fig. 1, there is first an evolution of one-dimensional order starting with periodic alternating amorphous blocks of materials a and b. With increasing order, first one and then both materials are polycrystalline with varying degrees of texture. The different materials can have similar or different structures. These films have one-dimensional structural order in that the position and average composition of each layer along the growth direction is defined

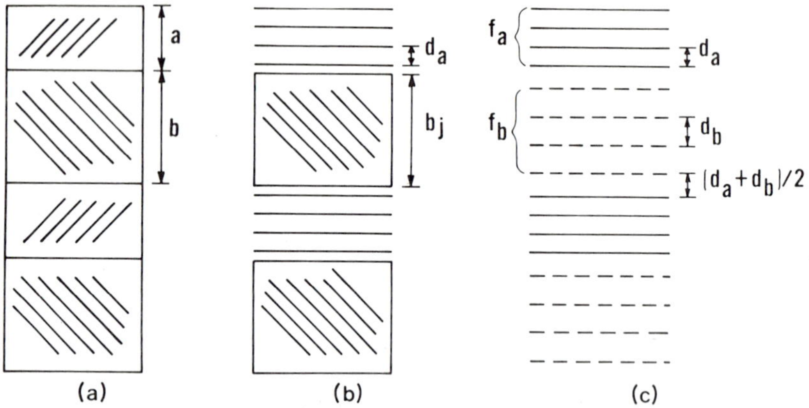

Fig. 1. Evolution of one-dimensional order: (a) alternating amorphous regions, (b) textured polycrystalline regions alternating with amorphous regions, and (c) alternating textured polycrystalline regions with interplanar spacings d_a and d_b.

by a translation vector and a modulation wavelength. Within a given plane there can be long-range order, but there is not necessarily any order between the lateral position of a plane in one block of one material and the lateral position of a plane in a different block of the same material. The Cu_nNb_m system is an example of this type of multilayer with polycrystalline blocks of Cu with [111] texture interspersed between polycrystalline blocks of Nb with [110] texture. The resulting film has a high degree of chemical order and of one-dimensional structural order (Schüller, 1980; Lowe *et al.* 1981). The film also has good coherence along the growth direction but has poor lateral coherence between blocks. As indicated on the right of Fig. 1, in the middle of each block the structure and lattice parameters correspond to those of the pure bulk materials.

The evolution of three-dimensional order is best thought about in terms of the interface. If the difference between the lattice parameters of the two materials is sufficiently small, then the layers can be constrained to match perfectly by introducing a coherency strain into each layer. The strain energy is balanced by the energy needed to introduce a misfit dislocation that would relieve part of the coherency strain (Hilliard, 1979). The two extremes are illustrated in Fig. 2. In Fig. 2a, a coherent structure is formed by putting one block under uniaxial tension and the other under uniaxial compression. In Fig. 2b, an incoherent interface results when some of the strain is relieved by dislocations. There is a phase transition when dislocations develop and their number increases continuously until all the coherency strain is relieved and the interfaces are incoherent. This is an example of a commensurate–incommensurate (CI) transition. The misfit dislocations are similar to the domain walls in rare gases physisorbed on graphite or to the discommensurations in the charge density wave state of 2H–$TaSe_2$ (Frank and van der Merwe, 1949; McMillan, 1976). In the case of multilayers the transition occurs because the balance between coherency strain

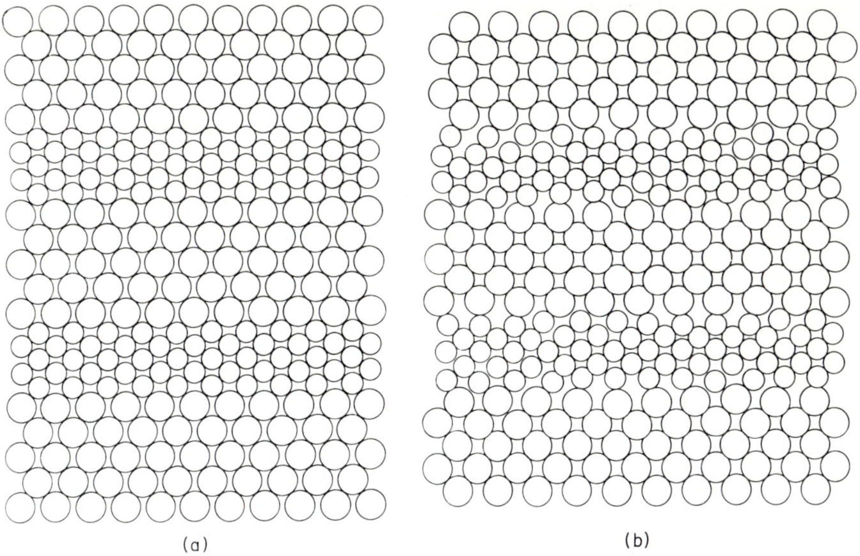

Fig. 2. Comparison of coherent and incoherent structures. (a) In a coherent structure there is a single lattice parameter perpendicular to the modulation direction. Alternating regions are under uniaxial tension and compression. (b) In an incoherent structure the coherency strains are relieved by misfit dislocations.

energy and dislocation energy changes with modulation wavelength. The coherency strain energy per unit volume is constant, whereas the dislocation energy varies as the number of interfaces per unit volume. Since the dislocation energy decreases with increasing wavelength, A CI transition occurs in all multilayer systems, but the critical wavelength λ_c will depend on the mismatch between the two components and on their elastic properties. For the Cu_nNb_m system λ_c is of the order of 10 Å (Lowe *et al.*, 1981), and for the $(GaAs)_n(GaAs_{0.5}P_{0.5})_m$, λ_c is ~250 Å (Matthews and Blakeslee, 1974). If the symmetry of the layers on each side of the interface is different, then the CI transition is anisotropic. For a (110) plane of a bcc structure to become commensurate with a (111) plane of a fcc structure one or both of the planes must have an anisotropic strain. The Nb_nAl_m system has an anisotropic CI transition as a function of λ (McWhan *et al.*, 1983).

The stabilization of new crystallographic phases in multilayer systems is an area that is only just beginning to be explored. For example, in the Nb_nZr_m system, bulk Nb is bcc and Zr is hcp at room temperature. At short wavelengths a bcc phase of Zr is stabilized in the multilayer (partly as a result of interdiffusion) and at longer wavelengths the Zr regions revert to the hcp structure (Lowe and Geballe, 1984).

Of more importance to the electronic and magnetic properties of multilayers is the lowering of the local symmetry that results from the coherency strains if the multilayer is coherent. In many cases the coherency strains are equivalent to putting the pure elements under a uniaxial strain of 1–2 GPa (10–20 kbar) as illustrated in Fig. 2a. For a metal such as Pd, in which the Fermi energy lies near a peak in the density of states, the resulting 1–2% tetragonal or rhombohedral distortion can drastically change the electronic properties (Brodsky and Freeman, 1980). Similarly, in a semiconductor such a stress could lift degeneracies in the band structure. The coherency strains result principally in uniaxial strains rather than hydrostatic strains; i.e., there is little if any change in density.

III. X-RAY DIFFRACTION FROM MULTILAYERS

X-ray diffraction provides the most easily interpreted structural data characterizing a broad range of existing multilayers. This technique is nondestructive, and most of the measurements can be done in a reflection geometry leaving the sample on the substrate. The evolution of one-dimensional order along the growth direction can be studied on a standard diffractometer, but the detailed evolution of three-dimensional order necessitates scans in general directions in reciprocal space. The positions and line shapes of the Bragg reflections of the average lattice and of the surrounding satellite reflections are used to calculate the average lattice constants and the average modulation wavelength and to obtain a measure of fluctuations in both quantities that may be caused by composition gradients and by loss of coherence. The intensities of the reflections are used to obtain the composition and the strain wave profiles.

There are several detailed developments of diffraction from one-dimensional modulated structures in the literature that are aimed principally at the problem of spinoidal decomposition in alloys (DeFontaine, 1966). Also there are models, which are applicable in the limit of small strain modulation, in standard textbooks (Guinier, 1963). The object of this section is to give an overview, based on model calculations, of how the diffraction data evolve with increasing order and to point out their implications for the structure of multilayers. This section is divided into three subsections dealing with the evolution of one-dimensional order, three-dimensional order, and crystal perfection.

A. Evolution of One-Dimensional Order

Figure 1 shows a sequence of films in which there is a progression from films with alternating amorphous regions to films in which one and then both regions are polycrystalline with texture. The diffraction from these structures is calculated for idealized models that have sharp interfaces and blocks containing an integral number of layers. The expressions for the structure factor squared are presented below for each type of multilayer and then the diffraction patterns are compared in Fig. 3.

The film represented by Fig. 1a has high chemical order but no structural order. The intensity of N alternating amorphous blocks with height $d = a + b$

can be approximated by regions of constant scattering power g_a and g_b. If $a = b$, then the intensity as a function of the scattering vector along the modulation direction q is given by (Lynn *et al.*, 1976)

$$I(q) \propto \frac{4}{q^2} \sin^2\left(\frac{qd}{4}\right)\left[\frac{\sin^2(Nqd/2)}{\sin^2(qd/2)}\right]\left[g_a^2 + g_b^2 + 2g_a g_b \cos\left(\frac{qd}{2}\right)\right].$$

Reflections occur at values of $q = 2\pi l/d$ for odd integer values of l with an intensity of $I(l) = (Nd/\pi l)^2(g_a - g_b)^2$. The scattering function, normalized to $I(l = 0)$, is shown in Fig. 3a. For comparison with the multilayers described below, $N = 10$ and g_a and g_b correspond to Nb and Al.

The multilayer shown in Fig. 1b has regions that are polycrystalline with texture alternating with amorphous regions. This type of structure occurs in multilayers of WC in which the tungsten deposits with [110] normal to the film and in which the carbon is amorphous (Underwood *et al.*, 1979). This system can be modeled as sets of n_a planes with spacing d_a that are separated by a distance b_j and repeated N times. To illustrate the effect of fluctuations in b_j on the diffraction, the distance b_j is assumed to have a gaussian distribution around an average value $\bar{b}$ with a width of c^{-1} (McWhan *et al.*, 1983). Defining $q = 2\pi l/d_a$ and $p = \pi/(cd_a)$, the intensity is

$$I(l) = \frac{f^2 \sin^2(\pi n_a l)}{\sin^2(\pi l)}\left[\frac{1 + e^{-2Np^2l^2} - 2\cos(2\pi l N(n_a + (\bar{b}/d_a)))e^{-N^2p^2l^2}}{1 + e^{-2p^2l^2} - 2\cos(2\pi l(n_a + (\bar{b}/d_a)))e^{-l^2p^2}}\right].$$

The diffraction pattern was calculated for $N = 10$, $n_a = 5$, $c^{-1} = 0.05$, and $\bar{b}/d_a = 5$. The results are shown in Fig. 3b. The new features are the appearance of reflections that correspond to the spacing between the planes plus satellites at $l \pm m/10$, which result from the periodicity of $5d_a + \bar{b} = 10d_a$ of blocks of polycrystalline material. The effect of the gaussian distribution of spacings b_j is to introduce an effective Debye–Waller factor that progressively reduces the higher-order reflections but does not affect the linewidth. This is a useful measure of the degree of order achieved during the growth process. If (after correcting for the Lorentz polarization, absorption, and scattering factors) the intensity of the higher-order reflections is greatly reduced, then better control of deposition rates and shutter motion are needed.

Progressing to Fig. 1c, in which there are alternating textured regions of materials a and b, the intensity is given by (Segmüller and Blakeslee, 1973)

$$I(l) = \left(\frac{\sin \pi l n N}{\sin \pi n}\right)^2 \left\{ f_a^2 \left(\frac{\sin \pi l n_a d_a/d_0}{\sin \pi l d_a/d_0}\right)^2 + f_b^2 \left(\frac{\sin \pi l n_b d_b/d_0}{\sin \pi l d_b/d_0}\right)^2 \right.$$
$$\left. + 2 f_a f_b \cos \pi l n \left(\frac{\sin \pi l n_a d_a/d_0}{\sin \pi l d_a/d_0}\right)\left(\frac{\sin \pi l n_b d_b/d_0}{\sin \pi l d_b/d_0}\right)\right\},$$

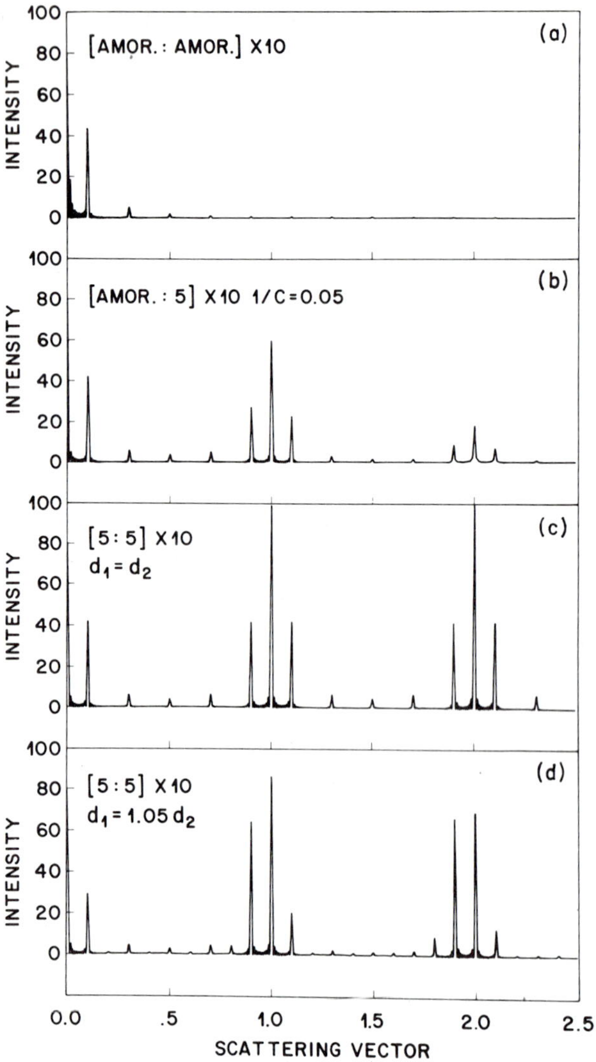

Fig. 3. Scattering intensity $[I(0) = 100]$ versus wave vector in units of the reciprocal lattice vector **c**. (a)–(d) correspond to the idealized models shown in Fig. 1. (a) Alternating amorphous regions of Nb and Al. (b) Blocks of 5 layers of Nb with $d_a = 2.33$ Å separated by distances of $\bar{b} = 5d_a$. (The solid line assumes a 5% width to a gaussian distribution of b values around $\bar{b}$.) (c) Multilayer of 5 layers of Nb followed by 5 layers of Al with $d_a = d_b$ or to a multilayer as in (b) with $b_i = 5d_a$. (d) Effect on the multilayer in (c) of adding a strain modulation with $d_a = 1.05d_b$.

where $n = n_a + n_b$ and $nd_0 = n_a d_a + n_b d_b$. The calculated diffraction pattern setting $N = 10, f_a = 41, f_b = 14$, $d_a = d_b$, and $n_a = n_b = 5$ is shown in Fig. 3c. A similar pattern is obtained for the model of alternating amorphous and polycrystalline regions if $c^{-1} = 0$. A diffraction pattern obtained with the scattering vector normal to the film does not by itself establish that both regions are polycrystalline. The diffraction pattern for $d_a = 1.05 d_b$ is shown in Fig. 3d. For $d_a = d_b$, there is only scattering power modulation that results in satellites at $l \pm m/10$ of equal intensity, but for $d_a \neq d_b$ there is a competition between scattering power modulation and a displacement wave modulation. This leads to a marked asymmetry in the intensity of the plus and minus intensity. At the same time the intensity of the reflections corresponding to the average lattice decrease on going from $l = 0 \rightarrow 1 \rightarrow 2$. This can be viewed as a frozen phonon in which intensity is taken from the average Bragg peak and put into the harmonics at $l \pm m/10$. A similar shift in the weight of the intensity of the satellite results if the ratio of d_a/d_b is held fixed and the values of n_a and n_b are increased. This has been demonstrated for the Cu_nNb_m multilayer system (Fig. 4) (Schüller, 1980). As the blocks are made thicker, the envelope function [in brackets in the expression above for $I(l)$] changes from a single central peak around the value of the average lattice spacing to a bimodal distribution that is peaked around the d spacings of the two constituents.

The preceding idealized model with a square-wave modulation is only achieved in systems in which the two components are immiscible under the conditions of growth. More generally, there is interdiffusion, and the resulting composition and displacement waves are intermediate between a square wave and a sine wave. In this general case the prefactor and the exponents in the expression for the scattering amplitude are approximated by Fourier series expansions of the scattering power and the displacements, respectively. The exponents can be expanded as series of Bessel functions J_p, but the expressions become sufficiently complex that attempts to use them to fit experimental data have been limited to truncating both the Fourier expansions and the Bessel function expansions to terms up to second order. If the position and scattering power of an atom in the mth plane are given by

$$z_m = d_0[m + (n\Delta_1/2\pi)\sin(2\pi m/n) + (n\Delta_2/4\pi)\sin(4\pi m/n)],$$

$$F_m = F_0[1 + \phi_1 \cos(2\pi m/n) + \phi_2 \cos(4\pi m/n)],$$

where F_0 and d_0 are the average structure factor and spacing of a layer and Δ_i and ϕ_i are the coefficients of the Fourier expansion of the displacement and scattering power waves, respectively, then Segmüller and Blakeslee (1973a) give the following expressions for the amplitude of the main Bragg peak at $l = 0$, 1,

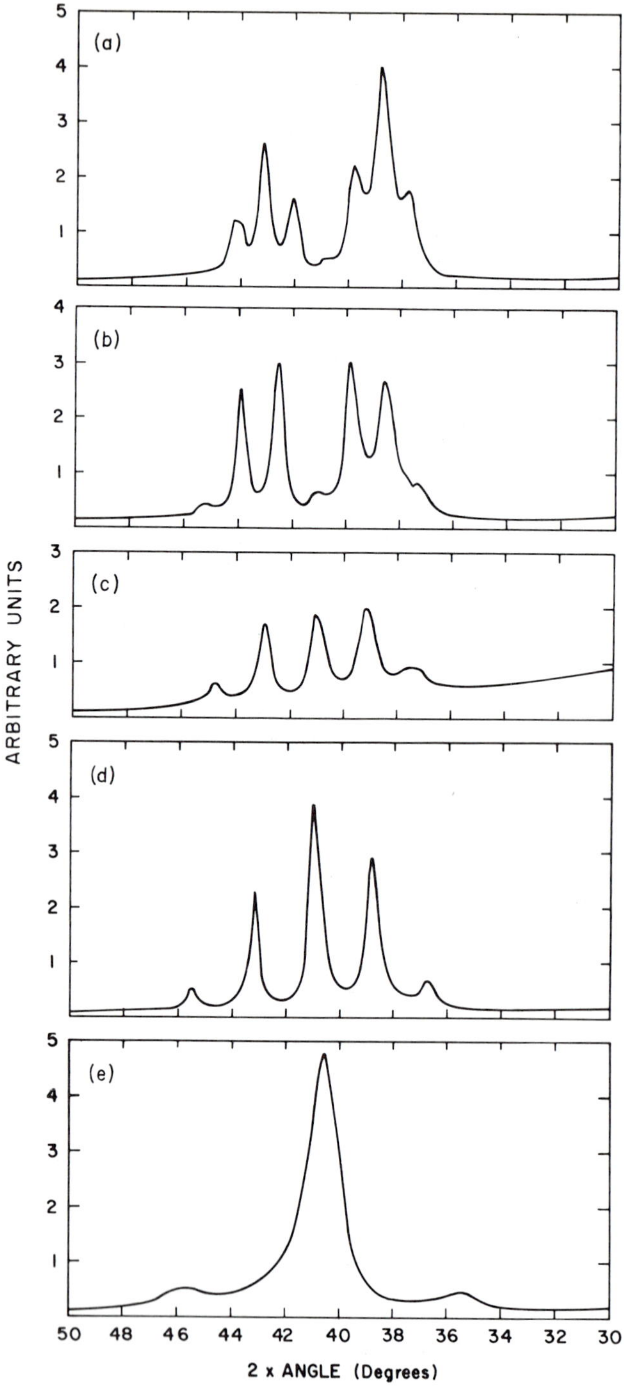

Fig. 4. X-ray scattering observed in Nb_nCu_m multilayers as a function of modulation wavelength. (a) $\lambda/2 = 46$ Å, (b) $\lambda/2 = 34$ Å, (c) $\lambda/2 = 25$ Å, (d) $\lambda/2 = 22$ Å, and (e) $\lambda/2 = 10$ Å. [From Schüller (1980).]

2, . . . , and the first- and second-order satellites at $l \pm 1/n$ and $l \pm 2/n$:

$$A(l) = F_0 G_0[J_{01}J_{02} - \phi_1 J_{11}J_{12} + \phi_2 J_{21}(J_{02} + J_{22})],$$

$$A\left(l \mp \frac{1}{n}\right) = F_0 G_0 \left\{ \pm J_{11}J_{02} - J_{11}J_{12} + \frac{\phi_1}{2}[J_{01}(J_{02} \pm J_{12}) + J_{21}(J_{02} + J_{22})] \mp \frac{\phi_2}{2} J_{11}(J_{02} + J_{22}) \right\},$$

$$A\left(l \mp \frac{2}{n}\right) = F_0 G_0 \left\{ \pm J_{01}J_{12} + J_{21}(J_{02} + J_{22}) \pm \frac{\phi_1}{2}[J_{11}(J_{02} - J_{22})] + \frac{\phi_2}{2}[J_{01}(J_{02} + J_{22}) \pm J_{21}J_{12}] \right\},$$

where $J_{p1} = J_p(ln\Delta_1)$, $J_{p2} = J_p(ln\Delta_2/2)$, and $G_0 = \sin \pi lN/\sin \pi l$. The Bessel function arguments vary with l, and a complete set of intensities for the satellites around $l = 0$, 1, and 2 can be fitted to the preceding equations to give values for ϕ_1, ϕ_2, Δ_1, and Δ_2. In practice the coefficients are highly correlated, and a unique least-squares fit to a given set of data is elusive. High-quality samples and intensity measurements are needed to address detailed questions about the interfaces, such as whether the displacements at the interface simply scale with the composition or show deviations that reflect the different bonding conditions in the presence of a composition gradient rather than those in a random alloy of a given composition.

In the limit of a displacement wave of small amplitude that scales with the composition such as exists in $(GaAs)_n(AlAs)_m$, the intensity of the satellites approaches a simple limiting form. If the atomic scattering power and position vary between $f(1 \pm \eta)$ and $d(1 \pm \epsilon)$, respectively, then the satellite intensities are given by (Guinier, 1963; Fleming *et al.*, 1980b)

$$I\left(\frac{l}{d} \pm \frac{m}{\lambda}\right) \Big/ I\left(\frac{l}{d}\right) \approx \frac{C_m^2}{4}\left[\left(\frac{l}{d} \pm \frac{m}{\lambda}\right)\lambda\epsilon \mp \eta\right]^2.$$

The Fourier coefficients $|C_m|$ and ϵ can be obtained from the intensities of the satellites. However, it is necessary to assume a simple waveform such as a rectangular or trapezoidal wave in order to assign phases to C_n because only $|C_n|$ is measured. For satellites around $l = 0$, the strain term is small and $|C_m|$ can be obtained directly and then Fourier transformed to give an average composition wave in the multilayer. In magnetic systems, polarized neutron-scattering studies yield similar results for the magnetization wave in the multilayer (Felcher *et al.*, 1980). The evolution of the composition wave can be traced as a function of wavelength as illustrated in Fig. 5, in which the profile in $(GaAs)_n(AlAs)_m$ is compared for $n:m = 1:1$ and $7:4$ (Dernier *et al.*, 1977). Similarly, as discussed in Chapter 11 on diffusion, the intensities can be studied as a function of

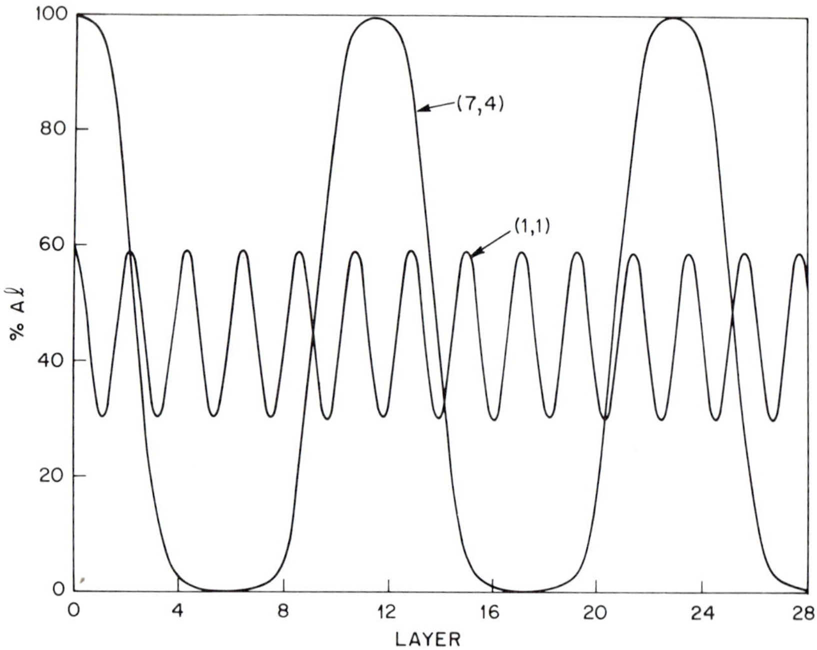

Fig. 5. Composition modulation wave obtained by a Fourier transformation of the measured scattering amplitudes of the harmonics around the (002) reflection of $(GaAs)_n(AlAs)_m$, with $n:m = 1:1$ and $7:4$. [After Dernier *et al.* (1977).]

annealing time and detailed pictures of the composition with various degrees of interdiffusion can be obtained as illustrated for $(GaAs)_{12}(AlAs)_9$ in Fig. 6 (Fleming *et al.*, 1980b).

A number of other probes have been used to measure the composition wave in multilayers. These include high-energy helium backscattering and channeling, transmission electron microscopy, and Auger profiling. Backscattering and channeling are particularly attractive when the multilayer is composed of elements with similar average x-ray atomic scattering factors, such as $Ge_n(GaAs)_m$, and they provide not only composition wave information but also evidence for dislocations, stacking faults, and so on (Chang *et al.*, 1981). Detailed information on the growth mechanism was obtained on Au_nAg_m multilayers by depositing the film, layer by layer, and by making backscattering and channeling measurements between layer depositions (Culbertson, 1981). Lattice imaging transmission electron microscopy techniques give a real space image of a multilayer, and by making densitometer traces of the resulting films, a qualitative composition profile can be obtained, as illustrated in Fig. 7 for $(GaAs)_{39}(Ga_{0.6}Al_{0.4}As)_{39}$ (Petroff, 1977). It is difficult to relate the changes in film contrast to absolute changes in concentration. The power of the electron

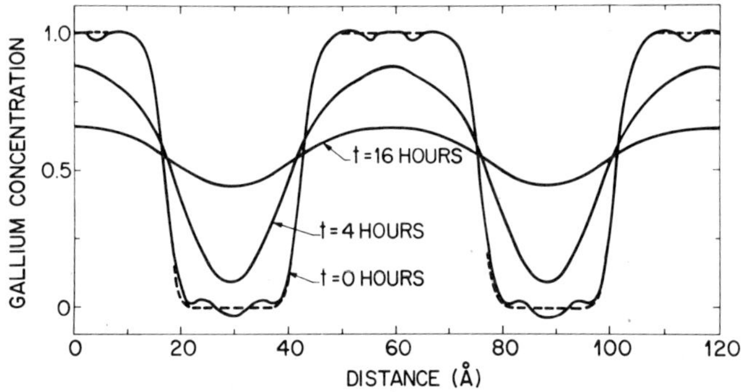

Fig. 6. Change in the composition modulation with annealing at $T = 860°C$ for $(GaAs)_{12}(AlAs)_9$. [From Fleming *et al.* (1980).]

microscope lies in being able to give a global picture of a multilayer and to detect various kinds of defects. Another qualitative probe of the composition wave is the combined use of argon sputter etching and Auger electron spectroscopy (Ludeke *et al.*, 1974). The composition profile obtained for 50-Å layers of GaAs and $Ga_{0.75}Al_{0.35}As$ is shown in Fig. 8. This technique relies on the uniformity of the sputtering rate and assumes that minimum damage or chemical changes result from the sputtering process. This technique gives composition profiles for wavelengths of ~ 100 Å and above.

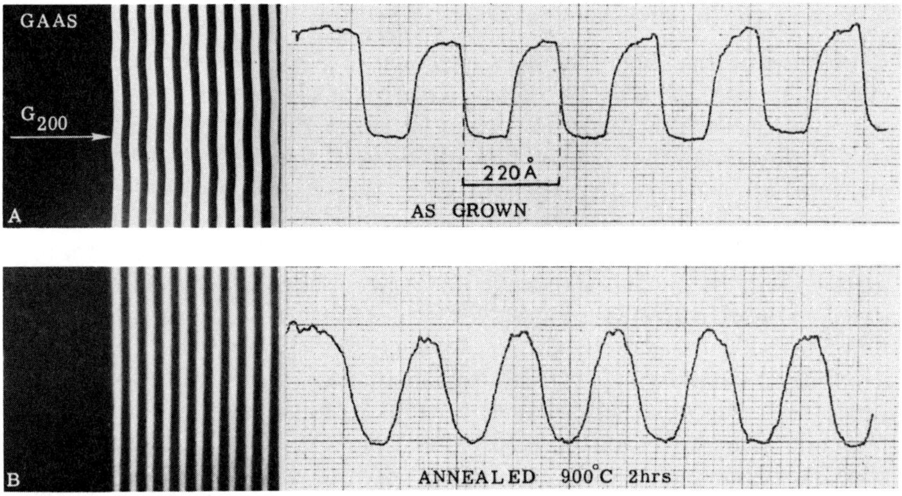

Fig. 7. Dark-field transmission electron microscopy micrograph of a cross section through a $(GaAs)-(Ga_{0.6}Al_{0.4}As)$ multilayer with $\lambda = 220$ Å. (A) As-grown film and (B) annealed film. [From Petroff (1977).]

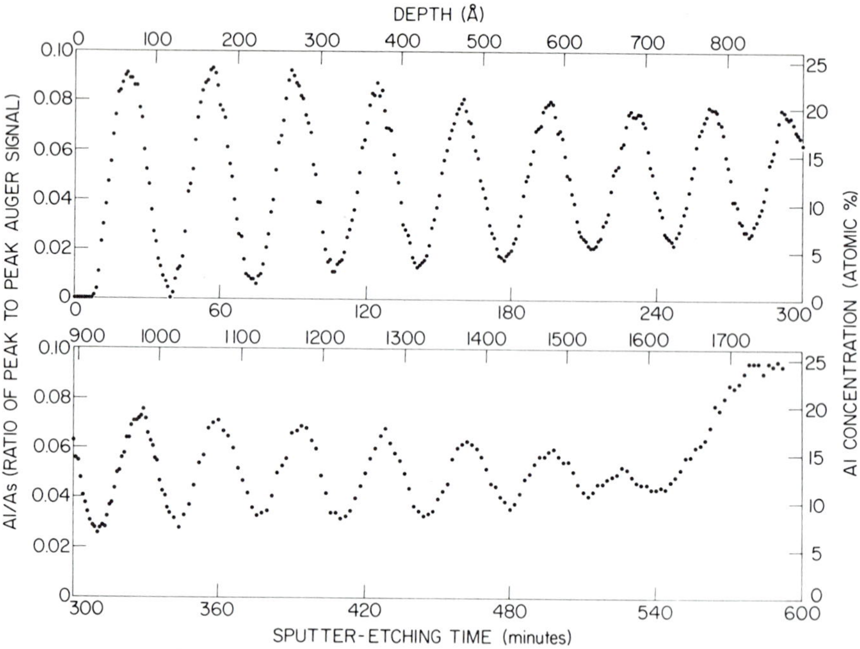

Fig. 8. Composition modulation obtained by Auger electron spectroscopy of a 16.5-period superlattice consisting of 50-Å blocks of GaAs and $Ga_{0.75}Al_{0.25}As$. [From Ludeke *et al.* (1974).]

B. Evolution of Three-Dimensional Order

In the preceding section the stacking sequence of layers of atoms was developed without regard to the lateral coherence of the planes with respect to each other. As discussed in Section II, when the mismatch across the interface is small (below some critical modulation wavelength), the interfaces are coherent; i.e., there is a single lattice parameter that describes the interplanar spacings perpendicular to the growth direction. This results in a coherency strain as the material with the larger lattice parameter is compressed and vice versa for the material with the smaller parameter. The presence of coherency strains can be observed directly by measuring the *d* spacings of planes perpendicular to the film. This is illustrated for the Cu_nNi_m system in Fig. 9 (Gyorgy *et al.*, 1982). For a multilayer sample with [111] texture of alternating 8 Å of Cu and 8 Å of Ni, a single $(2\bar{2}0)$ reflection is observed between the positions expected for the $(2\bar{2}0)$ reflection of pure Cu and pure Ni. With increasing wavelength part of the coherency strain is relieved by the formation of misfit dislocations and the $(2\bar{2}0)$ broadens as shown for the 30-Å Cu – 30-Å Ni sample in Fig. 9. If all the coherency strain is relieved, then the Cu and Ni regions each have the *d* spacing found in the respective pure elements, and there are two separate reflections instead of

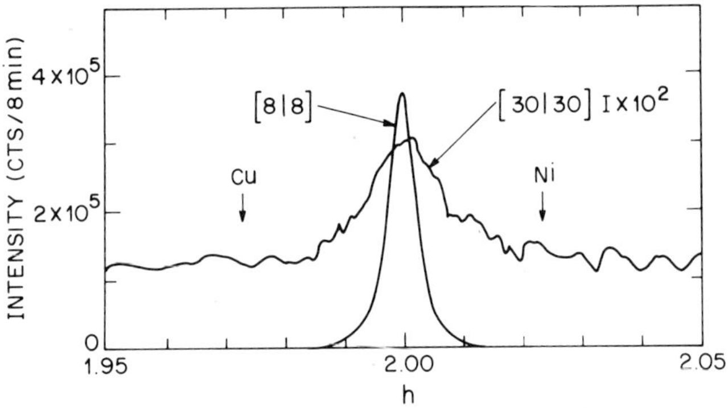

Fig. 9. Intensity versus h for the $(2\bar{2}0)$ of coherent Cu_nNi_m films for two different modulation wavelengths. A single reflection midway between the positions expected for pure Cu and pure Ni is direct evidence for a coherent structure. The broadened reflection for the sample with $\lambda = 60$ Å suggests a partial loss of coherency. [From Gyorgy *et al.* (1982).]

a single average reflection. At the interface the $(2\bar{2}0)$ planes of Cu and Ni, which are perpendicular to the film, are continuous in a coherent structure, whereas each misfit dislocation corresponds to adding an extra plane of Ni atoms (or alternatively deleting an extra plane of Cu atoms) until there are ~200 planes of Cu for every 205 planes of Ni. The coherency strain varies inversely with the number of planes between misfit dislocations, and the average density of dislocations can be inferred from the broadening of the $(2\bar{2}0)$ reflection of the CuNi[30|30] sample. The broadening of the $(2\bar{2}0)$ of the [30|30] sample over that of the [8|8] sample implies an average distance between dislocations of 500 and 1000 Å. This is a lower limit as there are other causes of line broadening.

In the limit of widely spaced misfit dislocations, the measurement of a line shape by x-ray diffraction becomes insensitive, and although x-ray topography can be used to probe the dislocation array, transmission electron microscopy is the best probe. Detailed studies have been made, especially on semiconductor systems that include the determination of the Burgers vector, correlation of dislocations from layer to layer, and occurrence of dislocation arrays. For example, in $(GaAs)_n(GaAs_{0.5}P_{0.5})_m$ grown on GaAs (001), the critical wavelength for the appearance of dislocations is of the order of 250 Å and the data are consistent with {111} slip planes and Burgers vectors of the form $\frac{1}{2}a\langle 110\rangle$ rather than a simple edge dislocation with a (100) or (010) slip plane (Matthews and Blakeslee, 1974).

Another more indirect way to probe the coherency strain by x-ray diffraction is to determine the strain wave along the growth direction by analyzing the intensities of the harmonics. The resulting value of the strain amplitude ϵ is compared with that expected for the pure elements in the absence and presence of coherency strains. The coherency strain energy varies as the biaxial modulus, which for cubic crystals is given by (Cahn, 1962)

$$Y(hkl) = \frac{1}{2}(C_{11} + 2C_{12}) \times \left[3 - \frac{C_{11} + 2C_{12}}{C_{11} + 2(2C_{44} - C_{11} + C_{12})(h^2k^2 + h^2l^2 + h^2l^2)}\right],$$

where C_{ij} are the elastic moduli and hkl the Miller indices. In the absence of coherency strains

$$\epsilon = (1/a)\, da/dc,$$

where a is the lattice parameter and c the concentration of the binary alloy. When the modulation direction is [111], a coherent structure will have a variation in d spacing along [111] given by

$$\epsilon_{\text{coh}} = (Y(111)/2C_{44})\epsilon_{\text{incoh}}.$$

For a system like CuNi, $\epsilon_{\text{incoh}} = 0.024$ and $\epsilon_{\text{coh}} = 0.039$ (Hilliard, 1979), so that the amplitude of the strain wave is substantially larger for the coherent structure. A qualitative measure of the transition from coherent to incoherent interfaces as a function of modulation wavelength can be obtained from the ratio of the plus to the minus harmonic. In the limit of small ϵ this ratio is

$$R = \frac{I^-(l)}{I^+(l)} \approx \left[\left(\frac{l\lambda}{d} - 1\right)\epsilon + \eta\right]^2 \Big/ \left[\left(\frac{l\lambda}{d} + 1\right)\epsilon - \eta\right]^2.$$

The ratio approaches 1.0 at large λ. If the heavier component has the larger-lattice parameter, then R increases more rapidly with decreasing λ for a coherent structure than for an incoherent structure. A nonmonotonic increase in R with decreasing λ may indicate an incoherent to a coherent transition. However, an anomaly could also result from a change in the structure of one of the components of the multilayer or from one of the components becoming amorphous. It is difficult to draw unambiguous conclusions about the degree of coherency solely on the basis of diffraction data taken with the scattering vector along the modulation direction.

In recent years there have been numerous attempts to grow multilayers in which the structure of each component is different (Schüller, 1980). The most common example is a combination of bcc and fcc metals. The corresponding close-packed planes at the interface are (110) and (111), and one or both of the planes will have an anisotropic coherency strain if the structure is coherent. Most of the metallic materials studied to date are polycrystalline in the plane of the film, and the only evidence for coherency strains comes from observed deviations in the positions of the powder diffraction rings from the position expected for the pure elements. For example, in Nb_nCu_m multilayers (Lowe *et al.*, 1981) no evidence for coherency strains was found while in the Nb_nAl_m

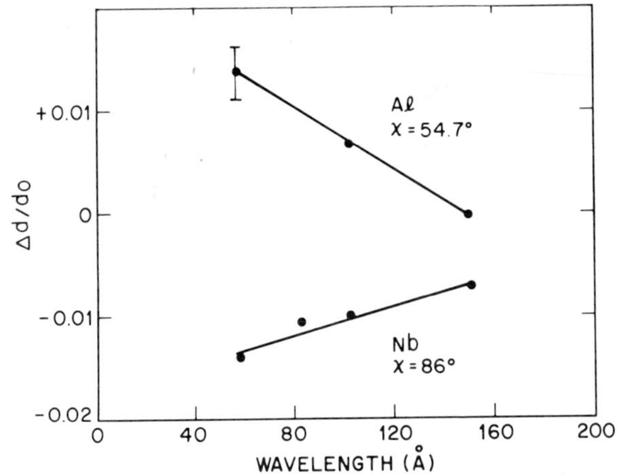

Fig. 10. The variation in the *d* spacing of Al(200) and Nb(002) with wavelength for a multilayer ($\lambda = 58$ Å) with [110] and [111] texture in the Nb and Al regions, respectively. The deviations from the values expected for films of the pure elements are evidence for a coherent (commensurate) to incoherent (incommensurate) transition as a function of λ. The angle x is from the film normal.

system (McWhan *et al.*, 1983), the Nb (110) planes contract ~0.5% along [1$\bar{1}$0] and ~1.5% along [002] and the Al expands. With increasing modulation wavelength, the anisotropic coherency strain decreased toward zero, as illustrated in Fig. 10. The Nb_nAl_m multilayers are at least partially coherent at short wavelengths and then become incoherent at longer wavelengths.

C. Crystal Perfection

A number of different types of fluctuations from the average structure can be assessed using x-ray scattering. If the multilayer is a textured polycrystalline film, then the structure consists of many grains that have (1) a range of orientations with respect to the growth direction, (2) an average grain size, (3) a degree of local strain resulting from the stresses applied by neighboring grains, and (4) a range of modulation wavelengths in different grains resulting from chemical variations across the film. If the multilayer is coherent and approaches a single crystal, then more subtle deviations can be studied. For example, the growth may be anisotropic because of the chemical nature of the growth surface or because the growth surface is not cut parallel to a major crystallographic plane, and the resulting surface is stepped. Other possible fluctuations are clustering effects at the interface if the alloy system has a miscibility gap and the film is grown below the critical temperature for phase separation.

The distribution of grain orientations or texture in metallic multilayers is a very sensitive function of substrate temperature and growth rate. The simplest way to screen samples while searching for the optimum growth conditions is to measure the *rocking* curve. The film is mounted on a diffractometer so as to measure the intensity of a Bragg reflection normal to the film and the film is rotated through the reflection. Figure 11 shows a comparison of the rocking curves for two samples of Cu : Ni [20 Å|10 Å] deposited on freshly cleaved mica and prepared on successive days with the only apparent difference being the deposition rate (Gyorgy *et al.*, 1982). In one case the texture is reasonably good, while in the other the texture is very poor. Comparisons of elastic properties such as the biaxial modulus as a function of modulation wavelength are only meaningful if the microstructure as reflected (among other things) by the rocking curves is similar.

With increasing wavelength as the films become incoherent, there is a tendency for the texture to become poorer and for the films to break up into grains that show a range of orientations as a result of local strains. As the average grain size decreases, there is a broadening of the Bragg peaks along the scattering vector as opposed to the rocking curve that samples the width of the Bragg spot almost perpendicular to the scattering vector (Guinier, 1963). If the film is composed of a distribution of grains that are incoherent, then the diffraction spot will have a constant longitudinal width when expressed in units of the scattering vector. The size of the average domain L is related to the full width at half-maximum Δq by $L \approx 5.6/\Delta q$. In units of the scattering angle this is the Scherrer formula

$$\Delta(2\theta) = 0.9\lambda_0/L \cos\theta_0,$$

where θ_0 is the Bragg angle for x rays of wavelength λ_0. In addition to the grain-size broadening ($\Delta q = \text{const}$), there is strain broadening for which $\Delta q/q$ is constant for different orders of a reflection. As discussed in the literature, it is difficult to separate these two effects uniquely.

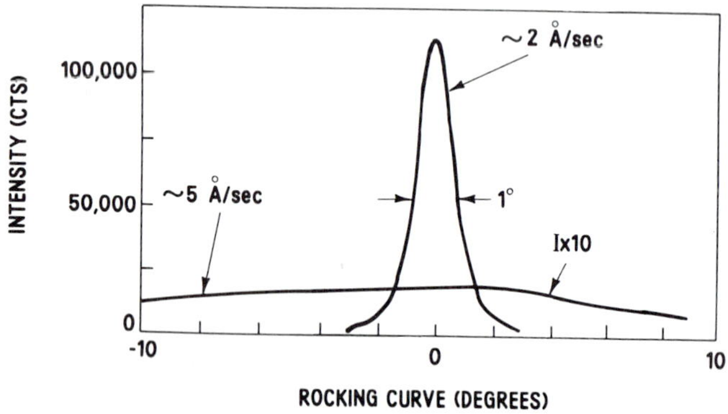

Fig. 11. Rocking curves for two samples of Cu/Ni [20 Å/10 Å] illustrating the sensitivity to the growth conditions in metallic superlattices.

If there is a chemical gradient across the sample, then there will be a distribution of modulation wavelength around some average value, and this is reflected in the widths of the harmonics. The position of the harmonics varies as $mk = 2\pi m/\lambda$, so that the relative width of the harmonics along the growth direction increases linearly with the order of the harmonic. This is illustrated in Fig. 12 for several different multilayers (Dernier *et al.*, 1977; Gyorgy *et al.*, 1982). The slopes of the lines indicate that there is a spread in modulation wavelengths of $\Delta q/q = \Delta\lambda/\lambda \approx 3$ and 5% for $(GaAs)_7(AlAs)_4$ and Cu_4Ni_4, respectively. In addition to chemical gradients, it has been reported that there are marked improvements in the quality of Cu_nNi_m films as n and m approach integer values (Flevaris *et al.*, 1981).

For coherent single-crystal films such as $(GaAs)_n(AlAs)_m$ that are grown epitaxially on a single-crystal GaAs substrate, it is more appropriate to consider

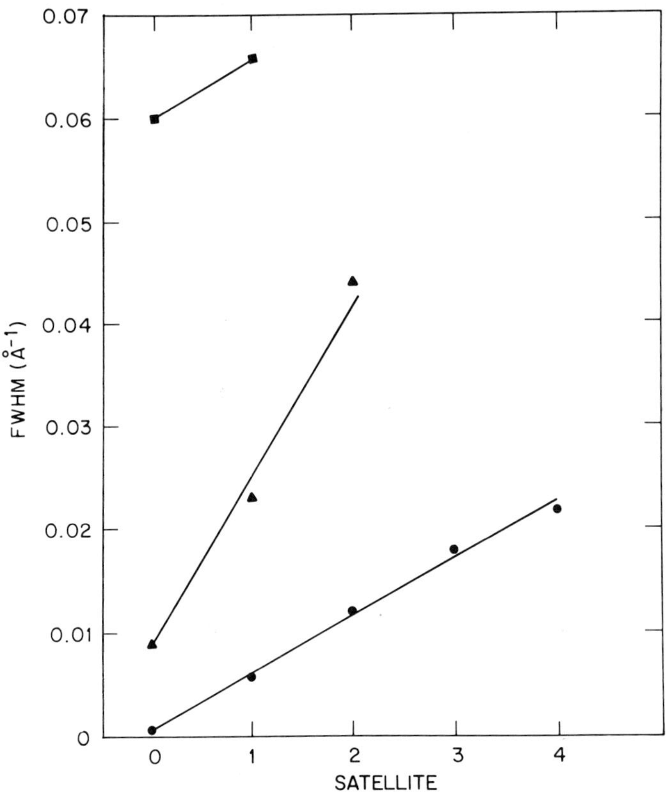

Fig. 12. Comparison of the longitudinal widths of the harmonics in $(GaAs)_7(AlAs)_4$ and Cu_4Ni_4 multilayers. [■, Cu_4Nb_4; ▲, Cu_4Ni_4; ●, $(GaAs)_7(AlAs)_4$.] The slope of the linear increase in the width with order of the harmonics indicates 3 and 5% fluctuations, respectively, in the modulation wavelength across the films.

the volume over which the multilayer is coherent rather than the model of an ensemble of misoriented grains. As illustrated in Fig. 13, the harmonics around each main Bragg reflection fall on a line perpendicular to the growth direction that passes through the Bragg reflection. The slope of this line relative to the crystallographic axis gives the angle δ shown in Fig. 13. The intensity of the reflections can be approximated by an ellipsoid with the full-widths at half-maximum in each direction being inversely proportional to the coherence in that direction. As indicated already, the widths of the harmonics along the growth direction increase linearly with order as a result of fluctuations in the modulation wavelength. Similarly, the widths of the harmonics along **a*** and **b*** increase linearly as a result of fluctuations in the direction of the chemical modulation. This limits the average chemical coherence in the plane of the film in that the interface between GaAs and AlAs regions wanders.

The coherence length in the plane can be anisotropic because the growth surface is anisotropic or because the surface is cut off-axis. For example, a sample of $(GaAs)_{2.4}(AlAs)_{0.5}$ was grown on a GaAs (100) surface that had been

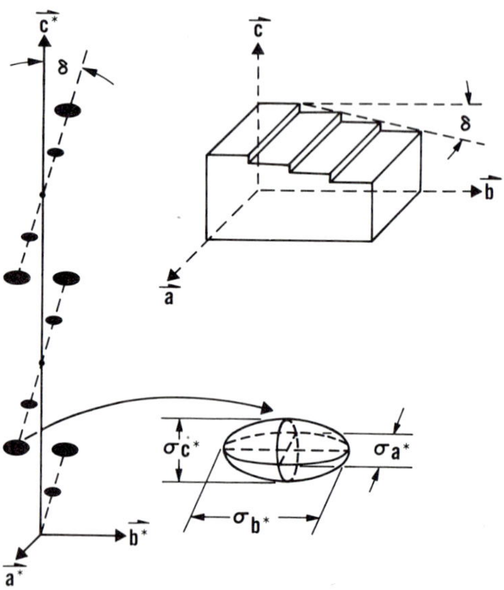

Fig. 13. Scattering from a multilayer grown on a stepped surface. If the surface is cut at an angle δ, then the harmonics are on a line that makes an angle δ with the crystallographic axis of the substrate. As discussed in the text, σ_{a^*}, σ_{b^*}, and σ_{c^*} are related to the volume in real space over which the multilayer is chemically coherent. (Arrows indicate vectors in figures; boldface indicates vectors in text.)

cut ~2° off-axis (McWhan and Gossard, 1978). The measured modulation wave vector was $\mathbf{k} = (\pm 0.025\mathbf{b}^* \mp 0.66\mathbf{c}^*)$, which corresponds to an off-axis cut of 2.1°. Measurement of σ_{a^*} and σ_{b^*} for the plus and minus first harmonic around the (002) and (004) reflections yields correlation lengths (after correcting for instrumental resolution) of $\Gamma_a = 2/\sigma_{a^*} = 131$ Å and $\Gamma_b = 77$ Å. The interplanar spacing in GaAs is 2.83 Å, and the calculated average distance between steps is 77.1 Å for an angle of 2.1°, which is fortuitously close to $\Gamma_b = 77$ Å. In a number of samples that were cut within 1° of the [001], a definite anisotropy was observed with the longest correlation lengths being along the tetrahedral bonding direction $[1\bar{1}0]$ of a (001) surface of the GaAs structure as illustrated in Fig. 14 (Fleming *et al.,* 1980b). The solid curve is

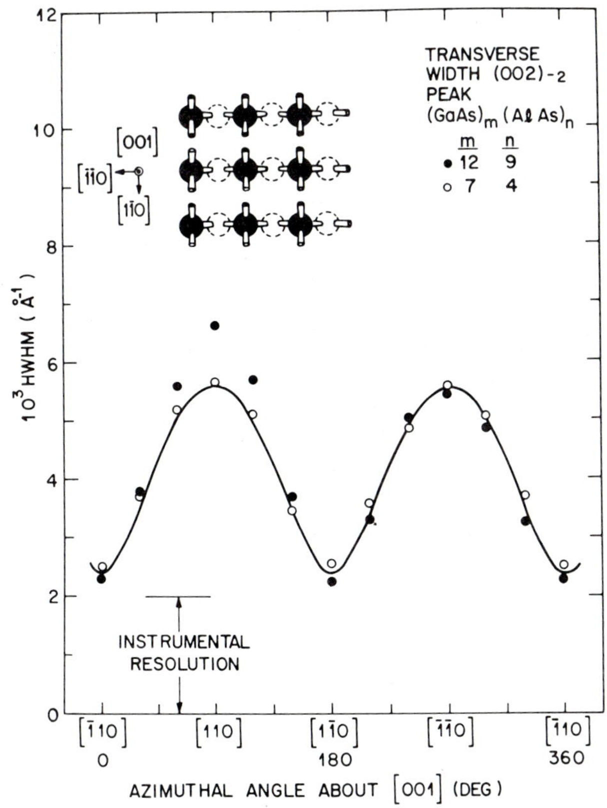

Fig. 14. Variation in the width of the second harmonic around the (002) in two $(GaAs)_n(AlAs)_m$ multilayers (i.e., variation in the σ_{a^*}–σ_{b^*} plane of Fig. 13). This illustrates a growth anisotropy that may be related to the asymmetry in the bonding at the surface, which is illustrated in the inset. [From Fleming *et al.* (1980b).]

calculated by assuming a correlation length in real space of

$$\Gamma(\phi) = [500^2 \sin^2 \phi + 175^2 \cos^2 \phi]^{1/2} \quad (\text{Å}).$$

The anisotropy may be an extrinsic effect that is related to the overall surface smoothness because at least one sample with an optically smooth surface did not show any anisotropy.

In some cases the growth direction may differ from the crystallographic axis because of the intrinsic atomic nature of the substrate. Multilayers of Nb_nTa_m have been grown on a number of different faces of sapphire (Durbin *et al.*, 1982). Films grown on the $(01\bar{1}2)$ plane of sapphire at substrate temperatures above ~825°C have a [100] texture (Hertel *et al.*, 1982). However, the [100] direction of the Nb_nTa_m multilayer is found to make a reproducible angle of 2.7° with the normal to the sapphire substrate.

In addition to the fluctuations in the interface already discussed, there may be clustering at the interface in systems that form solid solutions. The alloy system may have a real or an incipient miscibility gap under the growth conditions (both temperature and coherency strain). At the interface where the composition falls inside the miscibility gap, there is a tendency for phase separation. If there is sufficient surface mobility during growth, some degree or short-range order (SRO) may exist in the interfacial layers. If there are no correlations between interfaces, then the SRO gives rods of diffuse scattering through the Bragg spots along **c***. The width of the rod gives the average cluster size at the interface, and any modulation in intensity along **c*** is related to the thickness of the interfacial region over which SRO exists. Interfacial spinoidal decomposition has been observed in liquid-phase epitaxial grown $In_xGa_{1-x}As_yP_{1-y}$ layers that are lattice matched to InP (Quillec *et al.*, 1982), but to date no conclusive evidence for interfacial clustering in multilayers has been reported. Rods of diffuse scattering have been observed in $(GaAs)_n(AlAs)_m$ and Cu_nNi_m multilayers. In Fig. 15 transverse scans through [001] for a GaAs substrate, a random $Ga_{0.5}Al_{0.5}As$ layer grown on GaAs and a $(GaAs)_7(AlAs)_4$ multilayer are compared (Dernier *et al.*, 1977). The transverse scan for the 7 : 4 multilayer was taken midway between harmonics. There is evidence for rods of diffuse scattering in both the multilayer and the nominally "random" alloy, and this suggests clustering. This weak rod of diffuse scattering was observed in a series of multilayers ranging from $n:m = 1:1$ to 10 : 4. Much more pronounced phase separation has been observed in $Ga_{0.75}Al_{0.25}As$ grown on the (110) instead of the (100) surface of GaAs (Petroff *et al.*, 1982).

In multilayers of poorer quality, such as Cu_nNi_m, the interpretation of the diffuse scattering is more ambiguous because of the effects of finite coherence lengths on the scattering (Gyorgy *et al.*, 1982). The standard scattering function for a one-dimensional lattice of N atoms, $\sin^2 \pi Nl/\sin^2 \pi l$, has subsidiary weak maxima that decreased in intensity as l deviates from integral values. For a distribution of crystallite sizes, there will be long tails around the Bragg peaks that appear like modulated rods of diffuse scattering if the coherence along the

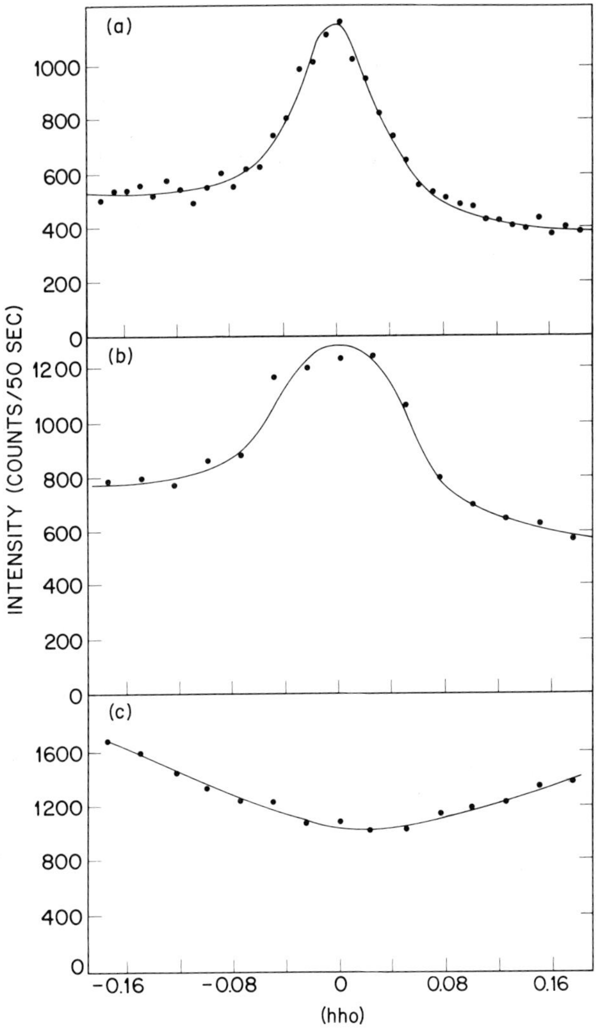

Fig. 15. Scans made transverse to the growth direction at various values of *l*. The peaks in the 7 × 4 multilayer and in the "random" $Ga_{0.5}Al_{0.5}As$ alloy indicate rods of diffuse scattering that imply clustering in the plane of the film. (a) (7, 4), $l = 2.40$; (b) alloy, $l = 2.20$; and (c) GaAs, $l = 3.60$.

growth direction is substantially smaller than the coherence in the plane of the multilayer. Finally, there are possible phonon anomalies that could contribute thermal diffuse scattering. Potentially important information about each of these effects can be obtained from x-ray diffraction measurements, but detailed studies as a function of wave vector and temperatures will be needed to separate the different contributions to the scattering.

IV. SURVEY OF MULTILAYERS

The amount and quality of structural information vary widely from system to system. The most extensive studies, using the largest number of probes, have been reported for the semiconductor systems. In contrast to data reported for the semiconductors, only diffraction data along the growth direction has been reported for most of the metallic systems. At the present time, only a few of the metallic multilayers have been made with a reasonable degree of three-dimensional structural coherence.

A. Semiconductor Multilayers

1. $(GaAs)_n(GaAs_{1-x}P_x)_m$

X-ray measures have been made throughout reciprocal space for this first semiconductor multilayer system to be grown (Esaki *et al.*, 1970; Segmüller and Blakeslee, 1973a,b). Samples of $(GaAs)_n\,(GaAs_{0.5}P_{0.5})_m$ with modulation wavelengths of $\lambda \approx 200$ Å were analyzed by using both a step model and a frozen-phonon model to obtain both compositions and strain profiles. Electron microscopy studies show that the structures are completely coherent for $\lambda \leq 250$ Å (Matthews and Blakeslee, 1974).

2. $(GaAs)_n(AlAs)_m$

The $(GaAs)_n(AlAs)_m$ system has been studied more extensively than any other multilayer system. X-ray (Dernier *et al.*, 1977; Fleming *et al.*, 1980b; Segmüller *et al.*, 1977), transmission electron microscopy (Petroff, 1977; Chang *et al.*, 1973; Petroff *et al.*, 1978; Gossard *et al.*, 1976), auger profiling (Ludeke *et al.*, 1974; Chang and Koma, 1976), ion backscattering (Chang *et al.*, 1973; Mayer *et al.*, 1973), and optical measurements (Weisbuch *et al.*, 1981a; Dingle, 1977; Dingle *et al.*, 1976) have been made as a function of wavelength, growth, and annealing conditions and composition to give a detailed picture of the composition wave and the interfacial smoothness. Most of the samples have been grown on (001) GaAs, but a few have been grown on (110) or (111) GaAs. The difference in lattice parameters is small, and a coherent structure is formed. The lattice parameter in the (001) plane is constrained by the GaAs substrate to $a = b = 5.563$ Å. The pure AlAs regions are under uniaxial tension; i.e., there is a tetragonal distortion with $e_{xx} = e_{yy} = -0.0014$ and $e_{zz} = +0.0012$ (Segmüller *et al.*, 1977). X-ray studies of $(GaAs)_n(AlAs)_m$ samples with $(n:m) = (1:1)$, $(2:2)$, $(5:4)$, $(7:4)$, $(10:4)$, and $(12:9)$ are consistent with the interfacial regions between 100% GaAs and 100% AlAs being ~2 monolayers (5.7 Å) thick (Dernier *et al.*, 1977; Fleming *et al.*, 1980b). As illustrated in Fig. 5, this means that a nominally alternating monolayer sample has a 30% modulation of the composition. [The actual composition of the sample shown in Fig. 5 was $(GaAs)_{1.2}(AlAs)_{1.0}$ so that the Al content varied from 60 to 35%.] Optical measurements, based on the luminescence, absorption, and excitation spectra with decreasing layer thickness, are also consistent with an interfacial fluctuation of 1–2 monolayers. More qualitative measurements of the sharpness of the interface are illustrated in Figs. 7 and 8 for transmission electron microscopy (TEM) and Auger profiling. The sharpness of the composition wave is sensitive to substrate temperature, and there appears to be a balance between mobility of the atoms impinging on the surface and a roughening of the surface resulting in a loss of a coherent modulation. This has been demonstrated in electron diffraction studies of the satellites in alternating monolayer samples (Petroff *et al.*, 1978) and in optical linewidth measurements in samples with $\lambda = 350-400$ Å (Weishuch *et al.*, 1981b). The critical temperature varies with wavelength from $T_s = 570°C$ for ultrathin structures to $T_s \sim 690°C$ for the longer-wavelength samples. However, other factors such as As : Ga ratio and As_4/As_2 conversion as well as subtle differences in the MBE chambers may strongly influence these results.

The macroscopic uniformity of the multilayer acros the film is quite high (see Fig. 7). As indicated in Fig. 12, a measure of the microscopic variation in modulation direction is given by the rate of increase in σ_{b*} with the order of the harmonic. For a (7, 4) sample the widths of the harmonics increase linearly with a slope of $\sigma_{b*}(n)/n\mathbf{k} = 0.023$, and this corresponds to a fluctuation of ± 11 mrad in the growth direction. In many samples the linewidth depends on the direction in the plane (i.e., σ_{b*} or σ_{a*}), and this is illustrated in Fig. 14. There is a marked growth anisotropy that may be related to the anisotropic bonding on the (001) surface as illustrated in the inset in Fig. 14.

The homogeneity of the layering period across the film has been determined by x-ray measurements both at small angles (Chang *et al.*, 1976) and at large angles (Dernier *et al.*, 1977). In Fig. 12, the linear increase in the longitudinal width of the harmonics with order indicates a range of periods and the slope gives a spread of $\Delta\lambda/\lambda = 3\%$. This spread results mainly from the geometry of the MBE apparatus in which the Al oven is displaced relative to the center of the substrate and the Ga oven. This has been mapped out by making x-ray measurements with a finely focused beam on different parts of the sample (Dernier *et al.*, 1977; Chang *et al.*, 1976). Finally, there is some evidence (Fig. 15) for clustering in the interfacial regions between GaAs and AlAs.

3. $(Ga_{1-x}Al_xAs)_nGe_m$

Gallium arsenide and Ge have a lattice mismatch comparable to that in $(GaAs)_n(AlAs)_m$ with $\Delta a/a \approx 0.8 \times 10^{-3}$ to 1.4×10^{-3} in the temperature range $400 \leq T_s \leq 800°C$, but the growth of good multilayers is strongly dependent on x, T_s, and λ. Smooth interfaces were achieved for substrate temperatures of 400–500°C and modulation wavelengths above 25–30 Å (Petroff *et al.*, 1979; Chang *et al.*, 1981). At higher temperatures and/or shorter λ, columnar structures were observed in TEM studies. At short λ the interdiffusion of Ge and GaAs leads to island growth and rough interfaces. High epitaxial quality and low-defect-density films were only obtained for layer thicknesses above ~ 35 Å.

4. Other Semiconductor Systems

Several other systems have been shown to form good multilayers but have not been studied as extensively: $(GaAs)_n(InAs)_m$ has a 7% lattice mismatch and the resulting multilayers have large concentrations of misfit dislocations (Gossard, 1979). Similarly, Ge_nSi_m multilayers have a heavily misfit-dislocated structure (Kasper *et al.*, 1975). Also $(InAs)_n(GaSb)_m$ multilayers have been studied by ion-beam crystallography, and evidence for interface relaxation has been presented (Chu *et al.*, 1982).

B. Metals

1. Both Metals with Same Structure

The quality of multilayers made from different metals that have the same crystal structure depends principally on the degree of lattice mismatch. Interdiffusion occurs during growth in most cases, so that at shorter wavelengths the amplitude of the chemical modulation is reduced. The Ag_nAu_m, Cu_nNi_m, and Nb_nTa_m systems have been studied in the most detail, and they are discussed individually, with brief reference made to other systems.

Ag_nAu_m. The mismatch of these fcc metals is <0.2%, and high-quality multilayers have been grown both on mica (Cook and Hilliard, 1969; Cook, 1969; Purdes, 1976) and Ag single-crystal surfaces (Culbertson *et al.*, 1981). In the latter study a multilayer was grown on Ag (111) layer by layer, and the growth followed by high-energy ion-scattering studies. These measurements established that the growth mode is the Frank and van der Merwe layer by layer mode (Frank and van der Merwe, 1949). The films grown on mica were studied by x-ray diffraction with the scattering vector parallel with the growth direction. An epitaxial strain resulting from the mica substrate complicates the determination of the coherency strain, but Ag_nAu_m multilayers appear to be coherent over the range of wavelengths studied ($\lambda < 100$ Å).

Cu_nNi_m. Because of their interesting magnetic and elastic properties, the Cu_nNi_m system has been studied more extensively than any other metallic superlattice (Hilliard, 1979; Felcher *et al.*, 1980; Gyorgy *et al.*, 1982; Flevaris *et al.*, 1981; Purdes, 1976; Tsakalakos, 1977). The most successful growth has been achieved on mica that was first covered with a Cu buffer layer. After supporting the film on a glass or other substrate, both the mica and the Cu buffer layer can be removed before measuring the properties of the multilayer.

Transmission electron microscopy studies establish that the films grow with [111] texture and that the [1$\bar{1}$0] or [$\bar{1}$10] directions of the film are parallel to the [010] of mica (Purdes, 1976). Microtwining has been observed in both TEM (Purdes, 1976) and x-ray measurements (Gyorgy *et al.*, 1982). The composition wave is asymmetric because of interdiffusion with the faster diffusion occurring in the Cu-rich regions (Tsakalakos, 1977), so that in a sample with $\lambda = 16$ Å the Cu concentration varies from 9 to 74% (Gyorgy *et al.*, 1982). Only for wavelengths above ~40 Å does the amplitude reach 100%.

The lattice mismatch is ~2.4% and a loss of coherency is expected at longer wavelengths. Transmission electron microscopy measurements of 500-Å Cu–tÅ Ni–500-Å Cu [100] sandwiches show the appearance of misfit dislocations at $t \approx 15$ Å [Nakahara (1980) and private communication]. In an alternating multilayer of equal thicknesses of Cu and Ni, the critical wavelength may be substantially longer than $\lambda = 30$ Å because both the Cu and Ni regions are strained. X-ray measurements with the scattering vector both parallel and perpendicular to the growth direction show that substantial coherency strains exist even for $\lambda = 122$ Å (Gyorgy *et al.*, 1982). (See Subsection III.B and Fig. 9.) There is evidence that at wavelengths below 40 Å the amplitude of the interplanar spacing modulation does not simply scale with the amplitude of the chemical modulation A but instead approaches the value expected for a coherent structure as A^2; i.e., $\epsilon = \epsilon_{\text{inc}} + c\,A^2$ (Purdes, 1976). Above $\lambda \approxeq 40$ Å, ϵ saturates at ~0.039 corresponding to a coherent structure.

The texture is strongly dependent on the growth parameters (see Subsection III.C and Fig. 11), and in most studies the texture has been estimated by measuring the rocking curve or by comparing the integrated intensity of a Bragg peak with that of a standard. Under typical growth conditions, the fluctuations in the modulation wavelength are ~5% (see Section III.C and Fig. 12), and the coherence along the growth direction exceeds 500 Å.

Nb_nTa_m. Both Nb and Ta are body-centered cubic and the lattice mismatch is <0.2%. High-quality multilayer crystalline films have been grown on heated sapphire substrates by using a molecular-beam-epitaxy growth chamber with electron-beam sources (Durbin *et al.*, 1981, 1982) and by sputtering (Hertel *et al.*, 1982). For substrate temperatures of 700–900°C, good epitaxial single-crystal growth is observed. The growth direction in the bcc structure was controlled by the crystallographic plane of the sapphire or MgO substrate. The growth directions and the substrate planes are [100] on cleaved (100) MgO or (01$\bar{1}$2) sapphire; [110] on (11$\bar{2}$0) sapphire; [111] on (0001) sapphire; and [211] on (10$\bar{1}$0) sapphire. As discussed in Subsection III.C, the [100] axis is 2.7° off the normal to the sapphire (10$\bar{1}$2). Also there is a critical temperature for good [100] epitaxial growth on (10$\bar{1}$2) sapphire. Below ~825°C a poor [110]-oriented film results, but above 825°C the texture is [100] and the rocking curve is <1°. The best films are those with [110] texture grown on (11$\bar{2}$0) sapphire, where the rocking curves are $<0.1°$. At the substrate temperatures required for good epitaxial growth, there is some interdiffusion. Two independent estimates of the width of the interfacial region give ~ 9 Å for the spread over which the composition goes from 10 to 90% (Durbin *et al.*, 1982; Hartel *et al.*, 1982). As a result of interdiffusion, a sample with $\lambda = 18.6$ Å shows a composition modulation of $\sim 37-63\%$.

Other Systems. A number of multilayers have been made to study interdiffusion, and in most of these the x-ray measurements have been $\theta-2\theta$ scans that were used to estimate the amplitude of the composition wave as a function of annealing. A number of these are collected in Table I along with those discussed earlier.

TABLE I

Metallic Multilayers

System	References
AgAu	Culbertson *et al.*, 1981; Cook and Hilliard, 1969; Cook, 1969; Purdes, 1976
AgPd	Henein, 1979
AgTi	Zheng *et al.*, 1982
AlBe	Underwood *et al.*, 1979; Underwood and Barbee, 1981
AlNb	McWhan *et al.*, 1982
AuC	Spiller, 1976
AuCu	DuMond and Youtz, 1940; Paulson and Hilliard, 1977
AuMg	Dinklage, 1967
AuNi	Yang, 1970
CCu	Spiller, 1976
CW	Underwood *et al.*, 1979; Underwood and Barbee, 1981
CoMn	Stearns, 1982
CuNb	Schüller, 1980; Lowe *et al.*, 1981
CuNi	Hilliard, 1979; Felcher *et al.*, 1980; Gyorgy *et al.*, 1982; Flevaris *et al.*, 1981; Purdes, 1976; Tsakalakos, 1977
CuPd	Philofsky and Hilliard, 1969
FeGe	Lynn *et al.*, 1976
FeMg	Dinklage, 1967
FeSb	Hosoito *et al.*, 1982
GeMn	Schoenborn *et al.*, 1974; Saxena and Schoenborn, 1977
GeNb	Ruggiero *et al.*, 1982
GePb	Willens *et al.*, 1982
MgPb	Dinklage and Frerichs, 1963
MnNi	Stearns, 1982
Mn-stearate	Pomerantz and Segmüller, 1980
NbTa	Durbin *et al.*, 1982; Durbin *et al.*, 1981; Hertel *et al.*, 1982
NbTi	Zheng *et al.*, 1981
NbZr	Lowe and Geballe, 1984

2. Metals with Different Structures

The structures of five multilayer systems have been reported in which the two elements have different structures: one involving fcc–hcp elements (AgTi) (Zheng *et al.*, 1982), two involving bcc–fcc elements [NbCu (Schüller, 1980; Lowe *et al.*, 1981) and NbA (McWhan *et al.*, 1983)] and two involving bcc–hcp elements [NbTi (Zheng *et al.*, 1981) and NbZr (Lowe and Geballe, 1984)].

Ag_nTi_m. Both of these elements have close packed structures, and therefore it is the stacking sequence of close-packed layers that changes. A careful study of the modulation amplitude as a function of substrate temperature established that the optimum temperature was ~90°C. Samples grown at this temperature approach single crystals with an fcc stacking sequence in the Ag regions and an hcp stacking in the Ti regions. However, each region in essence picks up the appropriate sequence where the previous region left off. For example, a multilayer of repeated four-layer sequences of Ag and Ti might be: |abca|caca|bcab|abab|cabc|bcbc|. . . . This leads to a microscopically twinned crystalline film as successive fcc (on hcp) regions have different orientations: abc followed by cba (or caca followed by abab).

Nb_nCu_m. These elements are not mutually soluble, and so there is no interdiffusion and sharp interfaces are expected. X-ray studies show one-dimensional coherence with [110] and [111] texture in the Nb and Cu regions, respectively (see Fig. 4). There is no evidence of coherency strains. The quality of the films deteriorates at short wavelengths, and the grain size decreases.

Nb_nAl_m. The Nb regions show an orthorhombic distortion that indicates the presence of anisotropic coherency strains. The Nb contracts in the plane of the film and the Al expands. The coherency strains decrease with increasing wavelength, as shown in Fig. 10. The composition modulation is approximately trapezoidal with interfacial regions that are ~11 Å thick. The interdiffusion is asymmetric with the Al concentration varying from 10 to 100% in a film with $\lambda = 58$ Å.

Nb_nTi_m *and* Nb_nZr_m. These systems are similar in that both have regions of the phase diagram where solid solutions are bcc. The x-ray measurements indicate that there is interdiffusion in both systems and that a bcc phase is stabilized in the Ti and Zr regions. The lattice parameters of bcc Ti, which is stable at high temperature, and bcc Nb differ by only 0.2%, so that the misfit is reasonably small and a coherent structure is expected. The misfit is larger in the Nb_nZr_m system and the bcc phase is stabilized for $\lambda \leq 40$ Å. At longer wavelengths the Zr regions revert to the hcp structure, and the Zr and Nb have [0001] and [110] textures, respectively. The Nb_nZr_m films are coherent in the bcc–bcc phase and incoherent in the bcc–hcp phase.

C. Polycrystalline–Amorphous Multilayers

There are a number of applications in which "two-dimensional" regions of a metal separated by amorphous regions of a second material can be used. The crossover from three- to two-dimensional behavior could be studied by preparing samples with different thicknesses of the metallic region. Examples of properties that can be studied are magnetism, superconductivity, and melting; the corresponding systems (which are listed in Table I) are Mn–stearate, Nb–Ge, and Pb–Ge. Another major application of multilayers is for x-ray mirrors and neutron polarizers. In the x-ray case, the appropriate choice of components is a heavy metal followed by weakly absorbing material, and systems such as W–C, Au–C, and Cu–C have been used successfully (See Table I). In the neutron case, multilayers of Fe–Ge and Mn–Ge have been used.

ACKNOWLEDGMENTS

The author thanks his colleagues both at AT&T Bell Laboratories and Brookhaven National Laboratory for many helpful discussions and especially thanks J. D. Axe, R. C. Dynes, J. F. Dillon, Jr., R. M. Fleming, A. C. Gossard, F. M. Gyorgy, D. E. Moncton, J. M. Rowell, and L. R. Walker.

REFERENCES

Brodsky, M. B., and Freeman, A. J. (1980). *Phys. Rev. Let.* **45,** 133.

Cahn, J. W. (1962). *Acta Metall.* **10,** 179.

Chang, C.-A., Chu, W.-K., Mendez, E. E., Chang, L. L., and Esaki, L. (1981). *J. Vac. Sci. Technol.* **19,** 567.

Chang, L. L., and Koma, A. (1976). *Appl. Phys. Lett.* **29,** 138.

Chang, L. L., Esaki, L., Howard, W. E., Ludeke, R., and Schul, G. (1973). *J. Vac. Sci. Technol.* **10,** 655.

Chang, L. L., Segmüller, A., and Esaki, L. (1976). *Appl. Phys. Lett.* **28,** 39.

Cho, A. Y. (1970). *J. Appl. Phys.* **41,** 2780.

Chu, W. K., Saris, F. W., Chang, C.-A., Ludeke, R., and Esaki, L. (1982). *Phys. Rev. B.* **26,** 1999.

Cook, H. E. (1969). *J. Phys. Chem. Solids* **30,** 1097.

Cook, H. E., and Hilliard, J. E. (1969). *J. Appl. Phys.* **40,** 2191.

Culbertson, R. J., Feldman, L. C., Silverman, P. J., and Boehm, H. (1981). *Phys. Rev. Lett.* **47,** 657.

DeFontaine, D. (1966). In "Local Atomic Arrangement Studied by X-Ray Diffraction (J. B. Cohen and J. E. Hilliard, eds.), p. 51, Gordon and Breach, New York.

Dernier, P. D., Moncton, D. E., McWhan, D. B., Gossard, A. C., and Wiegmann, W. (1977). *Bull. Am. Phys. Soc.* **22,** 293.

Dingle, R. (1977). *J. Vac. Sci. Technol.* **14,** 1006.

Dingle, R., Gossard, A. C., and Wiegmann, W. (1976). *Bull. Am. Phys. Soc.* **21,** 367.

Dinklage, J. B. (1967). *J. Appl. Phys.* **38,** 3781.

Dinklage, J. B., and Frerichs, R. (1963). *J. Appl. Phys.* **53,** 2436.

DuMond, D. W. M., and Youtz, J. P. (1935). *Phys. Rev.* **48,** 703.

DuMond, D. W. M., and Youtz, J. P. (1940). *J. Appl. Phys.* **11,** 357.

Durbin, S. M., Cunningham, J. E., Mochel, M. E., and Flynn, C. P. (1981). *J. Phys. F* **11,** L223.

Durbin, S. M., Cunningham, J. E., and Flynn, C. P. (1982). *J. Phys. F* **12,** L75.

Esaki, L., Chang, L. L., Tu, R. (1970). *Proc. Int. Conf. Low Temp. Phys., 12th* (E. Kanda, ed.), p. 551. Keigaku Publ., Tokyo.
Felcher, G. P., Cable, J. W., Zheng, J. Q., Ketterson, J. B., and Hilliard, J. E. (1980). *J. Magn. Magn. Mater.* **21,** L198.
Fleming, R. M., Moncton, D. E., McWhan, D. B., and Disalvo, F. J. (1980a). *Phys. Rev. Lett.* **45,** 576.
Fleming, R. M., McWhan, D. B., Gossard, A. C., Wiegmann, W., and Logan, R. A. (1980b). *J. Appl. Phys.* **51,** 357.
Flevaris, N. K., Baral, D., Hilliard, J. E., and Ketterson, J. B. (1981). *Appl. Phys. Lett.* **38,** 992.
Frank, F. C., and van der Merwe, J. H. (1949). *Proc. R. Soc. London* **198,** 205.
Gossard, A. C. (1979). *Thin Solid Films* **57,** 3.
Gossard, A. C., Petroff, P. M., Wiegmann, W., Dingle, R., and Savage, A. (1976). *Appl. Phys. Lett.* **29,** 323.
Guinier, A. (1963). "X-ray Diffraction." Freeman, San Francisco California.
Gyorgy, E. M., McWhan, D. B., Dillon, J. R., Jr., Walker, L. R., and Waszczak, J. V. (1982). *Phys. Rev. B* **25,** 6739.
Henein, G. E. (1979). Ph.D. thesis, Northwestern Univ., Evanston, Illinois.
Hertel, G., McWhan, D. B., and Rowell, J. M. (1982). *Proc. Conf. Superconductivity in d and f band Metals, Karlsruhe, Germany,* p. 299.
Hilliard, J. E. (1979). In "Modulated Structures" (J. M. Cowley, J. B. Cohen, M. B. Salamon, and B. J. Wuensch, eds.), p. 407. AIP Conf. Proc. 53,Amer. Inst. Phys. New York.
Hosoito, N., Kawaguchi, K., Shinjo, T., and Takada, T. (1982). *J. Phys. Soc. Jpn.* **51,** 2701.
Iizumi, M., Axe, J. D., Shirane, G., and Simaoka, K. (1977). *Phys. Rev. B* **15,** 4392.
Kasper, E., Herzog, H. J., and Kibbel, H. (1975). *Appl. Phys.* **8,** 199.
Lowe, W. P., and Geballe, T. H. (1984). *Phys. Rev. B* **29,** 4961.
Lowe, W., Barbee, T., Jr., Geballe, T. H., and McWhan, D. B. (1981). *Phys. Rev. B* **24,** 6193.
Ludeke, R., Esaki, L., and Chang, L. L. (1974). *Appl. Phys. Lett.* **24,** 417.
Lynn, J. W., Kjems, J. K., Passell, L., Saxena, A. M., and Schoenborn, B. P. (1976). *J. Appl. Cryst.* **9,** 454.
McMillan, W. L. (1976). *Phys. Rev. B* **14,** 1496.
McWhan, D. B., and Gossard, A. C. (1978). Unpublished results.
McWhan, D. B., Gurvitch, M., Rowell, J. M., and Walker, L. R. (1983). *J. Appl. Phys.* **54,** 3886.
Matthews, J. W. (ed.) (1975). "Epitaxial Growth." Academic Press, New York.
Matthews, J. W., and Blakeslee, A. E., (1974). *J. Cryst. Growth* **27,** 118.
Mayer, J. W., Ziegler, J. F., Chang, L. L., Tsu, R., and Esaki, L. (1973). *J. Appl. Phys.* **44,** 2322.
Nakahara, S. (1980). *Thin Solid Films* **69,** 241.
Paulson, W. M., and Hilliard, J. E. (1977). *J. Appl. Phys.* **48,** 2117.
Petroff, P. M. (1977). *J. Vac. Sci. Technol.* **14,** 973.
Petroff, P. M., Gossard, A. C., Wiegmann, W., and Savage, A. (1978). *J. Cryst. Growth* **44,** 5.
Petroff, P. M., Gossard, A. C., Savage, A., and Wiegmann, W. (1979). *J. Cryst. Growth* **46,** 172
Petroff, P. M., Cho, A. Y., Reinhart, F. K., Gossard, A. C., and Wiegmann, W. (1982). *Phys. Rev. Lett.* **48,** 170.
Philofsky, E. M., and Hilliard, J. E. (1969). *J. Appl. Phys.* **40,** 2198.
Pomerantz, M., and Segmüller, A. (1980). *Thin Solid Films* **68,** 33.
Purdes, A. J. (1976). Ph.D. thesis, Northwestern Univ., Evanston, Illinois.
Quillec, M., Daguet, C., Benchimol, J. L., and Launois, H. (1982). *Appl. Phys. Lett.* **40,** 325.
Ruggiero, S., Barbee, T. W., Jr., and Beasley, M. R. (1982). *Phys. Rev. B* **26,** 4894.
Saxena, A. M., and Schoenborn, B. P. (1977). *Acta Cryst. A* **33,** 805.
Schoenborn, B. P., Casper, D. L. D., and Kammerer, O. F. (1974). *J. Appl. Cryst.* **7,** 508.
Schüller, I. K. (1980). *Phys. Rev. Lett.* **44,** 1597.
Segmüller, A., Blakeslee, A. E. (1973a). *J. Appl. Cryst.* **6,** 19.

Segmüller, A., and Blakeslee, A. E. (1973b). *J. Appl. Cryst.* **6,** 413.
Segmüller, A., Krishna, P., and Esaki, L. (1977). *J. Appl. Cryst.* **10,** 1.
Sinha, S. K. (ed.) (1980). "Ordering in Two Dimensions." North-Holland, Publ., New York.
Spiller, E. (1976). *Appl. Optics* **15,** 2333.
Stearns, M. B. (1982). *J. Appl. Phys.* **53,** 2436.
Tsakalakos, T. (1977). Ph.D. thesis, Northwestern Univ. Evanston, Illinois.
Underwood, J. H., and Barbee, T. W., Jr. (1981). *Appl. Optics* **20,** 3027.
Underwood, J. H., Barbee, T. W., Jr., and Keith, D. C. (1979). *Proc. Soc. Photo-Opt. Instrum. Eng.* **184,** 123.
Weisbuch, C., Dingle, R., Gossard, A. C., and Wiegmann, W. (1981a). *Solid State Commun.* **38,** 709.
Weisbuch, C., Dingle, R., Petroff, P. M., Gossard, A. C., and Wiegmann, W. (1981b). *Appl. Phys. Lett.* **38,** 840.
Willens, R. H., Kornblit, A., Testardi, L. R., and Nakahara, S. (1982). *Phys. Rev. B* **25,** 290.
Woodall, J. W. (1972). *J. Cryst. Growth* **12,** 32.
Yang, W. M. C. (1970). Ph.D. thesis, Northwestern Univ., Evanston, Illinois.
Zheng, J. Q., Ketterson, J. B., Falco, C. M., and Schüller, I. K. (1981). *Physica* **108B,** 945.
Zheng, J. Q., Ketterson, J. B., Felcher, G. P. (1982). *J. Appl. Phys.* **53,** 3624.

PART II

SEMICONDUCTORS

3 ASPECTS OF THE THEORY OF SUPERLATTICES

J. N. Schulman*

Department of Physics and Astronomy
University of Hawaii
Honolulu, Hawaii

T. C. McGill

T. J. Watson, Sr., Laboratory of Applied Physics
California Institute of Technology
Pasadena, California

I. INTRODUCTION

Superlattices present the theorist with an opportunity to apply some of the basic principles of solid-state physics. The basic question of the role of periodicity in determining the properties of solids is put to a new test in the artificially created world of the superlattice. These new man-made materials offer possibilities for providing solids with new electronic spectra, lattice vibrations, collective excitations, transport properties, and dielectric properties.

* Present address: Hughes Research Laboratories, Malibu, California.

ISBN 0-12-170470-X

Much of the theory of the superlattice depends on adequately applying our understanding of the bulk properties of the constituent materials in the new periodic structure. In this approach to the problem the superlattice is made up of layers of bulk materials with a nasty, hopefully unimportant, interface between adjacent layers. A full understanding of the superlattice requires that the interface region also be understood. The relationship between bulk and superlattice properties is discussed in Section II. Given the assumption that the interface does not cause undue disruption, there are several theoretical techniques that are useful in calculating superlattice electronic properties. The methods, which are generalizations to the superlattice of techniques that have previously been used to examine bulk semiconductors, are discussed in Section III. Section IV concentrates on some of the results of the calculations, illustrating typical characteristics of superlattice electronic structure. Section V surveys the physics of phonons in superlattices. Phenomena similar to those exhibited by the electronic spectra occur also for phonons. Section VI surveys a variety of physical phenomena in superlattices, especially those that are just starting to be explored both theoretically and experimentally. These include optical properties, impurities, transport properties, and collective excitations. Conclusions are presented in Section VII.

II. GENERAL CONCEPTS

On the most fundamental level, any particular superlattice consisting of alternating layers of a specified pair of semiconductors with specified layer thicknesses can be viewed as a completely new material. A more useful approach recognizes that each alternating layer is made up of a bulk semiconductor, such as GaAs or AlAs, whose bulk properties may be well known. Although the following point of view may prove inaccurate when the alternating layers are very thin, it is often useful in obtaining a qualitative understanding of superlattice properties to consider it as consisting of alternating layers of semiconductors whose bulk properties are left somewhat intact. This does not imply, however, that the superlattice properties are necessarily a simple superposition of the properties of its two constituents. Although the bulk properties, such as the electrostatic potential or ionic positions, are envisioned as remaining unchanged, the boundary conditions satisfied by the electron or phonon wave functions are influenced by the thickness of the layers in the superlattice. This section discusses two concepts that follow from this point of view, band folding and mixing of bulk solutions with complex wave vectors.

A. Band Folding

The band folding concept is a simple method for obtaining an approximate understanding of electron or phonon dispersion curves in a superlattice. Defining the z direction to be perpendicular to the alternating layers, the electron or phonon energy E versus wave vector k_z in the z direction can be roughly predicted from the known E versus k_z dispersion curves for the bulk materials. This can be done by displaying the bulk dispersion curves while assuming an extended bulk unit cell in the z direction. For a superlattice with a total of L atomic layers of its constituent semiconductors per superlattice unit cell, the bulk unit cell should also consist of L layers. In the extended zone scheme, the superlattice potential would open up gaps in the bulk dispersion curves at intervals of $4\pi/La$ (where a is the conventional zincblende unit cell size) in k_z. Folding these curves back into a single Brillouin zone of width $4\pi/La$ would give the superlattice dispersion curves. Thus, folding back the bulk dispersion curves in the same manner as if there were an extended unit cell gives an approximation to the superlattice dispersion curves.

Although this approximation seems crude, it is useful in sorting out the multiplicity of superlattice bands produced due to the large superlattice unit cell as compared to the bulk unit cell, and it can lead to qualitative results for the dispersion curves. Figure 1 shows a clear example. The GaAs–AlAs (100) superlattice band structure in the [100] direction is plotted for three GaAs layer thicknesses (Ivanov and Pollmann, 1979). The folded bulk GaAs band structures (dotted lines) provide good approximations to the superlattice curves

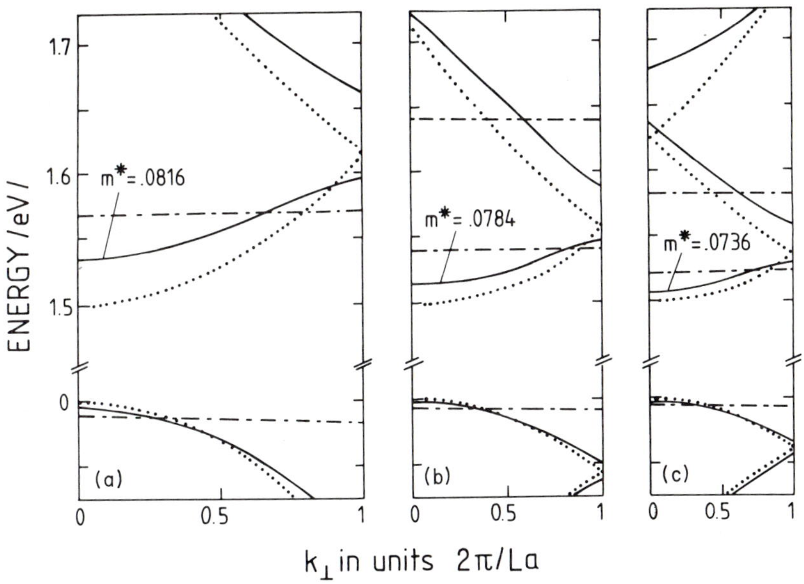

Fig. 1. GaAs–AlAs superlattice bands in the [100] direction for GaAs slab thicknesses of (a) 57, (b) 85, and (c) 114 Å. (Solid lines: 11-Å AlAs slab thickness; dash-dotted lines: 85-Å AlAs slab thickness; dotted lines: folded bulk GaAs band structure.)

(solid lines). The main differences are shifts in energy and a splitting of bulk degeneracies at zone edges.

B. Bulk Solutions with Complex Wave Vectors

Another useful concept that relates the superlattice properties to those of its constituent bulk materials is that of describing the superlattice electron or phonon wave function in terms of bulk states with complex wave vectors. By these are meant solutions to the bulk Schrödinger equation with complex values of k_z.

$$H_i \psi_n^{(i)}(\mathbf{k}_t, k_z^{(n)}, \mathbf{r}) = E \psi_n^{(i)}(\mathbf{k}_t, k_z^{(n)}, \mathbf{r}), \tag{1}$$

with $i = 1, 2$ labeling the two bulk semiconductors, $\mathbf{k}_t$ the wave vector parallel to the superlattice planes, and n the allowed k_z values for fixed $\mathbf{k}_t$, i, and E. A simple and efficient method for finding the complex k_z solutions to this equation has been developed within the tight-binding, $\mathbf{k} \cdot \mathbf{p}$, and pseudopotential schemes (Chang and Schulman, 1982).

The usefulness of the complex wave vector states is that the superlattice Hamiltonian away from any disruption near the interfaces is equal to the bulk Hamiltonians of one or the other of the constituent materials. The superlattice wave function in these regions, therefore, can be written as a linear combination of bulk complex solutions with the same energy and $\mathbf{k}_t$ (Schulman and McGill, 1981; Schulman and Chang, 1981a):

$$\psi(\mathbf{r}) = \sum_n C_n^{(i)} \psi_n^{(i)}(\mathbf{r}), \tag{2}$$

within material i. The solutions with complex values of k_z represent evanescent states with decay lengths equal to $(\mathrm{Im}(k_z))^{-1}$.

In general, a full calculation is needed to determine the exact linear combinations in Eq. (2). There are energy ranges of interest, however, in which symmetry or other considerations limit the number of complex k_z bulk states that contribute substantially to the superlattice wave function. In these cases much can be learned about the superlattice wave function from an analysis of the complex wave vector bulk states alone. In the more general case, the coefficients $C_n^{(i)}$ in Eq. (2) give the mixing of the complex wave vector bulk states by the superlattice potential. This will be discussed in more detail in Subsection IV.C.

III. CALCULATION OF ELECTRONIC PROPERTIES

This section will discuss the methods that have been used to calculate the electronic properties of semiconductor superlattices. In order of increasing computational complexity, these are the quantum-well model, the two-band approximation, the full tight-binding method, and the pseudopotential method. The different methods rely to varying degree on empirical parameters that are related to the bulk or atomic properties of the two constituent semiconductors. The effects of this and other assumptions in the various models on the calculated electronic properties will be discussed here.

A. Band Offsets

The first three methods share a common empirical parameter whose value must be found—the band offset. In all three methods the bulk band structure of each constituent material, or at least that part of it which is of interest, is independently determined with an arbitrary energy zero. The parameters that produce the bulk band structures are then to be used in the superlattice calculation, but first the relative energy lineup, the band offset, must be determined.

Attempts to calculate the offsets theoretically have been inconclusive. Several approaches have been taken. The simplest guide to the offset value is the electron affinity rule (Milnes and Feucht, 1972). This method locates the conduction band edge of each material relative to the vacuum by using the experimentally determined bulk electron affinity. Inaccuracies arise from experimental uncertainties (typically tenths of an electron volt), and the omission of surface and interface effects. Harrison (1980) used a unified tight-binding model to predict electron affinities, and thus offsets, theoretically for several materials. Trends seem to be qualitatively described, but the accuracy is limited by the omission of interface effects and self-consistency. Two more detailed theoretical attempts to determine the offsets were the electrostatic potential matching scheme of Frensley and Kroemer (1977) and the application of the self-consistent pseudopotential method (Pickett *et al.*, 1978; Ihm *et al.*, 1979) to the problem. As will be shown, these methods also had limited success. It is

TABLE I

GaAs–AlAs Heterojunction Band Offsets Expressed as $\Delta E_v/\Delta E_g$[a]

Source	$\Delta E_v/\Delta E_g$
Frensley and Kroemer (1977)	0.56
Pickett *et al.* (1978)(110)	0.17
Harrison (1980)	0.03
Dingle *et al.* (1975)(100)	0.15

[a] ΔE_v is the GaAs valence maximum energy minus the AlAs valence band maximum energy and ΔE_g is the AlAs direct band gap minus the GaAs direct band gap. Specific orientations are indicated.

Table II

InAs–GaSb Heterojunction Band Offsets Expressed as E_s[a]

Source	E_s
Frensley and Kroemer (1977)	−0.14
Ihm *et al.* (1979)(100)	−0.21
Harrison (1980)	0.10
Madhukar *et al.* (1979)(100)	0.06
Sai-Halasz *et al.* (1978a)(100)	0.15
Sai-Halasz *et al.* (1977)	0.14

[a] Specific orientations are indicated.

possible that a more detailed structural and chemical knowledge of the atomic layers very close to the interface is needed to produce more accurate values.

Tables I–III demonstrate the variation in the values of the offsets as calculated by several methods. Also listed are experimentally determined values for the GaAs–$Ga_{1-x}Al_xAs$ and InAs–GaSb superlattices. For example, Table I lists values for the GaAs–AlAs valence band maxima discontinuity as a fraction of the total difference in direct band gaps between the two semiconductors (see Fig. 2). The generally accepted value is that of Dingle *et al.* (1975), $\Delta E_v/\Delta E_g = 0.15$. This value was originally determined by using a simple well model to fit features in the optical absorption spectra of the GaAs–$Ga_{1-x}Al_xAs$ super-

TABLE III

Ge–GaAs Heterojunction Band Offsets[a]

Source	ΔE_v	ΔE_c
Frensley and Kroemer (1977)	0.71	0.07
Pickett *et al.* (1978)(110)	0.35	0.40
Harrison (1980)	0.41	0.35
Baraff *et al.* (1977)(100)	(0.75)	0
Kunc and Martin (1980a,b)(100)	−0.03	0.78
Perfetti *et al.* (1978)(100)	0.25	0.50
Kraut *et al.* (1980)(110)	0.53	(0.22)
Mönch and Grant (1980)(110)	(0.36)	0.39
Bauer (1981)(110)	0.4	(0.35)
Shay *et al.* (1976)	0.69	0.06

[a] ΔE_v is the Ge valence band maximum minus the GaAs valence band maximum energy. ΔE_c is the GaAs conduction band minimum minus the Ge conduction band minimum energy. Specific orientations are indicated. Values in parentheses assume $\Delta E_v + \Delta E_c = 0.75$ eV.

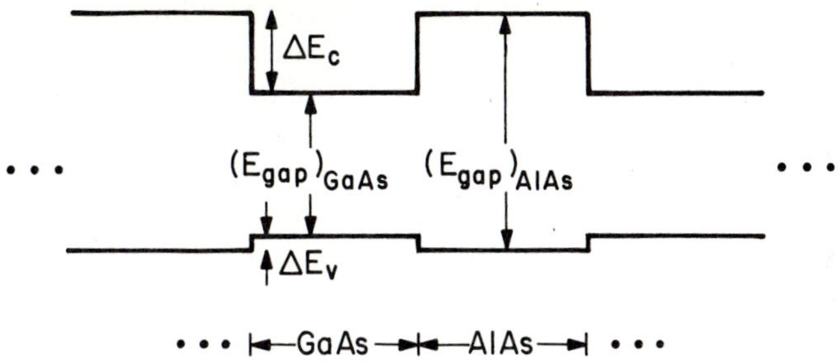

Fig. 2. Line-up of the GaAs and AlAs conduction and valence band edges in the GaAs–AlAs superlattice. ΔE_v and ΔE_c are the valence and conduction band edge discontinuities.

lattice. It has since then been used successfully in analyzing optical spectra from superlattices, quantum wells, and two-dimensional electron gases at the GaAs–$Ga_{1-x}Al_xAs$ interface. Only the theoretical calculation of Pickett *et al.*, who used a self-consistent pseudopotential method, is close to that of Dingle *et al.* The electrostatic potential matching scheme of Frensley and Kroemer (1977) and the unified tight-binding model of Harrison (1980) both produce discontinuities that are substantially different from Dingle's value.

The situation is similar for the InAs–GaSb heterojunction (Table II). The critical parameter here is E_s, the GaSb valence band maximum energy minus the InAs conduction band minimum energy (see Fig. 3). Several experiments have been successfully analyzed by using values of E_s between 0.06 and 0.15 eV. Using electron affinities to locate the conduction band edges relative to the vacuum level seems to be accurate in this case (Sai-Halasz *et al.*, 1977). It

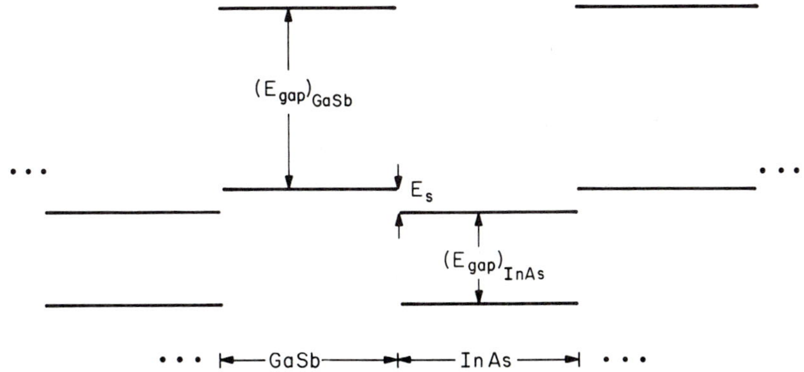

Fig. 3. Lineup of the InAs and GaSb conduction and valence band edges in the InAs–GaSb superlattice. E_s is the energy separation between the GaSb valence band maximum and the InAs conduction band minimum.

produces a value for E_s of 0.14 eV, which compares well with the 0.15-eV value deduced from analyzing experimental superlattice energy gaps with a simple two-band model (Sai-Halasz *et al.*, 1978a). Madhukar *et al.* (1979) fit the same data with a tight-binding model and found a value of 0.06 eV for E_s. In this case Harrison's method agrees roughly with the experimental findings, while Frensley and Kroemer's does not. The self-consistent pseudopotential method that seemed successful in the GaAs–AlAs case is less accurate for InAs–GaSb, perhaps due to the neglect of the spin-orbit interaction (Ihm *et al.*, 1979).

The band offsets that have been most extensively studied experimentally are those of the Ge–GaAs system (Table III). The consensus of the more recent experiments (Kraut *et al.*, 1980; Mönch *et al.*, 1980; Bauer, 1981) is that the valence band discontinuity lies between 0.36 and 0.53 eV. The electron affinity rule (Shay *et al.*, 1976) fails here, while the theoretical approaches give a large range of values (Frensley and Kroemer, 1977; Pickett *et al.*, 1978; Harrison, 1980; Baraff *et al.*, 1977; Kunc and Martin, 1980a,b).

The CdTe–HgTe heterojunction and superlattice are at an early stage of experimental investigation. Measurements by Kuech and McCaldin (1982) imply a valence band offset in this system that is not as small as that predicted by electron affinity arguments. They speculate, however, that this may be caused by inversion at the interface and that the true offset may be small.

B. Quantum-Well Model

The electronic structure of superlattices of certain compositions and in certain energy ranges can be obtained from a simple quantum-well model. For example, the band offsets for the GaAs–$Ga_{1-x}Al_xAs$ superlattice are such that the bulk conduction band minima and valence band maxima line up as shown in Fig. 2. The band gap of the $Ga_{1-x}Al_xAs$ forms a barrier to the electrons in the GaAs conduction well and the holes in the GaAs valence well. The simplest calculation based on this model uses the effective mass approximation and writes the electronic wave function as a plane wave, just as in the Kronig–Penney problem (Esaki and Tsu, 1970). The energy solutions form a series of minibands within the conduction and valence wells.

It is clear that for values of x such that the $Ga_{1-x}Al_xAs$ alloy has a direct band gap, the band edges indicated in Fig. 2 should be the direct edges. For x greater than 0.4, however, $Ga_{1-x}Al_xAs$ has an indirect conduction band minimum at the X point. The quantum-well model gives no guidance as to which minimum to use. In general, a more detailed calculation that includes more of the band structure is needed.

This model has been widely used to analyze experimental data, especially optical experiments. As long as the widths of the alternating layers are large, >50 Å, the quantum-well model should give an adequate description of the electronic structure of the lowest conduction bands and highest valence bands.

This has indeed been the case for most of the superlattices on which experiments have been performed.

Several steps can be taken within this framework to improve agreement with experiment over the simple Kronig–Penney calculation and to extend the validity of the model over a larger energy range. A major defect in the model is the assumed parabolic bulk energy dispersion curves at all energies resulting from the effective mass approximation. Since the accuracy in the determination of the superlattice energy levels can be no better than that for the bulk semiconductors, it is important to use bulk dispersion curves that are as accurate as possible. Figure 4 illustrates the deviation from parabolicity of the bulk GaAs and AlAs energy bands in the fundamental band gap region. It shows parabolic dispersion curves superimposed on dispersion curves determined from a tight-binding calculation with the same effective masses at the valence and conduction band edges (Schulman and Chang, 1981a). In order for a superlattice state energy to be correctly computed from a Kronig–Penney model calculation, it is necessary that the energy lie within the parabolic regions of the GaAs and AlAs bulk bands.

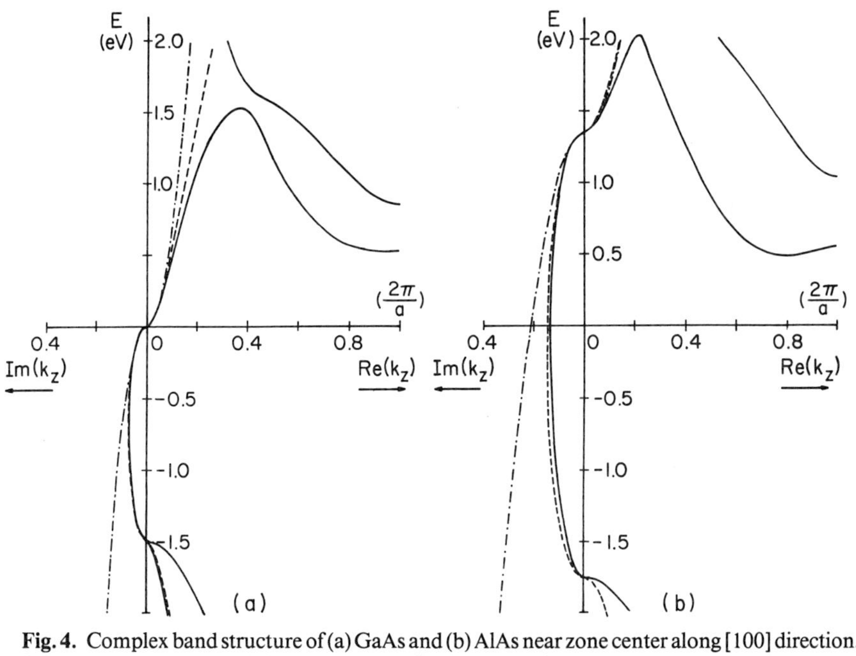

Fig. 4. Complex band structure of (a) GaAs and (b) AlAs near zone center along [100] direction. $\mathrm{Re}(k_z)$ is plotted to the right of each figure, $\mathrm{Im}(k_z)$ is to the left. (Solid lines: tight-binding model; dashed lines: two-band model; dash-dotted lines: Kronig–Penney model.) The GaAs conduction band minimum is the energy zero. The AlAs valence band maximum is close to −1.7 eV. [From Schulman and Chang (1981a).]

It can be seen that the GaAs conduction band parabolic curve begins to depart significantly from the results of the tight-binding calculation for energies greater than roughly 0.5 eV above the conduction band minimum. Fortunately, experiments on GaAs–$Ga_{1-x}Al_xAs$ superlattices have investigated electronic states that fall within this range of energy. For example, a GaAs–AlAs superlattice made from alternating layers with widths of 50 Å each has a lowest conduction band energy of only about 0.1 eV above the GaAs edge (Mendez *et al.*, 1981). Thus, the departure from parabolicity in superlattices with thicker GaAs layers or for superlattices with $Ga_{1-x}Al_xAs$ layers replacing the AlAs (which can also have small lowest conduction band energies) is moderate.

Also important is the effect of nonparabolicity in the $Ga_{1-x}Al_xAs$ layers. The lowest superlattice conduction band energy, while quite close to the GaAs conduction band edge, is, because of the band offsets, far from the $Ga_{1-x}Al_xAs$ conduction band edge at the zone center, except for small values of x. Thus, the deviation from parabolicity in the $Ga_{1-x}Al_xAs$ can be substantial. The extreme case, shown in Fig. 4 for $x = 1$, results in a large difference in bulk k_z between the tight-binding and parabolic dispersion curves for energies close to the GaAs conduction band edge. Again, many of the experimental results reported in the literature that were analyzed by using the Kronig–Penney model pertained to GaAs–$Ga_{1-x}Al_xAs$ superlattices with small values of x, so that large errors would not be expected in those cases. Also, since the superlattice state in this energy range is highly localized in the GaAs layers, the $Ga_{1-x}Al_xAs$ dispersion is less important in influencing the superlattice state energy.

An illustration of the error resulting from the use of a Kronig–Penney calculation for the quantum well model is shown in Fig. 5 (Schulman and Chang, 1981a). Superlattice conduction minimum dispersion curves in the z direction, calculated by using the Kronig–Penney and tight-binding methods, are shown for superlattices with two layers of AlAs alternating with a varying number of GaAs layers. Agreement between the two methods decreases for superlattices with smaller numbers of GaAs layers because the resulting energies are farther from the GaAs conduction band minimum and the effects of nonparabolicity become more important. The two methods have also been compared for higher-lying superlattice conduction states. These correspond to higher states in the quantum wells. The discrepancies between the two methods follow the same pattern—higher energies have greater discrepancies.

Figure 5 also shows that for energies above ≈ 0.52 eV the predictions of the two methods diverge. The tight-binding curves flatten out at this energy, whereas the Kronig–Penney curve continues to increase with increasing perpendicular wave vector. Its significance is that it is the indirect conduction band minimum energy at the X point in both GaAs and AlAs (with the specified band offsets). Near it, the superlattice state is more closely related to the bulk X-point states than it is to the zone center states. The Kronig–Penney model ignores

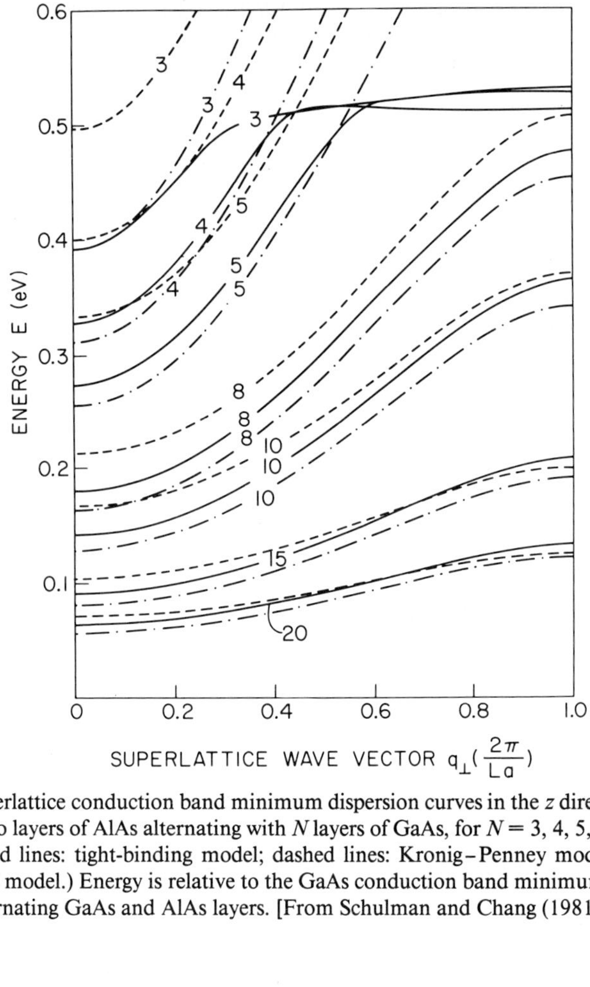

Fig. 5. Superlattice conduction band minimum dispersion curves in the z direction for superlattices with two layers of AlAs alternating with N layers of GaAs, for $N = 3, 4, 5, 8, 10, 15, 20$ as indicated. (Solid lines: tight-binding model; dashed lines: Kronig–Penney model; dash-dotted lines: two-band model.) Energy is relative to the GaAs conduction band minimum. L is the total number of alternating GaAs and AlAs layers. [From Schulman and Chang (1981a).]

these states and thus it fails completely here. Note that this energy is far below the AlAs direct conduction band minimum.

There have been several attempts to improve the Kronig–Penney model while keeping within the same basic framework. Mukherji and Nag (1975, 1978) modified the very simplest model by using different effective masses for the two-constituent semiconductors. They also artificially imposed a modification of the parabolic dispersion curve by using Kane's relationship for an energy-dependent effective mass. Conduction band energies of GaAs–$Ga_{1-x}Al_xAs$ were calculated for superlattices with x values of 0.2, 0.5, and 0.7. The GaAs layer width used was 50 Å, while the $Ga_{1-x}Al_xAs$ width varied from 10 to 100 Å. They found that including nonparabolicity decreased the superlattice conduction state energies (measured relative to the GaAs conduction band minimum) by as much as 15% for $x = 0.2$ and 25% for $x = 0.5$ and 0.7.

In addition, Mukherji and Nag made the first attempt to evaluate the effect of the indirect $Ga_{1-x}Al_xAs$ X-point conduction band minimum on the superlattice conduction band states. They found that the lowest energy states were not related to the X-point minimum, even for x values such that $Ga_{1-x}Al_xAs$ is indirect. They predicted additional narrow bands above the conduction band minimum that would be derived from the X-point states. This result is supported by the tight-binding calculation, as shown in Fig. 5.

In general, the simple Kronig–Penney model must be applied with special care when analyzing experiments in which optical transitions between superlattice states composed of bulk states away from the zone center are important. This was attempted by Mendez *et al.* (1981) in their analyses of electroreflectance spectra of $In_{1-x}Ga_xAs$–$GaSb_{1-y}As_y$ and GaAs–$Ga_{1-x}Al_xAs$ superlattices. Not surprisingly, they showed that the variation of the optical transition energies for transitions away from the zone center as a function of layer thickness in the $In_{1-x}Ga_xAs$–$GaSb_{1-y}As_y$ system did not follow Kronig–Penney behavior. The more complicated behavior they did observe must be analyzed by using Mukherji and Nag's method or the more detailed tight-binding or pseudopotential methods that can describe bulk band structure away from the zone center.

C. Two-Band Models

The two-band model is the simplest model that realistically incorporates some elements of the band structure of the constituent semiconductors. The two bands of interest are the lowest conduction band and the light-hole band. It provides two major improvements over a Kronig–Penney type calculation. First, the bulk dispersion curves resulting from the model are parabolic near the zone center but contain the corrections to parabolicity for larger values of the wave vector. In fact, an excellent estimate of the departure from parabolicity of the bulk bands, and thus of the accuracy of a superlattice energy calculated using the Kronig–Penney model, can be estimated by using simple formulas derivable from a Kane type two-band model (Kane, 1957). The bulk conduction and valence band dispersion curves in this model can be written as (Schulman and McGill, 1981)

$$E = E_v \pm \frac{\hbar^2 \gamma \sqrt{k^2 + \gamma^2}}{m_\pm}, \tag{3}$$

where

$$\gamma = \sqrt{E_g \Big/ \hbar^2 \left(\frac{1}{m_+} + \frac{1}{m_-}\right)}.$$

Here, E_v is the energy of the valence band edge, m_- the valence band effective mass, m_+ the conduction band effective mass, and E_g the band gap. Other two-band models would produce slightly different forms for the dispersion relation as compared to Eq. (3), depending on the basis set used (Sai-Halasz *et al.*, 1978b).

The second improvement over the Kronig—Penney model is that the two-band model recognizes that Bloch states are matched at the interfaces. In the Kronig—Penney model the states are plane waves with different normal components of the wave vectors in the two alternating layers. No account is taken of the differences in the periodic part of the bulk Bloch functions, which are ignored. The two-band model allows the Bloch functions in each material to differ from each other. This makes the boundary conditions at the interfaces more realistic, although more complicated.

Also, the use of Bloch functions in the two-band model includes the possible mixing of bulk states from different bands in forming the superlattice state. This is an especially important effect in the InAs–GaSb superlattice in which the band offsets are such that the InAs lowest conduction band is near in energy to the GaSb valence band maximum. Superlattice states that mix the InAs conduction band state and the GaSb light-hole state would be expected (Sai-Halasz *et al.*, 1977). A similar effect occurs in the GaAs–$Ga_{1-x}Al_xAs$ superlattice valence bands in which the light-hole and spin-orbit split-offs bands can interact. Section IV contains examples of this mixing.

There have been two types of two-band model calculations carried out for superlattices: those based on a Kane type analysis (Sai-Halasz *et al.*, 1977; White and Sham, 1981; Bastard, 1981, 1982b) and those based on the tight-binding method (Sai-Halasz *et al.*, 1978b; Ivanov and Pollmann, 1979). Sai-Halasz *et al.* (1977) used Bloch functions obtained from Kane's two-band Hamiltonian to replace the plane waves of the Kronig–Penney model. The superlattice dispersion relation was then obtained by matching the wave function and its derivative at the interfaces. The resulting superlattice dispersion relation is

$$\cos(kd_0) = \cos(k_1 d_1)\cos(k_2 d_2) - f \sin(k_1 d_1)\sin(k_2 d_2), \tag{4}$$

with

$$f = \tfrac{1}{2}[w + (1/w)] \qquad \text{and} \qquad w = [ik_1 + (u_1'/u_1)]/[ik_2 + (u_2'/u_2)].$$

Here, k is the superlattice wave vector in the direction perpendicular to the interface, d_0 is the superlattice period, d_1 and d_2 are the thicknesses of the alternating slabs, k_1 and k_2 are the wave vectors of the Bloch states in the alternating layers, and u_1, u_2, u_1', and u_2' are the values of the periodic parts of the Bloch states and their first derivatives in the alternating slabs at the interfaces. The Kronig–Penney formula is recovered if the logarithmic derivatives are set equal to zero.

The effect of including two bands in their calculation is clearly demonstrated in Fig. 6. The superlattice dispersion curves for three $In_{1-x}Ga_xAs-GaSb_{1-y}As_y$ superlattices with three sets of x, y values: (a) $x = 0$, $y = 0$, (b) $x = 0.75$, $y = 0.15$, (c) $x = 0.67$, $y = 0.70$ are shown. The bulk conduction and valence band edges and the band offsets are also indicated. In case (c), the dispersion curves calculated using the Kronig–Penney and two-band models are almost identical. This follows from the large energy separation between the bulk $In_{1-x}Ga_xAs$ conduction band and the bulk valence band of the $GaSb_{1-y}As_y$. Substantial deviation from the Kronig–Penney model is found for alloy superlattices (a) and (b) in which the conduction and valence bands of the different materials are close in energy.

Bastard (1981, 1982b) made use of the same Kane formalism and also obtained Eq. (4). Investigating the boundary conditions in more detail, he expressed the correction to the Kronig–Penney boundary conditions for the quantum-well model in a simple way. First, the effective mass envelope function is assumed continuous across the interface. Then, the derivative of the

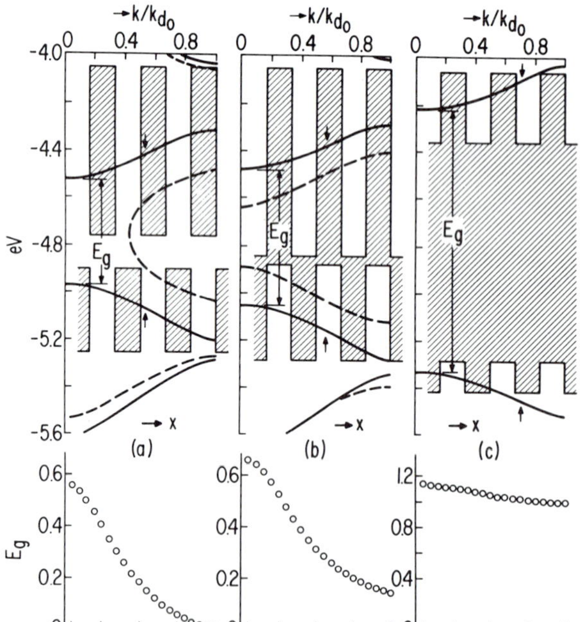

Fig. 6. On the upper graphs, E versus k_z relationships for three $In_{1-x}Ga_xAs-GaSb_{1-y}As_y$ superlattices consisting of four layers each of $In_{1-x}Ga_xAs$ and $GaSb_{1-y}As_y$ together with their potential profiles. The x and y values are given in the text. $k_{d_0} = \pi/d_0$, where d_0 is the bulk lattice constant. (Solid lines: two-band model; dashed lines: Kronig–Penney model; lower curves: energy gaps versus layer thickness.) [From Sai-Halasz *et al.* (1977).]

envelope function divided by a simple function of the band energy is matched. This reduces to matching the derivative divided by the effective mass at interfaces in which band mixing is small, such as in the GaAs–$Ga_{1-x}Al_xAs$ superlattice. For the interacting band case, the InAs–GaSb superlattice, the full expression is used.

The other two-band calculation based on the Kane model is that of White and Sham (1981). White and Sham derived boundary conditions that matched light-hole band and conduction band wave functions across the interfaces, but not their derivatives. Bastard (1982b) has demonstrated the equivalence of these boundary conditions with his own. Since both Bastard's and White and Sham's methods produce reasonable agreement with experimental data, it is probable that the inaccuracies in their boundary conditions are not significant within the range in which the two-band model is applicable. As shown in Subsection IV.C, more than two bands can contribute significantly to the superlattice wave functions close to the interface, so that their boundary conditions would not be expected to be accurate in all cases. Altarelli (1982) has devised a new, very efficient method for calculating superlattice electronic properties within a similar, partially three-band, Kane model framework.

There have been two two-band model calculations on superlattices based on the tight-binding method, one for the InAs–GaSb superlattice (Sai-Halasz *et al.*, 1978b) and the other for the GaAs–$Ga_{1-x}Al_xAs$ superlattice (Ivanov and Pollmann, 1979). Both use a basis set consisting of one s orbital on the cation and one p orbital on the anion. The empirical parameters were chosen to reproduce the experimental bulk data on band gaps, effective masses, and band offsets. Thus, these models could be expected to describe well the lowest conduction band and the light-hole valence band. In addition, Sai-Halasz *et al.* included the dispersion in the heavy-hole band by allowing p orbitals on the cation to interact with the p orbitals on the anion. Both calculations produced superlattice dispersion curves and thickness-dependent energy levels. Also, Sai-Halasz *et al.* used a self-consistent Thomas–Fermi approximation to investigate band bending in the InAs–GaSb superlattice.

A notable feature of both of these calculations is that, unlike the conventional tight-binding model, the complexity of these calculations does not increase with the thickness of the alternating superlattice layers. Sai-Halasz *et al.* replaced the Hamiltonian method with a transfer matrix approach. The superlattice bands are found by examining the trace of the transfer matrix. Ivanov and Pollmann used an equivalent approach based on a resolvent technique within a Green's function formulation of scattering theory.

The limitation of the two-band model is that it is valid only within a limited energy range. A major improvement of the two-band model over the quantum-well model is in its ability to describe the bulk dispersion curve for imaginary values of the wave vector k_z. As shown in Fig. 4, a GaAs–AlAs superlattice state with an energy slightly above the GaAs conduction band minimum has an energy that lies well within the AlAs band gap. The two-band model produces an excellent approximation to the full tight-binding dispersion curve that connects the conduction band minimum and valence band maximum across the AlAs gap. The two curves fall almost on top of each other in Fig. 4. The superlattice dispersion curve that results is in closer agreement with that of the full tight-binding calculations than with that of the quantum-well calculation (Fig. 5) in both energy location and shape of the curve. The two-band model also fails, however, when the superlattice energy approaches the X-point energy and for the same reason: the bulk structure away from $\mathrm{Re}(k_z) = 0$ is ignored.

The two-band model is obviously inadequate in describing superlattice states in which more than two bulk states may be expected to contribute significantly. For example, the band offsets in the InAs–GaSb superlattice are such that certain superlattice valence band state energies are near the InAs conduction band minimum, the GaSb light-hole band maximum, and the GaSb spin–orbit split-off band maximum. A more detailed model is thus necessary to include all three bulk states.

D. Full Tight-Binding Models

The simplest model that incorporates a realistic description of the bulk band structures is the tight-binding model. First introduced for the GaAs–AlAs superlattice (Schulman and McGill, 1977, 1979c), it has also been applied to the InAs–GaSb (Nucho and Madhukar, 1978; Madhukar *et al.*, 1979; Madhukar and Nucho, 1979), CdTe–HgTe (Schulman and McGill, 1979a,b, 1981), Si–GaP (Madhukar and Delgado, 1981), and GaAs–GaAsP (Osbourn, 1982) superlattices. As most commonly employed, the tight-binding method makes use of one s orbital and three p orbitals per anion and cation. Thus, a total of eight bands, four valence and four conduction, can be described. As in the two-band model, the empirical parameters in the theory are chosen to reproduce features in the bulk band structures such as optical transition energies and effective masses. The valence bands are quite accurately fit in this way (Chadi and Cohen, 1975). The conduction bands are more troublesome but can also be adequately described for the purposes of a superlattice calculation by including second neighbor overlap parameters (Schulman and McGill, 1977) or more orbitals (Schulman and Chang, 1981a).

The overlap parameters between orbitals on opposite sides of the interface between the two alternating semiconductors must be considered separately. In superlattices with a common anion, such as GaAs–AlAs, the As cation overlaps have been taken to be the same as for bulk materials. The second neighbor cation matrix elements and the on-site As matrix elements were taken to be the average of those for the two bulk materials (Schulman and McGill, 1977). The InAs–GaSb calculations made use of bulk InSb and GaAs parameters at the interfaces (Nucho and Madhukar, 1978).

These assumptions point out the main limitation of the tight-binding method: the approximate description of the interface electronic structure. This model, like the quantum-well and two-band models, assumes a bulklike potential right up to the interface. No self-consistency in the electronic density or atomic rearrangement, which would modify the potential near the interface, is included. Fortunately, as will be discussed subsequently, calculations that include a self-consistent redistribution of electronic charge show that its effect is small and of short range. Also, for lattice-matched semiconductors with similar valence electronic structure little atomic rearrangement is expected.

As mentioned previously, the original formulation of the tight-binding method for superlattices made use of a Hamiltonian matrix whose dimension grows linearly with the superlattice unit cell length in the z direction. This was a serious limitation on the usefulness of the method, since superlattices of only about 40 atoms per unit cell in the z direction could be handled. An equivalent reformulation of the method has been developed (Schulman and Chang, 1981a, 1983) that employs a Hamiltonian of constant dimension. In addition, this method allows a clear analysis of the superlattice electronic state in terms of the bulk Bloch states of which it is composed.

In this method the superlattice state of energy E and specified $\mathbf{k}_t$ is written as in Eqs. (1) and (2) as a sum of bulk Bloch states with complex values of k_z. The number of Bloch states needed is, for instance, 16 for each material for a (100) superlattice when four orbitals per atom and up to second neighbor interactions are included (Osbourn and Smith, 1979). They can be found by using the method of Chang and Schulman (1982) or Schulman and Chang (1983). The 32 Bloch states thus form a basis in which the Hamiltonian matrix is written. Diagonalizing this Hamiltonian gives the superlattice state energy and the correct linear combination of Bloch states needed to describe it. Since the Bloch states that form the basis are energy dependent, it is necessary to iterate this method so that the energy resulting from diagonalizing the Hamiltonian is the same as that used in finding the Bloch states. In practice, very few (two to five) iterations are needed and the computer time involved is insignificant.

E. Pseudopotential Method

The pseudopotential method is the most sophisticated of the methods so far described. Its principal advantage is that it incorporates detailed, albeit pseudowave functions. With the pseudowave functions, the valence charge density and the resulting electrostatic potentials can be estimated.

The pseudopotential model, as it has been applied to superlattices, also has empirical parameters that must be determined. The parameters in this case determine the form and intensity of the atomic pseudopotentials. They have been fit by attempting to reproduce features of the atomic or the bulk electronic structures or both. The empirical pseudopotential method (Caruthers and Lin-Chang, 1977, 1978; Andreoni *et al.*, 1978) uses these parameters to create the superlattice potential by superimposing the atomic pseudopotentials. A basis set (usually plane wave) is chosen and the electronic properties are calculated by diagonalizing the Hamiltonian. The self-consistent pseudopotential method continues and calculates a new electrostatic potential based on the charge distribution given by the wave functions.

The self-consistent pseudopotential method allows a direct determination of the band offsets between the two semiconductors. The most common method for achieving this (Pickett *et al.*, 1978; Ihm *et al.*, 1979; Kunc and Martin, 1980) is to determine the average electrostatic potentials in the separate bulk materials and then to determine them again on either side of, and away from, an interface formed by joining them. The shift in the difference between the potentials of the two materials determines the offsets.

The main disadvantage of the pseudopotential method is the computational difficulties involved for superlattices with large alternating layer thicknesses. For example, Caruthers and Lin-Chung in their non-self-consistent calculation found that for a general wave vector the largest GaAs–AlAs superlattice for which it was practical to carry out their calculation consisted of just three layers of GaAs alternating with three layers of AlAs (Caruthers and Lin-Chung, 1978). At the zone center, they could treat a total of 18 layers of GaAs and AlAs per superlattice unit cell. It is possible that this limitation could be eliminated by expanding the superlattice state in terms of bulk states with complex values of k_z, as was done within the tight-binding formalism. A method that finds the complex k_z states within the pseudopotential and $\mathbf{k} \cdot \mathbf{p}$ formalisms has been developed (Chang and Schulman, 1982), but it has not yet been applied to the superlattice problem.

Another difficulty of the pseudopotential method is one it shares with the tight-binding method. Similar calculations carried through on the basis of somewhat different empirical pseudopotential parameters give conflicting results. This is discussed in more detail in Section IV.

The non-self-consistent pseudopotential method has an additional difficulty when applied to the superlattice problem. As it has been applied so far, the band offsets between the bulk semiconductors are a product, not an input, of the calculation (Caruthers and Lin-Chung, 1978). However, Pickett *et al.* (1978) have shown for the (110) Ge–GaAs and GaAs–AlAs interfaces that the offsets can change substantially when self-consistency is included. It is unclear whether the offsets produced by the non-self-consistent calculations agree with accepted values.

IV. ELECTRONIC PROPERTIES

In this section we shall discuss a few of the electronic properties of semiconductor superlattices that have been investigated theoretically by using the methods described in Section III. Of special interest is the electronic structure in the energy range of the band gap, including band-gap dependences on superlattice parameters and the character of the electronic states. Also to be discussed are the related problems of the confinement of electronic states and the mixing of bulk states by the superlattice potential.

The focus in this section will be on the two superlattice systems that have been the most thoroughly investigated theoretically. These are the GaAs–AlAs and InAs–GaSb superlattices. Also to be mentioned is the CdTe–HgTe superlattice. This system has only recently been fabricated in the laboratory (Faurie, 1981). Prior theoretical calculations (Schulman and McGill, 1979a,b) predict unique properties for this superlattice.

A. Interface Electronic Structure

The electronic properties of semiconductor interfaces have been studied extensively and reviewed elsewhere (Pollmann, 1980). There are two important results from these calculations that are most relevant to superlattices. The first is the values of the band offsets. This has been discussed in Section II. The second and related result is the effect of the interface on the electrostatic potential and

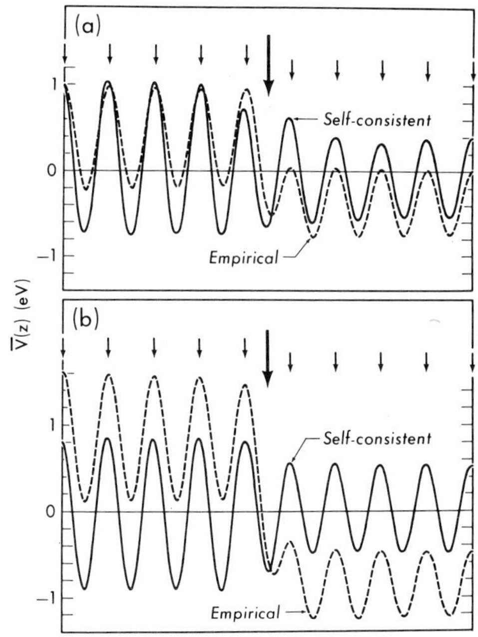

Fig. 7. Empirical and self-consistent potentials, averaged parallel to the interface, for the (110) interfaces of (a) Ge–GaAs and (b) AlAs–GaAs. The large arrow denotes the geometric interface, while the smaller arrows show the positions of atomic planes. [From Pickett *et al.* (1978).]

thus the superlattice Hamiltonian. If the presence of the interface disrupts the potential at large distances away from the interface, the approximation employed by the quantum-well, two-band, and tight-binding models, consisting of using parameters characteristic of the bulk materials close to the interface, is inadequate.

Fortunately, detailed studies of the interface potential show that the interfacial disruption can be slight and highly localized within two or three atomic planes from the interface. For example, Fig. 7 shows the potential averaged over planes parallel to the interface near the GaAs–AlAs and Ge–GaAs interfaces (Pickett *et al.,* 1978). It can be seen that the potential is indistinguishable from bulklike within two layers. The main effect of the joining of the two materials is the shift in average bulk potentials brought about by the self-consistent iteration. The effect is incorporated in the value of the band offset and thus can be included empirically in the less sophisticated models.

The short range of the disruption near the interface is not surprising for two lattice-matched semiconductors with similar chemical character such as GaAs–AlAs, InAs–GaSb, and CdTe–HgTe. Other pairs of materials may not be so simple and could exhibit as yet unknown reconstruction and charge transfer.

B. Band Gaps

Since the alternating materials out of which a superlattice is constructed are semiconductors, it is not surprising that the superlattice itself is also usually a semiconductor. Of practical significance is the fact that the superlattice band gap is adjustable by varying the thicknesses of its alternating layers.

There have been several calculations of the superlattice band gap dependence on layer thickness for the GaAs–AlAs, the InAs–GaSb, and the CdTe–HgTe systems. The GaAs–AlAs superlattice will be discussed first.

For large GaAs and AlAs alternating layer thicknesses, the band-gap dependence is smooth and monotonically decreasing with GaAs layer thickness, just as would be the case for a series of square wells separated by finite barriers. Both the conduction electrons and the valence holes in the GaAs layers near the band edges are in these effective wells, as seen in Fig. 2. The Kronig–Penney, two-band, and tight-binding models all show this simple behavior down to very thin GaAs and AlAs layer thicknesses—about 10 Å. So do two of the three pseudopotential calculations performed on this system (Andreoni *et al.,* 1978; Pickett *et al.,* 1978). The exception is the pseudopotential calculation of Caruthers and Lin-Chung (1978). They find an almost flat variation of band gap with layer thickness for up to nine atomic layers each of GaAs and AlAs (as compared with Kronig–Penney behavior). Consistent with this result is that the lowest conduction band state in their calculation does not resemble the state for a particle in a well. Instead, it has a large charge density very close (within two atomic layers) to the GaAs–AlAs interfaces. They conclude that this is caused by an abruptly changing potential at the interface. Andreoni *et al.* (1978), and implicitly all others who have calculated the band-gap variation, claim that Caruthers and Lin-Chung overestimated the cationic potential difference and thus the abruptness of the change in the interface potential.

For very thin layers, the quantum-well model does break down. Andreoni *et al.* find that the thin-layer superlattices have band structures very close to that of the alloy with the same composition. The tight-binding study of Schulman and McGill (1979c) is in qualitative agreement but predicts a faster decrease in band gap with increasing thickness. This may be due to the inaccuracies in the effective masses produced by the particular choice of parameters in the tight-binding model. Indeed, subsequent studies using the tight-binding model have been careful to ensure that experimental effective masses were obtained. This has resulted in the more successful reproduction of experimental band gaps.

The InAs–GaSb superlattice band-gap variation cannot be analyzed in quite so simple a manner, but the well model still leads to a qualitative understanding. Figure 3 shows that the bottommost conduction states should be localized in the InAs layers while the topmost valence states would be mostly in the GaSb layers. What is not obvious is the relative energy positioning of the lowest InAs and highest GaSb-like state. For large layer thicknesses, in fact, the states cross. The InAs state has a lower energy than the GaSb state. When this occurs, the superlattice becomes a semimetal (Chang *et al.*, 1979; Guldner *et al.*, 1980). Rough agreement for the thickness at crossover exists among theoretical calculations (Sai-Halasz *et al.*, 1978b; Madhukar *et al.*, 1979; Chang, 1980a,b) and experiment (Chang *et al.*, 1979)—about 100Å each of InAs and GaSb.

The problem with the quantum-well model as applied to the InAs–GaSb superlattice is that the well limits are not well defined. For example, the bottommost conduction state, which is mostly localized in the InAs layers, will decay into the GaSb layers. However, the effective well edge that governs the decay is not simply the GaSb conduction band edge, because the InAs state can join onto the GaSb light-hole state as well as the GaSb conduction minimum state.

A similar situation exists for the CdTe–HgTe superlattice. Due to their common anions, band offsets for this superlattice have been conjectured to be such that the valence band maxima of bulk CdTe and HgTe are very close in energy (Schulman and McGill, 1979a), Fig. 8. The zero band gap of HgTe and the small value of the valence band discontinuity lead to unique band-gap behavior. The state at the conduction band minimum in the superlattice is made up of conduction band states in the HgTe layers, as the simple quantum-well model would lead one to expect. However, the large (1.6-eV) conduction band discontinuity does not confine the electron. The close proximity of the CdTe valence band maximum to the HgTe conduction band minimum allows the electrons to tunnel through the CdTe layers with relative ease (Schulman

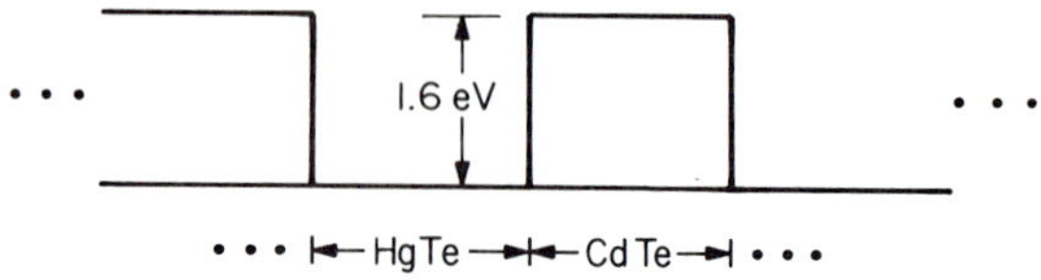

Fig. 8. Proposed lineup of the CdTe–HgTe conduction and valence band edges in the CdTe–HgTe superlattice. HgTe is a zero-band-gap semiconductor.

and McGill, 1981). This superlattice also exhibits an interesting transition from an alloylike conduction minimum electronic structure for small layer thicknesses to quantum-well behavior as the layer thickness increases (Schulman and McGill, 1979b). This is manifested by a transition in symmetry of the conduction band minimum state as the layer thickness increases. The overall dependence of band gap on layer thickness has been calculated (Schulman and McGill, 1979a; Bastard, 1982b). The gap goes to zero for large HgTe layer thickness and increases with decreasing HgTe layer thickness or increasing CdTe ("barrier") thickness.

C. Superlattice Wave Functions

As discussed previously, the short range of the interfacial disruption of the otherwise bulklike potential allows the superlattice state to be accurately described as a linear combination of bulk states in each material with complex wave vector values. In this section the decomposition of the superlattice state into its bulk state components will be illustrated. As a particular example, several valence band superlattice states at the Brillouin zone center of the (100) GaAs–AlAs superlattice, consisting of 50 Å (17 atomic double layers) of GaAs alternating with the same amount of AlAs, will be examined in detail. The theoretical framework used is that of the empirical tight-binding method (Schulman and Chang, 1981b).

Typical types of superlattice states are the following. The simplest of the states are similar in form to Kronig–Penney well-type states. These are states in which just one bulk state from near the bulk zone center dominates in each material. Spatial plots of these states have distinctive single- or multiple-peaked structures that are basically discrete lattice versions of sine waves. Other superlattice states have more than one bulk state contributing significantly. Their spatial plots are correspondingly more complicated. Bulk states with nonzero real and imaginary parts play an important role in some superlattice states. Even when they decay rapidly away from the interface [large Im(k_z)], they can represent a significant contribution close to the interface itself and, therefore, are important in determining the matching of bulk states across the interface. This implies that models that omit these states, such as the Kronig–Penney and two-band models, may not describe the boundary conditions at the interface realistically.

The complex wave vectors of the bulk states that make up the superlattice state are shown in Fig. 9. The complex band structures of GaAs and AlAs are displayed in the same manner as in Fig. 4. The spin-orbit interaction is included here. The complex wave vectors for a superlattice state with a given energy can be seen by finding the bulk complex states with the same energy. For example, the state in Table IV with energy −0.06 eV is composed of bulk complex states at that energy in Fig. 9. It shows that there are four distinct bands at energy −0.06 eV in GaAs and four in AlAs. There are actually a total of 20 bands for each material, but they are related to the 4 shown by negating or complex conjugating the wave vector, or by other symmetries peculiar to the (100) direction in zinc blende materials (Chang, 1982). Two of the four GaAs wave vectors are purely real—those in the light- and heavy-hole bands. The spin-orbit split-off band wave vector is purely imaginary, and the state emanating

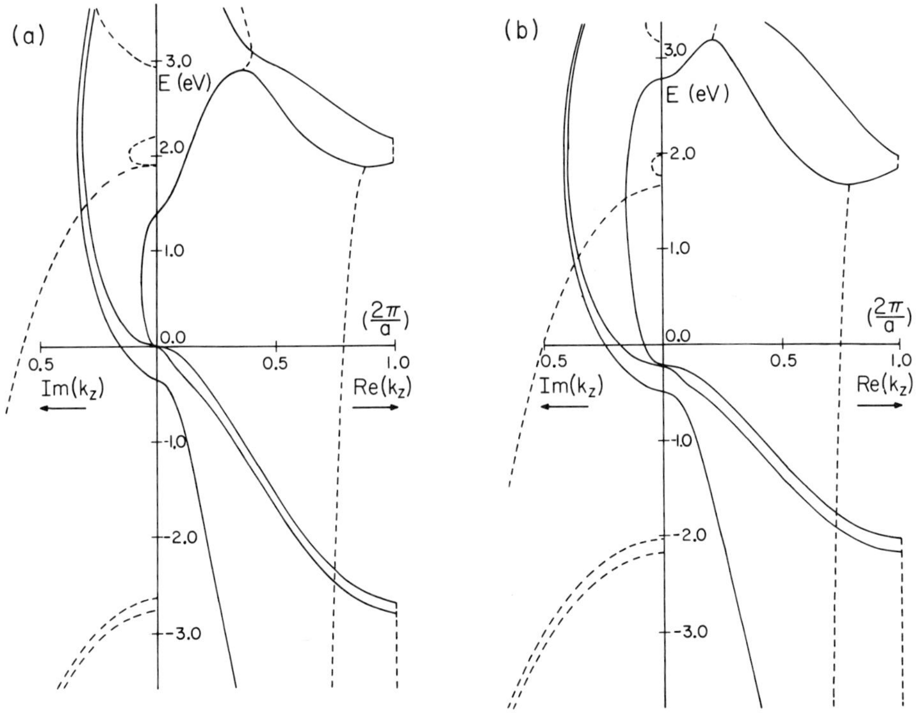

Fig. 9. Complex band structures of (a) GaAs and (b) AlAs along the [100] direction. Both the real bands [$\mathrm{Im}(k_z) = 0$] and the purely imaginary bands [$\mathrm{Re}(k_z) = 0$] are denoted by solid lines. The complex bands [$\mathrm{Re}(k_z) \neq 0$ and $\mathrm{Im}(k_z) \neq 0$] are denoted by pairs of dashed lines. The lines associated with $\mathrm{Re}(k_z)$ and $\mathrm{Im}(k_z)$ are plotted in the right and left portions of each figure, respectively. Energy is relative to the GaAs valence band maximum.

TABLE IV

GaAs–AlAs (100) Superlattice States at the Brillouin Zone Center for 50 Å of GaAs Alternating with 50 Å of AlAs[a]

Energy (eV)	GaAs	AlAs
1.56	c-(0.29,0)	c-(0,0.93)
−0.02	hh-(0.28,0)	hh-(0,1.12)
−0.06	lh-(0.19,0)	lh-(0,0.44)
−0.08	hh-(0.57,0)	hh-(0,0.96)
−0.17	hh-(0.82,0)	hh-(0,0.64)
−0.22	lh-(0.48,0)	lh-(0,0.15)
−0.26	hh-(1.04,0)	hh-(0.39,0)
−0.27	hh-(1.05,0)	hh-(0.45,0)
−0.29	so-(0,.28), lh-(0.64,0)	lh-(0.26,0)
−0.35	hh-(1.22,0), lh-(0.79,0)	hh-(0.87,0), so-(0,0.79)
−0.38	hh-(1.28,0)	hh-(1.00,0)
−0.39	hh-(1.29,0), lh-(0.88,0)	lh-(0.51,0)
−0.40	so-(0.24,0)	so-(0,0.63)
−0.41	lh-(0.94,0)	so-(0,0.57)

[a] The complex wave vectors of the dominant bulk states are given in units of a^{-1}, where a is the conventional lattice spacing. The GaAs valence band maximum is the zero of energy. The symbols c, hh, lh, and so stand for the bulk conduction, heavy-hole, light-hole, and split-off valence bands.

down from the conduction band minimum near the X-point is complex. The light- and heavy-hole AlAs wave vectors are purely imaginary.

Table IV lists all of the superlattice state energies at the zone center between −0.41 and 1.56 eV (the energy zero is the bulk GaAs valence band maximum). Also indicated are the bulk GaAs and AlAs states that give the dominant contribution to each superlattice state. They are labeled by the bulk complex bands to which they belong: conduction, heavy-hole, light-hole, or spin-orbit split-off bands. There is some ambiguity in assigning these labels to the complex bands because there are no band gaps in complex wave vector space. For example, a band originating from the conduction band minimum connects with the light-hole band maximum. Points on this curve with energies closer to the conduction band minimum are labeled as being conduction band states in Table IV.

An example of a superlattice state in which one bulk state is dominant is the state with energy −0.06 eV. Bulk GaAs and AlAs light-hole states contribute substantially more than the others in this case. A spatial plot of the wave function shows that the state resembles closely a bottommost light-hole quantum-well state. The same state plotted with the coefficients of the bulk GaAs states other than the light-hole state set equal to zero is only slightly different. This demonstrates that the one-band quantum-well model would be a reasonable approximation in describing the state. Similarly, the state with energy −0.02 eV is found to be the bottommost heavy-hole quantum-well state. The states with energies between −0.02 and −0.22 eV all resemble heavy- or light-hole quantum-well states, including excited states.

The next state, with energy −0.26 eV, is low enough in energy that it is below the AlAs valence band edge. It has a dominant heavy-hole band contribution in the AlAs layers with a purely real AlAs wave vector. The wave function thus has a substantial amplitude in the AlAs as well as the GaAs layers. This is true of all the states with energies between −0.26 and −0.39 eV. In addition, the states with energies of −0.29, −0.35, and −0.39 eV have more than one bulk state with substantial coefficients, as indicated. The state with energy −0.40 eV, as shown in Fig. 10, is interesting because it corresponds to the ground-state spin-orbit split-off band quantum-well state, although this is not obvious from examining its spatial plot. Only when the coefficients of the bulk states other than for the split-off state are set equal to zero does the characteristic well state

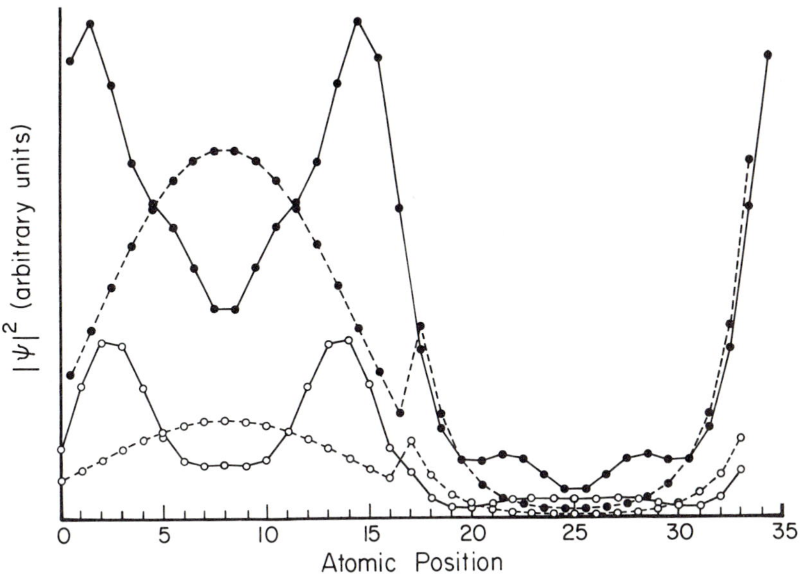

Fig. 10. Zone center electronic state with energy −0.40 eV relative to the GaAs conduction band minimum in a GaAs–AlAs superlattice with 50-Å GaAs and AlAs layer thicknesses. (Filled circles: anion; open circles: cation; solid lines: total state; dashed lines: light-hole bulk state contribution only.) The left half is in the GaAs layers and the right half is in the AlAs layers. [From Schulman and Chang (1981b).]

appear. Thus, the one-band, quantum-well model would not be expected to describe this state accurately. The state with energy −0.41 eV is an excited light-hole-band well state.

V. PHONONS

Lattice vibrations in superlattices exhibit many of the same phenomena associated with the new superlattice periodicity as electrons do. The additional periodicity shows up as a folding of the phonon dispersion curves onto a Brillouin zone decreased in size along the superlattice repeat direction. This folding process results in additional modes at the zone center.

This phenomenon has been explored theoretically by Tsu and Jha (1972) and Barker *et al.* (1978) and in a continuum model for the acoustic modes by Colvard *et al.* (1980). The effects of phonon folding can be illustrated by considering the results of Barker *et al.* (1978). They considered a one-dimensional model of a (100) GaAs–AlAs superlattice as illustrated in Fig. 11.

The force constants for AlAs and GaAs were taken to be the same, only the masses were assumed to be different. The force constants were chosen to give a "good" fit to the phonon dispersion curves of GaAs. The resulting phonon

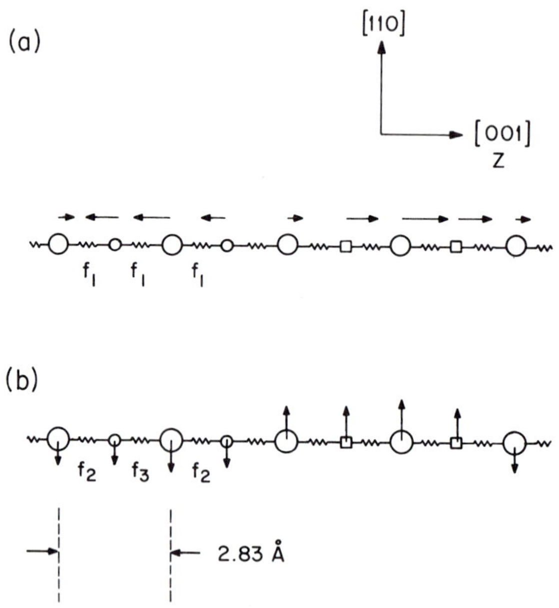

Fig. 11. The linear chain model for the GaAs–AlAs superlattice. The springs and appropriate spring constants are indicated. The values of these parameters are given in Barker *et al.* (1978). (Large open circles: As; small open circles: Ga; open squares: Al.)

dispersion curves for the monolayer and bilayer GaAs–AlAs superlattice are presented in Fig. 12. The additional dispersion curves—those over the four expected for bulk GaAs or AlAs—are due to the phonon folding. Of course, some small shifts in the frequencies also occur.

An important quantity determining infrared optical spectra is the infrared absorption strength S. It is given by

$$S = \frac{2}{\pi} \int \frac{\epsilon_2(\omega)}{\omega}\, d\omega,$$

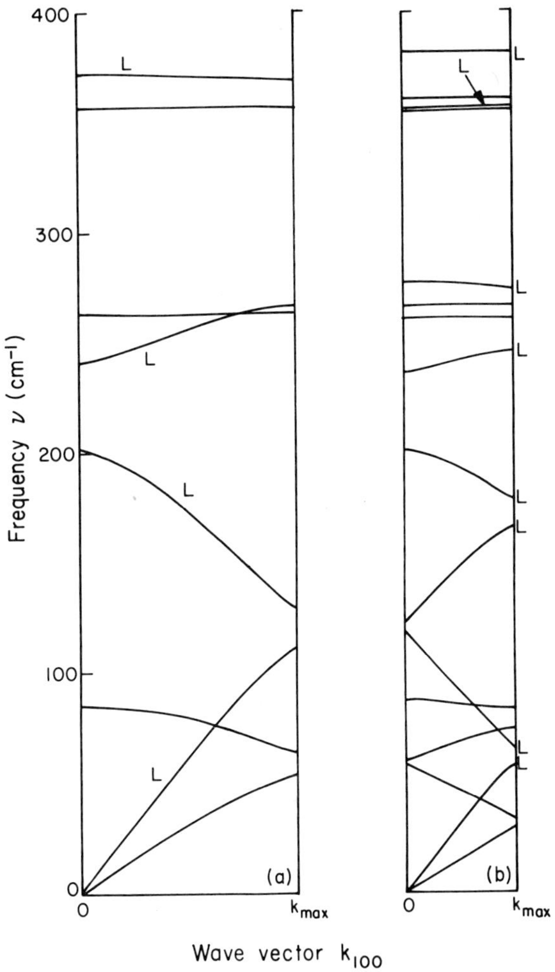

Fig. 12. The phonon dispersion curves for the (a) one-on-one and (b) two-on-two GaAs–AlAs superlattices obtained in the linear chain approximation. [After Barker *et al.* (1978).]

where $\epsilon_2(\omega)$ is the imaginary part of the dielectric constant and the integral is taken over the absorption from a given transition. Using their one-dimensional model for the lattice vibrations and an appropriate value of the effective charge, Barker *et al.* (1978) calculated S for all the $q = 0$ phonon modes. The results of this calculation are given in Table V. Some important points should be noted from the results in this table. First, the intensities of the modes that are analogous to the bulk modes are rather large. Second, modes that are produced by the folding are very weak, where they are allowed.

Further, these authors estimated the relative Raman intensities for modes that can produce scattering with the incident light incoming along the (100) direction, the z direction, and polarized along the x direction. The detected photons are outgoing in the z direction with polarization in either the x (R_{xx}) or y direction (R_{xy}). The results of these are presented in Table VI. In contrast to the case of the infrared absorption some of the additional modes are predicted to be rather strongly Raman active. This difference is due simply to the rather large difference in the bond polarizabilities for AlAs and GaAs that appear in the Raman scattering intensities in contrast to the rather small difference between the effective charges in GaAs and AlAs.

As noted by Merlin *et al.* (1980), the simple theory of Barker *et al.* (1978) did not take into account the long-range forces acting on the superlattice polar

TABLE V

Transverse Phonon Frequencies ν_j and Infrared Strengths S_j for Modes in the Model Bulk GaAs and AlAs; the One-on-One Superlattice; and the Two-on-Two Superlattice[a]

GaAs		AlAs	
ν_j (cm^{-1})	S_j	ν_j (cm^{-1})	S_j
269.0	1.93	363.0	2.0
Monolayer (1, 1)		**Bilayer (2, 2)**	
ν_j (cm^{-1})	S_j	ν_j (cm^{-1})	S_j
85.2	0.0087	57.5	0.0
262.8	0.82	59.0	0.0039
357.8	1.13	86.2	0.0
		260.4	0.0
		266.2	0.90
		354.6	0.0
		360.5	1.06

[a] Results from Barker *et al.* (1978).

TABLE VI

Longitudinal Phonon Frequencies ν_j and Raman Strengths for Bulk GaAs and AlAs, the One-on-One Superlattice, and the Two-on-Two Superlattice[a]

GaAs			AlAs		
ν_j (cm^{-1})	R_{xx}	R_{xy}	ν_j (cm^{-1})	R_{xx}	R_{xy}
289.0	0.0	6.7	391.4	0.0	0.72
Monolayer (1, 1)			Bilayer (2, 2)		
ν_j (cm^{-1})	R_{xx}	R_{xy}	ν_j (cm^{-1})	R_{xx}	R_{xy}
201.4	32.7	0.0	118.4	24.5	0.0
241.3	0.0	2.41	120.8	0.0	0.11
372.1	0.0	1.18	201.4	0.0	0.0
			236.4	3.02	0.0
			277.3	0.0	2.86
			356.1	1.03	0.0
			381.8	0.0	0.68

[a] The results are from Barker *et al.* (1978).

phonons. They note that one should observe additional phonon modes that result not just from the decrease in zone size, but also from the differences in the polar phonons propagating normal and parallel to the superlattice layers. They examined the problem by using a simple theory for the dielectric function of a layered medium (Rytov, 1956a). If ϵ_1 and ϵ_2 are the dielectric constants for layers with thickness d_1 and d_2, respectively, then

$$\epsilon_\perp = (d_1\epsilon_1 + d_2\epsilon_2)/(d_1 + d_2), \tag{5}$$

$$\epsilon_\parallel = \epsilon_1\epsilon_2(d_1 + d_2)/(d_1\epsilon_2 + d_2\epsilon_1). \tag{6}$$

For superlattices with the symmetry D_{2d}, there are two modes: a singly degenerate B_2 mode and a double degenerate E mode. The LO and TO phonons of both B_2 and E types are given by the zeros and poles of the appropriate dielectric functions: B_2(LO) is the zero of $\epsilon_\perp$; B_2(TO) is the pole of $\epsilon_\parallel$; E(LO) is the zero of $\epsilon_\perp$, and E(TO) is the pole of $\epsilon_\parallel$. This long-range effect alone can produce two phonons of each type, whereas in the case of the bulk III–V semiconductor, we would find only one mode of each type. The physics associated with these additional modes is different from that producing the dispersion curve folding. The first is due to the dielectric anisotropy, while the second is due to the changes in the local periodicity.

A model similar to the Kronig–Penney model has been applied to the long-wavelength acoustic phonon modes (Colvard *et al.*, 1980). In this calculation one considers a layered elastic medium characterized by densities and sound velocities appropriate to the direction under consideration. Using the results from Rytov (1956b), they find that the dispersion curve Ω versus q for LA phonons in the (100) direction is given by

$$\cos(qd) = \cos\frac{\Omega d_1}{C_1}\cos\frac{\Omega d_2}{C_2} - \frac{1+k^2}{2k}\sin\frac{\Omega d_1}{C_1}\sin\frac{\Omega d_2}{C_2},$$

where $k = (C_1\rho_1)/(C_2\rho_2)$, with C_1 and C_2 and ρ_1 and ρ_2 being the sound velocities and densities appropriate for the two layers, respectively. The widths of the individual layers are d_1 and d_2, and d is the total thickness. This formula for the dispersion curve results in a number of branches. They correspond to the LA acoustic branch of the bulk material folded into the smaller Brillouin zone of the superlattice.

Ren and Dow (1982) have treated the effect of the dispersion curve folding on umklapp processes that influence thermal conductivity. They find that the additional umklapp scattering processes can produce rather substantial decreases in the thermal conductivity.

Although most of the fundamental physics has been mapped out in these simple model calculations, detailed calculations of comparable sophistication to the tight-binding and pseudopotential calculations for the electronic spectra have not been carried out. These could provide the kind of detailed insight provided by similar calculations of the electronic spectra.

VI. OTHER PROPERTIES

The elementary excitations discussed in the preceding sections result in a number of observable properties of semiconductor superlattices. In this section, we shall address some of the properties that have not been discussed earlier.

Many of the useful properties of semiconductors are due to the properties of impurities in the semiconductors. In superlattices the properties of the standard shallow dopants as well as deep levels are modified substantially from their properties in the corresponding bulk materials. Bastard (1982a) and Mailhiot and co-workers (1982a,b) have considered the case of a coulombic center in a quantum well. Bastard assumed that the carrier was confined in an infinite well. He finds substantial modifications in the binding energy in that the binding is found to increase as the well thickness is decreased. Mailhiot *et al.* have treated the case of a donor in a GaAs layer sandwiched between two layers of $Ga_{1-x}Al_xAs$. The properties of the $Ga_{1-x}Al_xAs$ layers are included explicitly by taking account the finite band offset and the differences in effective masses and dielectric constants. The results of their calculation for an impurity at the center

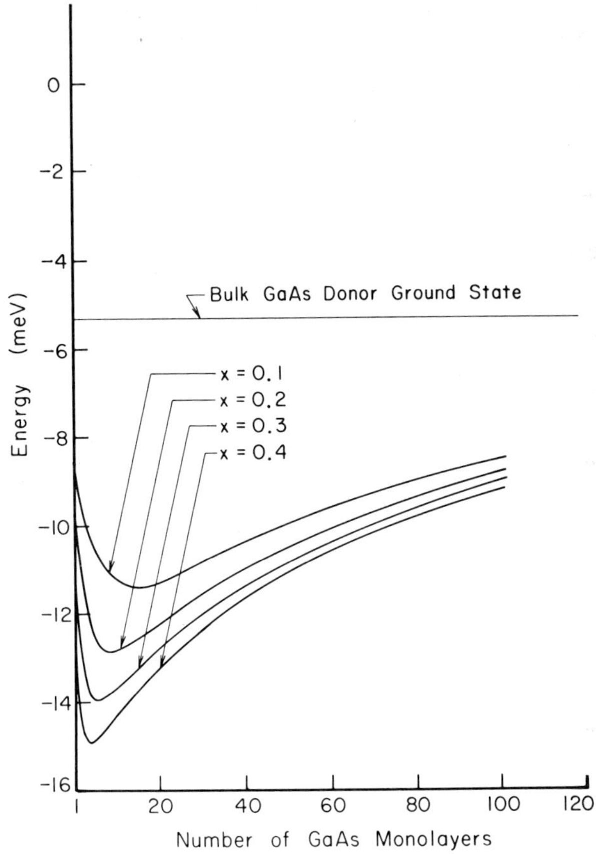

Fig. 13. The binding energy of a donor in the center of a GaAs well surrounded by two layers of $Ga_{1-x}AlAs$. The ordinate is the number of GaAs layers. The alloy concentration x is varied from 0.1 to 0.4. [From Mailhiot *et al.* (1982b).]

of the GaAs well are given in Fig. 13. They find that, while the binding energy at first increases owing to the confinement of the donor wave function in the layer, for thin layers the properties of the $Ga_{1-x}Al_xAs$ layers dominate, leading to binding energies that are the same as those for the $Ga_{1-x}Al_xAs$ alloy. Both Bastard and Mailhiot and co-workers studied the variation of the binding energy due to position in the layer. Both find that there is substantial variation in the binding energy due to moving the impurity about in the well.

The current theories make use of the effective mass approximation that is valid for the case of donors in GaAs–$Ga_{1-x}Al_xAs$ as long as x is such that the $Ga_{1-x}Al_xAs$ is direct (Ando, 1982). However, for acceptors or donors in other systems, it will be necessary to include explicitly the band structure of the quantum-well structure instead of simply applying the effective mass approximation.

Excitons in superlattices can be quite different from those in normal bulk semiconductors. For the case of GaAs – $Ga_{1-x}Al_xAs$ in which the electron near the conduction band edge and hole near the valence band edge are confined in the same layers, the behavior of the binding energy is very similar to that found in the case of impurities. In the limit of very thin layers and total confinement the binding energy should approach that of a two-dimensional exciton that binds four times as strongly as a three-dimensional exciton (Bassani and Parravicini, 1975). Calculations on the infinite-well model by Bastard *et al.* (1982) give the expected results. In superlattices, other cases are also possible. For example, in the case of InAs – GaSb (Bastard *et al.*, 1982), the electron and hole are in different layers and the binding energy of the exciton is reduced over the three-dimensional value.

One theoretical study suggests that nonlinear optical properties of superlattices could be quite large (Tsu, 1971). The additional dispersion due to the shortened Brillouin zone can produce a substantial nonlinear optical response. Although this phenomenon has been studied theoretically, it has not been explored in detail. In particular, the role of scattering that will tend to suppress this phenomenon has not been included.

Much of the transport theory for heterostructures has dealt with the transport along the interface between the GaAs and $Ga_{1-x}Al_xAs$ and hence does not involve true superlattice effects. The original suggestion by Esaki and Tsu (1970) of negative resistance for currents perpendicular to the superlattice layers due to the curvature in the minibands has not been explored extensively. The mean free path of the carriers are sufficiently small that they reduce this effect substantially.

Ktitorov and co-workers (1971) have suggested that one might observe Stark ladders (Wannier, 1960) in superlattices. However, the scattering of the carriers and the large intersubband matrix elements due to the applied electric field are likely to make it difficult to observe such a phenomenon.

Chin and co-workers (1980) have proposed that a superlattice structure would have widely different electron and hole impact ionization rates. This proposal has led to other suggestions for ways to control the impact ionization phenomenon (Blauvelt *et al.*, 1981; Tanoue and Sakai, 1982).

The collective excitations of the carriers in semiconductor superlattices have been explored by Das Sarma and Madhukar (1981) and Das Sarma and Quinn (1982). Das Sarma and Quinn have explored the optical plasmons, acoustic plasmons, magnetoplasmons, helicon, and Alfvén waves as the superlattice parameters are changed in such a way as to go from the two-dimensional to the three-dimensional limit.

VII. CONCLUSIONS

Techniques have been developed for treating the electronic spectra of ideal superlattices by using both first principles and empirical techniques. The empirical techniques (two-band model, tight-binding model) are very successful in treating the properties of superlattices since they allow the extrapolation of our knowledge of the band structure of the bulk semiconductor to the properties of superlattices. Calculations based on these techniques have provided results for the electronic band structure of abrupt superlattices. Perhaps more important, these techniques provide a method for treating some of the questions about transport, impurity states, excitonic states, deviations from an ideal abrupt interface (i.e., grading), alloys, and so on. Self-consistent calculations allow the examination of the charge densities at the interface and hence give some information about the interface region.

The lattice vibrations of superlattices have not been treated in as much detail as the electronic properties. The techniques for treating these problems at the same level as the electronic properties are available but have not been applied.

The properties that are of real interest are those that can lead to novel applications. A partial list of these novel applications would include linear and nonlinear optical properties and transport properties. As discussed in this chapter, some of the interesting transport properties include widely different impact ionization rates and negative resistance phenomena. Here a great deal remains to be done. The theorist can hope to make precise calculations based on experience with bulk semiconductors. Theoretical investigations of these properties should indicate which superlattice systems should be studied.

In this chapter, we have not addressed the very important question presented by realistic interfaces and their preparation. These include the character and stability of the interface in the superlattice and the mechanisms of growth. For these problems, theoretical work is difficult but essential in guiding the experimental studies in the field.

Since this chapter was written, the theoretical and experimental investigation of semiconductor superlattices has continued to undergo rapid development. Progress has been reported on several of the issues discussed here. R. Dingle's empirical rule that the GaAs–GaAlAs conduction band discontinuity is 85% of the band gap difference has been challenged. Thus, the one offset that had been thought to be relatively well established is no longer so. The valence band structure of GaAs–GaAlAs superlattices has come under intensive study, especially with applied magnetic fields. Agreement between theory and experiment does not exist concerning the nature and field dependence of the Landau levels. New types of superlattices have become prominent, especially those in which there is substantial lattice mismatch between the two constituents—strained layer superlattices. A few other areas in which much work has been accomplished include the CdTe–HgTe superlattice, detailed calculations of phonon properties, and the prediction of new many-body phenomena. In addition, proposals for the application of superlattices in optical and high-speed devices are becoming more common. It is clear that many theoretical issues remain to be investigated in this field.

ACKNOWLEDGMENTS

The authors wish to gratefully acknowledge the Army Research Office for supporting the preparation of this manuscript under contract DAAG29-80-C-0103. One of us (JNS) would like to acknowledge the support provided by the Office of Naval Research under contract N00014-82-K-0458.

REFERENCES

Altarelli, M. (1982). *Proc. Int. Conf. Phys. Semicond. 16th,* Montpellier, France (M. Averous, ed.), p. 747, North Holland Publ., Amsterdam.
Ando, T., and Mori, S. (1982). *Surf. Sci.* **113,** 124.
Andreoni, W., Baldereschi, A., and Car, R. (1978). *Solid State Commun.* **27,** 821.
Baraff, G. A., Appelbaum, J. A., and Hamann, D. R. (1977). *Phys. Rev. Lett.* **38,** 237.
Barker, A. S., Merz, J. L., and Gossard, A. C. (1978). *Phys. Rev. B* **17,** 3181.
Bassani, F., and Parravicini, G. P. (1975). "Electronic States and Optical Transitions in Solids." Pergamon, Oxford.
Bastard, G. (1981). *Phys. Rev. B* **24,** 5693.
Bastard, G. (1982a). *Surf. Sci.* **113,** 165.
Bastard, G. (1982b). *Phys. Rev. B* **25,** 7584.
Bastard, G., Mendez, E. E., Chang, L. L., and Esaki, L. (1982). *Phys. Rev. B* **26,** 1974.
Bauer, R. S. (1981). Private communication.
Blauvelt, H., Margalit, S., and Yariv, A. (1981). *Electron Lett.* **18,** 375.
Caruthers, E., and Lin-Chung, P. J. (1977). *Phys. Rev. Lett.* **38,** 1543.
Caruthers, E., and Lin-Chung, P. J. (1978). *J. Vac. Sci. Technol.* **15,** 1459.
Chadi, D. J., and Cohen, M. L. (1975). *Phys. Status Solidi B* **68,** 405.
Chang, L. L. (1980a). *Proc. Int. Conf. Phys. Semicond. 15th, Kyoto, Japan.*
Chang, L. L. (1980b). *J. Phys. Soc. Jpn. Suppl.* **49A,** 997.
Chang, L. L., Kawai, N. J., Sai-Halasz, G. A., Ludeke, R., and Esaki, L. (1979). *Appl. Phys. Lett.* **35,** 939.
Chang, Y. C. (1982). *Phys. Rev. B* **25,** 605.
Chang, Y. C., and Schulman, J. N. (1982). *Phys. Rev. B* **25,** 3975.
Chin, R., Holonyak, N., Jr., Stillman, G. E., Tang, J. Y., and Hess, K. (1980). *Electron. Lett.* **38,** 467.
Colvard, C., Merlin, R., Klein, M. V., and Gossard, A. C. (1980). *Phys. Rev. Lett.* **45,** 298.
Das Sarma, S., and Madhukar, A. (1981). *Phys. Rev. B* **23,** 865.
Das Sarma, S., and Quinn, J. J. (1982). *Phys. Rev. B* **25,** 7603.
Dingle, R., Gossard, A. C., and Wiegmann, W. (1975). *Phys. Rev. Lett.* **34,** 1327.
Esaki, L., and Tsu, R. (1970). *IBM J. Res. Dev.* **14,** 61.
Faurie, J. P. (1981). *J. Cryst. Growth* **54,** 577.
Frensley, W. R., and Kroemer, H. (1977). *Phys. Rev. B* **16,** 2642.

Guldner, Y., Vieren, J. P., Voisin, P., Voos, M., Chang, L. L., and Esaki, L. (1980). *Phys. Rev. Lett.* **45,** 1719.
Harrison, W. A. (1980). "Electronic Structure and the Properties of Solids." Freeman, San Francisco, California.
Ihm, J., Lam, P. K., and Cohen, M. L. (1979) *Phys. Rev. B* **20,** 4120.
Ivanov, I., and Pollmann, J. (1979). *Solid State Commun.* **32,** 869.
Kane, E. O. (1957). *J. Phys. Chem. Solids* **1,** 249.
Ktitorov, S. A., Simin, G. S., and Sindalovskii, V. Y. (1971). *Fiz. Tverd. Tela.* 13, 2230 [*Sov. Phys.-Solid State* **13,** 1872 (1972)].
Kraut, E. A., Grant, R. W., Waldrop, J. R., and Kowalczyk, S. P. (1980). *Phys. Rev. Lett.* **44,** 1620.
Kuech, T., and McCaldin, J. (1982). *J. Appl. Phys.* **53,** 3121.
Kunc, K., and Martin, R. M. (1980a). *Proc. Int. Conf. Phys. Semicond. 15th, Kyoto, Japan.*
Kunc, K., and Martin, R. M. (1980b). *J. Phys. Soc. Jpn. Suppl.* **49A,** 1117.
Madhukar, A., and Delgado, J. (1981) *Solid State Commun.* **37,** 199.
Madhukar, A., and Nucho, R. N. (1979) *Solid State Commun.* **32,** 331.
Madhukar, A., Dandekar, N. V., and Nucho, R. N. (1979). *J. Vac. Sci. Technol.* **16,** 1507.
Mailhiot, C., Chang, Y. C., and McGill, T. C. (1982a). *Surf. Sci.* **113,** 161.
Mailhiot, C., Chang, Y. C., and McGill, T. C. (1982b). *Phys. Rev. B* **26,** 4449.
Mendez, E. E., Chang, L. L., Landgren, G., Ludeke, R., Esaki, L., and Pollak, F. H. (1981). *Phys. Rev. Lett.* **46,** 1230.
Merlin, R., Colvard, C., Klein, M. V., Morkoç, H., Cho, A. Y., and Gossard, A. C. (1980). *Appl. Phys. Lett.* **36,** 43.
Milnes, A. G., and Feucht, D. L. (1972). "Heterojunctions and Metal-Semiconductor Junctions," pp. 34ff. Academic Press, New York.
Mönch, W., and Gant, H. (1980). *J. Vac. Sci. Technol.* **17,** 1094.
Mukherji, D., and Nag, B. R. (1975). *Phys. Rev. B* **12,** 4338.
Mukherji, D., and Nag, B. R. (1978). *Solid State Electron.* **21,** 555.
Nucho, R. N., and Madhukar, A. (1978). *J. Vac. Sci Technol.* **15,** 1530.
Osbourn, G. C. (1982). *J. Appl. Phys.* **53,** 1586.
Osbourn, G. C., and Smith, D. L. (1979). *Phys. Rev. B* **19,** 2124.
Perfetti, P., Denley, D., Mills, K. A., and Shirley, D. A. (1978). *Appl. Phys. Lett.* **33,** 667.
Pickett, W. E., Louie, S. G., and Cohen, M. L. (1978). *Phys. Rev. B* **17,** 815.
Pollmann, J. (1980). *Festkörperprobleme* [Advances in Solid State Physics] (J. Treusch, ed.), **20,** p. 117. Vieweg, Braunschweig.
Ren, S. Y., and Dow, J. D. (1982). *Phys. Rev. B* **25,** 3750.
Rytov, S. M. (1956a). *Sov. Phys. JETP Engl. Transl.* **2,** 466.
Rytov, S. M. (1956b). *Akust. Zh.* **2,** 71 [*Sov. Phys. Acoust. (Engl. Transl.)* **2,** 68 (1956)].
Sai-Halasz, G. A., Tsu, R., and Esaki, L. (1977). *Appl. Phys. Lett.* **30,** 651.
Sai-Halasz, G. A., Chang, L. L., Welter, J.-M., Chang, C.-A., and Esaki, L. (1978a). *Solid State Commun.* **27,** 935.
Sai-Halasz, G. A., Esaki, L., and Harrison, W. A. (1978b). *Phys. Rev. B* **18,** 2812.
Schulman, J. N., and Chang, Y. C. (1981a). *Phys. Rev. B* **24,** 4445.
Schulman, J. N., and Chang, Y. C. (1981b). Unpublished. [Parameters used were similar to those used in Schulman and Chang (1981a) with the addition of the spin-orbit interaction.]
Schulman, J. N., and Chang, Y. C. (1983). *Phys. Rev. B* **27,** 2346.
Schulman, J. N., and McGill, T. C. (1977). *Phys. Rev. Lett.* **39,** 1680.
Schulman, J. N., and McGill, T. C. (1979a). *Appl. Phys. Lett.* **34,** 663.
Schulman, J. N., and McGill, T. C. (1979b). *J. Vac. Sci. Technol.* **16,** 1513.
Schulman, J. N., and McGill, T. C. (1979c). *Phys. Rev. B* **19,** 6341.
Schulman, J. N., and McGill, T. C. (1981). *Phys. Rev. B* **23,** 4149.

Shay, J. L., Wagner, S., and Phillips, J. C. (1976) *Appl. Phys. Lett.* **28,** 31.
Tanoue, T., and Sakaki, H. (1982). *Appl. Phys. Lett.* **41,** 67.
Tsu, R., and Esaki, L. (1971). *Appl. Phys. Lett.* **19,** 246.
Tsu, R., and Jha, S. S. (1972). *Appl. Phys. Lett.* **20,** 16.
Wannier, G. H. (1960). "Elements of Solid State Theory," pp. 190–193. Cambridge Univ. Press, London and New York.
White, S. R., and Sham, L. J. (1981). *Phys. Rev. Lett.* **47,** 879.

4 COMPOSITIONALLY MODULATED SUPERLATTICES

Leroy L. Chang and E. E. Mendez

IBM Thomas J. Watson Research Center
Yorktown Heights, New York

I. INTRODUCTION

The existence of a long-period superlattice was first discovered in the Au–Cu alloy system in which antiphase boundaries were found to be spaced regularly (Johnson and Linde, 1936). This discovery promoted both structural investigations by x-ray and electron microscopy for the direct observation of the period (Ogawa and Watanabe, 1954) and theoretical considerations of electronic states that the period introduced (Jones, 1969). The presence of the additional periodicity was, in fact, not limited to this system alone, as subsequently shown in a variety of materials, including other metal alloys, polytype semiconductors, and layered, intercalated compounds.

The semiconductor superlattice (Esaki and Tsu, 1970) used in the present context refers to a compositionally modulated, periodic structure, consisting of alternate, thin layers of two semiconductors. With the period in thickness shorter than the electron mean free path, the potential provided by the layers modifies significantly the band structure. Quantum states are created, and the electron system associated with the states becomes restricted in its three-dimensional motion, leading to unusual properties not present in the host materials. The layers are to be introduced during epitaxial deposition with tailored prescriptions in terms of their thickness and composition. Thus, unlike superlattices that exist in nature, the present structures are synthesized or man-made.

Early efforts and developments for the experimental realization of the super-

ISBN 0-12-170470-X

lattice are described in Chapter 1 of this volume. After attempts were made with different techniques for different materials, the formation of the superlattice was achieved and demonstrated in GaAs–GaAlAs by molecular beam epitaxy (Chang *et al.*, 1973a). Work during the initial period was reviewed, covering both the deposition process (Chang and Ludeke, 1975; Cho and Arthur, 1975) and the superlattice properties (Dingle, 1975; Esaki and Chang, 1976). Since then rapid progress has been made, in scope as well as in depth, with investigations in transport, optical, and magneto properties, covering a variety of semiconductor compounds and alloys. Several review articles have appeared in the literature (Chang and Esaki, 1979 and 1980; Gossard, 1982; L. L. Chang, 1983). We list in Table I the different systems that, to our knowledge, have been studied, including the polytype superlattice of three components (Esaki *et al.*, 1981), the doping superlattice, the subject matter of Chapter 5, in which the periodic potential is provided by impurities (Esaki and Tsu, 1970; Döhler, 1972), and the noncrystalline superlattice made of hydrogenated amorphous materials (Abeles and Tiedje, 1983).

In this chapter, we describe the experimental observations and our understanding of them in superlattices and related heterostructures such as single and

TABLE I

Semiconductor Superlattices

Systems	Techniques	Remarks
GaAs–GaAlAs(AlAs)	MBE, MOCVD, LPE	Extensively studied
InAs(InGaAs)–GaSb(GaSbAs)	MBE	Extensively studied
GaSb–AlSb	MBE	Luminescence and absorption
InAs–AlSb	MBE	Transport and magnetotransport
GaAs(GaP)–GaAsP	MOCVD, CVD	Strained superlattice, luminescence
GaAs–InAs(InGaAs)	MBE, MOCVD	Metallurgical and light-emitting properties
InP–InGaAs(InAlAs)	MBE,MOCVD	Magneto properties
InP–InGaAsP	LPE	Light-emitting properties
InSb–GaSb	Sputtering	Ordered structure, interdiffusion
Ge–GaAs(GaAlAs)	MBE	Metallurgical properties, defects
Si–SiGe	MBE, CVD	Dislocations, mobility enhancement
CdTe–HgTe	MBE	Zero-finite gap, magnetoabsorption
PbTe–PbSnTe	Hot-wall	Optical and magneto properties
PbTe–PbGeTe	MBE	Dislocations, Auger profiling
InAs–GaSb–AlSb	MBE	Polytype superlattice and heterostructures
GaAs(*n*)–GaAs(*p*)	MBE	Doping superlattice, tunable gap
Si(H)–SiN(H)	CVD	Amorphous superlattice, optical properties

multiple quantum wells. Our interest lies primarily in the electronic properties in the quantum regime, as opposed to the mere formation of metallurgically alternating layers. In terms of materials, we focus on the GaAs–GaAlAs and InAs–GaSb systems that have been investigated most extensively, but we will touch on others in Table I to the extent that they present new features. Theoretical considerations, treated thoroughly in Chapter 3, will be limited to those essential to the presentation of the experimental results. Fabrication and characterization, the latter being the subject of Chapter 2, will be covered only as they relate directly to the superlattice structure. Needless to say, we confine ourselves to semiconductors; metallic superlattices, which have experienced similar progress in recent years, as described fully in Part III of this volume.

II. THEORETICAL CONSIDERATION

From a simplistic point of view, the superlattice can be represented as a series of potential barriers and wells along the z direction perpendicular to the surface. For energies near the fundamental energy gap, the periodic potential is given by the variations of band edges at the Γ point, as illustrated in Fig. 1a for GaAs–GaAlAs. The barrier height or well depth for electrons is provided by the conduction band discontinuity, $\Delta E_c = E_c(\text{GaAlAs}) - E_c(\text{GaAs})$ and that for holes, by the valence band edge discontinuity, $-\Delta E_v = E_v(\text{GaAlAs}) - E_v(\text{GaAs})$. Both heavy and light holes are present and their degeneracy is lifted by the potential. The heavy holes usually play a dominant role and will be referred to as holes unless specified otherwise. The period of the superlattice is the sum of the two layer thickness, $d = d_1(\text{GaAs}) + d_2(\text{GaAlAs})$.

The total energy of the electrons in this framework can be written as $E = (\hbar^2 k_\parallel^2/2m_e^*) + E_z(k_z)$. The first term represents the kinetic energy associated with motion parallel to the surface, assuming a parabolic and isotropic band. The second term is the confinement energy, which can be solved by the Kronig–Penney model using plane waves and matching boundary conditions at the interfaces. By requiring k_z to be real, a series of allowed quantum states or subbands E_{nz} are obtained, which are separated by minigaps at the zone boundaries, $k_{nz} = n\pi/d$. The dispersion relation and the density of states are shown schematically in Fig. 2 for the ground ($n = 1$) and the first excited ($n = 2$) states. The energy difference between the ground states of electrons and holes, which can be similarly obtained, is defined as the energy gap of the superlattice, $E_{gs} = E_{1e} - E_{1h}$. The subband energy as a function of layer thickness is shown in Fig. 3a, which uses $d_1 = d_2 = d_0$ and material parameters as listed in Table II. These calculated results from the simple model are quite adequate in explaining experimental observations in the GaAs–GaAlAs system, including early transport (Chang *et al.*, 1973a, 1974a; Esaki and Chang, 1974) and optical (Dingle *et al.*, 1974) measurements from which energy quantization was established.

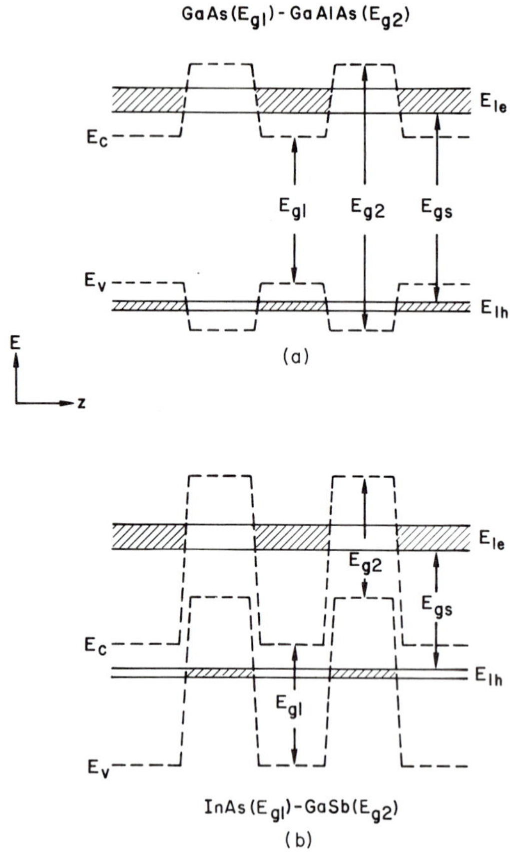

Fig. 1. Schematic energy diagrams of (a) GaAs–GaAlAs and (b) InAs–GaSb superlattices. E_{gs} is the energy gap of the superlattice, defined between the ground subbands of electrons and holes.

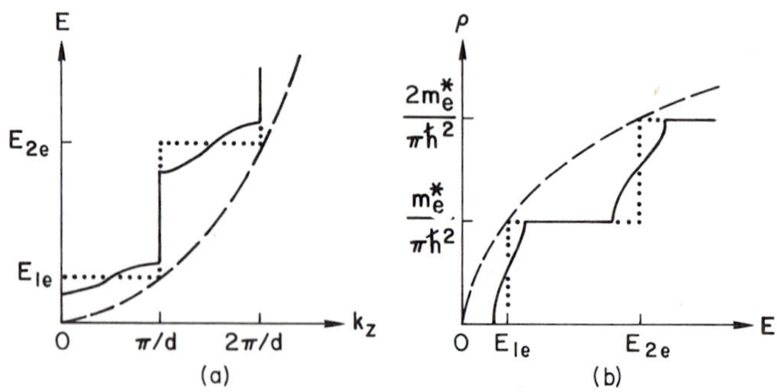

Fig. 2. Schematic diagrams of (a) energy dispersion and (b) density of states. Indicated are the two-dimensional (dotted), three-dimensional (dashed), and intermediate (solid) cases.

One of the most important features of the superlattice, that of controlling the dimensionality of the electron system by varying the superlattice configuration (Chang, 1978), can be seen from Figs. 2 and 3. Thin wells lead to high subband energies and thin barriers enhance the interaction between adjacent wells, which results in a finite bandwidth. The electron system in this limit approaches the three-dimensional case with $E = \hbar^2k^2/2m_e^*$ and $\rho = (2m_e^{*3}E)^{1/2}/\pi^2\hbar^3$, the familiar free-electron expressions. The subbands decrease in both energy and width in the opposite situation and eventually become two-dimensional (Ando

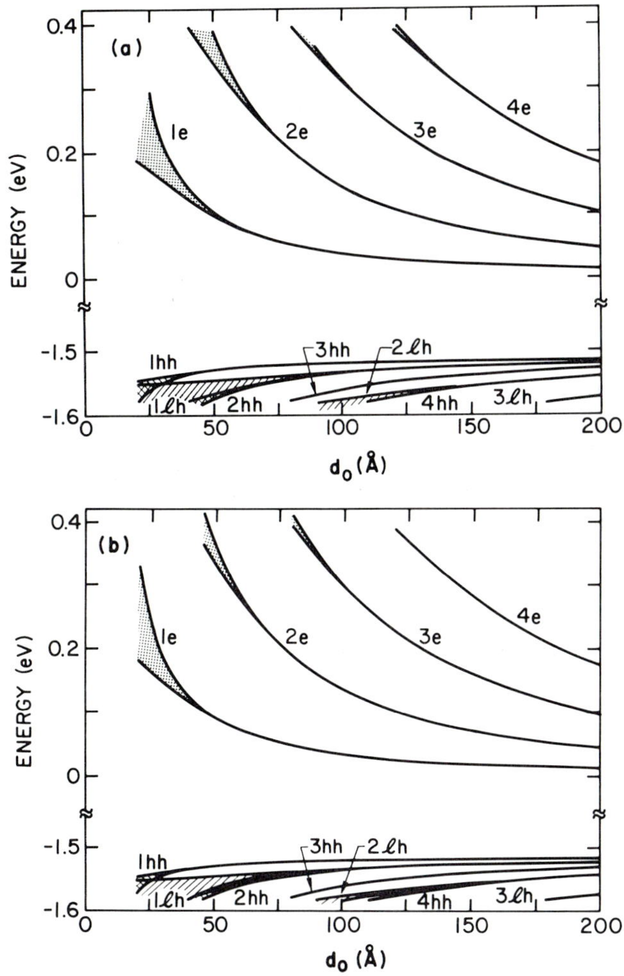

Fig. 3. Calculated subband energy positions as a function of (equal) layer thickness for GaAs–GaAlAs with (a) Kronig–Penney model and (b) envelope wave-function method and for InAs-GaSb with (c) Kronig–Penney model and (d) envelope wave-function method. *(Continued on next page)*

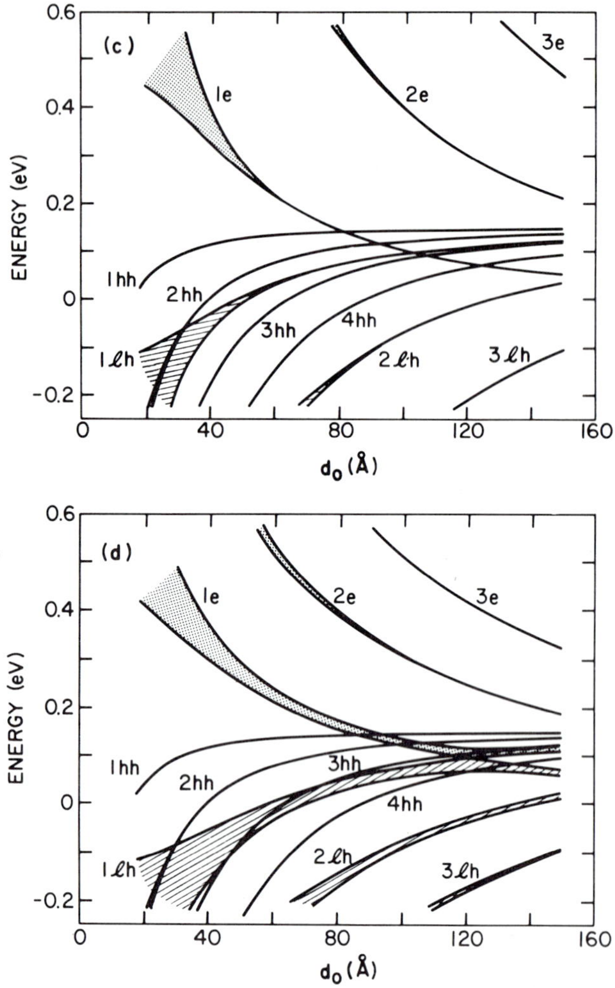

Fig. 3. *(Continued)*

et al., 1982), which is characterized by discrete bound states $E_n = n^2\pi^2\hbar^2/2m_e^*d^2$ and energy-independent density of states $\rho = m_e^*/\pi\hbar^2$. Both the extremes and the intermediate dimensionality between two and three are illustrated in Fig. 2. For most superlattices investigated experimentally, they fall in, or deviate not far from, the two-dimensional regime.

In determining the positions and characters of the subband structure, it is clear that the band edge discontinuities are important parameters, which are usually not known. The simplest guideline is the so-called electron affinity rule (Milnes and Feucht, 1972) in which the discontinuity is measured as the energy

TABLE II

Parameters Used in the Calculation of the Subband Structure Shown in Fig. 3[a]

	E_g	ΔE_c	ΔE_v	m_e^*	m_{lh}^*	m_{hh}^*
GaAs	1.52			0.067	0.082	0.45
		0.425	0.075			
$Ga_{0.6}Al_{0.4}As$	2.02			0.10	—	0.57
InAs	0.42			0.023	—	0.40
		0.96	0.57			
GaSb	0.81			0.042	0.042	0.43

[a] The energies are expressed in electron volts: E_g is the Γ-point energy gap at zero temperature; ΔE_c and ΔE_v are the conduction and valence band discontinuities, respectively. The effective masses are expressed in units of the free electron mass: m_e^* for electrons, m_{lh}^* for light holes and m_{hh}^* for heavy holes.

difference between the conduction band edge of the bulk and the vacuum level. A number of theoretical calculations have been attempted (Harrison, 1977; Frensley and Kroemer, 1977; Pickett *et al.*, 1977); the results from the method of linear combination of atomic orbits (Harrison, 1977 and 1980) appear to provide the best overall agreement with experimental values. For the GaAs–GaAlAs system discussed so far, both ΔE_c and ΔE_v have been found experimentally to be positive (Dingle, 1975; Dingle *et al.*, 1974), so that electron and hole states are quantized in the same GaAs layers. A different situation ensues when either of the two discontinuities is negative and, consequently, electrons and holes are confined in spatially alternate layers. An extreme case exists in InAs–GaSb. Here not only is $-\Delta E_v$, defined by E_v(GaSb) $-$ E_v(InAs), negative but its magnitude is even larger than the band gap of InAs (Sai-Halasz *et al.*, 1977, 1978a). In other words, E_v(GaSb) lies in energy above E_c(InAs), as shown in Fig. 1b; the difference having been measured to be 0.15 eV (Sai-Halasz *et al.*, 1978c). In addition to spatial separation of carriers, this superlattice exhibits a semiconductor-to-semimetal transition with an increase in the InAs layer thickness (Chang *et al.*, 1979). As can be seen in Fig. 3c, the relative position of E_{1e} and E_{1h} changes and the transition occurs when they cross each other, i.e., $E_{gs} = 0$. This type of superlattice is sometimes referred to as type II, as opposed to type I, which is represented by the GaAs–GaAlAs system.

The subband structure has thus far been considered from the Kronig–Penney model. In principle, the introduction of layers of two different materials creates a new lattice whose band structure can only be calculated in a fundamental way by treating the dissimilar atoms as a super cell. Both tight-binding (Schulman and McGill, 1977; Nucho and Madhukar, 1978) and pseudopotential (Caruthers and Lin-Chung, 1977; Ihm *et al.*, 1979) methods have been used for this purpose. In addition to the difficulty of self-consistency in handling the charge redistribution at the interface, the main obstacle in these calculations lies in the computational complexity associated with the size of the unit cell, which increases considerably with the layer thickness. A variety of techniques have been devised to overcome this problem, including modified tight-binding calculations (Ivanov and Pollman, 1979) and effective-mass approximations in the Kane framework (White and Sham, 1981; Bastard, 1981b). By use of envelope functions with proper matching conditions at the interface (Bastard, 1981b), improvement was made over the Kronig–Penny model by taking into account the effects of band nonparabolicity and mixing.

Results of the subband energies obtained from the envelope function method are shown in Fig. 3b and d to compare with those from the Kronig–Penny model. For GaAs–GaAlAs, for which the conduction and valence bands are far apart, the two methods give nearly identical energies within the range under consideration. For InAs–GaSb, however, the comparison shows large discrepancies, because the E_c of InAs is highly nonparabolic and closely located to E_v of GaSb, as mentioned above. In addition, the envelope function calculations show the effect of anticrossing between electrons and light holes. The heavy hole subbands in the two cases are, of course, similar, for they both were obtained separately without coupling to other bands. The decoupling is valid in principle only for $k_{\parallel} = 0$, under which condition the calculations in Fig. 3d were made (Bastard, 1981b); the semiconductor–semimetal transition here occurs at $d_0 = 85$ Å. The superlattice would exhibit the behavior of a narrow-gap semiconductor rather than that of a semimetal under the opposite condition (Altarelli, 1983).

Aside from the band structure, other effects in superlattices, notably the binding energies of excitons and impurities, have attracted recent theoretical interest. Impurity states in a quantum well are different from those in the bulk in two aspects: The binding energies are enhanced and become dependent on the spatial location of the impurities. By using a variational calculation with an infinite potential well (Bastard, 1981a), the binding energy was shown to approach the bulk value for thick layers but increase to four times as large in the thin-layer limit. A maximum exists if a finite well is used, which is important in the region of thin layers (Mailhiot *et al.*, 1982; Greene and Bajaj, 1983b). As the impurity is moved from the center to the edge of the well, the energy decreases to form an impurity band, with peak densities usually present at the band extrema. The binding energies of excitons as a function of layer thickness have been treated by a similar formalism (Bastard *et al.*, 1982a). It is interesting to note that such energies are rather small in InAs–GaSb, as a result of the spatial separation of electrons and holes.

Theoretical considerations have covered other topics as well. The varying dimensionality of the electron system in the superlattice makes it ideal to study the collective modes of excitations with different degrees of coupling between the layers. Various modes were predicted, including plasmons, magnetoplasmons, helicon, and Alfven waves (Das Sarma and Madhukar, 1981; Bloss, 1982; Das Sarma and Quinn, 1982). In a different area dealing with lattice properties, the periodicity in the superlattice is expected to generate new and modified vibrational modes. Phonons at energies different from those in the bulk can arise as a result of zone-folding and anisotropic propagation (Barker *et al.,* 1978; Merlin *et al.,* 1980; Covard *et al.,* 1980). Selective transmission (Narayanamurti *et al.,* 1979) and electromagnetic generation (Quinn *et al.,* 1983) of phonons in a superlattice medium have also been pursued.

From the structural point of view, several specific schemes of superlattices have been considered, including modulation-doped, one-dimensional, and random heterostructures. In modulation doping, donors are introduced only in the region of potential barriers, while electrons, dissociated from their parent impurities and accumulated in neighboring wells, are capable of achieving high mobilities. This effect, the subject of Chaper 6, has been observed experimentally (Dingle *et al.,* 1978; Witkowski *et al.,* 1980) and analyzed theoretically in terms of various scattering and screening processes (Mori and Ando, 1979; Price, 1982). The one-dimensional structure, as proposed (Sakaki, 1980), is an ultrafine wire where high electron mobility can be reached as a result of suppressing elastic scattering. It has been attempted experimentally (Petroff *et al.,* 1982). In fact, by creating a surface potential well at the edge of the superlattice, an electron system with one-to-two dimensionality can quite generally be produced (Chang *et al.,* unpublished). The random superlattice with a controllable degree of randomness presents a number of interesting situations in density of states and energy gaps (Dow *et al.,* 1982). Heterostructures with a random potential may indeed be an ideal medium for the study of localization.

III. FABRICATION AND CHARACTERIZATION

With the advance in semiconductor thin-film technology in recent years, a number of processes have been used for the deposition of superlattices. Early attempts, which succeeded in the growth of relatively thick layers, included chemical vapor disposition (CVD) (Blakeslee and Aliotta, 1970), molecular beam epitaxy (MBE) (Cho, 1971), and liquid-phase epitaxy (LPE) (Woodall, 1972). The MBE technique was proved to be the most suitable and, as can be seen in Table I, was applied to a large variety of semiconductor compounds and alloys.

Molecular beam epitaxy refers to the process of epitaxial deposition from molecular or atomic beams in an ultrahigh-vacuum environment. The beams are ideally generated from Knudsen-type cells in which thermal quasi-equilibrium is maintained. They travel in rectilinear paths in the vacuum space with guidance from orifices and shutters and condense on the substrate under kinetically favorable conditions. The technique of MBE is basically one of vacuum evaporation that has long been used for the preparation of semiconductor films. Evolved through a series of early developments (Gunther, 1958; Schoolar and Zemel, 1964; Davey and Pankey, 1969), the foundation of MBE, as it is known today, was laid by the kinetic studies of Ga and As_2 on the surface of GaAs (Arthur, 1968). It was found that the As beam does not stick to a heated GaAs surface unless the surface is covered with Ga atoms. The growth of stoichiometric GaAs is kinetically controlled by the adsorption of As, while the growth rate is determined by the Ga beam, which has a sticking coefficient of unity. Further investigations of thermal desorption and surface reconstructions (Cho, 1970; Arthur, 1974) and of the kinetic behaviors of both As_2 and As_4 (Foxon and Joyce, 1975; Joyce and Foxon, 1975) resulted in a better understanding of the deposition process, leading to rapid progress in MBE. Films of high quality were produced, and structures and devices with great precision and performance were fabricated. In addition to the early work (Chang and Ludeke, 1975; Cho and Arthur, 1975), we refer to several review articles that have since appeared in the literature (Chang, 1980; Ploog, 1980; Wood, 1980; Foxon and Joyce, 1981).

A typical MBE system is shown schematically in Fig. 4, which is based on the original system used for the growth of the GaAs–GaAlAs superlattice (Chang *et al.,* 1973a,b). In addition to the sources of the constituents and dopant impurities, it contains a heated substrate holder, a set of individual shutters, and a number of analytic apparatuses for the purpose of *in situ* characterization and control. The growth of GaAs is usually done on (100) GaAs substrate at a temperature of 600°C and a growth rate of 2 Å/sec. The superlattice of GaAs–GaAlAs is produced by periodically opening and closing the Al shutter. In the original operation (Chang *et al.,* 1973a), the control unit as shown in Fig. 4 was a process control computer, which received its input of beam intensities sensed by the mass spectrometer, corrected the source temperatures whenever neces-

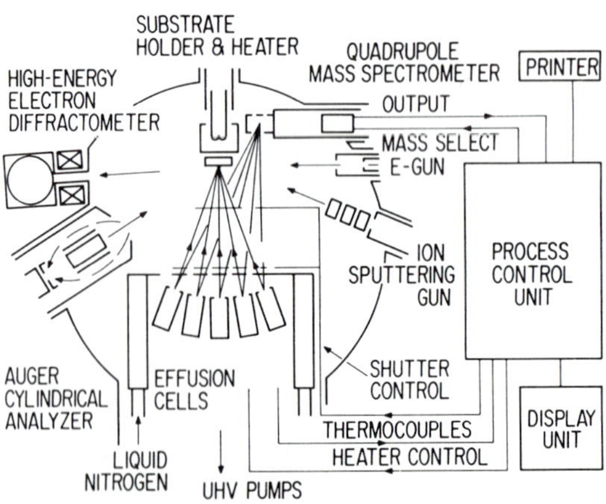

Fig. 4. A schematic diagram of the molecular beam epitaxy system with accessory apparatuses for the deposition of semiconductor superlattice.

sary to keep a constant growth rate, composition (Ga and Al), and a fixed surface environment (As to Ga and Al), and executed the shutters automatically at a predetermined layer thickness. With improvement in temperature stability of the cells, a simple timing circuit is usually sufficient for most control purposes. The design of the MBE systems has also improved in other aspects: the addition of an interlock loading chamber to facilitate sample introduction and withdrawal without contamination to the growth chamber and the incorporation of a rotational substrate holder to achieve uniform deposition, in particular, on large-size substrates.

The popularity of MBE for growing superlattices is, of course, not accidental, for it offers characteristic features uniquely satisfying the stringent requirements for such structures, as summarized in Table III. The slow growth rate is favorable for thickness control. The relatively low deposition temperature minimizes interdiffusion effects (Chang and Koma, 1976), which makes it possible to produce extremely abrupt interfaces. But the most important advantage lies in the capability of achieving atomically smooth surfaces. This arises from the two-dimensional growth nature of the MBE process (Cho and Arthur, 1975), which can be evaluated from the high-energy electron diffractometer (HEED) during deposition. A rough substrate surface exhibits a transmissionlike spotty pattern that undergoes a transition to become narrowly streaked as the surface is smoothed out (Dove *et al.*, 1973). For this reason, it is common practice to deposit a buffer layer on the substrate before starting the growth of the superlattice.

An example of the HEED pattern for a smooth GaAs surface is shown in the sequence in Fig. 5 to illustrate an important point, the effect of lattice matching on the growth of a superlattice (C.-A. Chang *et al.*, 1977). The patterns follow the heteroepitaxial deposition of GaSbAs on GaAs where the lattice mismatch exceeds 6% for the given alloy composition. It is clear that smooth surface can be achieved (Fig. 5d), but only after an intervening step of nucleation (Fig. 5c). Had the two materials been closely matched, such as GaAs–GaAlAs and InAs–GaSb, the transition from (a) to (d) would have proceeded directly and instantaneously. The critical mismatch from this criterion is in the vicinity of 2.5%. Recently, similar observations were reported for the GaSb–AlSb and InAs–AlSb systems (C.-A. Chang *et al.*, 1982). These results bear directly on

TABLE III

Superlattice versus Molecular Beam Epitaxy

Superlattice requirements	MBE characteristics
High-quality films	Ultrahigh-vacuum environment
Thin layers	Slow growth rate
Smooth interfaces	Planar growth process
Abrupt interfaces	Low-temperature deposition
Stringent structural demand	Precise control in composition, doping, and thickness

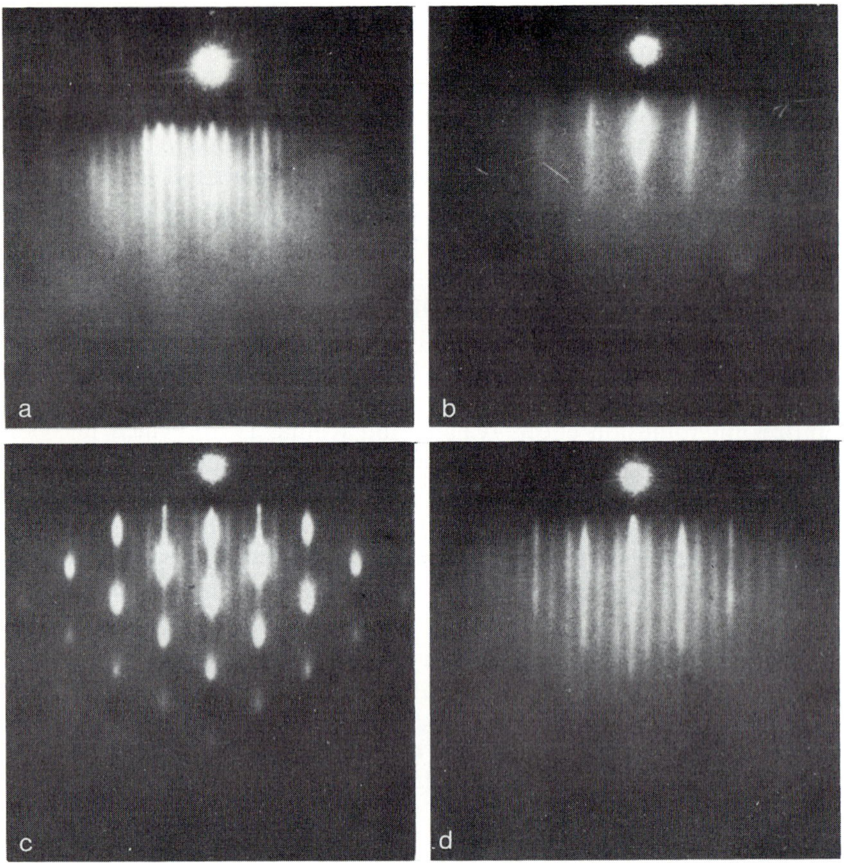

Fig. 5. Reflection high-energy electron diffraction patterns at 20 KeV and [110] azimuth in the growth of $GaSb_{0.8}As_{0.2}$ on GaAs (100): (a) the starting smoothed-out GaAs surface; (b) and (c) after the arrival of about one and three monolayers of $GaSb_{0.8}As_{0.2}$, respectively; and (d) after a total growth of 150 Å.

other mismatched systems of current interest such as GaP–GaAsP and GaAs–InGaAs (Osbourn, 1982; Gourley and Biefeld, 1982), where the effect of lattice strain is emphasized.

The use of HEED to extract surface information represents but one example of a variety of *in situ* characterization techniques. That MBE proceeds in an ultrahigh-vacuum environment makes it possible to install various modern instruments for studies in surface physics and chemistry. Moreover, such studies can be carried out on freshly prepared surfaces and interfaces for different orientations and under controllable dynamic conditions. For the superlattice, the analytic apparatuses shown in Fig. 4 can be used conveniently to provide structural characterization. For example, the profile of the structure in terms of composition versus distance can be obtained by combining the ion-sputtering facilities with the mass spectrometer (SIMS) or the Auger analyzer (Auger profiling). The latter in fact was the first technique used to resolve the periodicity with a layer thickness of 50 Å in the GaAs–GaAlAs superlattice (Ludeke *et al.*, 1974). It has been widely applied to other systems to demonstrate their alternating layers, including InSb–GaSb (Eltoukhy *et al.*, 1977), PbTe–PbSnTe or –PbGeTe (Kinoshita and Fujiyasu, 1980; Partin, 1982), and CdTe–HgTe (Faurie *et al.*, 1982).

Most structural characterizations, however, are performed outside, after the sample is withdrawn from the MBE system. Commonly used techniques include electron microscopy and x-ray diffraction. Both transmission (TEM) and scanning (SEM) modes have been used; the latter is easy to operate and has been routinely employed in nearly all the superlattice structures to reveal the periodicity for visual observation. The TEM requires the tedious procedure of sample thinning for a cross-sectional view, but it offers better resolution (Gossard, 1982) and is capable of providing structural information such as defects, as demonstrated in the case of Ge–GaAs (Petroff *et al.*, 1979; Kuan and Chang, 1982). Quantitative characterization can best be obtained from x-ray diffraction, both the large-angle Bragg reflection and the small-angle interference (Chang *et al.*, 1974b). The Bragg spectra exhibit the main reflection peak corresponding to the lattice of the superlattice as well as the satellite peaks as a result of lattice modulation. Many superlattice systems were evaluated by this technique: GaAs–GaAsP (Esaki *et al.*, 1970; Segmüller and Blakeslee, 1973), GaP–GaAsP (Gourley and Biefeld, 1983), InSb–GaSb (Eltoukhy *et al.*, 1977), GaSb–AlSb (Naganuma *et al.*, 1981; Mendez *et al.*, 1983), and Ge–GaAs (C.-A. Chang *et al.*, 1981). Most detailed analysis was done for GaAs–AlAs on GaAs substrates where reflections were taken along different reciprocal lattice rows (Segmüller *et al.*, 1977). It was concluded that AlAs was compressed

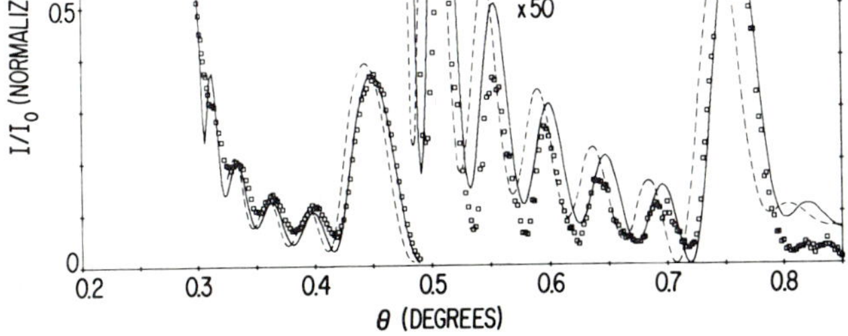

Fig. 6. X-ray interference pattern of a GaAs–AlAs superlattice with six periods. In comparison with the experimental data points, calculated results with different period thickness are plotted to demonstrate the agreement (solid curve with $d = 127.2$ Å) and the discrepancy (dotted curve with $d = 130.0$ Å).

parallel to the layer plane but extended in the perpendicular direction. The layers were fully strained without relief by the generation of dislocations.

An example of the small-angle interference spectra is shown in Fig. 6 for a GaAs–AlAs superlattice (Chang *et al.,* 1976). The pattern arises from the difference in refractive indices between the two materials, analogous to optical interference involving a series of regularly spaced slits. With a limited number of periods, both principal and secondary peaks become observable. The results, in terms of not only angular positions but also intensities and linewidths, are in excellent agreement with theoretical calculations with the assumption of abrupt interfaces. Of more significance is the fact that a change of one monolayer spacing is seen to result in an observable difference in the calculations. These observations indicate that the layers possess a degree of interfacial smoothness and periodic coherency on the scale of atomic dimensions. Similar measurements have been done across the sample surface to evaluate the structural uniformity (Chang *et al.,* 1976) and on structures with thickness in the range of a few monolayers (Esaki and Chang, 1976).

Another method of characterization that we would like to mention here is that of Rutherford backscattering and channeling, a technique that has not yet been commonly used for superlattices but is drawing interest. In fact, it was employed at the early stage to evaluate GaAs–GaAlAs (Mayer *et al.,* 1973). By projecting energetic ions against the target and energy analyzing the backscattered yield, the method constitutes a mass-sensitive, depth microscopy. Figure 7 shows the spectra of an InAs–GaSb structure (Chu *et al.,* 1982), where pronounced oscillations clearly demonstrate the periods. Furthermore, by orienting the ion beam along a particular crystal axis instead of a random direction, the backscattered intensity, or dechanneling yield, is decreased because of a reduction in scattering. This is shown for the ⟨100⟩; the yield is comparable to that of bulk GaSb, indicating the high quality of the superlattice. Although the

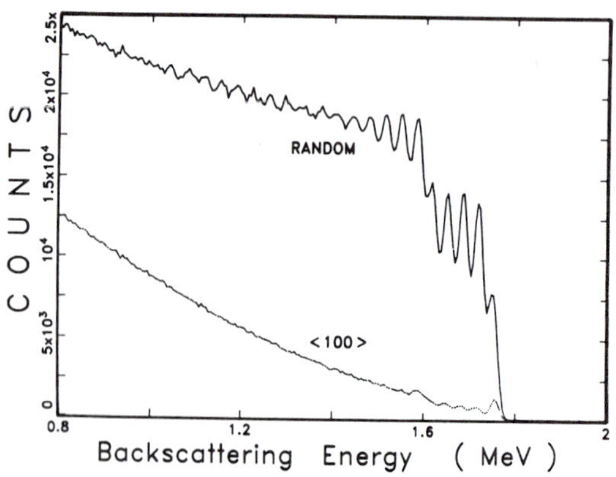

Fig. 7. Backscattering spectra for 2-MeV $^4He^+$ ions incident at random and ⟨100⟩ channeling directions of an InAs (200-Å)–GaSb (240-Å) superlattice. [After Chu *et al.* (1982).]

backscattering technique is not suitable for very thin layers in that oscillations cannot be well resolved, a comparison of the dechanneling yields for different orientations in InAs–GaSb has given information as to the bonding nature at the interfaces. The yields along ⟨110⟩ and ⟨111⟩ were found to be about a factor of two higher than along ⟨100⟩, leading to a lattice relaxation model of expanding and contracting bond-lengths at the heteroepitaxial boundaries (Chu *et al.*, 1982; C.-A. Chang, 1983).

As mentioned earlier, MBE is the most widely used but by no means the exclusive process for the fabrication of superlattices or fine heterostructures. Referring to Table I, other techniques include CVD, MOCVD, LPE, hot-wall, and sputtering. The CVD process in general refers to that of epitaxy resulting from chemical reactions of gaseous species, and layered structures are produced by alternatively introducing and removing different reactants in the growth chamber. Although this technique was used for GaAs–GaAsP (Blakeslee and Aliotta, 1970) and Si–SiGe (Manesevit *et al.*, 1982), the development of MOCVD by use of organometallic compounds greatly enhanced the capability of structure fabrication in terms of flexibility and control. Being quite competitive with MBE in structural quality, and, expectedly, more adaptable for production, it has found wide applications in GaAs–GaAlAs (Holonyak *et al.*, 1980; Maluenda and Frijlink, 1983), GaP–GaAsP (Osbourn, 1982; Gourley and Brefeld, 1982), GaAs–InGaAs (Ludowise *et al.*, 1983), and InP–InGaAs (Nicholas *et al.*, 1980; Razeghi and Duchemin, 1983). The growth of multilayers by LPE involves the contact of the substrate with various, saturated molten solutions. Structures of GaAs–GaAlAs (Woodall, 1972) and InP–InGaAsP (Rezek *et al.*, 1977) were successfully fabricated, although control at the interfaces seemed to pose an inherent problem. In both the hot-wall and sputtering techniques, a movable substrate configuration is usually used for making superlattices, whereby the substrate is periodically exposed through openings or windows to the depositing beams. The beams are generated by isothermal evaporations in the hot-wall method as in PbTe–PbSnTe (Kinoshita and Fujiyasu, 1980) and by multi-target sputtering as in InSb–GaSb (Eltoukhy *et al.*, 1977). The latter, although not popular in semiconductors, is one of the most commonly used in preparing metallic superlattices.

IV. ELECTRONIC PROPERTIES

A. Resonant Transport

Resonant transport refers to current conduction perpendicular to the heterostructure layers when quantum states are aligned in energy to enhance the flow of electrons. The conduction in a superlattice via the subband with a sufficient bandwidth, resulting from significant overlap of wave functions associated with states in the quantum wells, represents one of the extreme cases. The current characteristic is governed by the dispersion relation of E versus k_z shown in Fig. 2. The curvature change in this relation decelerates electrons under the application of an electric field, leading to a negative differential resistance, the original prediction on which early experimental efforts were focused (Esaki and Tsu, 1970; Chang *et al.*, 1973a). By using extremely thin GaAlAs barriers in GaAs–GaAlAs to broaden E_{1e}, the negative resistance was observed at threshold electric fields expected theoretically (Chang *et al.*, 1973a,b). Figure 8 shows an example of the current–voltage characteristic under pulsed measurements. The magnitude of the negative resistance was small, presumably suffering from leakage associated with defects. The effect was observable only in structures whose active area was minimized.

The other extreme of resonant transport deals with discrete quantum states in the two-dimensional electron system. Experimentally, the situation was realized in the simplest case by sandwiching a single well of GaAs between two GaAlAs barriers with outside GaAs electrodes. Figure 9 shows the current–voltage behavior with the structural configuration illustrated in the inset (Chang *et al.*, 1974a). When the energies of the incoming electrons coincide with the energies of the stationary states, the reflected waves from inside the well interfere destructively with those from outside so that only transmitted waves remain (Tsu and Esaki, 1973). The electron waves are peaked or "localized" at these energies inside the well. The resonance is manifested as current maximum at applied voltages equal to approximately twice the energies because of the

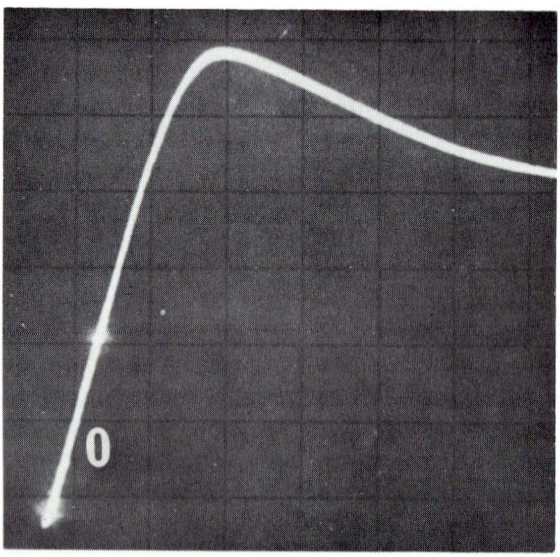

Fig. 8. Current–voltage characteristic of a GaAs (60-Å)–$Ga_{0.5}Al_{0.5}As$ (10-Å) superlattice. The measurement was made at room temperature with a 100-nsec pulse at a 10-kHz repetition rate.

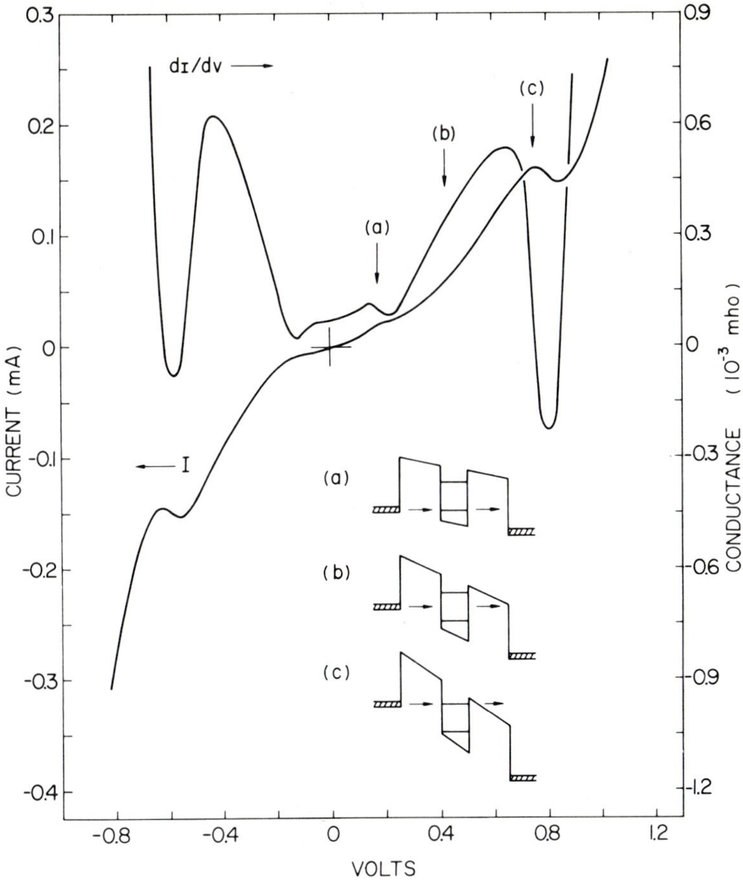

Fig. 9. Current and conductance characteristics in a double-barrier structure of $Ga_{0.3}Al_{0.7}As$ (80 Å)–GaAs (50 Å)–$Ga_{0.3}Al_{0.7}As$ (80 Å) at 77 K. Conditions at resonance (a and c) and off-resonance (b) are indicated on the curves and illustrated in the inset diagrams.

equality of the two barriers. The observed characteristics in Fig. 9, where conductance is also plotted to amplify the features, clearly demonstrate the resonance at (a) and (c). The resonant energies or voltages have been shown to be what are expected theoretically from structures with different well thickness (Chang *et al.*, 1974b). Similar observations were subsequently made in InGaAs–GaSbAs (Chang and Esaki, 1980) and again in GaAs–GaAlAs, which showed a sharpened resonance characteristic and a fast response time (Sollner *et al.*, 1983).

Superlattices with a narrow subband width provide yet another situation in which resonant transport can be investigated (Esaki and Chang, 1974). The current flow in this case is governed by a mixture of resonant tunneling through

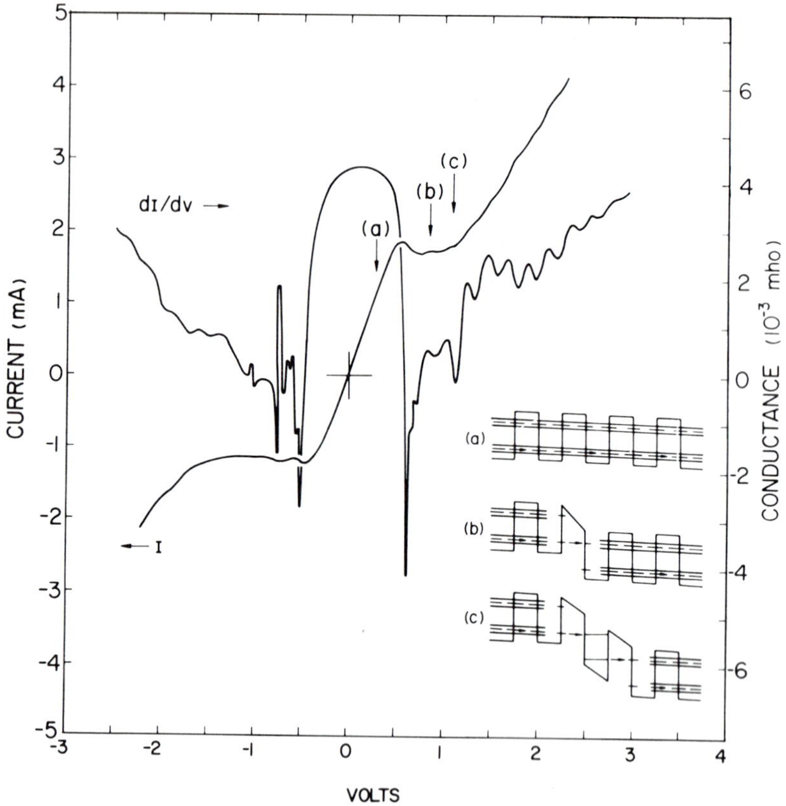

Fig. 10. Current and conductance characteristics of a GaAs (45-Å) – AlAs (40-Å) superlattice at 77 K. Conditions of band conduction (a) and localized tunneling (b and c) are indicated on the curves and illustrated in the inset diagrams.

stationary states and electron conduction via subbands, the two mechanisms just described. The band diagram and the current characteristics are shown in Fig. 10. With an increase in voltages so that the energy difference between adjacent wells approaches the bandwidth, the band conduction that is originally operative can no longer be maintained. A localized region is spontaneously created by fluctuations of nonuniformities where adjacent wells become essentially isolated. Current transport, limited by this region, is by a tunneling process, reaching a maximum when states of E_{1e} and E_{2e} in neighboring wells are aligned. The situation expands successively in space to result in a periodic variation of the current with respect to the applied voltage. The oscillations were clearly observed by the conductance plot as shown in Fig. 10. The crossover voltage, at which band conduction gives way to local tunneling, agrees with that calculated theoretically. That the period in voltage should be equal to the energy difference between the states has also been verified.

B. Optical Absorption

While transport measurements usually deal with electrons because of their high mobilities, optical experiments provide a means to study transitions involving states in both the conduction and valence bands. Among the different optical techniques, absorption gives the most direct information, as first demonstrated in isolated GaAs wells between thick GaAlAs layers (Dingle *et al.*, 1974). Figure 11 shows the absorption spectra in which peaks corresponding to excitonic transitions from E_{nh} to E_{ne} are observed. The spectrum resembles that of bulk GaAs for a thick well, and, in the case of a thin well, it shows in addition the contribution of light holes for the ground-state transition. Subsequent measurements were extended to various configurations (Dingle *et al.*, 1975) and, from a comparison of the transition energies with those calculated theoretically, the conduction band discontinuity ΔE_c was found to be 85% of the energy gap difference (Dingle, 1975). Essentially similar spectra were obtained from photoconductivity measurements (Tsu *et al.*, 1975). The capability of varying two independent parameters, the photon energy and the applied voltage, offers the advantage of investigating simultaneously the absorption and transport properties. Recently, well-resolved absorption spectra under hydrostatic pressures were reported (Kirchoefer *et al.*, 1982). The same pressure coefficient was obtained over a large range of energy, indicating the dominant role that the Γ-point plays in governing the absorption process.

The spatial separation of electron and hole states in the InAs–GaSb system

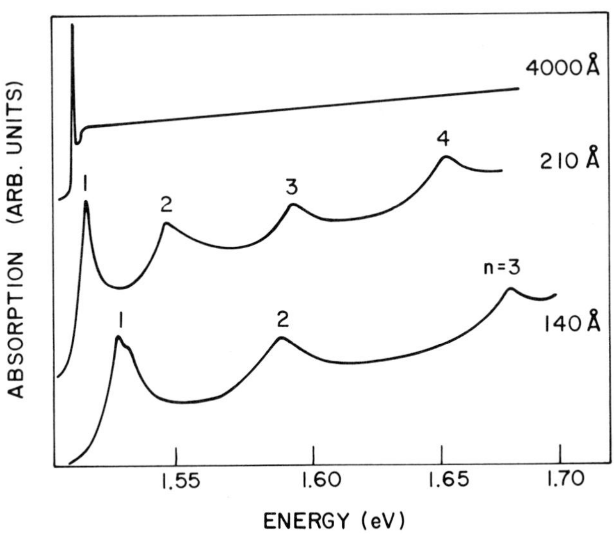

Fig. 11. Absorption spectra of GaAs quantum wells confined between thick $Ga_{0.8}Al_{0.2}As$ layers at 2 K. Indicated are the well thickness and the energy indices of the subbands involved in the transitions. [After Dingle *et al.* (1974).]

has a number of direct consequences on its absorption characteristics. The intensity of absorption is relatively weak, contributed only by the overlapping part of the wavefunctions of electrons and holes near the interfaces. The wave functions of these two carriers associated with subbands of the same energy indices ($\Delta n = 0$) are not necessarily in phase, so that transitions involving different subbands ($\Delta n \neq 0$) are equally probable. The spatial separation of carriers also causes a reduction in the exciton energy, as mentioned earlier. All these features are in direct contrast with those in GaAs–GaAlAs where both electrons and holes are confined spatially in the same GaAs regions. There is another difference from energy considerations, which has an impact on the experimental procedures. The transition energies of interest are smaller than the energy gap of the substrates typically used such as GaAs or GaSb. The tedious process of thinning or removing the substrate prior to taking the spectra is no longer required.

The absorption spectra were measured in the general alloy system InGaAs–GaSbAs, including the binary InAs–GaSb at different layer thicknesses (Sai-Halasz *et al.*, 1978c). The absorption edges were deduced, from which the E_v of GaSb was determined to be above the E_c of InAs by 0.15 eV, establishing the type II nature of this superlattice system. Subsequent experiments on samples with improved quality (L. L. Chang *et al.*, 1981b) revealed fine features, as shown in Fig. 12. A large number of transitions were seen and identified by comparing them with theoretical calculations of the subband energies as well as

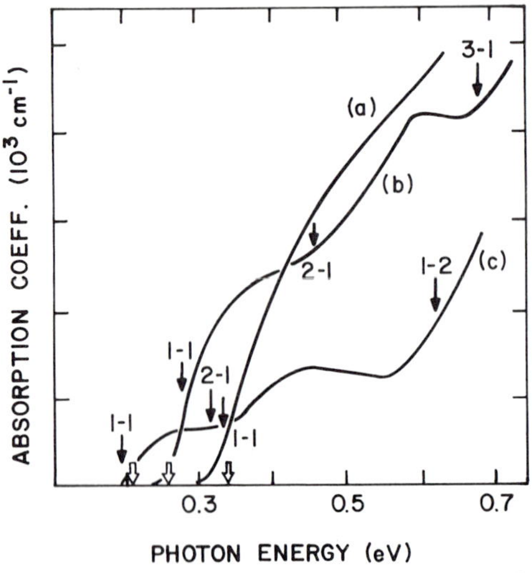

Fig. 12. Absorption spectra of InAs–GaSb superlattices (a) 21–21 Å, (b) 27–27 Å, and (c) 37–37 Å at 4.2 K. Indices of calculated subband transitions are marked by arrows. The block arrows indicate absorption edges obtained from magnetoabsorption.

the strengths of the overlapping integrals. It has been concluded that the ground-state transitions from heavy holes to electrons define the absorption edge, that light-hole contributions are usually unimportant, and that transitions of subbands involving different energy indices, as predicted, can be quite pronounced. Also, the excitonic effect is absent as expected (Bastard *et al.*, 1982a).

Absorption measurements have been carried out in GaSb–AlSb, the system that constitutes part of the proposed polytype superlattice (Voisin *et al.*, 1983). From the intensities, it was concluded that the system belongs to the type I category; that is, both electron and hole states are confined in GaSb. The deduced energy gap of GaSb was found to be invariably smaller than the known bulk value, and the discrepancy was explained on the basis of lattice strain. Results from subsequent experiments, shown in Fig. 13, have not only confirmed this strain effect quantitatively but showed its remarkable consequence in reversing the ground states of heavy and light holes (Voisin *et al.*, 1984). The lattice mismatch causes the GaSb lattice to stretch and the AlSb lattice to contract in the plane of the layers. For GaSb where the effect is of primary concern, this biaxial dilation results in a hydrostatic tensile stress which shrinks the gap, and a uniaxial compressive stress in the z direction that enlarges it. The latter, more importantly, splits the valence band, raising (in hole energy) the

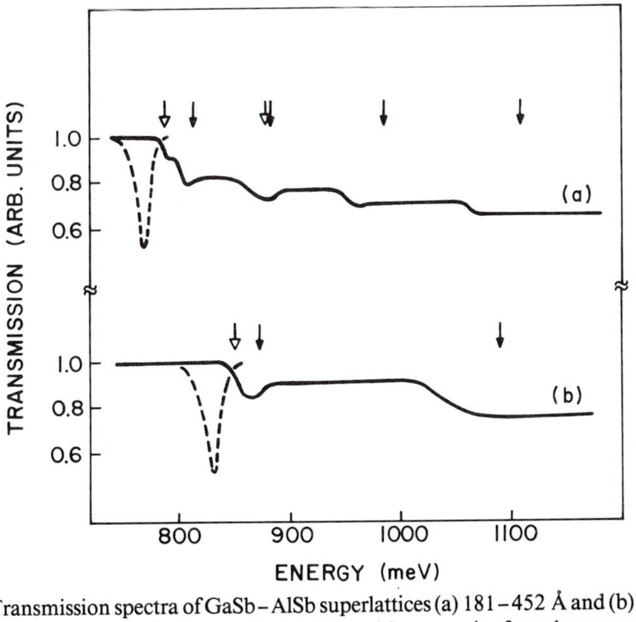

Fig. 13. Transmission spectra of GaSb–AlSb superlattices (a) 181–452 Å and (b) 84–419 Å at 4.2 K. Solid and open arrows indicate calculated transition energies from heavy and light holes, respectively, to electrons. Dotted curves represent luminescence.

subbands of the heavy holes with respect to those of the light holes and, consequently, leads to their reversal under proper layer thickness. The valence band discontinuity between AlSb and GaSb for the heavy holes, under this condition, is about 40 meV, a rather small value in consistency with the common-anion argument (Harrison, 1977, 1980). The experiments also exhibited excitonic absorption peaks (Voisin *et al.*, 1983, 1984), a further demonstration of the type I nature of the GaSb–AlSb system (Bastard *et al.*, 1982a).

C. Magnetotransport

Measurements under a magnetic field have played a vital role in the investigations of electronic properties in heterostructures, for such measurements are uniquely suitable for probing the low-dimensional electron system. As demonstrated in the Si-inversion layers (Ando *et al.*, 1982), the application of a perpendicular field B quantizes the motion of electrons in the layer plane, which together with the subband structure from electric quantization along z, makes the system completely quantized. Indeed, direct demonstration of the low-dimensionality of the superlattice came from transverse magnetoresistance experiments (Chang *et al.*, 1977). The results are shown in Fig. 14 for a GaAs–GaAlAs superlattice. The resistance is seen to oscillate with the field (Shubnikov–de Haas oscillations), which shifts progressively to higher fields as the field is tilted away from the surface normal ($\theta = 0°$).

In general, the constant energy surface in k-space perpendicular to the field is quantized according to $A_n = (2\pi eB/\hbar)(n + \frac{1}{2})$, where A_n is the cross-sectional area corresponding to the nth Landau level. Maxima in resistance occur whenever $A_n = A_f$, the Fermi cross section, so that the oscillation period in B^{-1} is given by $2\pi e/\hbar A_f$. It is thus generally possible to map the Fermi surface by measuring the periods as a function of the angle θ. Two important consequences of this relation are usually utilized in evaluating two-dimensional systems. One is that, since A_f at $\theta = 0°$ can be related directly to the electron density (n_s) or to the Fermi energy (E_f) if the effective mass is known, the period of oscillations can be used to determine these parameters. The other is that, as the Fermi surface at θ is related to that at zero by cos θ, such cosine dependence of the periods is commonly used to demonstrate two-dimensionality. It should be mentioned that the dependence of the oscillations merely on the perpendicular field component is only an approximation (Ando *et al.*, 1982). The parallel component may not be negligible and, in fact, at large angles and high fields, electric and magnetic quantizations are mixed and cannot be treated independently, as will be seen later.

Figure 15 shows the results from GaAs–GaAlAs superlattices with different configurations and electron densities. In one extreme, represented by either pure GaAs or a superlattice with a rather weak potential, the electron system is three-dimensional or essentially so with a spherical Fermi surface; the period of

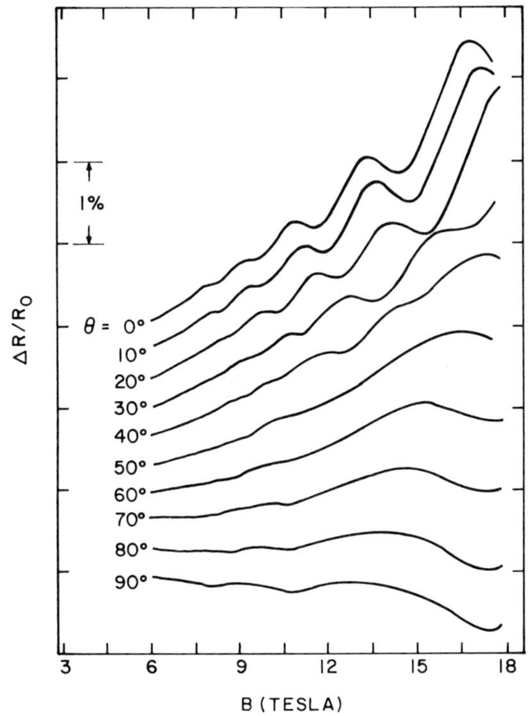

Fig. 14. Transverse magnetoresistance versus magnetic field of a GaAs (90-Å)–$Ga_{0.89}Al_{0.11}As$ (90-Å) superlattice at 4.2 K with an electron concentraion of 1.2×10^{18} cm^{-3}. Oscillations are shown to move to higher fields as the field is tilted away from the surface normal.

oscillations is independent of the field orientation. The other extreme is achieved with a strong potential where the system is two-dimensional, and its cylindrical surface gives rise to the cos θ variation as expected. The angular dependence falls between the two extremes for the intermediate case. The oscillations in Fig. 14, associated with electrons in the ground state E_{1e}, also follow the cosine relationship. The second subband is also occupied in this case, and the weak oscillation observed at parallel fields were interpreted as due to these electrons through magnetic breakdown at the zone boundaries. Mixed quantization may have already played a role under this condition (Ando, 1981). Oscillations under parallel fields were also reported when cyclotron orbits, shrinking with increasing fields, became confined in the GaAs layers (Yoshino *et al.*, 1983).

Shubnikov–de Haas oscillations have been used extensively in the study of the semiconductor–semimetal transition in the InAs–GaSb superlattice (Sakaki *et al.*, 1978; L. L. Chang *et al.*, 1981a). As mentioned earlier, this superlat-

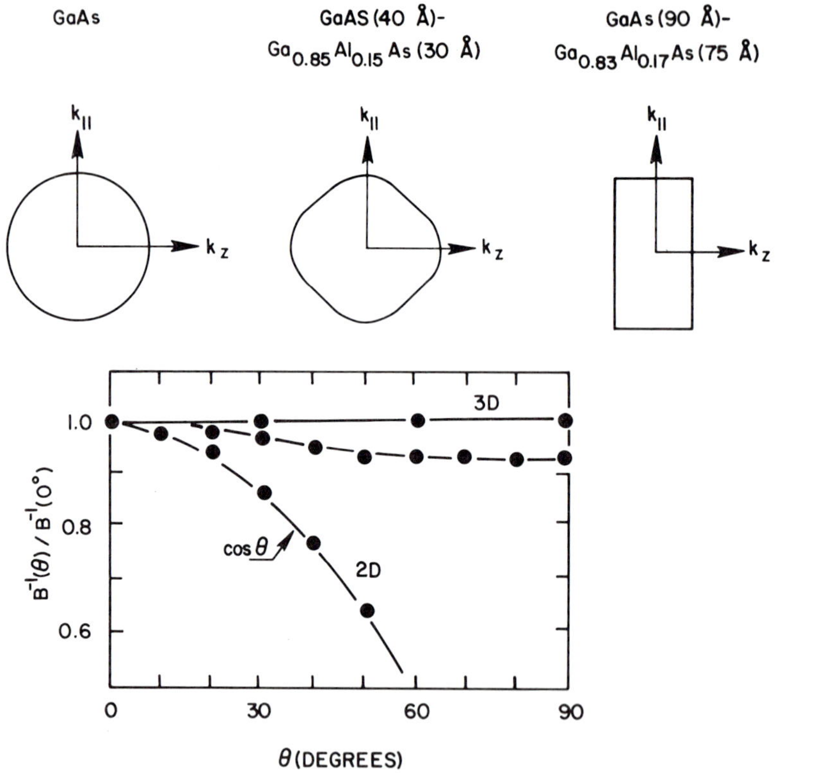

Fig. 15. Angular dependence of the period of Shubnikov–de Haas oscillations for three configurations from the three-dimensional GaAs through an intermediate case of superlattice to the two-dimensional superlattice of GaAs (90 Å)–$Ga_{0.83}Al_{0.17}As$ (75 Å). Corresponding Fermi surfaces from calculations are also shown.

tice system becomes semimetallic when the InAs layer exceeds 85 Å, beyond which carriers increase dramatically from charge transfer, as first demonstrated by Hall measurements (Chang *et al.,* 1979). With a series of configurations of different layer thickness, the subband structures were calculated, and E_f, defined with respect to E_{1e}, were compared with those obtained experimentally as shown in Fig. 16. The value of E_f increases first with the thickness as electron transfer starts. The space charge thus created leads to band bendings that tend to push toward each other the subbands of electrons and holes. This effect, together with the occupation of multiple subbands as their energy spacings are narrowed with increasing thickness, reduces E_f. Eventually, the band bendings localize electrons near the interface regions; the interaction between adjacent interfaces is no longer important and the superlattice behaves as a series of isolated heterojunctions. The Fermi energy and, in fact, the entire subband structure become saturated, independent of any further increase with thick-

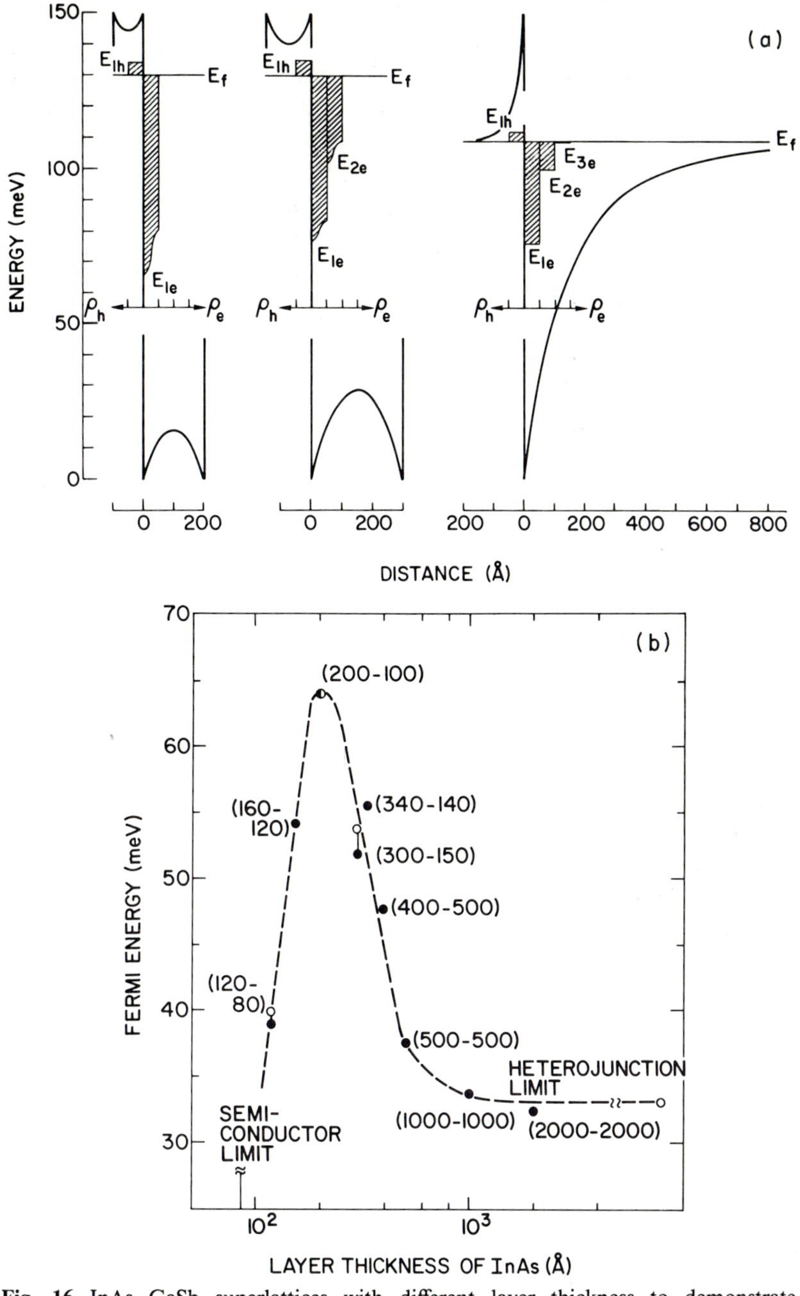

Fig. 16. InAs–GaSb superlattices with different layer thickness to demonstrate the semiconductor–semimetal transition: (a) the calculated subband energies and density of states and (b) the Fermi energy versus InAs layer thickness, the GaSb layer thickness is indicated by the second value in parenthesis.

ness. The band structures shown in Fig. 16 were originally calculated by the tight-binding method, taking into account the effect of band bending semiclassically by solving the Poisson equation within the Thomas–Fermi approximation (L. L. Chang *et al.*, 1981a). Recent self-consistent calculations for single wells in the range when only the ground state was occupied gave essentially the same results (Bastard *et al.*, 1982b). The data shown in Fig. 16 cover the entire semimetallic regime, from the semiconductor limit on the one end to the heterojunction limit on the other.

The semiconductor-to-semimetal transition occurs as a function of layer thickness in different samples, originating from the relative subband positions of electrons and holes $E_{1e} - E_{1h}$. The situation can be reversed in the same sample under magnetic fields. In the semimetallic state, the fields cause the sweeping of Landau levels of both electrons and holes across E_f, resulting in a reverse transfer of electrons from InAs back to GaSb. In the limit of crossing of the ground Landau levels, the semimetallic carriers are depleted and the superlattice behaves like a semiconductor. This, in principle, is expected to be quite a general property in semimetals (Brant and Svistova, 1970). The criterion for its observation requires that $\hbar\omega_c/2 = E_f$ can be reached with available fields. For superlattices with controllable configurations, this condition can be readily satisfied. A drastic increase in the magnetoresistance was observed experimentally at the critical field of carrier depletion, which was ascribed to the field-induced semimetal-to-semiconductor transition (Kawai *et al.*, 1980).

Since the Shubnikov–de Haas effect reflects the extremal cross sections in the Fermi surface, additional oscillations may arise from multiple subbands as well as from different points in the Brillouin zone within the single band. Such oscillations were observed for upper subbands (Sakaki *et al.*, 1978; Dingle *et al.*, 1980) that showed strong beating interference if the subband spacing was small (Dingle *et al.*, 1980). Of more interest is the existence of a maximum at $k_z = 0$ and a minimum at $k_z = \pi/d$ in the Fermi surface of a superlattice with substantial subband dispersion (L. L. Chang *et al.*, 1982). The observed magnetoresistance shows two distinct sets of oscillations and their periods themselves vary with the tilting angle in an oscillatory fashion (L. L. Chang *et al.*, 1982). The results are shown in Fig. 17 where the two sets are identified with the extrema at the zone center and edge. The oscillations with θ arise as the cyclotron orbits traverse into the extended zone of the superlattice, switching their roles as being a maximum or minimum in the Fermi surface periodically.

In a magnetic field, the energy not only exhibits Landau levels but shows spin splittings. The g factor in most semiconductors is small and decreasing further with electron energy in the conduction band (Pidgeon *et al.*, 1967). In a two-dimensional electron system, the spin splittings, unlike the Landau splittings, depend only on the magnitude of the field regardless of its orientation. This was used commonly to determine the g factor by properly tilting the field to space the Landau and spin levels (Fang and Stiles, 1968). The value of the g factor is

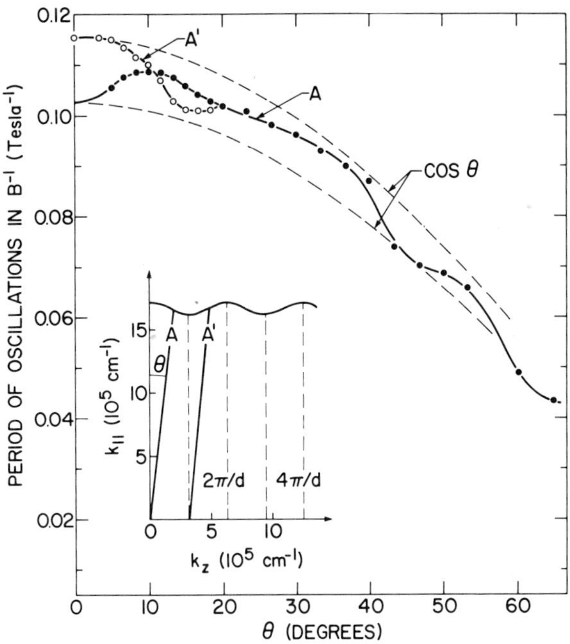

Fig. 17. Angular dependence of the period of Shubnikov–de Haas oscillations of an InAs (500-Å)–GaSb (500-Å) superlattice. The two sets of oscillations correspond to the two extrema of the Fermi surface at the reduced zone center (A) and edge (A′) as shown in the inset diagram.

also significantly enhanced by the self-energy exchange interaction (Janak, 1969; Ando and Uemura, 1974). Since such interaction is maximum when E_f lies between spin levels, the g value actually oscillates as a function of B, approaching its fully enhanced value with decreasing level broadening. Such an oscillation was observed (Englert *et al.*, 1982) with an enhancement of the order of 10 in GaAs–GaAlAs (Englert *et al.*, 1982; Narita *et al.*, 1982). Enhancement was also reported in other heterostructures such as InAs–GaSb (L. L. Chang *et al.*, 1982) and InP–InGaAs (Nicholas *et al.*, 1982).

Of great interest is the quantum Hall effect, first observed in Si (von Klitzing *et al.*, 1980) and subsequently in GaAs–GaAlAs (Tsui and Gossard, 1981). In such two-dimensional systems, the Hall resistance exhibits pronounced plateaus and the magnetoresistance vanishes, when E_f is located between magnetic levels, including both Landau and spin levels. The Hall resistance is given by B/en_s, and n_s is simply the product of level occupation (i) and level degeneracy (eB/h). The resistance value is precisely h/ie^2, which provides an accurate measure of the fundamental constant. Further experiments at low temperatures led to the striking observation that i could assume fractional numbers (Tsui *et al.*, 1982; Störmer *et al.*, 1983a), notably $i = \frac{1}{3}$, as shown in Fig. 18.

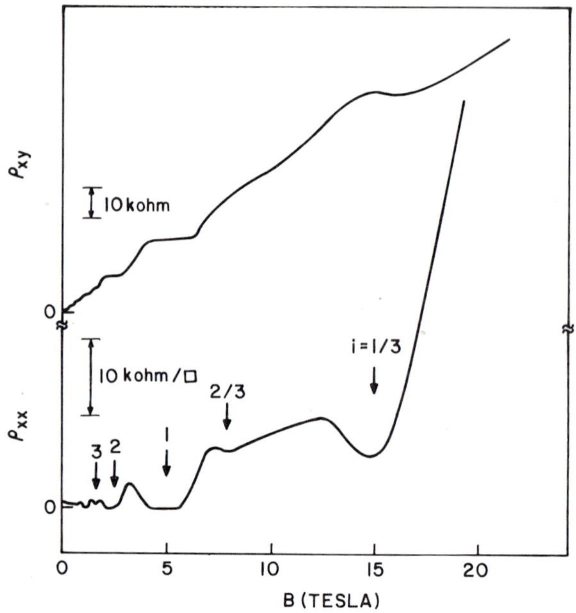

Fig. 18. Magnetoresistance ρ_{xx} and Hall resistance ρ_{xy} in a GaAs–$Ga_{0.7}Al_{0.3}As$ heterojunction at 0.48 K with an electron concentration n_s of 1.23×10^{11} cm^{-2} and a mobility of 9×10^4 $cm^2/Vsec$. The value i indicates the Landau level occupation, defined by $n_s h/eB$. [After Tsui *et al.* (1982).]

These results have attracted a large number of theoretical investigations (Laughlin, 1981, 1983; Fukuyama and Platzman, 1982; Yoshioka *et al.,* 1983; Thouless, 1984). While the integral quantum Hall effect can be understood on the basis of localized states between magnetic levels caused by impurities or defects, the fractional quantum Hall effect requires the formation of new, coherent many-particle states with gaps in the energy spectrum. Such states can be obtained by use of new electron wavefunctions characterizing an incompressible quantum liquid, which give rise to excitations involving fractionally charged electrons (Laughlin, 1983). The theory predicts a series of excitations, starting at $\frac{1}{3}$ and progressing to other fractions with increasingly large, odd denominators, and ending eventually to a crystalline state. Recent measurements with a dilute electron density have provided evidence to this trend (Mendez *et al.,* 1983c). The magnetoresistance showed a rather weak structure corresponding to $i = \frac{1}{5}$, but otherwise featureless characteristic down to $\frac{1}{11}$. In a separate development, the quantum Hall effect was also reported for holes in the GaAs–GaAlAs system (Störmer *et al.,* 1983a,c).

The quantum Hall effect has been observed in other heterostructure systems such as InP–InGaAs (Nicholas *et al.,* 1982; Guldner *et al.,* 1982a), InAs–GaSb

(Mendez *et al.*, 1984), and InAs–AlSb (Chang *et al.*, 1984). The generally high electron densities in these systems prevented reaching the extreme quantum limit ($i < 1$) with available fields. The resistances of a single InAs quantum well between two GaSb layers are shown in Fig. 19. In addition to the regular features, what can be described as doublets in the magnetoresistance appear to exist, notably the one in the vicinity of $i = \frac{5}{2}$, where a well-defined Hall step also develops. With decreasing temperatures, furthermore, the high-field peaks of the doublets were found to reduce strongly in intensity and shift to lower fields. These observations apparently cannot be explained by electrons alone. It was speculated that the presence of holes in the system, through their interactions with electrons in a way that was not understood, might be responsible for the additional features.

Other heterostructures, in which transport measurements under magnetic fields were performed, include GaAs–InGaAs (Schirber *et al.*, 1983), InGaAs–InAlAs (Kastalsky *et al.*, 1982), and PbTe–PbSnTe (Kinoshita *et al.*, 1982). Two-dimensional electron states were observed in all cases. We should mention that, in using the Schubnikov oscillations to probe the subband structure, the focus has been primarily on the oscillatory period. It is known that the amplitude of oscillations as functions of temperature and magnetic field provides a measure of the effective mass and the collision broadening. This was done, for example, in InAs–GaSb where the large nonparabolicity effect in the conduc-

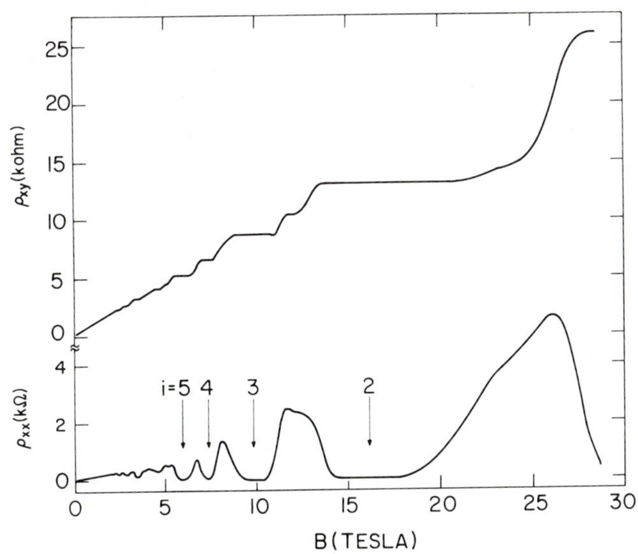

Fig. 19. Magnetoresistance ρ_{xx} and Hall resistance ρ_{xy} of a single well of InAs (150 Å) between GaSb at 0.55 K with an electron concentration of 8.5×10^{11} cm^{-2} and a mobility of 2.1×10^5 cm^2/V sec. The value i is similarly defined as in Fig. 18.

tion band of InAs resulted in a much enhanced electron mass (Sakaki *et al.*, 1978). However, the cyclotron resonance to be described in the next subsection provides a direct determination of this parameter.

D. Magnetoabsorption

Magnetoabsorption refers to optical transitions involving Landau levels in a magnetic field. Typically, different radiation sources are used, and for a given wavelength, the field is scanned to obtain the absorption or transmission spectrum. Two types of transitions are generally of interest. One is intersubband absorption arising from Landau levels between the valence and conduction bands; the energy of absorption peaks extrapolated to zero field provides information about the band structure. The other is cyclotron absorption or resonance between Landau levels across the Fermi energy within the same subband of electrons or holes. The variation of absorption energies with field, in this case, gives a direct determination of the effective mass. In considering the various energies involved, the effect of band nonparabolicity has to be taken into account. The simple two-band approximation with the replacement of E by $E(1 + E/E_g)$ has been found to be adequate for most experimental situations. The factor of mass enhancement, according to this approximation, is then given by $1 + 2E_f/E_g$. Early observations of cyclotron resonance were made both in GaAs–GaAlAs heterojunctions (Störmer *et al.*, 1979) and in InAs–GaSb superlattices (Bluyssen *et al.*, 1979), where enhanced electron masses, as expected, were reported.

In the InAs–GaSb system, extensive magnetoabsorption experiments have been performed, leading to a number of important observations. Results of intersubband transitions in the semiconductor regime (L. L. Chang *et al.*, 1981b) yielded absorption edges, or energy gaps of the superlattice E_{gs}, in agreement with those from simple absorption measurements, as shown in Fig. 12. But it is more interesting in the semimetallic regime in which such absorption is not possible until the Landau levels of electrons are lifted in energy across E_f above those of holes. The situation is illustrated in the inset of Fig. 20, in which the absorption peak energies are plotted versus field in the standard fashion (Guldner *et al.*, 1980). Both cyclotron resonance (CR) associated with E_{1e} and interband transitions between E_{1e} and E_{1h} are seen. The latter exhibit a series of branches, each corresponding to a specific Landau index, and converge to give a negative value of E_{gs}, a direct indication of the semimetallic nature. Subsequent measurements to higher fields and with more radiation sources provided more data points to corroborate these results, and, in addition, detected another branch of transitions (Maan *et al.*, 1981). This branch was identified as due to absorption of the same subbands but at the zone edge ($k_z = \pi/d$) at which the density of states also peaked. The energy difference at $B = 0$ between this and that shown in Fig. 20 for $k_z = 0$ measures directly the

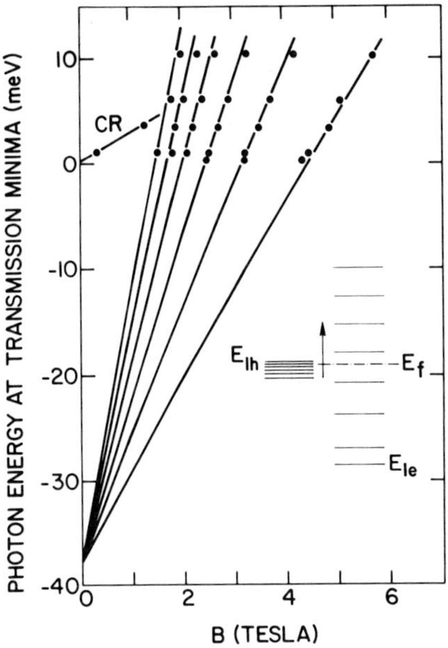

Fig. 20. Energies of magnetoabsorption in an InAs(120-Å)–GaSb(80-Å) superlattice. Solid lines are calculated Landau-level involving the ground state of electrons and holes as shown in the inset. CR indicates cyclotron resonance of electrons.

bandwidth of E_{1e}. The superlattice under consideration deviates significantly from strict two-dimensionality.

Another case of interest is the superlattice in the heterojunction limit shown in Fig. 16, in which more than one electron subband exists. The magnetoabsorption spectra showed a large number of peaks which are plotted in Fig. 21 in a similar fashion (Guldner *et al.*, 1982b). The interband series of the ground states $E_{1h} - E_{1e}$ designated as IS_1 is similar to that in Fig. 20. But additional transitions are present. The IS_2 branch corresponds to the ground Landau level transition between the same hole subband to the second electron subband $E_{1h} - E_{2e}$; the difference of IS_2 and IS_1 in energy at $B = 0$, in this case, equals the spacing between the two electron subbands. Two cyclotron resonances of electrons are visible, reflecting the two different effective masses at E_f for the two subbands as a result of the highly nonparabolic conduction band of InAs. The last branch, denoted CR_h, is attributed to the cyclotron resonance of holes whose mass is given correctly by the slope. The results shown in Fig. 21 demonstrate the ability of magnetoabsorption for elucidating the various transitions arising from the band structure of the superlattice.

As mentioned previously, while a perpendicular field B along z can be treated independently of the electric quantization, this cannot be done for B in an arbitrary direction where mixed subbands are formed. In the extreme case

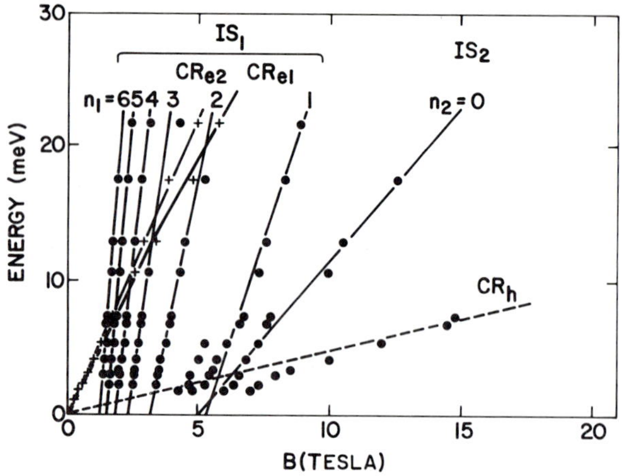

Fig. 21. Energies of magnetoabsorption in an InAs (1000-Å)–GaSb (1000-Å) superlattice in the heterojunction limit. Indicated are intersubband transitions (IS_1 and IS_2) and cyclotron resonances of electrons (CR_{e1} and CR_{e2}) and holes (CR_h).

such as B along x, for example, the equivalent magnetic contribution to quantization is given by $(\hbar^2/2m^*)(z'^2/l^2)$, where $z' = z + (\hbar k_y/eB)$ and $l = (\hbar/eB)^{1/2}$, the cyclotron radius. If this term is small in comparison with $V(z)$, the superlattice potential, it can be treated as a perturbation (Ando, 1981); but otherwise the Schrödinger equation must be solved in general to obtain the quantum states. Physically, this means that the energy depends on the center coordinate position of the cyclotron orbit $\hbar k_y/eB$. Calculations at a specific field are shown in Fig. 22a for the structure of Fig. 21, using the InAs layer thickness z_0 as the characteristic length (Maan *et al.*, 1982). With increasing B, the magnetic field contribution becomes gradually dominant, leading to Landau-like states with energy spacings $\hbar\omega_c$ near the center $k_y = 0$ and $2\hbar\omega_c$ at the interface $\hbar k_y = eBz_0$, the familiar magnetic surface states (Wanner *et al.*, 1975). Experimentally, well-defined transitions, which were not resolvable at low fields when magnetic and electric quantizations were completely mixed (Maan *et al.*, 1982), were observed as the field was increased, as shown in Fig. 22b. The transitions correspond to those near the center with an energy approaching $\hbar\omega_c$.

While cyclotron resonance has been commonly used to determine the effective mass and to reveal the two-dimensional nature of the electron system with tilted fields (Bluyssen *et al.*, 1982) similar to that in transport, it is a rather complex phenomenon depending on the field, the frequency of excitation, and the electron density, as demonstrated in Si inversion layers (Ando *et al.*, 1982). Investigations in heterostructures have begun to focus on cyclotron interactions with impurities and phonons. The resonance linewidth as a function of

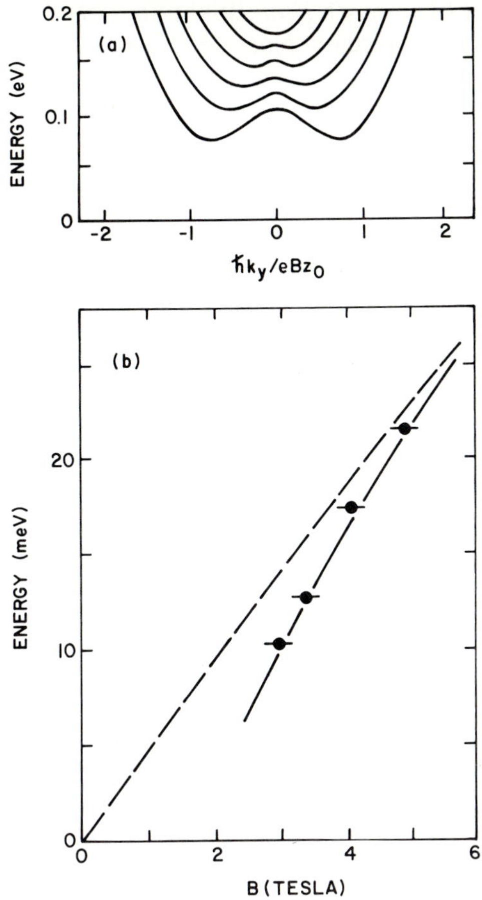

Fig. 22. InAs–GaSb superlattice of the same configuration as in Fig. 21 under parallel magnetic field. (a) Calculated subband dispersion where $\hbar k_y/eBz_0$ is the normalized center coordinate of the cyclotron orbit ($B = 4$ Tesla). (b) Observed energies of magnetoabsorption with calculated hybrid subband transition (solid line) and cyclotron resonance (dotted line).

the field was studied in GaAs–GaAlAs (Muro *et al.,* 1982). It was found to vary with $\sqrt{B}$ at low fields in accordance with the short-range scattering theory (Ando, 1975; Prasad and Fujita, 1978) but deviate significantly below this dependence as the field was increased (Voisin *et al.,* 1983). An oscillatory linewidth correlated with the occupation of the Landau levels was also reported (Englert *et al.,* 1983). The polaron effect was considered theoretically (Madhukar and Das Sarma, 1980; Das Sarma, 1984), predicting its enhancement in a two-dimensional system (Das Sarma, 1983). The electron screening, however, probably plays a more important role, which has the opposite effect of reducing the polaron mass (Seidenbusch *et al.,* 1984). In the InP–InGaAs system, the cyclotron linewidth was recently reported to depend on the field in a similar fashion but to increase dramatically at fields corresponding to the LO phonons in InAs (Brummell *et al.,* 1984).

E. Photoluminescence

Photoluminescent intensity and electron mobility are perhaps the two criteria most commonly used to evaluate the quality of a semiconductor, for they are both related to the defect states. While the latter measures the influence of defects on the scattering process in electron transport, the former reflects their presence through the recombination process, as defects usually result in luminescence at energies below the band gap or nonradiative recombination. That luminescence properties bear directly on lasers has provided additional stimulus for their extensive investigation. Indeed, stimulated emission was reported in quantum-well structures with different configurations and pumping schemes (van der Ziel *et al.,* 1975; Holonyak *et al.,* 1978; Tsang *et al.,* 1979). Activities in this area, which started with GaAs–GaAlAs and spread to other materials, are reviewed in Chapter 7 of this volume.

We are interested in this subsection in photoluminescence properties from spontaneous emission, related to the subband structure and the confinement of both carriers and impurities. The main recombination features in the luminescence spectrum of GaAs–GaAlAs at low temperatures and low excitations are those between the ground states of electrons and heavy holes and, to a much less extent, light holes (Miller *et al.,* 1980; Weisbuch *et al.,* 1981). Figure 23 shows a typical spectrum (Weisbuch *et al.,* 1981). The identification of these two transitions and their excitonic character were verified from measurements of polarizations by the technique of spin orientation. By populating upper subbands, additional transitions were observed as shown in the excitation spectrum that was taken by detecting the luminescence at a fixed wavelength while scanning the excitation. The transitions satisfy the usual selection rule and involve subbands of electrons and holes of the same energy index ($\Delta n = 0$). But weak forbidden transitions, including both parity conserving ($\Delta n =$ even) and nonconserving ($\Delta n =$ odd) recombinations, were also observed, particularly at high excitations (Miller *et al.,* 1981a).

The binding energy of the free excition in superlattices was initially estimated from absorption measurements (Dingle *et al.,* 1974). More systematic evaluation was derived from the energy difference between the ground and first excited state of the exciton, obtained from excitation spectroscopy (Miller *et al.,* 1981b). For the range of layer thickness under consideration, the enhancement of the binding energy with decreasing well width generally followed from initial (Miller *et al.,* 1981b) as well as subsequent, more refined, calculations (Bastard *et al.,* 1982a; Greene and Bajaj, 1983a). A precise quantitative determination is

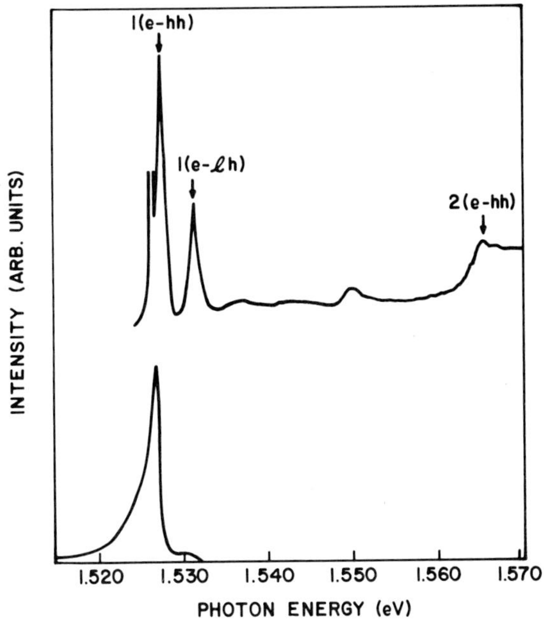

Fig. 23. Luminescence (lower) and excitation (upper) spectra of a GaAs(188-Å)–$Ga_{0.7}Al_{0.3}As$(19-Å) superlattice at 1.8 K. Indicated are transitions involving ground and first excited subbands of electrons and heavy holes, and ground subbands of electrons and light holes. The unmarked peak is due to a parity-conserving, forbidden transition. [Reprinted with permission from *Solid State Commun.* **37,** Weisbuch *et al.,* Intrinsic radiative recombination from quantum states in GaAs–AlGaAs multi-quantum well structures, copyright (1981), Pergamon Press, Ltd.]

difficult, however, since the experimentally observed transition of the excited state was not sharply defined. The binding energy enhancement was used advantageously to observe optical bistability in a superlattice etalon (Gibbs *et al.,* 1982). Also, transitions attributed to biexcitons were reported (Miller *et al.,* 1982). The binding energy was found to be significantly enhanced from the bulk value and to exhibit a rather weak thickness dependence, contrary to what was expected theoretically (Kleinman, 1983).

That the photoluminescence in a superlattice is predominantly intrinsic in nature offers a sharp contrast to that in bulk GaAs, in which extrinsic luminescence involving impurities or impurity-bound excitons usually prevail. The reason is not clear but may be related to the smearing of the impurity bands and the decreasing of the oscillator strength of their transitions (Weisbuch *et al.,* 1981). The binding energies obtained for the extrinsic luminescence from undoped GaAs–GaAlAs superlattices as well as single and double wells (Miller *et al.,* 1983) are shown in Fig. 24. The transition corresponds to that of electrons to neutral acceptors, presumably residual carbon. The results for the superlattice

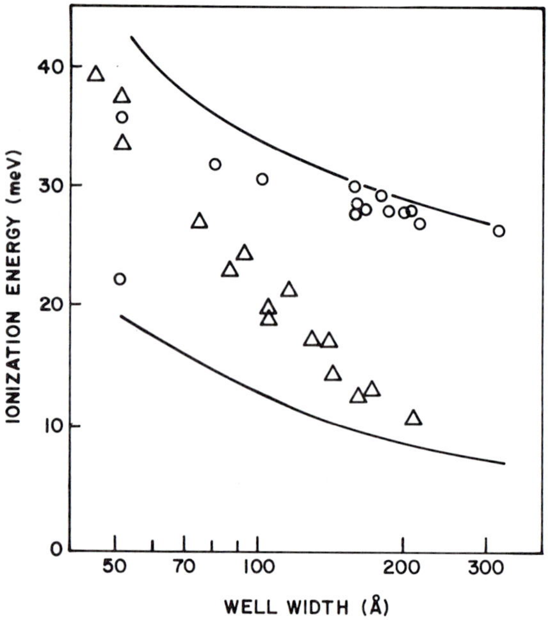

Fig. 24. Acceptor binding energy versus well width for GaAs–$Ga_{0.7}Al_{0.3}As$ superlattices (circles) and single and double quantum wells (triangles). The upper and lower curves are calculations for the acceptor to be located at the center and edge of the well, respectively. [After Miller *et al.*, (1983).]

are quite consistent with theoretical calculations that predict a maximum density of states at the well center and a second maximum at the well edge for thin wells (Bastard, 1981a). For cases with a limited number of wells, the data can be explained on the assumption that carbon is present mainly near the interface and essentially fills the well as its width is decreased (Miller *et al.*, 1983). Also reported were observations of exciton bound to neutral acceptors in purposely doped *p*-type materials (Miller *et al.*, 1983) and a series of transitions involving electron–acceptor, hole–donor, and donor–acceptor (Lambert *et al.*, 1983).

Compared to GaAs–GaAlAs, luminescent studies in other superlattice structures have been pursued to a much less extent. The spectrum of InAs–GaSb was shown to be dominant by band-to-band transitions of the ground states of electrons and holes (Voisin *et al.*, 1981). The lack of excitonic feature is consistent with the small binding energy as a result of spatial separation of the carriers (Bastard *et al.*, 1982a). Luminescent peaks were also observed at low energies and were attributed to impurities or thickness fluctuations. The GaSb–AlSb superlattice has attracted some recent interest, for which several luminescence investigations were reported (Naganuma *et al.*, 1981; Mendez *et al.*, 1983a; Griffiths *et al.*, 1983). That the luminescence originated from the ground subbands in GaSb was determined from the temperature dependence of the transition energy (Mendez *et al.*, 1983) and was subsequently confirmed from absorption experiments. One of the interesting features of this system lies in the fact that the L and X minima in the conduction band of GaSb are not far above the direct Γ point. The light electron mass associated with this point may push its subbands above those of the other minima with a narrowing of the well width, leading to a direct–indirect transition. This was not observed experimentally, however, as illustrated in Fig. 25 (Griffiths *et al.*, 1983). The transition enrgies vary with the well thickness in accordance with theoretical calculations (Bastard, 1981b), indicating the dominance of the Γ-point in the luminescence process throughout the energy range of consideration. But it was found that the intensity decreased, and an additional broad peak at low energy appeared as the well width was thinned beyond what was expected for the Γ–X crossover (Griffiths *et al.*, 1983).

Another superlattice system for which luminescence properties have been studied is GaP–GaAsP (Gourley and Biefeld, 1982). This system presents several interesting points: a relatively large lattice mismatch; a spatial separa-

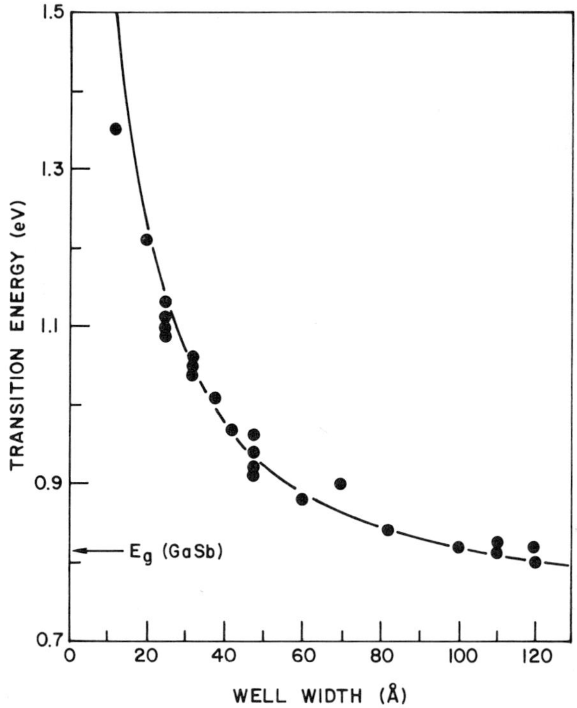

Fig. 25. Luminescence energy versus well width for GaSb–AlSb superlattices at 1.6 K. The solid curve shows a theoretical fit. [After Griffiths *et al.* (1983).]

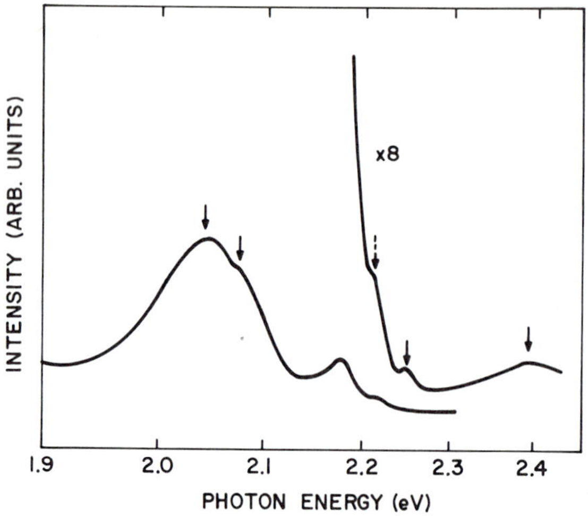

Fig. 26. Luminescence spectrum of a GaP–$GaAs_{0.4}P_{0.6}$ superlattice at 78 K. Arrows indicate transitions associated with the superlattice, of which four (solid arrows) can be identified with the calculated subband structures. [After Gourley and Biefeld (1982).]

tion of carriers, electrons in GaP, and holes in GaAsP; and a direct energy gap as a result of zone folding, although the two components are indirect materials themselves. The luminescence spectrum is shown in Fig. 26, in which arrows are marked to indicate transitions associated with the superlattice (Gourley and Biefeld, 1982). With the exception of the one (shown with a broken arrow) whose origin is unknown, the other four can find their counterparts in energy from theoretical calculations. In ascending order of energy, the first three were ascribed to transitions from the ground state of electrons at the folded X point to the heavy-hole, light-hole, and the split-off states in the valence band, and the last one to transition from the Γ-point conduction band to the ground heavy-hole state. In these calculations, the effect of strain was taken into account (Osbourn, 1982), which lifted the bulk degeneracy at the valence band and the valley degeneracy at the X point in the conduction band.

F. Other Electronic Properties

A large number of other measurement techniques have been used to probe specific properties of superlattices and heterostructures. Among them, inelastic light scattering is one of the most powerful spectroscopic tools. Early work dealt with resonant Raman scattering involving LO phonons of GaAs in GaAs–GaAlAs superlattice (Manuel *et al.*, 1976). The enhancement in scattering intensity at energies corresponding to intersubband transitions between the conduction and valence bands provides information about the subband energies as well as the two-dimensional density of states. This and subsequent investigations (Barker *et al.*, 1978; Merlin *et al.*, 1980; Sai-Halasz *et al.*, 1978b), with an apparent interest in the folding effect of phonon dispersion by the superlattice periodicity, led to the observation of anisotropy of phonon propagation in the layered medium (Merlin *et al.*, 1980). By using extremely thin layers, new modes arising from acoustic phonons folded from $2\pi/d$ onto the Brillouin zone center were obtained (Covard *et al.*, 1980), as shown in Fig. 27. The two sharp peaks, as indicated by arrows, occur at energies consistent with those calculated from a simple theory by treating the electrons in the Kronig–Penney model and the phonons in the elastic continuum limit. It should be mentioned, in this connection, that zone-boundary phonons were also observed through the selective transmission of the superlattice in tunnel junctions (Narayanamurti *et al.*, 1979).

Inelastic scattering has been used successfully in the investigation of electron excitations associated with discrete subbands in heterostructures, including heterojunctions (Abstreiter and Ploog, 1979) and superlattices (Pinczuk *et al.*, 1979) of GaAs–GaAlAs. With proper scattering geometry, this technique is capable of uncovering both single-particle and collective excitations to provide information on the energy levels and coulombic interactions (Pinczuk *et al.*, 1980). Tremendous progress has been made in this area which is reviewed in Chapter 6 of this volume. Figure 28 shows the spectra of a GaAs–GaAlAs quantum-well structure in the range of the ground state (Pinczuk *et al.*, 1980). The depolarized spectrum exhibits the single-particle transition assigned to the lowest spin-density excitation between subbands. In the polarized spectrum, coupled collective excitation with LO phonons are seen at higher energies. The effect of the depolarization field can thus be determined from the difference. Of

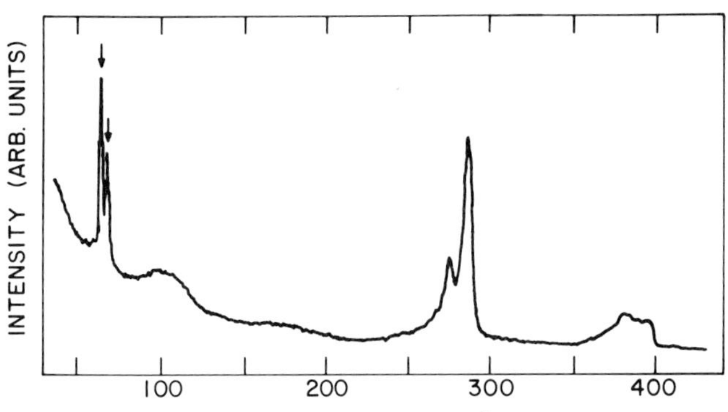

Fig. 27. Raman scattering spectrum of a GaAs (13.6-Å)–AlAs (11.4-Å) superlattice at 300 K and for $z(x, x)\bar{z}$. Arrows mark the folded phonon modes of interest. Peaks at high shifts are due to bulklike phonons. [After Covard *et al.* (1980).]

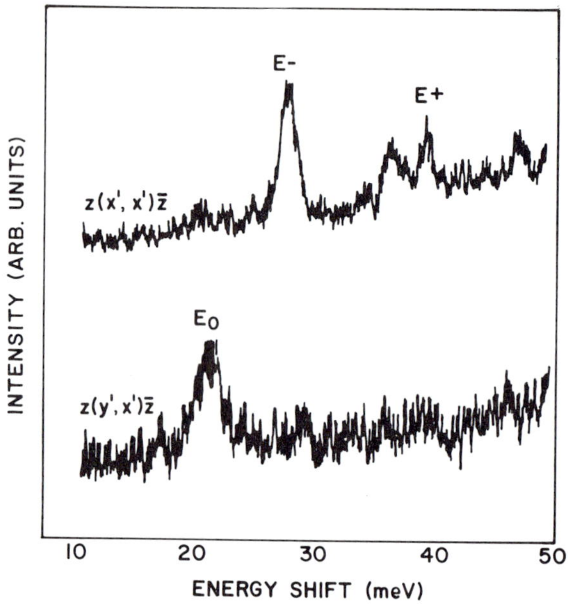

Fig. 28. Light-scattering spectra of a GaAs–$Ga_{0.82}Al_{0.18}As$ superlattice with a well width of 204 Å at 2 K. The depolarized spectrum (lower) shows the lowest intersubband transition of electrons. The polarized spectrum (upper) shows the phonon-coupled collective modes; other peaks due to phonons are not indicated. [Reprinted with permission from *Solid State Commun.* **36,** Pinczuk *et al.,* Spectroscopy of two-dimensional electron gases: Coulomb interactions, copyright (1980), Pergamon Press, Ltd.]

interest in inelastic scattering is the study of excitation associated with electron motion in the plane of the layers. The plasmon frequency dispersion, different from that in two or three dimensions, was observed (Olego *et al.,* 1982), which varied linearly with the parallel wavevector as predicted from electrodynamics for weakly coupled electron layers.

The interest in the subbands in a superlattice was largely confined to the region near the fundamental gap of the host semiconductors. The situation had remained until the question was raised as to the effect of the periodic potential on the band structure away from the Brillouin zone center. The use of electroreflectance spectroscopy addressed this question, focusing especially on the L point. Figure 29 shows the spectra of a series of GaAs–GaAlAs superlattices with increasing Al compositions (Mendez *et al.,* 1981). The transition energies at this point shift progressively to higher values, which have been shown to correspond to those of the ground states from the valence and its split-off bands to the conduction band by using Kronig–Penney calculations with proper effective masses along the ⟨111⟩ direction. In other words, the formation of subband at the L point is analogous to that at the Γ point. This is not the case in

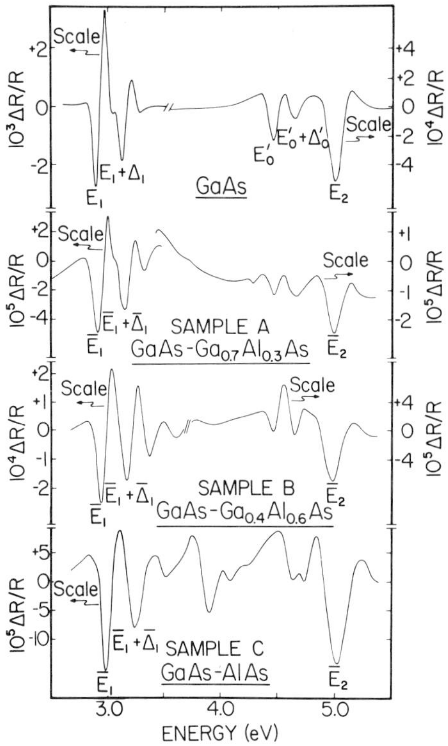

Fig. 29. Electroreflectance spectra of GaAs and GaAs–GaAlAs superlattices. The layer thickness in the superlattices are 77–66 Å, 50–50 Å, and 50–50 Å (top to bottom). Of main interest are the features marked E_1 and $E_1 + \Delta_1$ corresponding to transitions at the L point.

InGaAs–GaSbAs for which similar measurements have been performed (Mendez *et al.*, 1981). The observed dependence of the transition energies on alloy composition is complicated, which may reflect complex band mixing in this system, requiring a refined band-structure calculation. An alternative explanation involving alloy phase separation was also suggested (Phillips, 1981).

Optical properties have been pursued under the influence of specific, external parameters, such as electric and magnetic fields, hydrostatic pressure, and intense radiation. The dependence of photoluminescence on electric fields perpendicular to the layers in GaAs–GaAlAs quantum wells is illustrated in Fig. 30 (Mendez *et al.*, 1982). The increase of the field was achieved by placing the wells in the space-charge region of a Schottky barrier and decreasing the external voltage. The emission peaks were found to shift slightly to lower energies but reduce drastically in intensity, particularly for the exciton peak that became completely quenched. These observations were interpreted as being due to the field-induced separation of electrons and holes and modification of

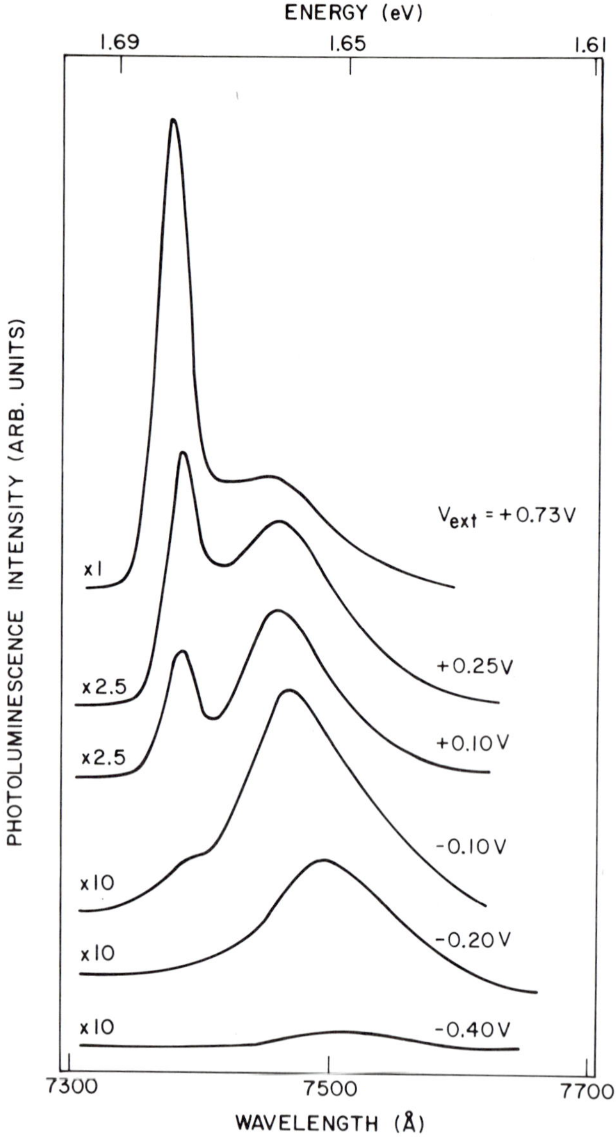

Fig. 30. Photoluminescence spectra of a GaAs (35-Å)–$Ga_{0.65}Al_{0.35}As$ (100-Å) superlattice at 6 K to illustrate the decrease in intensities with increasing electric fields.

the subband energies. Theoretical calculations based on isolated wells predicted the observed trend. But coupling between the wells, which is expected in the experimental structure, would be needed to explain quantitatively the energy lowering and the polarization of the carrier distribution (Bastard *et al.*, 1983).

The application of an electric field parallel to the layers of GaAs–GaAlAs superlattices has been investigated for its effect on both optical absorption

(Chemla *et al.*, 1983) and plasmon emission (Hopfel *et al.*, 1982). The absorption characteristics were found to be significantly modified by the electric field, which shifted and broadened the sharp excitonic absorption edge. For plasmon emission, the excitation was provided by hot electrons and coupling was achieved by a grating gate structure, which determined the emission frequency. The use of pressure to affect the electronic properties in heterostructures was pursued in two areas. It changes the band structure that in turn changes the optical absorption (Kirchoefer *et al.*, 1982), as mentioned earlier, and consequently the laser emission properties in GaAs–GaAlAs (Camras *et al.*, 1983). The pressure also affects the energetics of impurities in GaAlAs and, therefore, affects the electron density in GaAs at its interface (Mercy *et al.*, 1984). Other experiments include luminescence under a magnetic field (Worlock *et al.*, 1984) and absorption after strong light excitation (Shank *et al.*, 1982). The former resulted in an oscillatory spectra with a field dependence reflecting the complex valence-band structure. The latter probed the electron dynamics with the time-resolved picosecond technique, showing the effects of exciton screening and band filling.

Transport measurements under a magnetic field have been performed beyond those related directly to the subband structure. Magnetophonon oscillations were reported in GaAs–GaAlAs (Tsui *et al.*, 1980), GaInAs–AlInAs (Portal *et al.*, 1984), and GaInAs–InP (Portal *et al.*, 1983a). While the dominant phonons involved in the former two systems are those associated with GaAs and InAs, respectively, both GaAs and InP phonons may be present in the latter, depending on the structural configuration. Experiments with hot-electron excitation, similar to those in plasmon emission but with the application of magnetic fields, have led to the observation of cyclotron emission (Hopfel *et al.*, 1982). Magnetic fields were also used in connection with two-dimensional conductivity at low temperatures in the weak localization regime (Abrahams *et al.*, 1979). The magnetoresistance behaves differently when localization or correlation effects are dominant (Fukuyama, 1980; Hikami *et al.*, 1980; Lee and Ramakrishnan, 1982), as demonstrated in Si inversion layers (Bishop *et al.*, 1982). Work in this regime was extended to GaAs–GaAlAs (Poole *et al.*, 1981) and, recently, to InAs–GaSb (Washburn *et al.*, 1984), for which evidence of hole localization was reported. In a separate development, de Haas–van Alphen oscillations were observed in GaAs–GaAlAs superlattices (Störmer *et al.*, 1983b). Measurements of magnetization provide a direct means for evaluating the two-dimensional density of states.

ACKNOWLEDGMENTS

We would like to thank many of our collaborators who have contributed to the work covered in this chapter, in particular, L. Esaki and C.-A. Chang of this laboratory, M. Voos and G. Bastard of l'ENS, Paris, and J. C. Maan of MPI, Grenoble. This work was partially sponsored by the U.S. Army Research Office.

REFERENCES

Abeles, B., and Tiedje, T. (1983). *Phys. Rev. Lett.* **51,** 2003.

Abrahams, E., Anderson, P. W., Licciardello, D. C., and Ramakrishnan, T. V. (1979). *Phys. Rev. Lett.* **42,** 673.

Abstreiter, G., and Ploog, K. (1979). *Phys. Rev. Lett.* **42,** 1308.

Altarelli, M. (1983). *Physica* **117–118B–C,** 744.

Ando, T. (1975). *J. Phys. Soc. Jpn.* **38,** 989.

Ando, T. (1981). *J. Phys. Soc. Jpn.* **50,** 2978.

Ando, T., and Uemura, Y. (1974). *J. Phys. Soc. Jpn.* **37,** 1044.

Ando, T., Fowler, A. B., and Stern, F. (1982). *Rev. Mod. Phys.* **54,** 437.

Arthur, J. R. (1968). *J. Appl. Phys.* **39,** 4032.

Arthur, J. R. (1974). *Surf. Sci.* **43,** 449.

Barker, A. S., Merz, J. L., and Gossard, A. C. (1978). *Phys. Rev. B* **17,** 3181.

Bastard, G. (1981a). *Phys. Rev. B* **24,** 4714.

Bastard, G. (1981b). *Phys. Rev. B* **24,** 5693.

Bastard, G., Mendez, E. E., Chang, L. L., and Esaki, L. (1982a). *Phys. Rev. B* **26,** 1974.

Bastard, G., Mendez, E. E., Chang, L. L., and Esaki, L. (1982b). *J. Vac. Sci. Technol.* **21,** 531.

Bastard, G., Mendez, E. E., Chang, L. L., and Esaki, L. (1983). *Phys. Rev. B* **28,** 3241.

Bishop, D. J., Dynes, R. C., and Tsui, D. C. (1982). *Phys. Rev. B* **44,** 773.

Blakeslee, A. E., and Aliotta, C. F. (1970). *IBM J. Res. Dev.* **14,** 686.

Bloss, W. L. (1982). *Solid State Commun.* **44,** 363.

Bluyssen, H. J. A., Mann, J. C., Wyder, P., Chang, L. L., and Esaki, L. (1979). *Solid State Commun.* **31,** 35.

Bluyssen, H. J. A., Maan, J. C., Wyder, P., Chang, L. L., and Esaki, L. (1982). *Phys. Rev. B* **25,** 5364.

Brant, N. B., and Svistova, E. A. (1970). *J. Low Temp. Phys.* **2,** 1.

Brummell, M. A., Nicholas, R. J., Brunel, L. C., Huant, S., Baj, M., Portal, J. C., Razeghi, M., di Forte-Poisson, M. A., Cheng, K. Y., and Cho, A. Y. (1984). *Surf. Sci.* **142,** 380.

Camras, M. D., Holonyak, N., Jr., Coleman, J. J., Drickamer, H. G., Burham, R. D., Streifer, W., Scifres, D. R., Lindstrom, C., and Paoli, T. P. (1983). *J. Appl. Phys.* **54,** 4386.

Caruthers, E., and Lin-Chung, P. J. (1977). *Phys. Rev. Lett.* **38,** 1543.

Chang, C.-A. (1983). *J. Vac. Sci. Technol. B* **1,** 346.

Chang, C.-A., Ludeke,R., Chang, L. L., and Esaki, L. (1977). *Appl. Phys. Lett.* **31,** 759.

Chang, C.-A., Chu, W. K., Mendez, E. E., Chang, L. L., and Esaki, L. (1981). *J. Vac. Sci. Technol.* **19,** 567.

Chang, C.-A., Takaoka, H., Chang, L. L., and Esaki, L. (1982). *Appl. Phys. Lett.* **40,** 983.

Chang, C.-A., Mendez, E. E., Chang, L. L., and Esaki, L. (1984). *Surf. Sci.* **142,** 598.

Chang, L. L. (1978). *Surf. Sci.* **73,** 226.

Chang, L. L. (1980). In "Handbook of Semiconductors" (T. S. Moss, ed.), Vol. 3, p. 563. North-Holland Publ., Amsterdam.

Chang, L. L. (1983). *J. Vac. Sci. Technol. B* **1,** 120.

Chang, L. L., and Esaki, L. (1979). In "Prog. Crystal Growth and Characterization" (B. R. Pamplin, ed.), Vol. 2, p. 3. Pergamon, Oxford.

Chang, L. L., and Esaki, L. (1980). *Surf. Sci.* **98,** 70.

Chang, L. L., and Koma, A. (1976). *Appl. Phys. Lett.* **29,** 138.

Chang, L. L., and Ludeke, R. (1975). In "Epitaxial Growth" (J. W. Matthews, ed.), Part A, Chap. 2.2. Academic Press, New York.

Chang, L. L., Esaki, L., Howard, W. E., and Ludeke R. (1973a). *J. Vac.Sci. Technol.* **10,** 11.

Chang, L. L., Esaki, L., Howard, W. E., Ludeke, R., and Schul,G. (1973b). *J. Vac. Sci. Technol.* **10,** 655.

Chang, L. L., Esaki, L., and Tsu, R. (1974a). *Appl. Phys. Lett.* **24,** 593.

Chang, L. L., Esaki, L., Segmüller, A., and Tsu, R. (1974b). *Proc. Int. Conf. Phys. Semicond., 12th* (M. H. Pilkuhn, ed.) p. 688. Teubner, Stuttgart.
Chang, L. L., Segmüller, A., and Esaki, L. (1976). *Appl. Phys. Lett.* **28,** 39.
Chang, L. L., Sakaki, H., Chang, C.-A., and Esaki, L. (1977). *Phys. Rev. Lett.* **38,** 1489.
Chang, L. L., Kawai, N. J., Sai-Halasz, G. A., Ludeke, R., and Esaki, L. (1979). *Appl. Phys. Lett.* **35,** 939.
Chang, L. L., Kawai, N. J., Mendez, E. E., Chang, C.-A., and Esaki, L. (1981a). *Appl. Phys. Lett.* **38,** 30.
Chang, L. L., Sai-Halasz, G. A., Esaki, L., and Aggarwal, R. L. (1981b). *J. Vac. Sci. Technol.* **19,** 589.
Chang, L. L., Mendez, E. E., Kawai, N. J., and Esaki, L. (1982). *Surf. Sci.* **113,** 306.
Chang, L. L., Stiles, P. J., and Esaki, L. Unpublished.
Chemla, D. S., Damen, T. C., Miller, D. A. B., Gossard, A. C., and Wiegmann, W. (1983). *Appl. Phys. Lett.* **42,** 864.
Cho, A. Y. (1970). *J. Appl. Phys.* **41,** 2780.
Cho, A. Y. (1971). *Appl. Phys. Lett.* **19,** 467.
Cho, A. Y. Arthur, J. R. (1975). In "Prog. Solid State Chemistry" (J. O. McCaldin and G. A. Somorjai, eds.), Vol. 10, p. 157. Pergamon, New York.
Chu, W. K., Saris, F. W., Chang, C.-A., Ludeke, R., and Esaki, L. (1982). *Phys. Rev. B* **26,** 1999.
Covard, C., Merlin, R., Klein, K. V., and Gossard, A. C. (1980). *Phys. Rev. Lett.* **45,** 298.
Das Sarma, S. (1984). *Surf. Sci.* **142,** 341.
Das Sarma, S., and Madhukar, A. (1981). *Phys. Rev. B* **23,** 805.
Das Sarma, S., and Quinn, J. J. (1982). *Phys. Rev. B* **25,** 7603.
Davey, J. E., and Pankey T. (1969). *J. Appl. Phys.* **39,** 1941.
Dingle, R. (1975). In "Adv. Solid State Phys. " (H. J. Queisser, ed.), Vol. 15, p. 21. Verlag, Braunsweig.
Dingle, R., Wiegmann, W., and Henry, C. H. (1974). *Phys. Rev. Lett.* **33,** 827.
Dingle, R., Gossard, A. C., and Wiegmann, W. (1975). *Phys. Rev. Lett.* **34,** 1327.
Dingle, R., Störmer, H. L., Gossard, A. C., and Wiegmann, W. (1978). *Appl. Phys. Lett.* **33,** 665.
Dingle, R., Störmer, H. L., Gossard, A. C., and Wiegmann, W. (1980). *Surf. Sci.* **98,** 90.
Döhler, G. H. (1972). *Phys. Status Solidi B* **52,** 533.
Dove, D. B., Ludeke, R., and Chang, L. L. (1973). *J. Appl. Phys.* **44,** 1897.
Dow, J. D., Ren, S. Y., and Hess, K. (1982). *Phys. Rev. B* **25,** 6218.
Eltoukhy, A. H., Ziko, J. L., Wickersham, C. E., and Green, J. E. (1977). *Appl. Phys. Lett.* **31,** 156.
Englert, T., Tsui, D. C., Gossard, A. C., and Uihlein, C. (1982). *Surf. Sci.* **113,** 295.
Englert, T., Maan, J. C., Uihelin, C., Tsui, D. C., and Gossard, A. C. (1983). *Physica* **117–118B–C,** 631.
Esaki, L., and Chang, L. L. (1974). *Phys. Rev. Lett.* **33,** 495.
Esaki, L., and Chang, L. L. (1976). *Thin Solid Films* **36,** 285.
Esaki, L., and Tsu, R. (1970). *IBM J. Res. Dev.* **14,** 61.
Esaki, L., Chang, L. L., and Tsu, R. (1970). In *Proc. Int. Conf. Low Temp. Phys., 12th, Kyoto, Japan,* p. 551. (Keigaku, Tokyo.)
Esaki, L., Chang, L. L., and Mendez, E. E. (1981). *Jpn. J. Appl. Phys.* **20,** L259.
Fang, F. F., and Stiles, P. J. (1968). *Phys. Rev.* **174,** 823.
Faurie, J. P., Million, A., and Piaguet, J. (1982). *Appl. Phys. Lett.* **41,** 713.
Foxon, C. T., and Joyce, B. A. (1975). *Surf. Sci.* **50,** 434.
Foxon, C. T., and Joyce, B. A. (1981). In "Current Topics in Material Science" (E. Kaldis, ed.), Vol. 7, p. 1. North-Holland Publ., Amsterdam.
Frensley, W. R., and Kroemer, H. (1977). *Phys. Rev. B* **16,** 2642.
Fukuyama, H. (1980). *J. Phys. Soc. Jpn.* **48,** 2169.
Fukuyama, H., and Platzman, P. M. (1982). *Phys. Rev. B* **25,** 2934.

Gibbs, H. M., Tang, S. S., Jewell, J. L., McCall, D. A., Passner, A., and Wiegmann, W. (1982). *Appl. Phys. Lett.* **41,** 221.
Gossard, A. C. (1982). In "Treatise on Materials Science and Technology" (K. N. Tu and R. Rosenberg, eds.), Vol. 24, p. 13. Academic Press, New York.
Gourley, P. L., and Biefeld, R. M. (1982). *J. Vac. Sci. Technol.* **21,** 473.
Gourley, P. L., and Biefeld, R. M. (1983). *J. Vac. Sci. Technol. B* **1,** 383.
Greene, R. L., and Bajaj, K. K. (1983a). *Solid State Commun.* **45,** 831.
Greene, R. L., and Bajaj, K. K. (1983b). *J. Vac. Sci. Technol. B* **1-2,** 391.
Griffiths, G., Mohammed, K., Subbana, S., Kroemer, H., and Merz, J. (1983). *Appl. Phys. Lett.* **43,** 1059.
Guldner, Y., Vieren, J. P., Voisin, P., Voos, M., Chang, L. L., and Esaki, L. (1980). *Phys. Rev. Lett.* **45,** 1719.
Guldner, Y., Hirtz, J. P., Vieren, J. P., Voisin, P., Voos, M., and Razeghi, M. (1982a). *J. Phys.* **42,** L613.
Guldner, Y., Vieren, J. P., Voisin, P., Voos, M., Mann, J. C., Chang, L. L., and Esaki, L. (1982b). *Solid State Commun.* **41,** 755.
Gunther, H. G. (1958). *Z. Naturforsch.* **13A,** 1081.
Harrison, W. A. (1977). *J. Vac. Sci. Technol.* **14,** 1016.
Harrison, W. A. (1980). "Electronic Structure and the Properties of Solids." Freeman, San Francisco, California.
Hikami, S., Larkin, A. I., and Nagoaka, Y. (1980). *Prof. Theor. Phys.* **63,** 707.
Holonyak, N., Jr., Kolbas, R. M., Dupuis, R. D., and Dapkus, P. D. (1978). *Appl. Phys. Lett.* **33,** 73.
Holonyak, N., Jr., Laidig, W. O., Vojak, B. A., Hess, K., Coleman, J. J., Dapkus, P. D., and Bardeen, J. (1980). *Phys. Rev. Lett.* **45,** 1703.
Hopfel, R., Lindemann, G., Gornik, E., Stangl, G., Gossard, A. C., and Wiegmann, W. (1982). *Surf. Sci.* **113,** 118.
Ihm, J., Lam, P. K., and Cohen, M. I. (1979). *Phys. Rev. B* **20,** 4120.
Ivanov, I. and Pollmann, J. (1979). *Solid State Commun.* **32,** 869.
Janak, J. F. (1969). *Phys. Rev.* **178,** 1416.
Johnson, C. H., and Linde, J. O. (1936). *Ann. Phys., Leipzig.* **25,** 1.
Jones, H. (1969). *J. Phys. C,* **2,** 733.
Joyce, B. A., Foxon, C. T. (1975). *J. Cryst. Growth* **31,** 122.
Kastalsky, A., Dingle, R., Cheng, K. Y., and Cho, A. Y. (1982). *Appl. Phys. Lett.* **41,** 274.
Kawai, N. J., Chang, L. L., Sai-Halasz, G. A., Chang, C.-A., and Esaki, L. (1980). *Appl. Phys. Lett.* **36,** 369.
Kinoshita, H., and Fujiyasu, H. (1980). *J. Appl. Phys.* **51,** 5845.
Kinoshita, H., Takaoka, S., Murase, K., and Fujiyasu, H. (1982). *Int. Symp. MBE Related Clean Surface Tech., 2nd* (Jpn. Soc. Appl. Phys.), Tokyo, Japan, p. 6.
Kirchoefer, S. W., Holonyak, N., Jr., Hess, K., Gulino, D. A., Drickamer, H. G., Coleman, J. J., and Dapkus, P. D. (1982). *Appl. Phys. Lett.* **40,** 821.
Kleinman, D. A. (1983). *Phys. Rev. B* **28,** 871.
Kuan, T. S. and Chang, C.-A. (1982). *Proc. Int. Conf. Electron Microscopy, 10th,* Hamburg, Germany, p. 401.
Lambert, B., Deveaud, B., Regreny, A., and Talalaff, G. (1983). *Physica* **117-118B-C,** 717.
Laughlin, R. B. (1981). *Phys. Rev. B* **23,** 5632.
Laughlin, R. B. (1983). *Phys. Rev. Lett.* **50,** 1395.
Lee, P. A., Ramakrishnan, R. V. (1982). *Phys. Rev. B* **26,** 4009.
Ludeke, R., Esaki, L., and Chang, L. L. (1974). *Appl. Phys. Lett.* **24,** 417.
Ludowise, M. J., Dietze, W. T., Lewis, C. R., Holonyak, N., Jr., Hess, K., Camras, M. D., and Nixon, M. A. (1983). *Appl. Phys. Lett.* **42,** 257.

Maan, J. C., Guldner, Y., Vieren, J. P., Voisin, P., Voos, M., Chang, L. L., and Esaki, L. (1981). *Solid State Commun.* **39,** 683.
Maan, J. C., Uihlein, C., Chang, L. L., and Esaki, L. (1982). *Solid State Commun.* **44,** 653.
Madhukar, A., and Das Sarma, S. (1980). *Surf. Sci.* **98,** 135.
Mailhiot, C., Chang, Y. C., and McGill, T. C. (1982). *Surf. Sci.* **113,** 161.
Maluenda, J., and Frijlink, P. M. (1983). *J. Vac. Sci. Technol. B* **1,** 334.
Manesevit, H. M., Gergis, I. S., and Jones, A. B. (1982). *Appl. Phys. Lett.* **41,** 464.
Manuel, P., Sai-Halasz, G. A., Chang, L. L., Chang, C.-A., and Esaki, L. (1976). *Phys. Rev. Lett.* **37,** 1701.
Mayer, J. W., Ziegler, J. F., Chang, L. L., Tsu, R., and Esaki, L. (1973). *J. Appl. Phys.* **44,** 2322.
Mendez, E. E., Chang, C.-A., Chang, L. L., Esaki, L., and Pollak, F. H. (1980). *J. Phys. Soc. Jpn., Suppl. A* **4,** 1009.
Mendez, E. E., Chang, L. L., Landgren, G., Ludeke, R., Esaki, L., and Pollak, F. H. (1981). *Phys. Rev. Lett.* **46,** 1230.
Mendez, E. E., Bastard, G., Chang, L. L., Esaki, L., Morkoc, H., and Fischer, R. (1982). *Phys. Rev. B* **26,** 7101.
Mendez, E. E., Chang, C.-A., Takaoka, H., Chang, L. L., and Esaki, L. (1983a). *J. Vac. Sci. Technol. B* **1,** 152.
Mendez, E. E., Chang, L. L., Chang, C.-A., Alexander, L. F., and Esaki, L. (1984). *Surf. Sci.* **142,** 215.
Mendez, E. E., Heiblum, M., Chang, L. L., and Esaki, L. (1983c). *Phys. Rev. B* **28,** 4886.
Mercy, J. M., Bousquet, C., Robert, J. L., Raymond, A., Gregoris, G., Beerens, J., Portal, J. C., Frijlink, P. M., Delecluse, P., Chevrier, J., and Linh, T. (1984). *Surf. Sci.* **142,** 298.
Merlin, R., Covard, C., Klein, M. V., Morkoc, H., Cho, A. Y., and Gossard, A. C. (1980). *Appl. Phys. Lett.* **135,** 43.
Miller, R. C., Tsang, W. T., and Nordland, W. A., Jr., (1980). *Phys. Rev. B* **21,** 1569.
Miller, R. C., Kleinman, D. A., Munteanu, D., and Tsang, W. T. (1981a). *Appl. Phys. Lett.* **39,** 1.
Miller, R. C., Kleinman, D. A., Tsang, W. T., and Gossard, A. C. (1981b). *Phys. Rev. B* **25,** 1134.
Miller, R. C., Kleinman, D. A., Gossard, A. C., and Munteanu, D. (1982). *Phys. Rev. B* **25,** 6545.
Miller, R. C., Gossard, A. C., and Tsang, W. T. (1983). *Physica* **117–118B–C,** 714.
Milnes, A. G., and Feucht, D. L. (1972). "Heterojunctions and Metal-Semiconductor Junctions," Academic Press, New York.
Mori, S., and Ando, T. (1979). *Phys. Rev. B* **19,** 6433.
Muro, K., Narita, S., Hiyaurizu, S., Nanbu, K., and Hashimoto, H. (1982). *Surf. Sci.* **113,** 321.
Naganuma, M., Suzuki, Y., and Okamoto, H. (1981). *Int. Sym. GaAs and Related Compounds* (T. Sugano, ed.), p. 125. Inst. of Physics, University of Reading, Berkshire.
Narayanamurti, V., Störmer, H. L., Chin, M. A., Gossard, A. C., and Wiegmann, W. (1979). *Phys. Rev. Lett.* **43,** 2012.
Narita, S., Takeyama, S., Luo, W. B., Hiyamizu, S., Nanbu, K., and Hasimoto, H. (1982). *Surf. Sci.* **113,** 301.
Nicholas, R. J., Session, S. J., and Portal, J. C. (1980). *Appl. Phys. Lett.* **37,** 178.
Nicholas, R. J., Brummell, M. A., Portal, J. C., Razeghi, M., and Poisson, M. A. (1982). *Solid State Commun.* **43,** 825.
Nucho, R. N., and Madhukar, A. (1978). *J. Vac. Sci. Technol.* **15,** 1530.
Ogawa, S., and Watanabe, D. (1954). *Acta Crystallogr.* **7,** 377.
Olego, D., Pinczuk, A., Gossard, A. C., and Wiegmann, W. (1982). *Phys. Rev.* **25,** 7867.
Osbourn, G. C. (1982). *J. Appl. Phys.* **53,** 1536.
Partin, D. L. (1982). *J. Vac. Sci. Technol.* **21,** 1.
Petroff, P. M., Gossard, A. C., Savage, A., and Wiegmann, W. (1979). *J. Crystal Growth* **46,** 172.
Petroff, P. M., Gossard, A. C., Logan, R. A., and Wiegmann, W. (1982). *Appl. Phys. Lett.* **41,** 635.

Phillips, J. C. (1981). *Phys. Rev. B* **24,** 3620.
Pickett, W. E., Louis, S. G., and Cohen, M. L. (1977). *Phys. Rev. B* **17,** 815.
Pidgeon, C. R., Mitchell, D. L., and Brown, R. N. (1967). *Phys. Rev.* **154,** 737.
Pinczuk, A., Störmer, H. L., Dingle, R., Worlock, J. M., Wiegmann, W., and Gossard, A. C. (1979). *Solid State Commun.* **32,** 1001.
Pinczuk, A., Worlock, J. M., Störmer, H. L., Dingle, R., Wiegmann, W., and Gossard, A. C. (1980). *Solid State Commun.* **36,** 43.
Ploog, K. (1980). In "Crystals; Growth, Properties and Applications," (H. C. Freyhardt, ed.), Springer-Verlag, Berlin and New York, Vol. 3, p. 73.
Poole, D. A., Pepper, M., and Glew, R. W. (1981). *J. Phys. C* **14,** L995.
Portal, J. C., Cisowski, J., Nicholas, R. J., Brummell, M. A., Razeghi, M., and Poisson, M. A. (1983). *J. Phys. C* **16,** L573.
Portal, J. C., Gregoris, G., Brummell, M. A., Nicholas, R. J., Razeghi, M., Poisson, M. A., Cheng, K. Y., and Cho, A. Y. (1984). *Surf. Sci.* **142,** 368.
Prasad, M., and Fujita, S. (1978). *Physica* **91A,** 1.
Price, P. J. (1982). *Surf. Sci.* **113,** 199.
Quinn, J. J., Strom, U., and Chang, L. L. (1983). *Solid State Commun.* **45,** 111.
Razeghi, M., and Duchemin, J. D. (1983). *J. Vac. Sci. Technol. B* **1,** 262.
Rezek, E. A., Holonyak, N., Jr., Vojak, B. A., Stillman, G. E., Rossi, J. A., Kenne, D. L., and Fairing, J. D. (1977). *Appl. Phys. Lett.* **31,** 288.
Sai-Halasz, G. A., Tsu, R., and Esaki, L. (1977). *Appl. Phys. Lett.* **30,** 651.
Sai-Halasz, G. A., Esaki, L., and Harrison, W. A. (1978a). *Phys. Rev. B* **18,** 2812.
Sai-Halasz, G. A., Pinczuk, A., Yu, P. Y., and Esaki, L. (1978b). *Solid State Commun.* **25,** 381.
Sai-Halasz, G. A., Chang, L. L., Welter, J. M., Chang, C.-A., and Esaki, L. (1978c). *Solid State Commun.* **25,** 935.
Sakaki, H. (1980). *Jpn. J. Appl. Phys.* **19,** L735.
Sakaki, H., Chang, L. L., Sai-Halasz, G. A., Chang, C.-A., and Esaki, L. (1978). *Solid State Commun.* **26,** 589.
Schirber, J. E., Fritz, I. J., Dawson, L. R., and Osburn, G. C. (1983). *Phys. Rev. B* **28,** 2229.
Schoolar, R. B. Zemel, J. N. (1964). *J. Appl. Phys.* **35,** 1848.
Schulman, J. N., and McGill, T. C. (1977). *Phys. Rev. Lett.* **39,** 1680.
Segmüller, A., and Blakeslee, A. E. (1973). *J. Appl. Crystallog.* **6,** 120.
Segmüller, A., Krishna, P., and Esaki, L. (1977). *J. Appl. Crystallogr.* **10,** 1.
Seidenbusch, W., Lindemann, G., Lassnig, R., Edlinger, J., and Gornik, E. (1984). *Surf. Sci.* **142,** 375.
Shank, C. V., Fork, R. L., Green, B. I., Weisbuch, C., and Gossard, A. C. (1982). *Surf. Sci.* **113,** 108.
Sollner, T. C. L. G., Goodhue, W. D., Tannenwald, P. E., Parker, C. D., and Peck, D. D. (1983). *Appl. Phys. Lett.* **43,** 588.
Störmer, H. L., Dingle, R., Gossard, A. C., Wiegmann, W., and Sturge, M. D. (1979). *Solid State Commun.* **29,** 705.
Störmer, H. L., Chang, A., Tsui, D. C., Hwang, J. C. M., Gossard, A. C., and Wiegmann, W. (1983a). *Phys. Rev. Lett.* **50,** 1953.
Störmer, H. L., Haavasoja, T., Narayanamurti, V., Gossard, A. C., and Wiegmann, W. (1983b). *J. Vac. Sci. Technol. B* **1,** 423.
Störmer, H. L., Schlesinger, Z., Chang, A., Tsui, D. C., Gossard, A. C., and Wiegmann, W. (1983c). *Phys. Rev. Lett.* **51,** 126.
Thouless, D. J. (1984). *Surf. Sci.* **142,** 14.
Tsang, W. T., Weisbuch, C., Miller, R. C., and Dingle, R. (1979). *Appl. Phys. Lett.* **35,** 673.
Tsu, R., and Esaki, L. (1973). *Appl. Phys. Lett.* **22,** 562.
Tsu, R., Chang, L. L., Sai-Halasz, G. A., and Esaki, L. (1975). *Phys. Rev. Lett.* **34,** 1509.
Tsui, D. C., and Gossard, A. C. (1981). *Appl. Phys. Lett.* **38,** 550.

Tsui, D. C., Englert, T., Cho, A. Y., and Gossard, A. C. (1980). *Phys. Rev. Lett.* **44,** 341.
Tsui, D. C., Störmer, H. L., and Gossard, A. C. (1982). *Phys. Rev. Lett.* **48,** 1559.
van der Ziel, J. P., Dingle, R., Miller, R. C., Wiegmann, W., and Nordland, W. A., Jr., (1975). *Appl. Phys. Lett.* **26,** 463.
Voisin, P., Bastard, G., Goncalves da Silva, C. E. T., Voos, M., Chang, L. L., and Esaki, L. (1981). *Solid State Commun.* **39,** 79.
Voisin, P., Guldner, Y., Vieren, J. P., Voos, M., Maan, J. C., Delescluse, P., and Linh, N. T. (1983). *Physica* **117–118B–C,** 634.
Voisin, P., Bastard, G., Voos, M., Mendez, E. E., Chang, C.-A., Chang, L. L., and Esaki, L. (1983). *J. Vac. Sci. Technol. B* **1,** 409.
Voisin, P., Dalalande, C., Voos, M., Chang, L. L., Segmüller, A., Chang, C.-A., and Esaki, L. (1984). *Phys. Rev. B* **30,** 2276.
von Klitzing, K., Dorda, G., and Pepper, M. (1980). *Phys. Rev. Lett.* **45,** 494.
Wanner, M., Doezema, R. E., and Strom, U. (1975). *Phys. Rev. B* **12,** 2883.
Washburn, S., Webb, R. A., Mendez, E. E., Chang, L. L., and Esaki, L. (1984). *Phys. Rev. B* **29,** 3752.
Weisbuch, C., Miller, R. C., Dingle, R., Gossard, A. C., and Wiegmann, W. (1981). *Solid State Commun.* **37,** 219.
White, S. R. and Sham, L. J. (1981). *Phys. Rev. Lett.* **47,** 879.
Witkowski, L. C., Drummond, T. J., Stanchak, C. M., and Morkoc, H. (1980). *Appl. Phys. Lett.* **37,** 1033.
Wood, C. E. C. (1980). In "Physics of Thin Films" (W. R. Hunter and G. Haas, eds.), Vol. 11, p. 35. Academic Press, New York.
Woodall, J. M. (1972). *J. Cryst. Growth* **12,** 32.
Worlock, J. M., Maciel, A. C., Petrou, A., Perry, C. H., Aggarwal, R. L., Smith, M., Gossard, A. C., and Wiegmann, W. (1984). *Surf. Sci.* **142,** 486.
Yoshino, J., Sakaki, H., and Hotta, T. (1984). *Surf. Sci.* **142,** 326.
Yoshioka, D., Halperin, B. I., and Lee, P. A. (1983). *Phys. Rev. Lett.* **50,** 1219.

5 DOPING (n–i–p–i) SUPERLATTICES

Gottfried H. Döhler and Klaus Ploog

Max-Planck-Institut für Festkörperforschung
Stuttgart, Federal Republic of Germany

I. INTRODUCTION

In previous chapters of this volume, the fascinating possibilities for tailoring the properties of artificial semiconductor superlattices have been discussed. It has been shown that the band gap may have any value between zero (semimetal) and those of familiar bulk semiconductors. Also, the distance between the subbands and their widths may be designed independently by an appropriate choice of the alloy composition of the components and the thickness of the constituent layers.

The subject of this chapter, the doping superlattice, is a periodic sequence of n- and p-doped layers that may have undoped (i–) zones of the same semiconductor material in between (n–i–p–i crystals). The electronic properties of these structures have been investigated theoretically (Döhler, 1972a,b, 1978, 1979; Döhler and Ploog, 1979) and the predicted peculiarities verified experimentally (Ploog *et al.*, 1981a,b; Döhler *et al.*, 1981b, 1982; Künzel *et al.*, 1982a,b; Jung *et al.*, 1982; Rehm *et al.*, 1983). These n–p junction or homojunction superlattices, in contrast to their previously discussed heterojunction counterparts, in principle, can be fabricated with any arbitrary semiconductor as the host material, since there are no restrictions on the choice of materials due to the requirements of lattice matching. More important than this technical advantage for the preparation of doping superlattices is a new dimension of flexibility of this class of superlattices. In these structures, band gap, subband

SYNTHETIC MODULATED STRUCTURES

ISBN 0-12-170470-X

structure, and carrier concentration are not only properties that can be predetermined by an appropriate choice of design parameters, but they become quantities that can be varied from outside over a wide range for a given sample. Thus, doping superlattices form a new class of semiconductors in which the electronic properties are no longer fixed material parameters but tunable quantities.

These unique features are consequences of an *indirect gap in real space* that results from a parallel modulation of the conduction and valence band edges by the periodic impurity space-charge potential. The spatial separation between electrons and holes is the origin of the strong increase in the electron–hole recombination lifetimes as compared with unmodulated bulk semiconductors. This lifetime enhancement depends on the period and the amplitude of the space-charge potential. Therefore, the excess carrier lifetime is a quantity that may vary over many orders of magnitude, depending on the choice of the design parameters. At very long recombination lifetimes, even large deviations from thermal equilibrium are metastable; this implies that the electron and hole concentrations become variable quantities. In addition to the lifetime enhancement, the spatial separation between the electrons and holes has another drastic effect. The space charge of excess electrons and holes partly screens the impurity space-charge potential. As a consequence, the effective band gap as well as the two-dimensional subband structure also become tunable. It is obvious that the strongly enhanced recombination lifetimes and the ability to vary such fundamental properties as band gap and carrier concentration result in a lot of peculiarities involving nearly all the electronic features of doping superlattices.

In the theoretical part of this chapter we shall first summarize the basic theory of self-consistent calculations of the electronic structure. Then we shall discuss some consequences of the unusual electronic structure of these semiconductors with indirect gaps in real space, with respect to their electrical transport properties, optical absorption, luminescence, and elementary excitations. The third section of the chapter is devoted to the key aspects of preparing GaAs doping superlattices by molecular beam epitaxy (MBE). In the fourth section the results of experiments on GaAs doping superlattices will be reviewed and compared with theoretical predictions. In our concluding remarks we shall discuss some other doping superlattices. In particular, we shall outline some extensions of the original concept (hetero-$n-i-p-i$, amorphous doping superlattices). We shall not talk about the wide variety of possible device applications, since this would lead us beyond the scope of this chapter.

II. THEORETICAL CONSIDERATIONS

A. Electronic Structure

The periodic potential in doping superlattices is exclusively space charge induced. This is in contrast to the compositional superlattices, in which the different band gaps of the components induce periodic variations of the band edges. For a theoretical treatment of the electronic structure, it is a particular advantage of doping superlattices that one is dealing with a homogeneous semiconductor that is only modulated by the superposition of a periodic superlattice potential. For most cases of interest, the effective mass approximation provides a very satisfactory description. The space-charge potential varies slowly over distances on the order of the bulk lattice constant. The considered energies in the bands are sufficiently close to the band edges such that the neglect of corrections due to nonparabolicity of the bands is justified.

Another simplifying approximation, which requires a more careful discussion (later in this chapter), is to replace the space-charge distribution of the individual doping atoms by homogeneous distribution functions $en_D(z)$ and $-en_A(z)$ for donors and acceptors, respectively, that vary periodically in superlattice direction z.

The contribution of the bare ionized impurities to the superlattice potential is then determined by the solution of Poisson's equation (see Fig. 1)

$$v_I(z) = \frac{4\pi e^2}{\kappa_0} \int_0^z dz' \int_0^{z'} dz'' \{n_D(z'') - n_A(z'')\}, \tag{1}$$

where κ_0 is the static dielectric constant of the bulk semiconductor. Except for the case of a compensated doping superlattice that contains the same amount of donors and acceptors per period d, i.e.,

$$\int_{-d/2}^{d/2} dz\, n_D(z) = \int_{-d/2}^{d/2} dz\, n_A(z), \tag{2}$$

the condition of (macroscopic) neutrality of the crystal requires a periodic electron or hole space-charge distribution $-en(z)$ or $ep(z)$, respectively. These mobile charges provide a Hartree contribution to the superlattice potential of

$$v_H(z) = \frac{4\pi e^2}{\kappa_0} \int_0^z dz' \int_0^{z'} dz'' \{-n(z'') + p(z'')\}. \tag{3}$$

The simplest way to take into account the exchange and correlation corrections of the free carriers is to use the local density approximation of the density functional formalism. This yields a contribution (Ruden and Döhler, 1983a) of

$$v_{xc}(z) \simeq v_x(z) \simeq 0.611(e^2/\kappa_0)(4\pi/3n(z))^{-1/3}. \tag{4}$$

The carrier distributions $n(z)$ and $p(z)$, of course, have to be calculated self-consistently. With these approximations the self-consistent, one-particle Schrödinger equation to be solved becomes

$$\{-(\hbar^2/2m_i)\nabla^2 + v_{sc}(z)\}\phi_{i,\mu,\mathbf{k}}(\mathbf{r}) = \epsilon_{i,\mu}(\mathbf{k})\phi_{i,\mu,\mathbf{k}}(\mathbf{r}). \tag{5}$$

The subscript i may stand for electrons c, light vl, or heavy holes vh or possibly

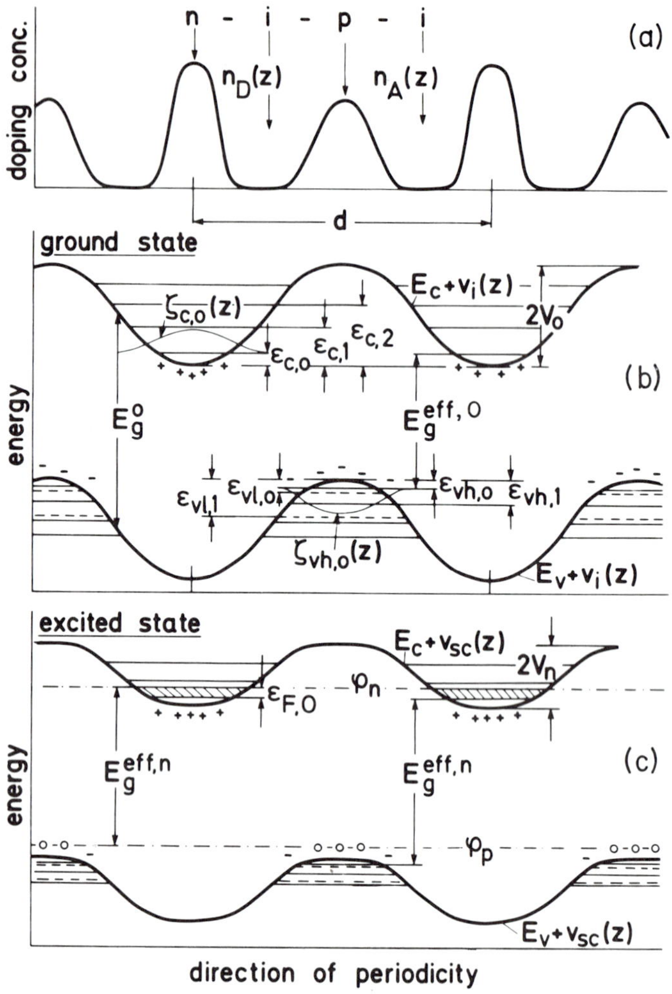

Fig. 1. (a) Periodic doping profile and real space band diagram in a compensated $n-i-p-i$ doping superlattice. (b) Ground state. Conduction and valence band edge are modulated by $v_I(z)$ from Eq. (1). (c) Excited state. The same superlattice with electrons in the n layers and holes in the p layers and separate quasi-Fermi levels ϕ_n and ϕ_p for electrons and holes, respectively. The effective band gap $E_g^{\text{eff},n}$ is larger than its value in the ground state, $E_g^{\text{eff},0}$. Plus signs indicate ionized donor levels near the conduction band edge and minus signs the negatively charged acceptor levels above the valence band edge.

for spin-orbit split-off holes vs. The index $\mu = 0, 1, 2, \ldots$, and the wave vector

$$\mathbf{k} = (\mathbf{k}_\parallel, k), \qquad \text{with} \quad |\mathbf{k}_\parallel| < |\mathbf{k}_{\parallel,\mathrm{BZ}}|, \qquad \text{and} \qquad -\pi/d < k_z \leqslant \pi/d \qquad (6)$$

are the quantum numbers of the subbands, similar to the previously discussed case of subbands in compositional superlattices. The self-consistent potential

$v_{sc}(z)$ is the sum of the bare ion contribution Eq. (1), the Hartree part Eq. (3), and the exchange and correlation potential Eq. (4)

$$v_{sc}(z) = v_I(z) + v_H(z) + v_{xc}(z). \tag{7}$$

If we neglect the mixing between different bands, caused by $v_{sc}(z)$, we may separate the quasi-free particle solution for propagation parallel to the doping layers, with kinetic energy $\hbar^2 k_\|^2/2m_i$, from the z-dependent motion in superlattice direction and write

$$\phi_{i,\mu,\mathbf{k}} = \exp(i\mathbf{k}_\| r_\|) u_{i,\mathbf{k}_0}(\mathbf{r}) \zeta_{i,\mu,k_z}(z), \tag{8}$$

where $u_{i,\mathbf{k}_0}(\mathbf{r})$ is the lattice-periodic part of the Bloch function of the i-band extremum at $\mathbf{k}_0$ and $\zeta_{i,\mu,k_z}(z)$ is the envelope function of the i, μ-subband with momentum k_z.

The Schrödinger equation reduces to

$$\{-\hbar^2/2m_i) d^2/dz^2 + v_{sc}(z)\} \zeta_{i,\mu,k_z}(z) = \epsilon_{i,\mu}(k_z) \zeta i,\mu,k_z(z) \tag{9}$$

and the subband energies become

$$\epsilon_{i,\mu}(\mathbf{k}) = (\hbar^2/2m_i) k_\|^2 + \epsilon_{i,\mu}(k_z). \tag{10}$$

In most cases of interest, it is justified to neglect the interaction between neighboring potential wells (multiquantum-well structures, see Chapter 3). This means that the k_z dependence of $\epsilon_{i,\mu}(k_z)$ in Eqs. (9) and (10) may be disregarded and the envelope function becomes

$$\zeta_{i,\mu,k_z}(z) = \sum_m \exp(ik_z md) \zeta_{i,\mu}(z - md), \tag{11}$$

where $\epsilon_{i,\mu}$ and $\zeta_{i,\mu}(z)$ are the eigenvalues and eigenfunctions for a single potential well.

A few general remarks about the space-charge potential and the solutions of Eq. (9) seem appropriate at this point.

(1) In order to give some idea of typical values we consider a compensated doping superlattice with equal thickness $d_n = d_p$ of the n- and p-doping layers, equal and constant doping concentration $n_D = n_A$, and zero thickness $d_i = 0$ of the undoped layers (see Fig. 1a). The periodic space-charge potential in this case contains $v_I(z)$ only, where $v_I(z)$ is parabolic and has an amplitude of

$$V_0 = (4\pi e^2/\kappa_0) n_D d_n^2/8. \tag{12}$$

Thus, V_0 is proportional to the doping concentration and to the square of the layer thickness. For GaAs ($\kappa_0 = 12.5$ at $T = 0$ K) with $n_D = n_A = 10^{18}\,\text{cm}^{-3}$ and $d_n = d_p = 50$ nm, for instance, the value of the amplitude is $V_0 = 905$ meV. From Fig. 1b we see that the effective energy gap is given by

$$E_g^{\text{eff},0} = E_g^0 - 2V_0 + \epsilon_{c,0} + |\epsilon_{vh,0}|, \tag{13}$$

where vh stands for heavy hole. The effective energy gap is strongly reduced below the value of the unmodulated bulk material E_g^0. (For the given example, $E_g^0 = 1.52$ eV.)

The subband energies and envelope wave functions are nearly exact solutions of the harmonic oscillator problem within the parabolic part of the respective potentials, with

$$\epsilon_{c,\mu} = \hbar(4\pi e^2 n_D/\kappa_0 m_c)^{1/2}(\mu + 1/2) \tag{14}$$

for the electrons and corresponding expressions for the light and heavy holes. For our present example, the subband separations are

$$\Delta\epsilon_c = 40.2 \text{ meV}, \qquad \Delta\epsilon_{vl} = 38.3 \text{ meV}, \qquad \text{and} \qquad \Delta\epsilon_{vh} = 17.3 \text{ meV},$$

respectively.

(2) If the thickness and/or the doping concentration of the n layers is increased, there will be a finite (two-dimensional) electron concentration in the n layers. Its value follows from the neutrality condition

$$n^{(2)} = n_D d_n - n_A d_p. \tag{15}$$

The spatial distribution of the electrons $n(z)$ depends on the wave functions of the populated subbands and the number of carriers in these subbands, i.e.,

$$n(z) = \sum_\mu n_\mu^{(2)} |\zeta_{c,\mu}(z)|^2. \tag{16}$$

The values of $n_\mu^{(2)}$ follow from the condition of equal Fermi levels in all populated bands,

$$\epsilon_{c,\mu} + (\hbar^2/2m_c)(2\pi n_\mu^{(2)}) = \epsilon_{c,0} + (\hbar^2/2m_c)(2\pi n_0^{(2)}) \tag{17}$$

and

$$\sum n_\mu^{(2)} = n^{(2)}, \tag{18}$$

if more than one subband is populated. The eigenfunctions $\zeta_{i,\mu}(z)$ are self-consistent solutions of Eqs. (9) and (11).

(3) The calculated value of E_g^{eff} from Eq. (13) would become negative for sufficiently strong doping and/or large layer thickness. For our example this will occur if $d_n = d_p$ exceeds 70 nm at $n_D = n_A = 10^{18}\,\text{cm}^{-3}$. In this case, Eq. (13) no longer holds, but the gap will be zero in the ground state and there will be electrons *and* holes in the n and p layers, respectively. This situation corresponds to the special case of a n-i-p-i semimetal.

1. Tunable Carrier Concentration

Figure 1 reveals an important feature of doping superlattices. The conduction subband wave functions are shifted by half a superlattice period with respect to those of the valence band. Therefore, the effective gap E_g^{eff} is an indirect gap in real space. Type II superlattices (see Chapter 3) also have such an indirect gap in real space. The spatial separation between electrons and holes, however, is less perfect in the latter case, and the consequences of the indirect gap are much less dramatic than in the doping superlattices.

The envelope wave functions of the lowest conduction and the uppermost valence subbands are indicated schematically in Fig. 1. Depending on the choice of the design parameters the electron–hole recombination lifetimes may become extremely long due to the small overlap between those states, as will be shown in Subsection B.1.

A consequence of such long lifetimes, exceeding those in familiar semiconductors by many orders of magnitude, is an unusually strong tunability of the carrier concentration. Very weak electron–hole generation rates or very small injection currents are required in order to induce large deviations of the steady-state carrier concentration from thermal equilibrium within the whole n–i–p–i crystal. Comparatively high nonequilibrium concentrations can be obtained only under extreme excitation conditions in conventional semiconductors.

The neutrality condition requires that the changes of the electron and the hole concentration are equal, since

$$n^{(2)} - p^{(2)} = \int_{-d/2}^{d/2} dz\,(n_{\mathrm{A}}(z) - n_{\mathrm{D}}(z)) = \mathrm{const} \tag{19}$$

also holds under nonequilibrium situations.

Equations (16)–(18) determine the subband population and the spatial distribution of the electrons also in the excited state of a doping superlattice. (We shall see later on that the situation is usually not exactly analogous for the holes in the p layers.)

2. Tunable Effective Band Gap and Subband Structure

The spatial separation between the electrons in the n layers and the holes in the p layers has, in addition to the lifetime enhancement, another unique effect, as mentioned in the introduction. The (negative) electron space charge compensates the (positive) donor space charge in the n layers and, vice versa, the hole space charge in the p layers. Thus, the self-consistent potential $v_{\mathrm{sc}}(z)$ in Eq. (9) flattens with increasing excess electron and hole concentration. This becomes evident in a more explicit way from inspection of Eqs. (1) and (3) for the bare impurity and the Hartree contribution to $v_{\mathrm{sc}}(z)$ [Eq. (7)].

It is clear that the effective band gap increases if the potential $v_{\mathrm{sc}}(z)$ becomes flatter as a function of increasing $n^{(2)}$ and $p^{(2)}$. The expression [Eq. (13)] for the gap in the ground state $E_{\mathrm{g}}^{\mathrm{eff},0}$ has to be generalized for the tunable effective band gap

$$E_{\mathrm{g}}^{\mathrm{eff},n} = E_{\mathrm{g}}^{0} - (v_{\mathrm{sc}}^{n}(d/2) - v_{\mathrm{sc}}^{n}(0)) + \epsilon_{\mathrm{c},0}^{n} + \epsilon_{\mathrm{vh},0}^{n}. \tag{20}$$

At this point we should note that there are some problems concerning a proper definition of the effective gap that are similar to those in doped bulk semiconductors. Instead of using Eq. (20), one might define an "optical gap," taking into account the Burstein shift (Burstein, 1954), due to the finite Fermi energies $\epsilon_{F,\mu}$ in the subbands. This definition, however, implies problems with respect to the momentum selection rule, which does not hold strictly. Finally, the effective band gap may be defined as the difference between the electron quasi-Fermi level

$$\phi_n = E_c + v_{sc}(0) + \epsilon_{c,0} + \epsilon_{F,0} \tag{21}$$

and the hole quasi-Fermi level ϕ_p that corresponds to the high-energy edge of the luminescence spectrum. It is a disadvantage of this definition that it does not represent a straightforward extension of familiar band gap conventions. In the next section we shall see that ϕ_p corresponds to the threshold of occupied acceptor impurity band states in typical $n-i-p-i$ crystals; i.e., on the low-energy side this definition of the gap

$$\tilde{E}_g^{\text{eff},n} = \phi_n - \phi_p = \Delta\phi_{np} \tag{22}$$

does not refer to normal valence (sub)band states, as one is used to.

The other important consequence of the carrier-concentration-dependent shape of the self-consistent superlattice potential $v_{sc}(z)$ is a decrease in the subband spacing as a function of increasing subband population. The decrease in the energy differences $|\epsilon_{c,\mu} - \epsilon_{c,\nu}|$ is a result of the flattening of the quantum well. A reduction by a factor of two to three can be achieved easily, as we shall learn from numerically calculated examples (Subsection II.A.4).

3. Influence of Discrete Impurity Potentials with Random Spatial Distribution

So far we have treated the space charge of the impurities as homogeneously smeared out in the respective doping regions. In doing so we have neglected (1) the spatial potential fluctuations that result from the random distribution of impurities in the doped regions and (2) the point charge character of the impurities that may lead to bound impurity states or to the formation of impurity bands. The significance of the deviations from the simplified homogeneous space-charge treatment has been discussed elsewhere (Ruden and Döhler, 1983a). The authors come to the conclusion that the random potential fluctuations are reasonably well screened by free carriers and that impurity band effects are usually negligible for the electrons. For holes, in contrast, it is found that they will normally populate an acceptor impurity band rather than valence subbands. This quite different behavior originates from the large difference between conduction electron and heavy-hole mass. This leads to a small donor binding energy, large subband distances, and Fermi energies in the former case, but a comparatively large acceptor binding energy and small heavy-hole subband distances in the latter one. It should be emphasized, however, that this argument applies mainly to typical direct gap III–V semiconductors. A discussion of the situation in indirect gap materials or in IV–VI semiconductors is beyond the scope of this contribution (Döhler and Ruden, 1984a).

The population of an acceptor impurity band implies a very simple but rather good approximation for the calculation of the self-consistent potential in the p layers (Ruden and Döhler, 1983a). The holes neutralize the acceptor potentials in the central part of the p layers. The width d_p^0 of the neutral zone, with $v_{sc}(z) = \text{const}$ for a given value of the hole concentration per p layer $p^{(2)}$, follows from

$$p^{(2)} = \int_{(d-d_p^0)/2}^{(d+d_p^0)/2} dz\, n_A(z), \tag{23}$$

whence

$$d_p^0 = p^{(2)}/n_A \tag{24}$$

in case of constant acceptor concentration.

The quasi-Fermi level for the holes becomes approximately

$$\phi_p \simeq E_v + v_{sc}(d/2) + E_A \tag{25}$$

if we assume that the Fermi level in the acceptor impurity band does not differ significantly from the isolated acceptor energy. In Eq. (25) E_v is the energy of the upper valence band edge at the center of the n-type layers.

4. Results of Self-Consistent Calculations

The first self-consistent calculations of the electronic subband structure in a $p-n-p$ three-layer structure system similar to the doping superlattice, as far as electrons are concerned, were reported by Döhler (1978). Exchange and correlation contributions were neglected in this work. At the same time results of similar calculations, for the ground state only and based on the Thomas Fermi approximation, were published by Ury and Holm-Kennedy (1978). A detailed study of the electronic structure of GaAs doping superlattices with different doping profiles has been carried out by Ruden and Döhler (1983a). In this study the influence of exchange and correlation interaction on the subband energies has also been investigated. The corrections have been found to increase the subband spacings by about 5 to 10%. Figure 2 shows the results for conduction subband edges $E_{c,\mu}$ and for the electron quasi-Fermi level ϕ_n as a function of the electron and hole concentration per layer $n^{(2)} = p^{(2)}$ for our former example of a GaAs doping superlattice without intrinsic layers and with $d_n = d_p = d/2 = 40$ nm, $d_i = 0$ nm, and $n_D = n_A = 10^{18}$ cm^{-3}. The reference energy is the quasi-Fermi level of the holes ϕ_p assumed at the position of a Be acceptor ($E_A = 28$ meV). The increase of the band gap from $E_g^{\text{eff},0} \simeq 0.95$ eV in the ground state to about $E_g^{\text{eff,max}} \simeq 1.5$ eV at the upper limit follows roughly the parabolic depen-

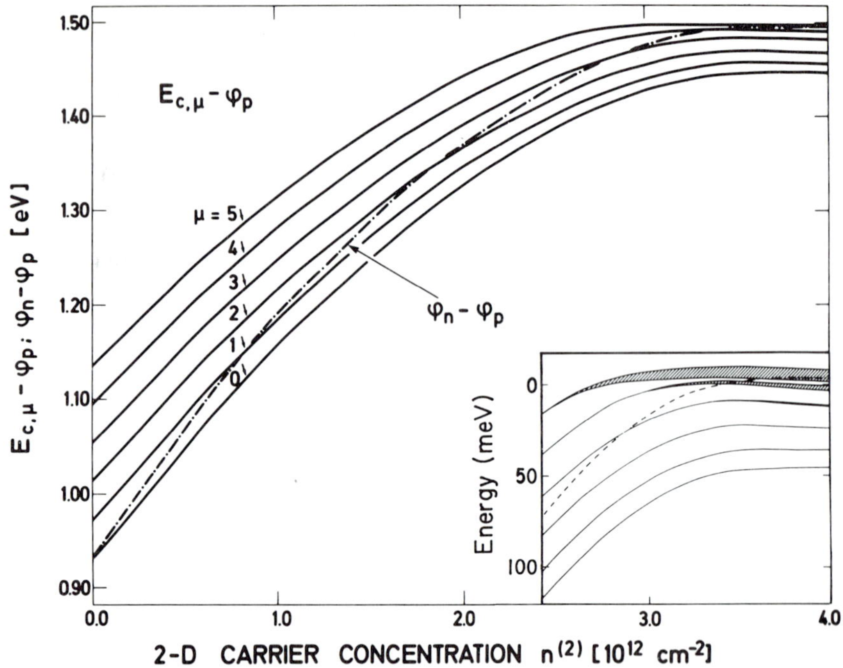

Fig. 2. Subband energies and quasi-Fermi level as a function of the electron concentration per period $n^{(2)}$ for a GaAs-doping superlattice. [From Ruden and Döhler (1983a).] The inset shows the same subbands for large values of $n^{(2)}$ on an expanded energy scale and with the maximum of the self-consistent potential chosen as zero.

dence that would be found from a classical treatment of the carriers neutralizing acceptors and donors. The maximum value of $n^{(2)} = p^{(2)} = 4 \times 10^{12}$ cm^{-2} in this case would correspond to a complete neutralization of the donors and the acceptors by the electrons and holes, respectively, with $E_g^{\text{eff,max}} = E_g^0$. Figure 2 also shows the decreasing subband distances and the increasing number of subbands populated (i.e., $\phi_n \geqslant E_{c,\mu}$) as a function of increasing $n^{(2)}$. The finite subband width due to the k_z dispersion becomes significant near zero energy, which corresponds to the classical free-particle threshold energy.

B. Peculiarities Resulting from the Unusual Electronic Structure

As a consequence of the tunable electronic structure, the electronic properties and characteristic material parameters are variable quantities in n–i–p–i superlattices. Apart from the ability to design a semiconductor with certain features, a property that compositional and doping superlattices have in common, the electrical and optical properties for a given n–i–p–i crystal may be modulated within a wide range of values. Because of this tunability n–i–p–i

crystals exhibit a lot of intriguing properties. Moreover, they represent a particularly interesting model substance for the study of dynamically two-dimensional many-body systems with variable carrier concentration. For the same reason, they also form a fascinating basic material for novel electronic and optoelectronic devices. In the following we will omit the discussion of properties that are qualitatively similar to those of heterojunction superlattices and restrict our considerations completely to properties that are specific to doping superlattices.

1. Lifetimes

In Subsection II.A the extremely long recombination lifetimes of excess electrons and holes were mentioned briefly as the origin of metastable deviations from thermal equilibrium, which, in turn, are a basic prerequisite for the tunability of band gap and carrier concentration. The direct recombination of an electron in the n layer with a hole in the p layer, whether radiative or nonradiative, e.g., by an Auger process, depends on the square of the overlap integral between the corresponding envelope wave functions (Döhler, 1972b). A detailed theoretical study of electron–hole recombination lifetimes in $n-i-p-i$ superlattices has not yet been performed. It is, however, possible to calculate analytically the overlap integral between the lowest conduction and the uppermost light-hole valence subband for the case of a compensated $n-p$ doping superlattice with constant values of n_D and n_A in the respective layers. It is a reasonably good approximation to use the appropriate harmonic oscillator ground-state wave functions.

$$\zeta_{i,0}(z) = \pi^{-1/4}\alpha_i^{-1/2} \exp\{-z^2/2\alpha_i^2\} \tag{26}$$

with

$$\alpha_c = \hbar^{1/2}(m_c\omega_c)^{-1/2} \quad \text{and} \quad \alpha_{vl} = \hbar^{1/2}(m_{vl}\omega_{vl})^{-1/2}, \tag{27}$$

where

$$\omega_c = (4\pi e^2 n_D/\kappa_0 m_c)^{1/2} \quad \text{and} \quad \omega_{vl} = (4\pi e^2 n_A/\kappa_0 m_{vl})^{1/2} \tag{28}$$

are the electron and light-hole (vl) bulk-plasma frequencies.

The square of the overlap integral is a measure for the reduction factor for the electron–hole recombination processes as compared to the corresponding unmodulated bulk values. In our case, we find

$$\begin{aligned}(\tau_{rec}^{nipi}/\tau_{rec}^{bulk})^{-1} &\simeq |\langle c, 0|vl, 0\rangle|^2 \\ &= [2\alpha_c\alpha_{vl}/(\alpha_c^2+\alpha_{vl}^2)] \exp[-(d/2)^2/(\alpha_c^2+\alpha_{vl}^2)].\end{aligned} \tag{29}$$

This result demonstrates that the lifetime enhancement depends on an exponential scale on the design parameters.

For our previous example ($n_D = n_A = 10^{18}$ cm^{-3}, $d_n = d_p = d/2 = 40$ nm,

$m_c = 0.067\ m_0$, $m_{vl} = 0.074\ m_0$, $\kappa_0 = 12.5$ for GaAs), we obtain

$$\tau_{rec}^{nipi}/\tau_{rec}^{bulk} \simeq 10^{13}. \tag{30}$$

Thus, the lifetime for a luminescent recombination is increased from the nanosecond range into the kilosecond range. If, however, the superlattice period is reduced by a factor of two, one finds

$$\tau_{rec}^{nipi}/\tau_{rec}^{bulk} \simeq 2 \times 10^3, \tag{31}$$

which amounts to optical lifetimes in the microsecond range. This example illustrates that any arbitrary lifetime between typical bulk values and quasi-infinite lifetimes may be achieved by appropriate tailoring within the range of easily accessible experimental design parameters.

It is clear that the recombination lifetimes for a given $n-i-p-i$ structure are, again, strongly tunable quantities. The overlap between electron and hole subbands increases if the effective gap $E_g^{\text{eff},n}$ increases. This is a consequence of the flattening of the self-consistent potential with increasing electron and hole concentration (see Fig. 1).

The lifetime enhancement as a function of the effective gap can be obtained from the overlap of the numerically calculated wave functions of the populated subbands. The following analytical, although rather crude, estimate provides some idea about the tremendous change in lifetime that can be observed in a given sample. We consider again, for the sake of simplicity, the case $n_D = n_A$ and $d_n = d_p$. We approximate the self-consistent potential by flat central portions of width

$$d_n^0 = n^{(2)}/n_D \tag{32}$$

and

$$d_p^0 = p^{(2)}/n_A, \tag{33}$$

respectively, and parabolically shaped bare impurity contributions of width

$$d_n^+/2 = d_p^-/2 = (d_n - n_D/n^{(2)})/2 \tag{34}$$

for the region in between.

The value from Eq. (34) inserted into Eq. (12) instead of d_n provides an estimate of $2V_n = v_{sc}(d/2) - v_{sc}(0)$ and via Eq. (20) of $E_g^{\text{eff},n}$ at finite carrier concentration. We assume that the self-consistent wave functions in the range $d_n^0 < z < d/2 - d_p^0/2$ do not differ too much from those in the ground state of a similar doping superlattice with d_n^+ and d_p^- from Eq. (34) as doping layer thicknesses. Then Eq. (29) with d_n^+ inserted for $d/2$ can be used as an estimate of the lifetime enhancement for the excited stated. If we neglect the difference between m_c and m_{vl} we can rewrite Eq. (29) by using our estimate for $2V_n$ and Eqs. (27) and (28) as

$$\tau_{rec}^{nipi}/\tau_{rec}^{bulk} \simeq \exp[2V_n/(\hbar\omega_c/2)]. \tag{35}$$

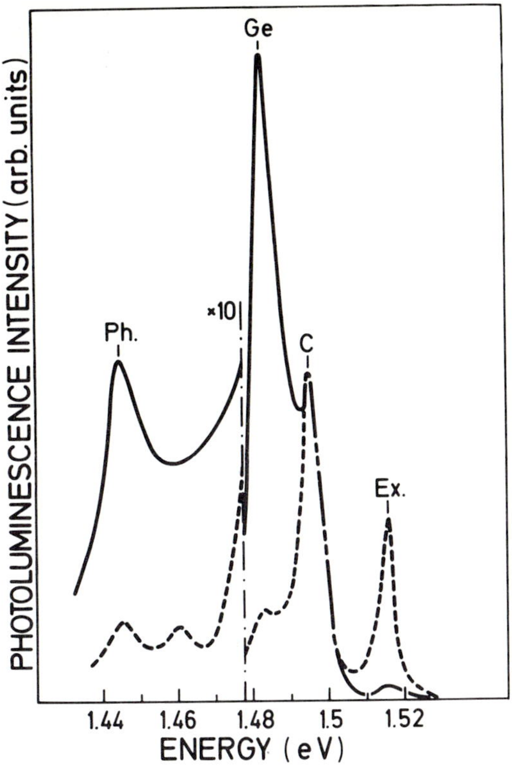

Fig. 16. Photoluminescence spectra (measured at 1.6 K, intensity 10 mW cm^{-2} of two Ge-doped n–GaAs films grown at different substrate temperatures, indicating the enhanced amount of Ge acceptors incorporated on As sites with increased substrate temperature. Dotted line: $T_S = 500°C$ (sample 1029). Solid line: $T_S = 595°C$ (sample 1033).

hole concentration at the $p-p^+$ transition was achieved by a step increase of the Be- or Ge-cell temperature.

The properties of these (Be, Si)- and (Ge, Ge)-doped $p-n$ junctions as derived from detailed $I-V$ and $C-V$ measurements on isolated mesa diodes are summarized in Fig. 17 (Ploog *et al.*, 1981a). The solid lines represent the theoretical values of breakdown voltage and depletion layer width for one-sided abrupt junctions as a function of carrier concentration in the p region as calculated by Sze and Gibbons (1966) under the assumption that avalanche multiplication dominates the breakdown characteristic. There is a good agreement between our experimental data and theory except for the sample labeled by double circles, where the 1-μm-thick p-type side of the junction is already totally depleted at 16 V. The average diode n values derived from the slope of the linear fit of the ln(I) versus V curve for $V \geq 3\ kT/e$ were found to be 1.8 ± 0.2 for all measured diodes indicating that the current flow across the

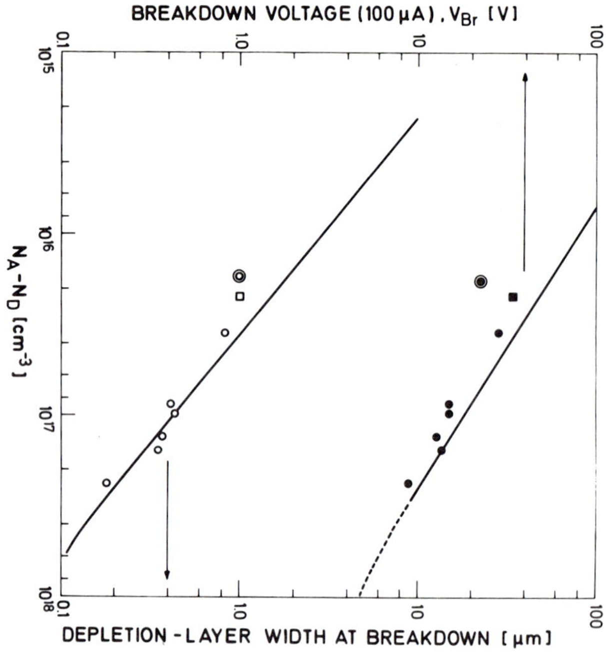

Fig. 17. Comparison of measured and calculated [solid lines (Sze and Gibbons, 1966)] characteristics of one-sided abrupt $p-n^+$ junctions in MBE-grown GaAs layers doped with (Be, Si) or (Ge, Ge) having different doping levels in the 1-μm-thick *p* region. The data points are derived from $I-V$ and $C-V$ measurements.

junction is dominated by recombination–generation processes (Sze, 1969).

The encouraging results obtained from isolated $p-n$ junctions in MBE-grown GaAs made feasible the fabrication of GaAs doping superlattices by combining many of such $p-n$ junctions in the same growth run. Figure 18 displays scanning electron micrographs of the (110) cleavage plane of three different periodic $n-p$ multilayer structures in GaAs grown by MBE. In Fig. 18a,b a total of 20 alternating *n*(Si)- and *p*(Be)-doped GaAs layers corresponding to 10 periods are clearly resolved. The thickness of the *i* layers in these samples was chosen to be close to zero by opening the Si shutter immediately after the Be shutter has been closed and vice versa.

Inspection of Fig. 18c reveals, however, that the use of Ge for *n*- as well as *p*-type doping of MBE GaAs in these doping multilayer structures has been less successful. On top of the 12 clearly observable alternate $n-p$ GaAs layers, 4 additional $n-p$ layers were deposited in the same growth run, which are no longer resolved due to the increasingly unstable Ga (4×2) growth conditions during growth of the *p*-type layers (Ploog *et al.*, 1979). Growth had to be stopped after 16 layers because of the onset of three-dimensional growth of Ga droplets.

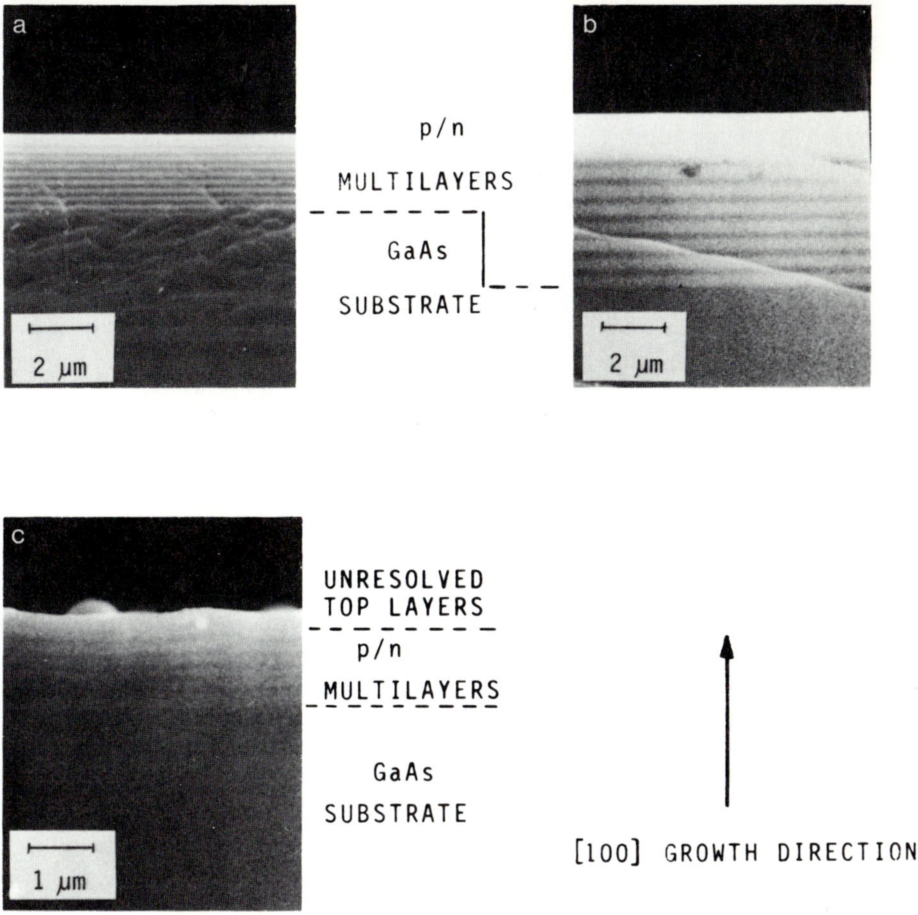

Fig. 18. Scanning electron micrographs of (110) cleavage plane of different periodic n–p doping multilayer structures in GaAs; (a) (Si, Be)-doped with $d_n = d_p = 0.12\ \mu$m; (b) (Si, Be)-doped with $d_n = d_p = 0.27\ \mu$m; (c) (Ge, Ge)-doped with $d_n = 0.16\ \mu$m and $d_p = 0.04\ \mu$m. No intrinsic (i) layer was interspersed between the intentionally doped layers.

In this particular structure, the n-type layers were obtained at an As_4/Ga flux ratio greater than 2, leading to the As-stabilized (2 × 4) surface structure, and the p-type material was grown at an As_4/Ga flux ratio of 1 resulting in a Ga-stabilized (4 × 2) surface structure. The width of the Ge incorporation on donor or acceptor sites was accurately determined by the time interval of placing the aperture into the As_4 beam path. It was monitored in the RHEED pattern. The occurrence of a wavelike bending in n–p GaAs *multilayer* structures, doped with Ge only, could also not be suppressed completely, when we performed the required variation of As surface population via a change of the substrate temperature from 530°C for n-type growth to 625°C for p-type growth at a constant As_4/Ga flux ratio of 2. Therefore, at present, only the application of Si as donor and Be as acceptor impurities yields a uniform doping within each individual layer and a controlled thickness with constant periodicity also for a large number of layers in sequence. Only in this case, the transition regions (interfaces) between the individual GaAs layers remain extremely smooth down to atomic steps also over large substrate areas (Ploog *et al.*, 1981a).

Ten periods of the doping multilayer structures as shown in Fig. 18a,b are sufficient to produce the novel superlattice effects in this material. In order to study these effects, we have now grown a large number of (Si, Be)-doped GaAs doping superlattices, in which we varied the layer thicknesses d_n, d_p, and d_i ($0 \leq d_i \leq 0.1$ μm) and thereby the periodicity d as well as the individual doping levels over wide ranges according to our theoretical considerations in Section II.

Finally, the question of the ultimate abruptness of the constituent (Be, Si)-doped $p-n$ junctions in these GaAs doping superlattices arises. Since the depth resolution combined with the required high elemental sensitivity of the available profiling techniques (e.g., SIMS) is not yet sufficient, we have combined the results obtained from $C-V$ profiling measurements of periodic doping profiles in n−GaAs:Si and in p−GaAs:Be, in order to derive the expected free-carrier profile of an abrupt $p-n$ junction in (Be, Si)-doped GaAs by point-by-point substraction of actually measured profiles (Ploog *et al.*, 1981a). With a final free-carrier concentration reaching 1.5×10^{17} cm^{-3} on both sides of the junction (not stable because of recombination!), the width necessary to change the free-carrier concentration from n-type 1×10^{17} cm^{-3} to the same level p type was found to be approximately 250 Å. In this simulation procedure, the problems arising from $C-V$ profiling measurements have not been taken into account. Since the measured free-carrier profile may differ by several Debye lenths [$\lambda_{\text{Debye}} \sim 450$ Å for $n = 1 \times 10^{16}$ cm^{-3} (Kennedy *et al.*, 1968)] from the actual *dopant* profile due to majority carrier diffusion (Johnson and Panousis, 1971), this value of 250 Å represents an upper limit, and the width required to change the actual dopant impurities Si and Be is much smaller. Furthermore, our experiments have shown that any noticeable interdiffusion of the dopants Si and Be does not occur at growth temperatures below 600°C.

IV. EXPERIMENTAL RESULTS

Experimental investigations of $n-i-p-i$ crystals were only started in 1980 when GaAs-doping superlattices grown by MBE became available. Within this short period the fundamental predictions of the theory, many of them made nearly 10 years earlier, could be verified. Also $n-i-p-i$ crystals were grown with semiconductors other than GaAs. Silicon-doping superlattices, for example, were grown by Sakamato (Shiraki, 1982) and PbTe-doping superlattices by Kinoshita *et al.* (1981). Detailed experimental studies of the electronic properties of those structures, however, have not yet been reported. The same is true for GaAs $n-i-p-i$ structures grown by other groups.

A. Tunable Bipolar Conductivity

The first experimental test of the tunability of the electron and the hole concentrations $n^{(2)}$ and $p^{(2)}$ in the n- and p-doped layers was made by measurements of the electron and hole conductivities $\sigma_{nn}(\Delta\phi_{np})$ and $\sigma_{pp}(\Delta\phi_{np})$ parallel to the layers. The modulation of $n^{(2)}$ and $p^{(2)}$ was achieved by injection and extraction of nonequilibrium carriers via selective electrodes and by photoexcitation as well.

1. Modulation by Selective Electrodes

The theoretical concept of selective electrodes to doping superlattices has been mentioned briefly in Subsection II.B.3.a. A region of strong n doping, penetrating all the doping layers from the top down to the bottom of the $n-i-p-i$ structure, forms ohmic contacts to all the n layers, whereas it provides blocking $n-p$ junctions with respect to the p layers. In an analogous manner, a strongly p-doped region represents a selective p-type contact. Ploog *et al.* (1981b) tested the feasibility of this concept by alloying small Sn or Sn/Zn balls from the surface as n^+ and p^+ electrodes, respectively, on GaAs-doping superlattices. They found that, indeed, perfect ohmic or excellent blocking contacts were achieved to the respective layers by this procedure. The results of layer conductance as a function of carrier injection and extraction in a $n-i-p-i$ semimetal (see Subsection I.A) obtained by Ploog *et al.* (1981b) are shown in Fig. 19 as an example. The electron and hole concentration in the crystal changes by injection or extraction of charge carriers if an external bias U_{np} is applied between selective n and p contacts until the quasi-Fermi level difference $\Delta\phi_{np} = \phi_n - \phi_p$ equals the external potential eU_{np} [Eq. (38)]. The relation between $n^{(2)}$, $p^{(2)}$, and $\Delta\phi_{np}$ in this $n-i-p-i$ semimetal differs quantitatively from the example shown in Fig. 2 by the negative value of $\Delta\phi_{np}^{\text{th}} = eU_{np}^{\text{th}}$ at which the hole concentration $p^{(2)}$ becomes zero. From the design parameters, together with Eq. (19), it follows that the electron concentration per layer $n^{(2)}$ exceeds the hole concentration $p^{(2)}$ by about $n_0^{(2)} = 2.1 \times 10^{12}\ \text{cm}^{-2}$. Thus, the condition of macroscopic charge neutrality requires a residual electron concentration of $n_0^{(2)} = 2.1 \times 10^{12}\ \text{cm}^{-2}$ if the hole layers are completely depleted at external voltages U_{np} below the threshold U_{np}^{th}. For $eU_{np} > eU_{np}^{\text{th}}$ one has $n^{(2)}(eU_{np}) = n_0^{(2)} + p^{(2)}(eU_{np})$.

The full lines in Fig. 19b correspond to the behavior expected from Eqs. (36) and (37) with the mobilities taken as constant fitting parameters to the experimental data points. The agreement between theory and experiment is good.

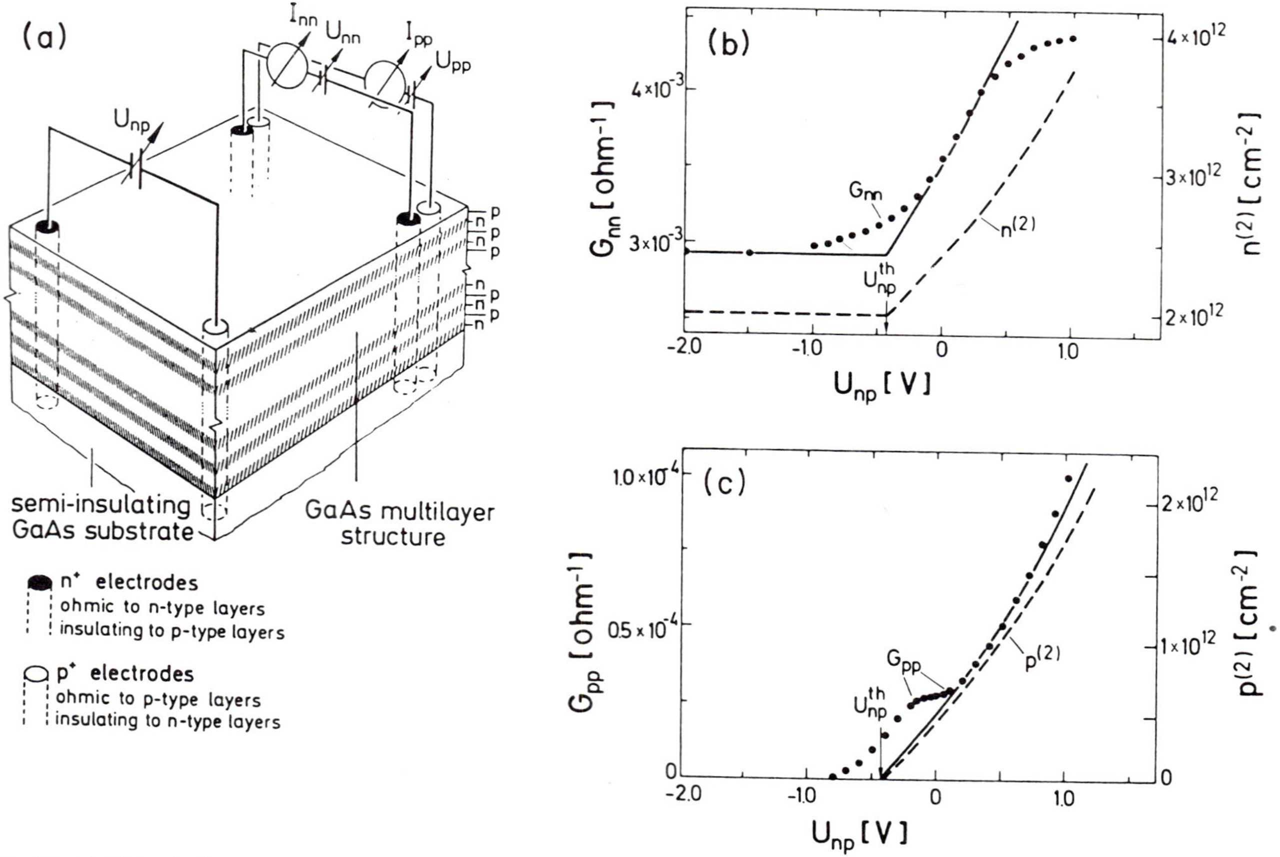

Fig. 19. (a) GaAs-doping superlattice provided with selective n^+ and p^+ electrodes for the modulation of the conductivities $\sigma_{nn}(\Delta\phi_{np})$ and $\sigma_{pp}(\Delta\phi_{np})$ in the respective layers. (b) and (c) Comparison between experimentally observed conductance G_{nn} and G_{pp} as a function of external potential $eU_{np} = \Delta\phi_{np}$ (filled circles) and calculated values (solid lines) with uniform mobility used as fitting parameter. Temperature 4.2 K, $d_n = d_p = 1900$ Å, $n = 3.0 \times 10^{17}$ cm^{-3}, and $p = 1.9 \times 10^{17}$ cm^{-3}.

Deviations of the experimental results from the ideal behavior were attributed mainly to the fact that the top p layer, which was not sandwiched between n layers, exhibited a threshold voltage, different from the p layers in the $n-i-p-i$ "bulk". Experiments on structures with improved design exhibit excellent agreement with calculated results (Künzel *et al.,* 1984). These results also demonstrate that the tunability via selective electrodes is not significantly affected by interlayer recombination currents j_{np} even if the doping concentration is increased into the 10^{18}-cm^{-3} range and d_n and d_p is decreased to below 100 nm.

2. Modulation by Photoexcitation

Whereas deviations from the ground state induced by an external bias include both carrier injection and extraction in the case of $n-i-p-i$ semimetals, it is clear that only $\Delta n^{(2)}, \Delta p^{(2)} > 0$ is possible in the case of photoexcitation. The advantage of photoconductivity experiments is a good homogeneity that can be achieved even at high excitation level. The interlayer recombination currents increase with $\Delta\phi_{np}$ because of the decreasing electron–hole lifetimes τ_{rec}^{nipi}, as discussed in Subsection II.B.1. As a consequence, the value of $\Delta\phi_{np}$ may become considerably less than eU_{np} in samples of large dimensions, and it may vary with the distance from the injecting electrodes, in particular, if eU_{np} approaches the value of E_g^0. The photoconductive response under homogeneous illumination of the sample, in contrast, only depends on the local balance between carrier generation and recombination rate. Thus, $\Delta n^{(2)} = \Delta p^{(2)}$ can be constant in the whole $n-i-p-i$ crystal, provided that the light intensity does not decrease significantly between the top and the bottom layers by absorption.

The photoconductive response was used for photo-Hall studies and for measurements of the tunable electron–hole recombination lifetime $\tau_{\text{rec}}^{nipi}(\Delta\phi_{np})$ by Künzel *et al.* (1982a), Döhler *et al.* (1982), and Ploog and Künzel (1982). The mobility obtained from photo-Hall measurements was found to increase as a function of carrier concentration (Künzel *et al.,* 1982a). It was somewhat lower than in bulk samples of the same doping level. This result is expected, since the carriers experience additional scattering by less perfectly screened impurity atoms in the uncompensated regions of the doping layers. The population of the conduction subbands was made too high in the studied samples in order to allow for the observation of two-dimensional subband effects on the mobility.

The decay of the photoconductive response $d\sigma_{nn}^{(2)}/dt$ after illumination was used for a determination of the recombination lifetimes. The relation between $\sigma_{nn}^{(2)}$, $n^{(2)}$, and $\Delta\phi_{np}$ allows us to deduce

$$dn^{(2)}/dt = n^{(2)}(\tau_{\text{rec}}^{nipi})^{-1} \tag{64}$$

as a function of $\Delta\phi_{np}$ directly from $d\sigma_{nn}^{(2)}/dt$. An example of τ_{rec}^{nipi} as a function of $\Delta\phi_{np}$ obtained in this way is displayed in Fig. 20. Note that the lifetimes are, indeed, in the kilosecond range (!) for $\Delta\phi_{np} \simeq 0.2$ eV.

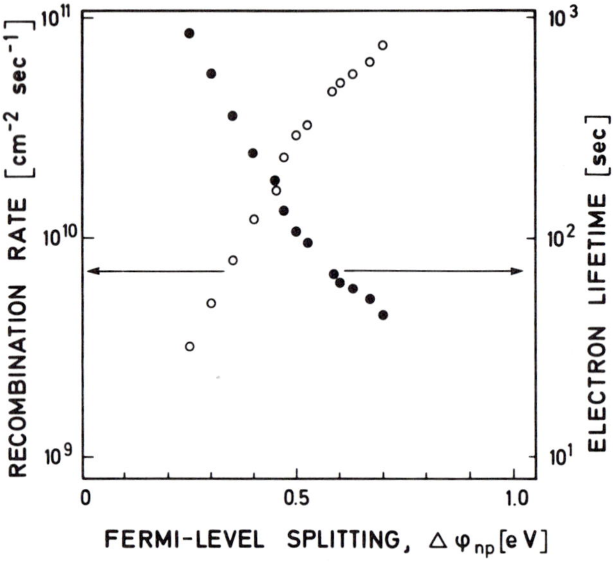

Fig. 20. Electron–hole recombination lifetimes as a function of quasi-Fermi level difference $\Delta\phi_{np}$ as deduced from the decay of photoconductivity for the sample shown in Fig. 19. [From Ploog and Künzel (1982).]

B. Tunable Absorption Coefficient

A very elegant way to measure the tunable absorption coefficient $\alpha^{nipi}(\omega; \Delta\phi_{np})$, discussed in Subsection II.B.4, is again provided by the giant photoconductive response of $n-i-p-i$ crystals. This method can be used conveniently within a range of $\Delta\phi_{np}$ values for which the generation rate, given by Eq. (40), is not much smaller than the recombination rate [Eq. (39)]. In the ground state a rather large photoresponse of $\Delta n^{(2)} > 10^{11}$ cm^{-2} is still obtained by a moderate excitation intensity of 1 W cm^{-2}, e.g., for an absorption coefficient as small as $\alpha(\omega; \Delta\phi_{np} = 0) \simeq 10^{-6}$ cm^{-1}(!) if $\tau_{\text{rec}}^{nipi} > 10^3$ s. This result demonstrates the extremely high sensitivity of this method.

So far, measurements of $\alpha^{nipi}(\omega; \Delta\phi_{np})$ have been performed only in doping superlattices with large period and relatively low doping level (Döhler *et al.*, 1982). In Fig. 21 the results of $\alpha^{nipi}(\omega; \Delta\phi_{np} = 0)$ as obtained from an analysis of $d\sigma_{nn}/dt$ under illumination with light of variable photon energy $\hbar\omega$ (assuming a quantum efficiency of unity) are compared with the theoretical curve. The agreement is very good over the full range. It should be noted that the reported results cover only a small fraction of the experimentally accessible part of the exponential absorption tail, since values down to $\alpha \simeq 10^{-6}$ cm^{-1} should be

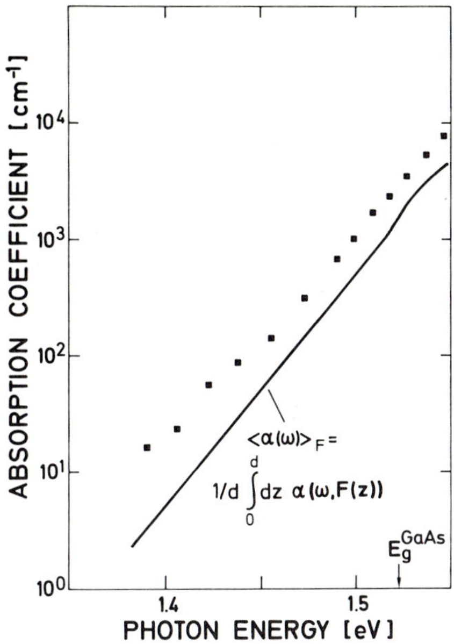

Fig. 21. Absorption coefficient in the ground state of a GaAs-doping multilayer superlattice where $T = 4.2$ K, $F_{max} = 2.2 \times 10^5$ V/cm, $d_n = d_p = 190$ nm, $n_D = 3 \times 10^{17}$ cm^{-3}, and $n_A = 1.9 \times 10^{17}$ cm^{-3}. Experimental points are obtained from the photoconductive response. Solid line calculated with the design parameters. [From Döhler *et al.* (1982).]

easily measurable, as we have just estimated. Döhler *et al.* (1982) also found that the variation of α^{nipi} as a function of $\Delta\phi_{np}$ agrees very well with the theory.

Investigations of the steplike structure of α^{nipi} as a function of ω or of the oscillatory behavior as a function of $\Delta\phi_{np}$, as discussed in Subsection II.B.4, have not yet been reported.

C. Tunable Luminescence

The observation of tunable luminescence due to radiative electron–hole recombination has provided the most striking evidence for the existence of a tunable band gap in $n-i-p-i$ crystals.

1. Photoluminescence

Steady-state photoluminescence studies have been performed on GaAs-doping superlattices with doping concentrations in the range from n_D, $n_A \simeq 10^{18}$ to 4×10^{18} cm^{-3} and layer thicknesses between 20 and 60 nm (Döhler *et al.*, 1981b; Jung *et al.*, 1982, 1983). Figure 22 displays a typical series

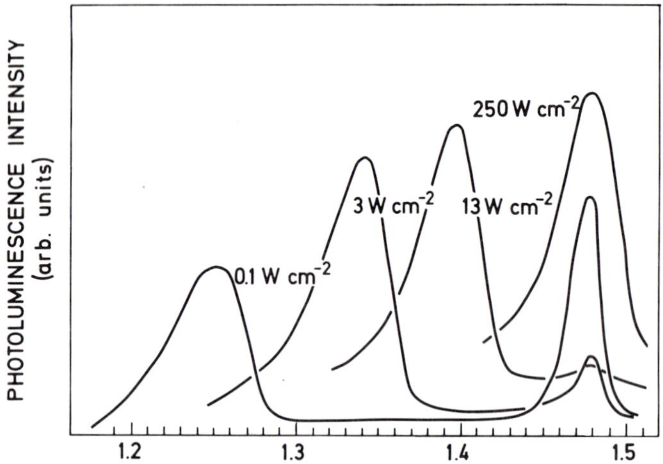

Fig. 22. Photoluminescence spectra of a GaAs-doping $n-i-p-i$ superlattice ($T = 2$ K, $d_n = d_p = 40$ nm, $n = p = 10^{18}$ cm^{-3}, and $\hbar\omega_{exc} = 1.916$ eV) at different excitation intensities showing the tunable, broad, and asymmetric bands due to recombination across the indirect gap in real space and the narrower near band edge emission line. (Reprinted with permission from *Solid State Commun.* **43,** Jung *et al.*, Copyright, 1982, Pergamon Press, Ltd.)

of PL spectra obtained at different excitation intensities I^{exc}. The broad asymmetric PL band that shifts by more than 200 meV as a function of excitation intensity is attributed to the specific $n-i-p-i$ luminescence mechanism originating from electron–hole recombination across the tunable indirect gap in real space. The shape of these PL bands corresponds to a broadened version of the steplike structured idealized spectra from Eq. (54) (see Fig. 9). The main reasons for this broadening are probably (1) the finite width of the acceptor impurity band, (2) potential fluctuations due to ionized impurities in the space-charge region, and (3) inhomogeneous excitation of the different $n-i-p-i$ layers due to the finite penetration depth of the laser light.

The dependence of the PL peak position on excitation intensity is shown for a series of samples with different design parameters in Fig. 23. In all the cases, it is found that the photoluminescence spectrum shifts roughly linearly to higher energies with the logarithm of the excitation intensity I^{exc}. This finding is in good agreement with the predicted exponential dependence between lifetime τ_{rec}^{nipi} and amplitude V_n of the superlattice potential. [See Eq. (35).] In the steady state the generation rate, which is proportional to I^{exc}, is balanced by the recombination rate

$$\dot{n}_{rec}^{(2)} = n^{(2)}(\tau_{rec}^{nipi})^{-1} \simeq n^{(2)}(\tau_{rec}^{bulk})^{-1} \exp[4V_n/\hbar\omega_c]. \qquad (65)$$

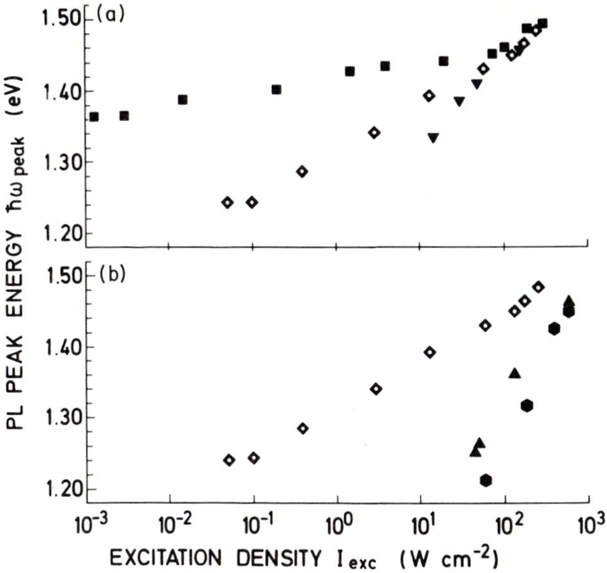

Fig. 23. Energies of the photoluminescence peak $\hbar\omega_{peak}$ as a function of excitation intensity I_{exc} for two sets of samples with different design parameters ($\hbar\omega_{exc} = 1.916$ eV). (a) Constant doping concentration ($n_D = n_A = 10^{18}$ cm^{-3}), variation of superlattice period. (■, $d_n = d_p = 20$ nm; ◇, $d_n = d_p = 40$ nm; ▼, $d_n = d_p = 60$ nm.) (b) Constant layer thickness ($d_n = d_p = 40$ nm), variation of doping concentration (◇, $n_D = n_A = 10^{18}$ cm^{-3}; ▲, $n_D = n_A = 2 \times 10^{18}$ cm^{-3}; ⬢, $n_D = n_A = 4 \times 10^{18}$ cm^{-3}.)

Thus, the relation

$$E_g^0 - \hbar\omega^{PL} \simeq 2V_n \propto \ln I^{exc} \tag{66}$$

follows.

The more symmetrical and narrower luminescence line in the PL spectra (see Fig. 22) with an almost intensity-independent position at about $\hbar\omega \simeq 1.48$ eV *cannot* be attributed to the luminescence of the semi-insulating substrate, since it is also observed with similar intensity in very thick doping superstructures of 2-μm thickness grown on either semi-insulating or n^+-doped substrates (Jung *et al.*, 1983). Therefore, it has to be interpreted as being due to vertical recombination of electrons at the (local) band edge [or, more probably, in shallow (residual) donors in the *p* layers] with holes in the acceptor impurity band, although the relative intensity of this luminescence line is considerably higher than expected from simple relaxation and recombination kinetics of the photoexcited carriers (Jung *et al.*, 1983). The peak position of this line agrees reasonably well with $\hbar\omega \simeq E_g^0 - E_A$. It is also compatible with $\hbar\omega \simeq E_g^0 - E_A - E_D + e^2/\kappa_0 r_0$, i.e., the energy for DA pair luminescence (r_0 stands for the average nearest acceptor distance). The other vertical process discussed in Subsection II.B.5.a, the electron–hole recombination in the n layers, has not been observed.

The observation of the decay of the luminescence by time-resolved measurements provides further support for the correctness of our interpretation of the tunable luminescence signal. As the carrier concentration decreases, the effective band gap $E_g^{\mathrm{eff},n}$ and the Fermi-level difference $\Delta\phi_{np}$ shrink because of the relation between these quantities and $n^{(2)}$ (see Fig. 2). Figure 24 shows an example for the shift of the luminescence spectra as a function of delay time after excitation (Rehm *et al.*, 1983). The relation between $\Delta\phi_{np}$ and $n^{(2)}$ allows us to deduce the total decay rate $\dot{n}^{(2)}$ and hence the total lifetime according to $\dot{n}^{(2)} = n^{(2)}(\Delta\phi_{np})(\tau_{\mathrm{rec}}^{nipi}(\Delta\phi_{np}))^{-1}$, directly from the observed shift of the luminescence signal as a function of delay time Δt, $d(\hbar\omega^{\mathrm{peak}})/d(\Delta t)$.

Total lifetimes of $\tau_{\mathrm{tot}}^{nipi} > 0.1$ msec were found in this way for $\Delta\phi_{np} < 1.2$ eV in the investigation by Rehm *et al.* (1983). The observed exponential increase of the lifetimes with decreasing effective band gap once more confirms the theoretically expected relation between $\tau_{\mathrm{rec}}^{nipi}$ and $E_g^{\mathrm{eff},n}$ as in the case of the steady-state photoluminescence.

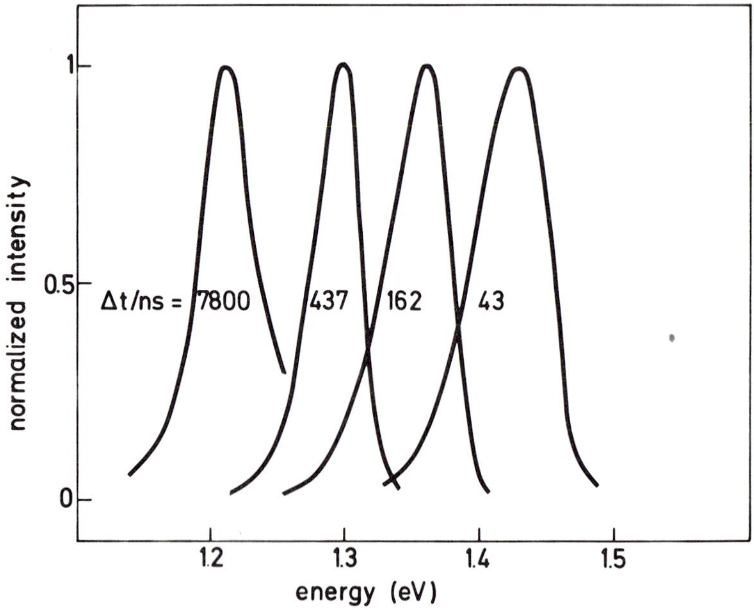

Fig. 24. Normalized luminescence spectra (at $T = 1.6$ K and $\hbar\omega_{\mathrm{exc}} = 1.92$ eV) of a GaAs-doping n–i–p–i superlattices with $d_n = d_p = 25$ nm and $n_D = n_A = 2 \times 10^{18}$ cm^{-3} for different delay times Δt after excitation. [From Rehm *et al.* (1983). Copyright 1983 North-Holland Publishing Co., Amsterdam.]

2. Electroluminescence

Studies of the tunable electroluminescence in GaAs-doping superlattices were carried out by Künzel *et al.* (1982b). The excitation was achieved by a variable external potential that was applied via selective electrodes (see Subsection IV.A.1). The electroluminescence spectrum could be shifted over 300 meV in a manner analogous to the case of photoluminescence (see Fig. 23), when the electrical power (instead of the light intensity) was varied over several orders of magnitude. The high-energy edge of the luminescence spectrum is determined by the quasi-Fermi level difference $\Delta\phi_{np}$ also in the case of electroluminescence. The applied external potential eU_{np}, however, was significantly larger than $\Delta\phi_{np}$ in the present case. The reason for this observation is a rather large distance between the selective n and p electrodes. Due to the nonnegligible series resistance of the layers, a considerable potential drop between the two electrodes exists. Hence, the *local* values of $\Delta\phi_{np}$, which actually determine the luminescence, are always smaller than eU_{np}.

D. Tunable Two-Dimensional Subband Structure

The observation of a two-dimensional subband structure in $n-i-p-i$ crystals and a study of its tunability so far have been reported only for the case of resonant inelastic light-scattering experiments (Döhler *et al.*, 1981b; Zeller *et al.*, 1982, 1983). In these experiments, two and in some cases three peaks could be detected in the spin–flip Raman spectra (see Fig. 25). These peaks are interpreted as being due to spin-density excitations between neighboring ($\Delta = 1$) and second- and third-nearest subbands ($\Delta = 2$ and 3). The spin-density intersubband excitation energies differ only slightly from the corresponding one-particle energy differences, as discussed in Subsection II.B.6. The decrease of the peak energies with increasing excitation intensity reflects the decrease of subband spacings associated with the enhancement of the steady-state carrier concentration. In order to make a quantitative comparison between measured and calculated intersubband excitation energies, the tunable luminescence signal was used as a tool to determine the steady-state carrier concentration $n^{(2)}$ at a given laser intensity via the calculated relation between $\Delta\phi_{np}$ and $n^{(2)}$. [The design parameters used in the self-consistent calculations for Fig. 2 correspond to those of the sample experimentally investigated by Döhler *et al.* (1981b).] The squares in Fig. 26 represent the observed Raman peak energies. The agreement of these energies with the calculated (appropriately averaged) intersubband excitations is excellent and provides definite evidence for the formation of two-dimensional subbands in these purely space-charge-induced quantum wells.

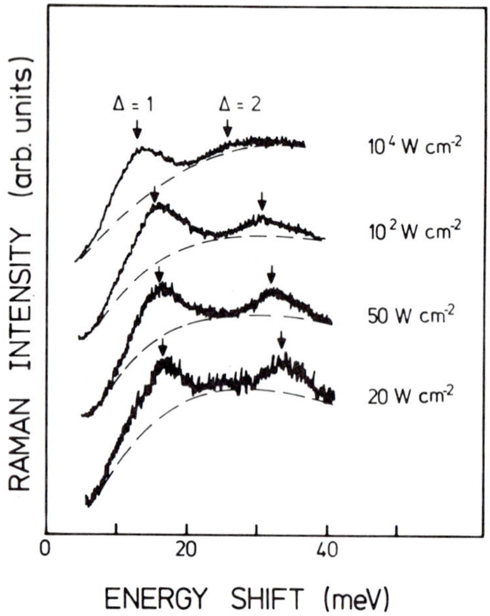

Fig. 25. Spin-density excitation spectra ($\lambda_L = 647.1$ nm and $T = 4.2$ K) observed with resonant Raman scattering at different laser intensities for a GaAs-doping $n-i-p-i$ superlattice with the design parameters $d_n = d_p = 40$ nm and $n_D = n_A = 10^{18}$ cm^{-3}. [From Döhler *et al.* (1981).]

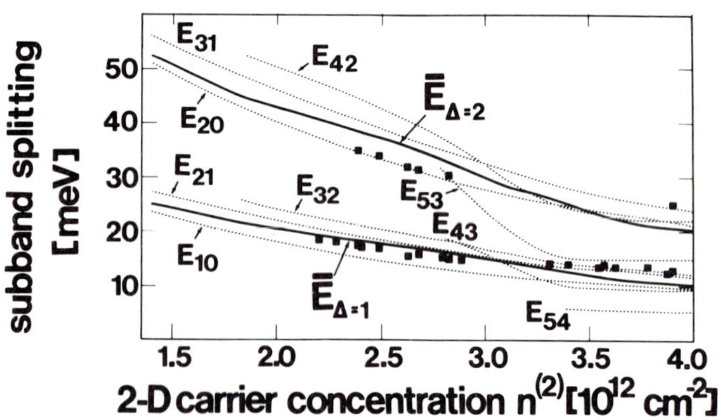

Fig. 26. Comparison between observed spin-density excitation peaks as a function of steady-state carrier concentration and calculated results for the sample from Fig. 25. Dotted lines show subband spacings between populated states in the n layers. The solid lines, labeled $\bar{E}_{\Delta=1}$ and $\bar{E}_{\Delta=2}$, represent averages for excitations between neighboring and second-nearest subbands, obtained by simple weighting of allowed phase space. [From Döhler *et al.* (1981).]

V. CONCLUDING REMARKS

In spite of the short period of investigation, the experimental work on GaAs doping superlattices has confirmed a large number of theoretically predicted and very unusual properties. There still remain, however, many interesting phenomena to be studied both in theory and experiment. In many of the cases, GaAs, on which we have mainly focused our attention, represents a quite ideal host material. For other purposes, semiconductors with lower or larger band gap will be more favorable. Effects related to two-dimensional subband formation or requiring high mobilities as well as absorption below the fundamental band gap E_g^0 are most easily studied in small-band-gap materials with low effective masses. For tunable luminescence in the visible range, of course, a large-band-gap material with the potential to achieve high donor and acceptor concentrations is desirable.

Not only the experimental but also the theoretical work has been restricted mainly to semiconductors with GaAs band structure (i.e., a direct gap at the Γ point). The interesting question of how far an indirect band gap in momentum space becomes efficient for optical transitions when the periodic space-charge potential of a doping superlattice transforms the crystal into a semiconductor with direct gap in momentum space [as in a (001)-Si $n-i-p-i$ crystal has not yet been studied. In the case of high doping concentrations in conjunction with small superlattice period, it seems likely that the net conduction-to-valence-band transition matrix elements may become rather large.

In addition, novel interesting features are to be expected in IV–VI $n-i-p-i$ crystals (Döhler and Ruden, 1984a) because of the special band structure of the bulk material with a direct gap in momentum space at the L points. The absence of bound impurity states and the variable dielectric constant make this group of materials particularly attractive.

Artificial superlattices do not rely on crystalline order. The properties of amorphous $n-i-p-i$ structures have been investigated theoretically (Döhler, 1982a), and it is expected that the experimental study of such exotic doping superlattices will provide interesting information about the bulk properties of amorphous Si.

We have emphasized that the homogeneity of the bulk material represents an advantage of doping superlattices from the technical point of view. For special purposes one may give up this particular advantage if, for instance, the goal is to combine the tunability of the electronic properties with high electron and hole mobilities. In a hetero-$n-i-p-i$ crystal with a bulk band gap, a normal E_g^0 doping superlattice, modified by the incorporation of undoped layers of a lower-band-gap material ($E_g^1 < E_g^0$) in the center of the doping layers (Döhler, 1981) (see Fig. 27), electrons and holes are spatially separated as in a conventional $n-i-p-i$ crystal. In addition, however, both types of carriers are also spatially separated from the respective impurity centers. Therefore, high elec-

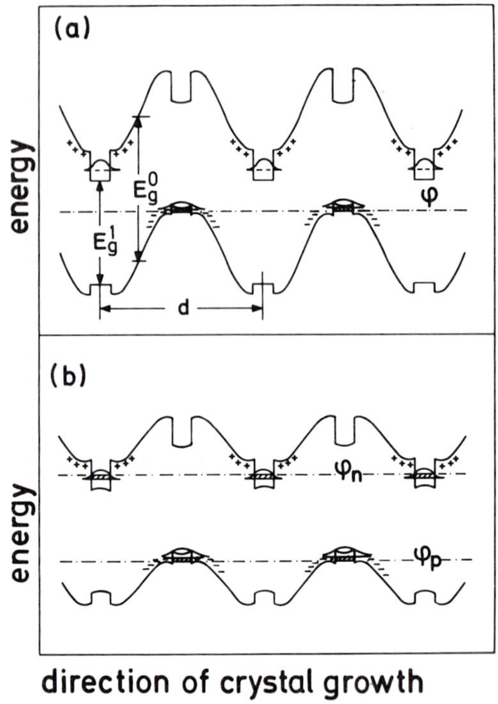

Fig. 27. Schematic real-space energy diagram of a hetero-$n-i-p-i$ crystal. p-type crystal in the (a) ground state ($n_0^{(2)} = 0$, $p_0^{(2)} \neq 0$), and (b) excited state ($n^{(2)} = \Delta$, $p^{(2)} = p_0^{(2)} + \Delta$). [From Döhler (1981).]

tron and hole mobilities are achieved, just as in the familiar modulation-doped compositional superlattice (see Chapter 3). This concept has been realized successfully by Künzel *et al.* (1983).

In this chapter, we have omitted completely the discussion of any device aspects of doping superlattices. It is evident that the tunability of electrical and optical properties implies a large variety of device applications. Depending on the design parameters, one may obtain devices of extremely high sensitivity or of extremely fast response times for the generation, amplification, detection, or storage of electrical or optical signals. Some of these ideas have been outlined by Döhler (1982b).

ACKNOWLEDGMENTS

We are grateful to P. Ruden for a critical reading of the manuscript and to Ms. G. Keck and Mrs. I. Zane for typing the manuscript.

REFERENCES

Ando, T., Fowler, A. B., and Stern, F. (1982). *Rev. Mod. Phy.* **43,** 437–672.
Burstein, E. (1954). *Phys. Rev.* **93,** 632–633.
Chandra, A., Wood, C. E. C., Woodard, D.W., and Eastman, L. F. (1979). *Solid-State Electron.* **22,** 645–650.
Chang, L. L. (1980). *In* "Handbook of Semiconductors" (T. S. Moss, ed.), Vol. 3, pp. 563–597. North-Holland Publ., Amsterdam.
Cho, A. Y. (1971). *J. Appl. Phys.* **42,** 2074–2081.
Cho, A. Y. (1976). *J. Appl. Phys.* **47,** 2841–2843.
Cho, A. Y., and Arthur, J. R. (1975). *Prog. Solid State Chem.* **10,** 157–191.
Cho, A. Y., and Hayashi, I. (1971). *J. Appl. Phys.* **42,** 422–4425.
Döhler, G. H. (1972a). *Phys. Status Solidi B* **53,** 79–92.
Döhler, G. H. (1972b). *Phys. Status Solidi B* **52,** 533–545.
Döhler, G. H. (1978). *Surf. Sci.* **73,** 97–105.
Döhler, G. H. (1979). *J. Vac. Sci. Technol.* **16,** 851–956.
Döhler, G. H. (1981). *Phys. Scripta.* **24,** 430–439.
Döhler, G. H. (1982a). *Verh. Dtsch. Phys. Ges. VI* **17,** 745.
Döhler, G. H. (1982b). *In* "Collected Papers of MBE-CST-2" (R. Ueda, ed.), pp. 20–23. Jpn. Soc. Appl. Phys., Tokyo.
Döhler, G. H., and Ploog, K. (1979). *Prog. Cryst. Growth Charact.* **2,** 145–168.
Döhler, G. H., and Ruden, P. (1984a). *Surf. Sci.* **142,** 474–485.
Döhler, G. H., and Ruden, P. (1984b) *Phys. Rev. B* **30,** 5932–5944.
Döhler, G. H., Ruden, P., Künzel, H., and Ploog, K. (1981a). *Verh. Dtsch. Phys. Ges. VI* **16,** 161.
Döhler, G. H., Künzel, H., Olego, D., Ploog, K., Ruden, P., Stolz, H. J., and Abstreiter, G. (1981b). *Phys. Rev. Lett.* **47,** 864–867.
Döhler, G. H., Künzel, H., and Ploog, K. (1982). *Phys. Rev. B* **25,** 2616–2626.
Esaki, L. (1958). *Phys. Rev.* **109,** 603–604.
Foxon, C. T., and Joyce, B. A. (1981). *In* "Current Topics in Materials Science" (E. Kaldis, ed.), Vol. 7, pp. 1–68. North-Holland Publ., Amsterdam and New York.
Heim, U., and Hiesinger, P. (1974). *Phys. Status Solidi B* **66,** 461–470.
Ilegems, M. (1977). *J. Appl. Phys.* **48,** 1278–1287.
Johnson, W. C., and Panousis, P. T. (1971). *IEEE Trans. Electron Devices* **ED-18,** 965–973.
Jung, H., Döhler, G. H., Künzel, H., Ploog, K., Ruden, P., and Stolz, H. J. (1982). *Solid State Commun.* **43,** 291–294.
Jung, H., Döhler, G. H., Künzel, H., and Ploog, K. (1983). *J. Appl. Phys.* **54,** 6965–6973.
Kennedy, D. P., Murley, P. C., and Kleinfelder, W. (1968). *IBM J. Res. Dev.* **12,** 399–40.
Kinoshita, H., Sakashita, T., and Fujiyasu, H. (1981). *J. Appl. Phys.* **52,** 2869–2875.
Künzel, H., and Ploog, K. (1980). *Appl. Phys. Lett.* **37,** 416–418.
Künzel, H., and Ploog, K. (1981). *Conf. Ser.-Inst. Phys.* **56,** 519–528.
Künzel, H., Fischer, A., and Ploog, K. (1980). *Appl. Phys.* **22,** 23–30.
Künzel, H., Graf, K., Hafendörfer, M., Fischer, A., and Ploog, K. (1981). *TM Tech. Messen* **48,** 295–300, 397–401, 435–440.
Künzel, H., Döhler, G. H., and Ploog, K. (1982a). *Appl. Phys. A* **27,** 1–10.
Künzel, H., Döhler, G. H., Ruden, P., and Ploog, K. (1982b). *Appl. Phys. Lett.* **41,** 852.
Künzel, H., Knecht, J., Fischer, A., and Ploog, K. (1983). *Appl. Phys. A* **30,** 73–77.
Künzel, H., Fischer, A., and Ploog, K. (1984). *J. Vac. Sci. Technol. B* **2,** 1–9.
Ploog, K. (1979). *J. Vac. Sci. Technol.* **16,** 838–846.
Ploog, K. (1980). In "Crystals: Growth, Properties, and Applications" (H. C. Freyhardt, ed.), Vol. 3, pp. 73–162. Springer-Verlag, Berlin and Heidelberg.
Ploog, K. (1981). *Annu. Rev. Mater. Sci.* **11,** 171–210.

Ploog, K. (1982). *Annu. Rev. Mater. Sci.* **12,** 123–148.
Ploog, K., and Fischer, A. (1977). *Appl. Phys.,* **13,** 111–121.
Ploog, K., and Künzel, H. (1982). *Microelectron. J.* **13** (3), 5–22.
Ploog, K., Fischer, A., and Künzel, H. (1979). *Appl. Phys.* **18,** 353–356.
Ploog, K., Fischer, A., and Künzel, H. (1981a). *J. Electrochem. Soc.* **128,** 400–410.
Ploog, K., Künzel, H., Knecht, J., Fischer, A., and Döhler, G. H. (1981b). *Appl. Phys. Lett.* **38,** 870–873.
Rehm, W., Künzel, H., Döhler, G. H., Ploog, K., and Ruden, P. (1983). *Physica* **117B/118B,** 732–734.
Ruden, P. (1982). Thesis, Univ. of Stuttgart, unpublished.
Ruden, P. (1983). *J. Vac. Sci. Technol. B* **1,** 285–288.
Ruden, P., and Döhler, G. H. (1983a). *Phys. Rev. B* **27,** 3538–3546.
Ruden, P., and Döhler, G. H. (1983b). *Phys. Rev. B* **27,** 3547–3553.
Shiraki, Y. (1982). In "Collected Papers of MBE-CST-2" (R. Ueda, ed.), pp. 179–182, Jpn. Soc. Appl. Phys., Tokyo.
Sze, S. M. (1969). *In* "Physics of Semiconductor Devices." Wiley, New York.
Sze, S. M., and Gibbons, G. (1966). *Appl. Phys. Lett.* **8,** 111–113.
Ury, I., and Holm-Kennedy, J. W. (1978). *Surf. Sci.* **73,** 179–183.
Wood, C. E. C. (1980). *In* "Physics of Thin Films" (W. R. Hunter and G. Haas, eds.), Vol. 11, 35–103. Academic Press, New York.
Wood, C. E. C., Woodcock, J., and Harris, J. J. (1979). *Conf. Ser./Inst. Phys.* **45,** 28–37.
Zeller, C., Vinter, B., Abstreiter, G., and Ploog, K. (1982). *Phys. Rev. B* **26,** 2124–2132.
Zeller, C., Vinter, B., and Abstreiter, G. (1983). *Physica* **117B/118B** 729–731.

6 MODULATION-DOPED SEMICONDUCTORS

A. C. Gossard and A. Pinczuk

AT&T Bell Laboratories
Murray Hill, New Jersey

I. INTRODUCTION

Much of the significance of semiconductors derives from the possibility of introducing charge carriers by incorporation of impurities. With the advent of techniques for growing finely modulated structures, the flexibility of distributions of impurities becomes very large, especially in comparison to the more limited configurations available with previously used bulk growth, diffusion, and implantation techniques. Such flexibility is important because the interfaces between doping regions or between doped regions and surfaces or interfaces are critical regions in control of semiconductor properties. In Chapter 5 by Döhler and Ploog the possibilities produced by fine variation in the doping profile in a single-semiconductor material are developed. In the present chapter, we consider finely varied dopant configurations in synthetically layered semiconductors. In these configurations, known as modulation-doped hetero-

ISBN 0-12-170470-X

structures, carriers are influenced both by the potentials created by transfer of charge and by potentials produced by the different electron affinities of the layers of dissimilar semiconductors. An important new element is provided by the heterostructures, in that their layers may be made so thin that the carrier states become confined and size quantized. Electrons will tend to move to layers of greater electron affinity, and holes to layers with larger hole affinity, where they can become confined in quantum states. The combination of these factors acting on the carriers and the flexibility and control in the growth technique creates a great deal of freedom in the design of structures with new charge distributions and transport characteristics.

A particularly rewarding use of the technique is the creation of conducting channels with high carrier concentrations but low impurity content. A major goal of solid-state physics and solid-state technology is the perfection of materials and structures in which charge carriers have long lifetimes, low scattering, high mobilities, and controlled densities. In materials such as semiconductors in which carriers are introduced by incorporation of impurities, however, the improvement of lifetimes and scattering rates is generally limited by scattering from impurities necessarily incorporated in order for carriers to be present or confined. The technique of modulation doping of semiconductor heterostructures provides a means for avoiding these limits on the performance of semiconductor materials and allows access to new regimes of physical effects and device performance.

The modulation doping (MD) technique traces its origins to semiconductor heterostructure and superlattice development that demonstrated charge carrier confinement by heterojunctions (Hayashi *et al.*, 1969) and showed that in layers that are sufficiently thin the carriers occupy quantized energy states (Chang *et al.*, 1974; Dingle *et al.*, 1974). The quantization occurs in the dimension perpendicular to the heterostructure, whereas the motion parallel to the interface is free. With only two degrees of free motion available, the charge carriers in each quantum state form a two-dimensional system. Deviations from two-dimensionality are produced only by scattering between quantum states or by tunneling between different layers.

Carriers can be introduced into the two-dimensional-quantum states by optical pumping of electrons between valence and conduction bands or by use of chemical doping. The first experimental observation of the two-dimensionality of free carriers was the detection of Shubnikov–de Haas magnetoresistance oscillations of the conduction electrons in uniformly Sn-doped molecular beam epitaxy (MBE)-grown GaAs–$Al_xGa_{1-x}As$ superlattices (Chang *et al.*, 1977). With thick (> 100-Å) barrier layers of $Al_xGa_{1-x}As$, motion of the electrons was demonstrated to be only in the planes of the layers, while with thinner barriers, evidence of barrier penetration was found. The observations required intense magnetic fields in order to make the cyclotron-orbit times of the carriers less than their scattering times. Although Esaki and Tsu (unpublished) had speculated that carrier mobilities might be enhanced by use of nonuniform doping, Chang *et al.* (1977) did not use modulation doping in preparation of the structures.

The first modulation-doped heterostructures were MBE-grown GaAs–$Al_xGa_{1-x}As$ superlattices with modulated silicon doping (Dingle *et al.*, 1978). The structures were designed to separate donors from mobile electrons in superlattices in order to increase the time between scattering events and thus observe more clearly any two-dimensional resistance effects. Silicon was selected as the dopant species because it could be incorporated with abrupt concentration profiles in MBE growth. The initial experiments showed that efficient introduction of carriers into the superlattice was produced by either uniform doping or doping of only barrier layers but that substantially increased mobilities were produced when only barrier layers were doped. The source of transfer of charge from the barriers to the GaAs wells is the lower energy of electron states in the wells relative to electrons residing in the barriers or bound to donors in the barriers. The principal mechanism for the mobility enhancement is the reduction in scattering of the carriers from impurities by means of the spatial separation of the impurities and the mobile electrons.

In a modulation-doped heterostructure superlattice, electrons confined to the quantum layers are bounded on each side of each well by the heterojunction potential step between the well materials and the barrier material. It is also possible with modulation doping to confine electrons at a single interface between an undoped channel material and a doped barrier material. This was achieved both in structures grown with MBE (Störmer *et al.*, 1979) and with LPE (liquid-phase epitaxy) (Tsui and Logan, 1979), although the MBE structures are more widely used and developed.

In the single-interface structures, the carriers in the channel are confined at the heterojunction potential step by the electric field of the ionized donors from which the carriers came. The resultant potential is essentially triangular in shape and confines carriers in quantum states much as in the superlattice case or in biased metal insulator semiconductor field effect transistor (MISFET) structures. Motion along the heterointerface occurs with enhanced mobility, much as in the modulation-doped superlattice. Superlattice and single-interface systems are illustrated schematically in Fig. 1. They have provided an excellent means for the realization of the two-dimensional electron gas and have shown many remarkable new physical phenomena. They also have provided an opportunity for exploring novel device phenomena and for developing high-performance device structures.

In Section II the principles of modulation doping are developed, particularly with regard to the various available parameters for design and construction of modulation-doped structures. Section III reviews the growth and measurement techniques. Section IV presents the experimentally observed conductivity and mobility behavior of modulation-doped structures. Section V treats the various phenomena of the two-dimensional electron gas in these structures, including light-scattering spectroscopy and the quantized Hall effect.

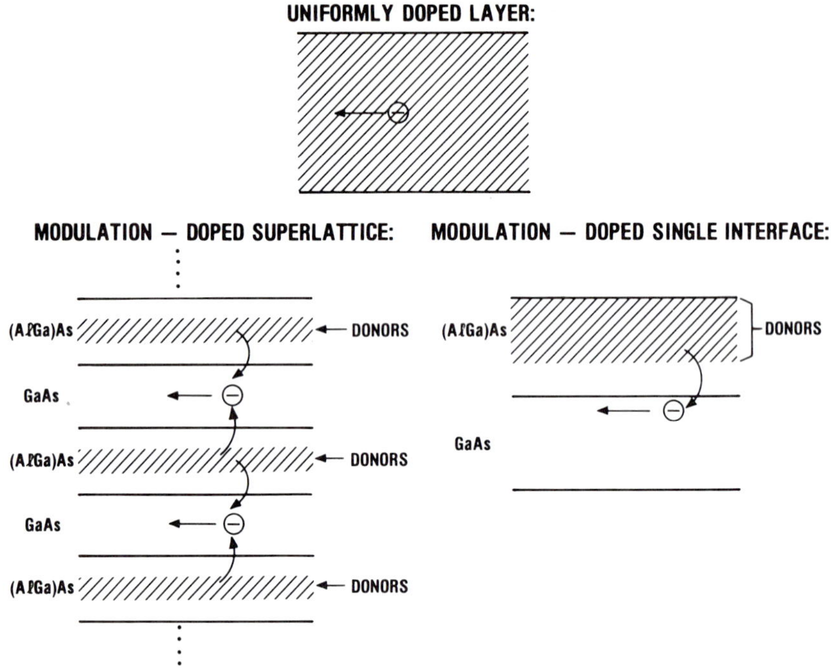

Fig. 1. Spatial arrangement of layers in GaAs modulation-doped structures. Crosshatched represent doped layers. Electrons transfer to lower energy mobile states in undoped GaAs layers.

II. PRINCIPLES OF MODULATION DOPING

A. Charge Transfer

The charge transfer and carrier confinement that form the basis for most modulation doping effects in heterostructures are produced by the step in band-edge energies occuring at the heterojunction. This step can be probed in several ways. Optical determination of confined particle energy levels permits a determination of confining potential barrier heights. The energy levels can be deduced from interband optical absorption spectra (Dingle *et al.*, 1974), photoluminescence excitation spectra (Weisbuch *et al.*, 1981), and intersubband Raman scattering spectra (Abstreiter and Ploog, 1979; Pinczuk *et al.*, 1979a). In the GaAs–$Al_xGa_{1-x}As$ system, both electrons and holes can be confined in the GaAs layers, whereas in the InAs–GaSb system, they might be confined in separate layers, with electrons contained in the InAs layers and holes in the GaSb layers (Sai-Halasz *et al.*, 1977). In the GaAs–$Al_xGa_{1-x}As$ system part of the difference in band gap between the two species occurs in the conduction band, corresponding essentially to the electron affinity difference between the

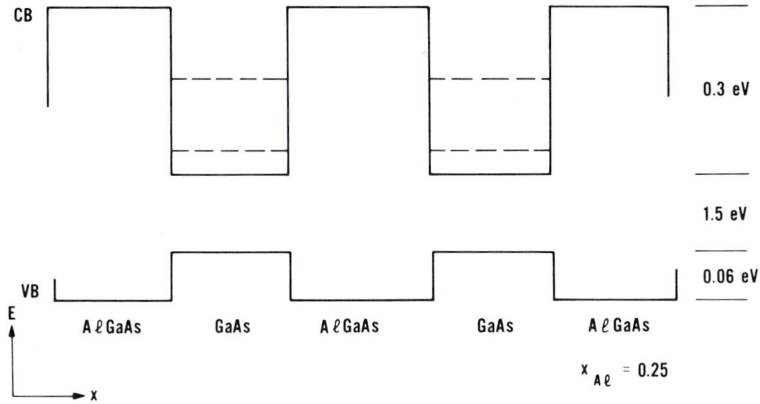

Fig. 2. Conduction-band and valence-band edges in undoped GaAs-(Al,Ga)As heterostructure. Electrons introduced into GaAs layers occupy quantum energies shown by dashed lines.

materials, and the remaining part occurs in the valence band (Dingle, 1975). The discontinuities are illustrated in Fig. 2.

The band-edge discontinuities form an upper limit for the difference in energy between an electron bound to a donor in the barrier layer near the interface and an electron in the (GaAs) channel layer. The actual energy difference will be reduced by several effects.

1. Donor Binding Energy

The electrons bound to donors in barrier material will lie below the bottom of the barrier conduction band. This binding energy is greater in $Al_xGa_{1-x}As$ than in GaAs, increasing with x to more than 100 meV in the range of $x = 0.25-0.40$ (Hiyamizu, 1983). It reduces the energy gained on dropping into the channel layer.

2. Conduction Electron Quantum Energy

The lowest conduction electron energy in the quantum confined state in the channel is raised by the size quantization effect. For thin wells, the quantum state energies E_n are approximated by the energy of a particle in a square box of the same width, d_1 with infinite barriers: $E_n = n^2h^2/8m^*d_1^2$, which has a value $E_1 = 56$ meV for electrons in the lowest state in GaAs ($m^*/m_0 = 0.0665$) at a well width $d_1 = 100$ Å. More exact values of the quantum energies that take into account the actual potential profiles may be calculated. This energy also reduces the energy gained by an electron transferring into the channel.

3. Conduction Electron Fermi Energy

For a nonzero concentration of electrons in the channel layer, the electron energies will occupy a Fermi distribution. At low temperatures, electron states up to a Fermi energy of $\pi\hbar^2 n/m^*$ above the lowest quantum energy E_1 will be occupied for a two-dimensional gas of concentration n carriers/cm^2 with only one confined quantum-well state occupied and spin degeneracy of 2. For $m^* = 0.0665m_0$ and $n = 1 \times 10^{12}$ cm^{-2}, this is 35 meV and also acts to reduce the energy gained by transferring additional electrons into the channel layer.

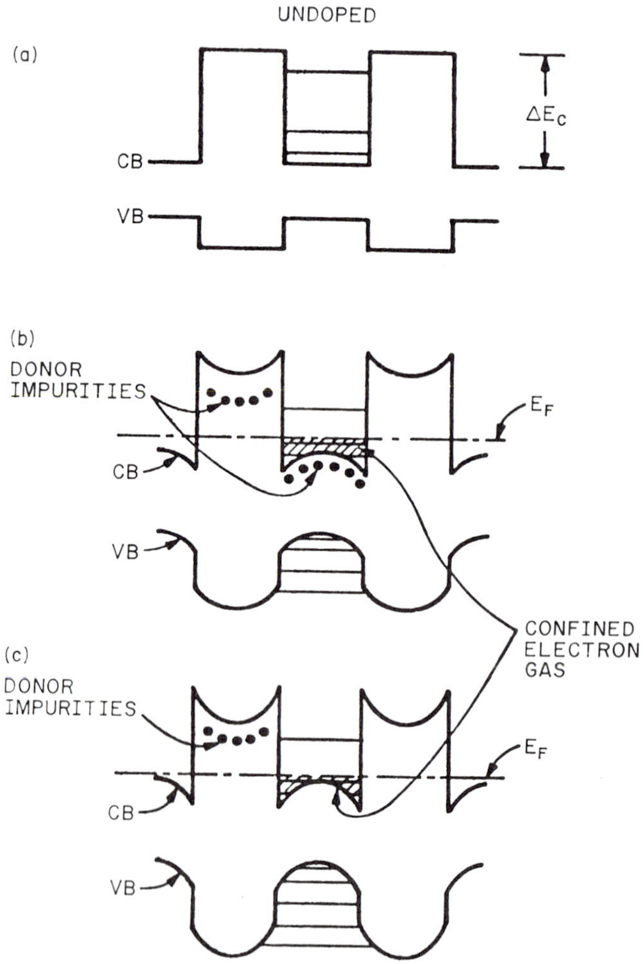

Fig. 3. Band edges, donor states, and electron states in (a) undoped, (b) uniformly doped, and (c) modulation-doped multilayer semiconductor system (Dingle *et al.*, 1978).

4. Spatial Transfer Coulomb Energy

Transfer of charge into the channel creates a negatively charged channel region and leaves behind a positive space charge of the ionized donors. The electric field and potential drop associated with this space charge are also appreciable. The potential drop for separation d of charge density σ is $\sigma d/\epsilon_s$, which is 138 meV for separation of 10^{12} electrons/cm^2 by a distance of 100 Å in GaAs of dielectric constant $\epsilon_s = 13.1$. Of course, for continuous or extended distributions of carriers and ions, appropriate integral expressions are to be employed. The coulomb potentials alter the shape of the potential wells and barriers and thus alter the quantum energies, so that an exact treatment requires a self-consistent calculation. The essential result of all of these considerations, however, is that it is energetically possible to transfer electrons up to densities of $\approx 10^{12}$ cm^{-2} over distances of ≈ 100 Å into wells as thin as 100 Å in modulation-doped GaAs–$Al_xGa_{1-x}As$ layers with $x \gtrsim 0.2$. These various effects are illustrated in the band diagram of Fig. 3 (Störmer *et al.,* 1981a).

B. Mobility-Determining Processes

The dominant sources of carrier scattering in uniformly doped semiconductors are phonons and impurities, with the phonon processes dominating scattering at high temperatures and the impurity processes predominant at low temperature. A plot of the limits to the mobility from these processes for very weakly doped GaAs ($\approx 10^{13}$ cm^{-3}) is shown in Fig. 4 (Wolfe *et al.,* 1970). In modulation-doped layers we may expect the phonon processes and rates to be roughly similar to the phonon scattering in uniformly doped material. The impurity scattering is drastically altered, however. When the impurity scatterers are separated from the carriers, the impurity scattering decreases both because of the falloff in Coulomb fields with distance and because the short-wavelength (large-wave-vector) potential fluctuations, which are the most important source of scattering by impurities, fall off with distance from the impurity distribution. These effects have been calculated in detail by Mori and Ando (1979), Price (1982), and Fishman and Calecki (1983). A simple expression for the resultant mobility is obtained (Price, 1982) for a degenerate two-dimensional sheet of n carriers/cm^2 separated by d from a two-dimensional sheet of n_I ions/cm^2,

$$\mu = 16\sqrt{2\pi}\, e d^3 n^{3/2}/\hbar\, n_I. \qquad (1)$$

The increase in mobility with carrier concentration n is especially noteworthy because it more than compensates the customary decrease of μ with n_I in uniformly doped cases where $n = n_I$. The increasing mobility with high n is a result of the increase in Fermi wave vector with n. With large Fermi wave vectors, small-k elastic ion scattering is less effective in relaxing the electron momentum distribution. The calculated mobility reaches 60,000 cm^2/V sec at $n = n_I = 10^{12}$ cm^{-2} and $d = 100$ Å. The mobility increase continues until a second quantum state becomes occupied, which occurs when the Fermi energy becomes equal to the difference between the first and second quantum states. The second quantum state forms the basis for a second subband of states and permits scattering of electrons between the subbands, which then produces a decrease in mobility, as first calculated by Mori and Ando (1979).

In the case of incomplete electron transfer to the quantum-well channels, electron states in both channel and barrier layers will be occupied. The conductivity will then consist of parallel conduction in the channels and barriers. For $Al_xGa_{1-x}As$ barriers of $x > 0.20$, however, donor states in the barriers are sufficiently deep that carriers are typically frozen out at cryogenic temperatures in the barriers and conduction is dominated by the quantum-well channel carriers.

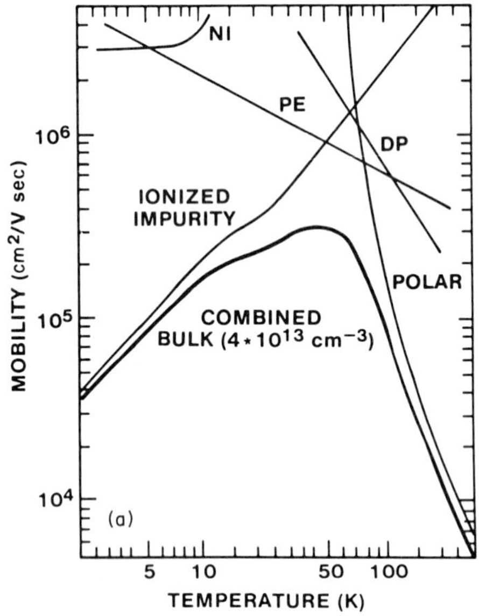

Fig. 4. (a) Theoretical mobility limits for main scattering processes in uniformly doped GaAs (Wolfe *et al.*, 1970). Neutral-impurity scattering is indicated by NI, PE indicates scattering by piezoelectrically active acoustic phonons, and DP indicates deformation-potential scattering by acoustic phonons. In modulation-doped material with no impurity scattering, the upper limit on mobility would be $\approx 5 \times 10^6$ cm^2/V sec. (b) Experimentally observed mobilities of high-mobility selectively doped GaAs–(Al,Ga)As single-interface structure (●, dark, $n = 2.4 \times 10^{11}$ cm^{-2}; ○, illuminated, $n = 3.6 \times 10^{11}$ cm^{-2}) and high-mobility, lightly uniformly doped bulk GaAs (Δ, $n = 4 \times 10^{13}$ cm^{-3}).

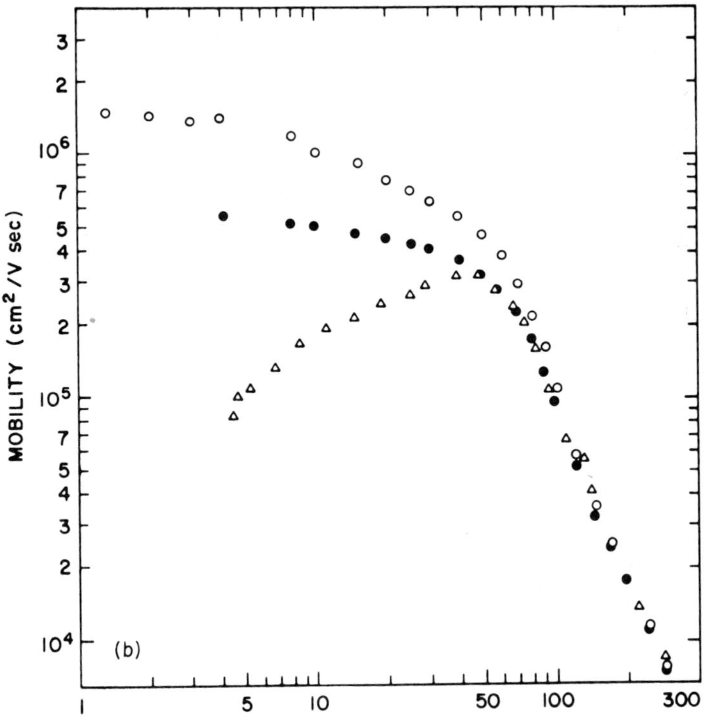

Fig. 4. *(Continued)*

III. GROWTH AND MEASUREMENT TECHNIQUES

A. Growth Techniques

Modulation-doped structures with enhanced mobilities have now been grown with liquid-phase epitaxy (Tsui and Logan, 1979), metal–organic chemical vapor deposition (Hersee *et al.,* 1982), and molecular beam epitaxy (Dingle *et al.,* 1978). In liquid-phase epitaxy, the structures are produced by successively sliding a seed crystal under gradually cooled melts containing doped and undoped barrier and well material, whereas in metal–organic chemical vapor deposition, the structures are obtained by successively introducing and purging the requisite reacting gasses from the reactor containing the crystal substrate. In molecular beam epitaxy, the modulation-doped structure is produced by shuttering on and off the constituent molecular beams needed for the respective layers. The smoothest and sharpest layers (Gossard, 1979; Petroff *et al.,* 1978) and the highest mobilities (Hiyamizu, 1982; Hwang *et al.,* 1982) have been demonstrated with the MBE technique, but the mobility enhancements observed with the other two techniques (Tsui and Logan, 1979; Hersee *et al.*, 1982) are also very appreciable. Consequently, all three techniques are potentially useful.

The essential factors that must be present to achieve a high mobility structure are (1) absence of scattering centers in the conducting channel regions, (2) smooth and pure interfaces bounding these regions, (3) minimization of traps and compensating centers in the barriers, and (4) separation of the dopant atoms from the channel regions. The absence of unintentional scattering centers in the channel and barriers requires a low dislocation density and a low impurity density, whereas the smooth and pure interfaces require growth under conditions giving a smooth crystal surface with low interdiffusion between layers. The separation of dopant atoms and channel regions requires dopants that do not diffuse and that incorporate directly in the lattice without requiring accumulation of an enriched surface layer before incorporation in the lattice.

In MBE of modulation-doped GaAs–(Al,Ga)As structures, the pure conducting channel regions are best obtained (1) by the use of substrate temperatures during epitaxial growth that are sufficiently high (600–700°C) to reduce incorporation of impurities, but not roughen or decompose the sample surface, (2) by the minimization of unintentional contamination from sources and hot surfaces, (3) by the substrate introduction by vacuum interlock to prevent contamination of the ultra high vacuum, and (4) by the use of arsenic-rich growth conditions. As emphasized in the next section, the smoothness and purity of interfaces generally depend on the order in which the elements of the interface are grown, with interfaces grown on GaAs surfaces being smoother and purer than interfaces grown on (Al,Ga)As. Absence of traps and compensating centers in (Al,Ga)As barrier layers is also favored by high substrate temperatures and good vacuum conditions, with optimal substrate temperatures being higher than for GaAs. The density of carriers measured in the channel for a selectively doped GaAs–(Al,Ga)As single interface is generally lower than that calculated for complete ionization of donor levels above the Fermi level and suggests that impurities or defect centers in (Al,Ga)As barriers trap charges in deep states, preventing their entry into the GaAs channels. This presumably accounts for much of the variability in charge transfer and mobility that is observed from sample to sample in MD structures grown in different systems or at different times. It also presents a limitation (in the high 10^{10}-cm^{-2} range) on the lowest amount of charge that can be reliably transferred to a conducting channel.

Silicon and beryllium are presently the most favorable *n*- and *p*-type dopants for creation of abrupt doping profiles in GaAs structures. Tin, although it can be incorporated at high concentrations, does not make a suitable dopant because of surface concentration effects that make it difficult to abruptly start and stop incorporation of tin in a GaAs structure. Diffusion of dopants occurs more rapidly between surface and near-surface layers than between layers farther beneath the surface. In order to avoid diffusion of dopants from the doped barrier toward the undoped channel during growth of a selectively doped single interface, the undoped channel layer may be grown before the doped barrier. In this way, dopant diffusion toward the growing surface will not move dopant atoms toward the channel.

B. Measurement Techniques

Basic characterization of modulation-doped structures has generally been based (1) on Hall effect and resistivity measurements, which together can determine the carrier concentration and mobility of a single charge layer or an assembly of identical charge layers, and (2) on oscillatory magnetoresistance data, which provide an alternative way for characterizing carrier densities and relaxation times. Each of the techniques requires ohmic contacts to the electron gas, which for the GaAs case can often be achieved with alloyed indium contacts or, for smaller contacts, gold–germanium–nickel-alloyed contacts. Capacitance measurements are also frequently useful and sometimes can be made with liquid-metal electrodes or electrolyte solutions. Optical methods, especially luminescence and inelastic light-scattering measurements, have also helped elucidate the behavior of modulation-doped structures and will be discussed in the section on excitations of two-dimensional electron gases in modulation-doped structures.

IV. CARRIER MOBILITIES

The maximum carrier mobilities and lifetimes that have been obtained with the MBE technique have increased steadily as growth techniques have improved and the relevant structural parameters have become understood (Dingle *et al.*, 1978; Hiyamizu, 1982; Hwang *et al.*, 1982). The highest carrier mobility that has been reported thus far is a value of 2.1×10^6 cm^2/V sec for electrons at a modulation-doped GaAs–$Al_3Ga_{1-x}As$ interface (Hiyamizu, 1982). This is approximately three orders of magnitude greater than in bulk structures of comparable local density. Comparison between the mobilities observed in an MD single interface with extremely high mobility and a very lightly uniformly doped bulk sample are shown in Fig. 4b (Hwang *et al.*, 1982). The similarity in the values at temperatures over 100 K illustrates the dominance of phonon processes in both systems at the higher temperatures. The superior mobility of the MD structure at low temperatures illustrates the reduced impurity scattering in MD structures, even when compared to bulk material of only 4×10^{13}-cm^{-3} carrier density. We now describe the experimental results that demonstrate the factors influencing mobilities.

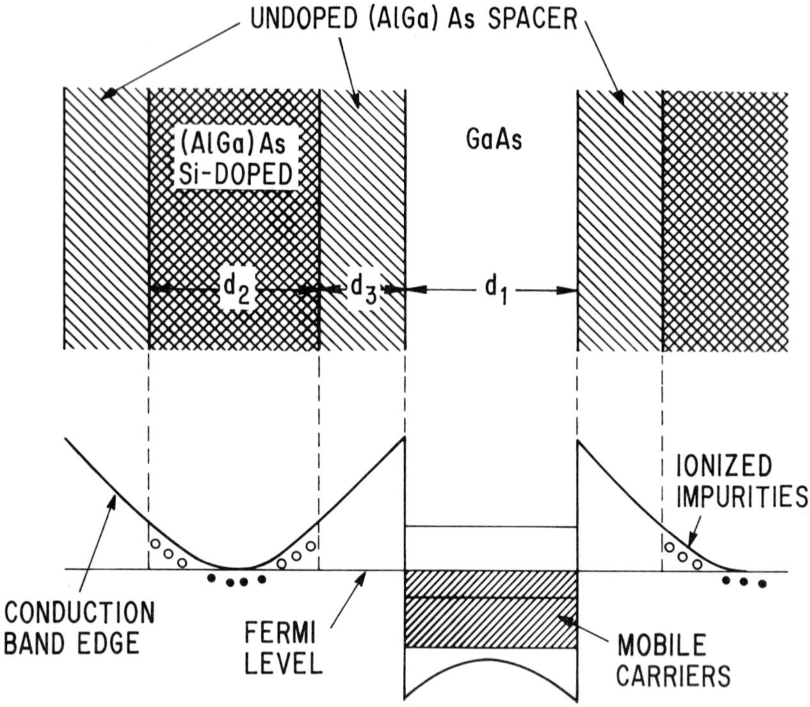

Fig. 5. Sequence of layers and relative position of the conduction-band edges in modulation-doped multi-quantum-well GaAs–(AlGa)As heterostructures.

A. Influence of Undoped Spacer Layer

The separation between mobile carriers in the conducting channels and ionized impurities in the barriers of modulation-doped structures can be increased by inserting undoped spacer layers in the barriers between the carriers and the ionized impurities as shown in Fig. 5 (Dingle *et al.*, 1978). Monotonically increased mobilities have been demonstrated in GaAs–$Al_xGa_{1-x}As$ with progressively increasing setback thicknesses d_3 from 0 to 150 Å (Drummond *et al.*, 1981a; Störmer *et al.*, 1981a). In order to demonstrate the effect, series of samples were grown in succession in which spacer layer thicknesses were increased. The temperature dependences of mobilities observed in a series of such samples are shown in Fig. 6. Throughout the range of temperature and spacings up to room temperature, the Hall mobilities increased with increasing width of undoped $Al_{0.12}Ga_{0.88}As$ spacer layers. The average Hall density decreased from 1.5×10^{17} cm^{-3} to 0.7×10^{17} cm^{-3} with the increase in spacer layer width. These results agree with the theoretical expectation that an undoped spacer layer progressively decreases ionized impurity scattering. However, the ob-

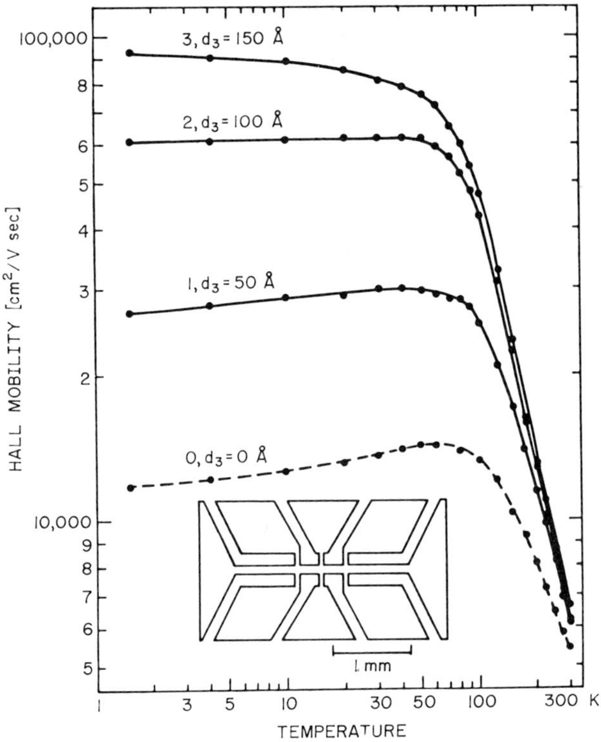

Fig. 6. Electron mobilities versus temperature for a series of modulation-doped multilayer samples with different undoped (Al,Ga)As spacer layer between GaAs channel and doped (Al,Ga)As regions (Störmer *et al.*, 1981a).

served dependence is less than the expected d_3^3 dependence, and samples studied with $d_3 > 150$ Å did not show further mobility increase. This occurs because other scattering mechanisms, perhaps associated with unintentional impurities or residual scattering from the heterojunction interface, become dominant at larger setback spacings of the dopant layer and because the carrier density in the layers was less when greater setbacks were present.

B. Effects of Carrier Density

Several methods can be used to vary the number of carriers per unit area in a given modulation-doped structure. These techniques, which permit variation of the carrier density without the need to produce series of different samples, are particularly useful in clarifying the role of the carrier density in the conduction process as well as in modifying conductivity as desired in devices. The techniques to be discussed involve photoexcitation of carriers, gate control of carriers by a metal gate electrode on the sample surface, and gate control by a backside gate beneath an insulating substrate on the opposite side of the substrate from the epitaxial conducting layers.

In the photoexcitation technique, a low-temperature persistent photoconductive effect can be used to vary the electron concentration continuously over a large range. The origin of the persistent photoconduction is not understood in detail but may be connected with deep traps in $Al_xGa_{1-x}As$ that can be optically activated (Störmer *et al.*, 1981b). Optically excited electrons transfer from the traps by tunneling to the GaAs channel, where they increase the carrier concentration.

By using this effect, a number of single-interface modulation-doped specimens have been studied. Light pulses emitted from light-emitting diodes onto GaAs–$Al_xGa_{1-x}As$ single-interface Hall bridges increased electron densities by amounts that persisted in the dark after the pulses ended. In this manner, charge densities in the 10^{11}–10^{12}-cm^{-2} range could be varied by up to a factor three. This produced increases in mobilities up to nearly one order of magnitude as measured by Hall effect, as shown in Fig. 7 (Störmer *et al.*, 1981b). The increases are monotonic in each case and actually more rapid than the $\mu \sim \sqrt{n}$ expected from the expression $\mu = 16\sqrt{2\pi}\, ed^3 n^{3/2}/\hbar n_{\mathrm{I}}$ for $n = n_{\mathrm{I}}$. It is possible that the optical technique modifies the distribution of ions near the channel as well as modifying the more distant ions and the carrier density and that this is also affecting the observed mobilities in a positive way with increasing total illumination.

The density of carriers can also be varied by the field effect induced by voltages on gate electrodes in proximity to the channel. An unusual but useful configuration involves a metal electrode fixed to the back of an insulating substrate (Fig. 8a) on which is grown a modulation-doped single interface (Störmer *et al.*, 1981c). The gate voltage may be applied between ohmic contacts to the channel electrons on the front and the back electrode, which may be a metal plate. By using the technique of Shubnikov–de Haas oscillations to determine carrier density and four-terminal resistance measurement to determine the resistivity, the carrier concentration was found to vary linearly with gate voltage by nearly a factor 5, while the conductivity varied by a factor 250 and the mobility by a factor 50, as shown in Fig. 8b (Störmer *et al.*, 1981c). The mobility increases with increasing positive gate voltage following an approximate $\mu \approx n^{5/2}$ law. The increased electron density improves screening of ionized impurity fields, while the positive backside bias increases the distance between the electrons in the channel and the impurities in the barrier by pulling electrons away from the interface. Both of these mechanisms increase mobility at positive biases and decrease mobilities under negative biases.

A more widely employed configuration for gate control of carrier concentration involves metal Schottky gates on the top surface of a modulation-doped single-interface structure. In this case, the gate voltage may be applied between

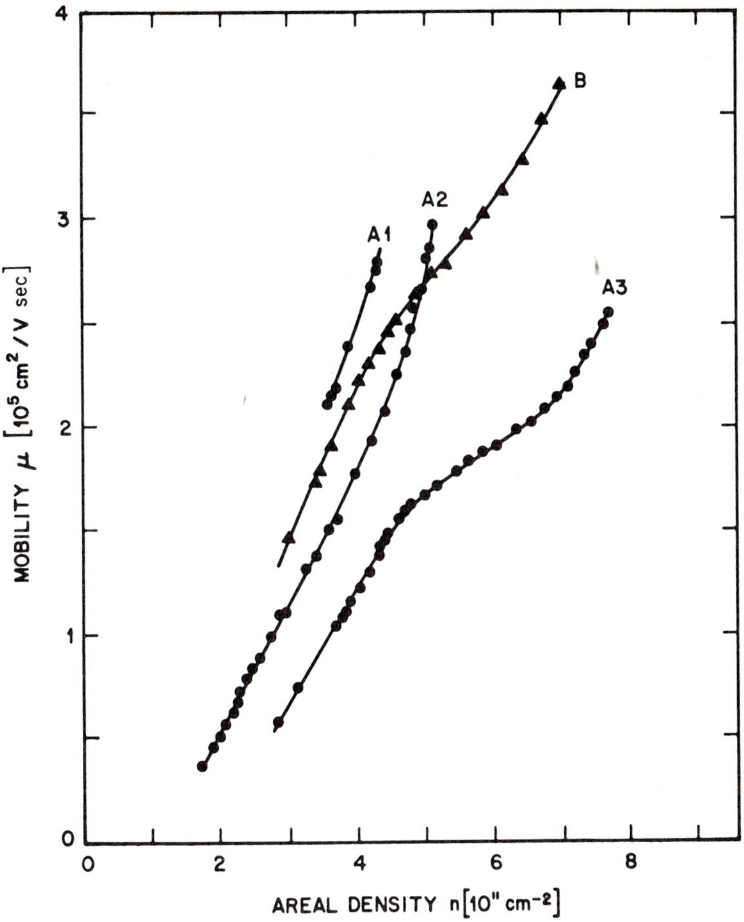

Fig. 7. Electron mobilities versus carrier density in several selectively doped single-interface GaAs–(Al,Ga)As samples. Carrier density is varied by persistent photoconductivity after illumination (Störmer *et al.,* 1981b).

the Schottky gate on the surface and the electrons in a channel approximately a depletion depth below the surface, which may contacted by alloyed ohmic contacts. This configuration forms the basis for selectively doped heterostructure transistors (Mimura, 1982). Strong variations in electron mobility with the gate-controlled carrier concentrations are also seen in this configuration, as shown in Fig. 9, which illustrates low-temperature mobilities in gated structures with and without an undoped spacer layer (Tsui *et al.,* 1981; Mimura, 1982). On devices covering a density range of 7×10^{10} cm^{-2} to 6×10^{11} cm^{-2}, mobilities increased with carrier concentrations according to power laws ranging from $\mu \approx n^{0.45}$ to $\mu \approx n^{1.4}$. The carrier density in these devices, with approximately 700 Å of $Al_xGa_{1-x}As$ and 200 Å of GaAs over the channel, varied linearly with gate voltage, in contrast to the density in uniformly doped layers, which varies as the square root of gate voltage. The linear change in carrier concentration is produced by the insulating action of the (Al,Ga)As barrier layers. The weaker dependence of mobility on carrier concentration, relative to the backside gate configuration, reflects the shift of carriers toward the interface on applying a positive voltage above the interface as opposed to displacement of carrier density away from the interface on application of positive voltage below the substrate.

The separation of carriers from the surface of the modulation-doped structure also provides the possibility of incorporating a layer of an insulator on the surface without having carriers in direct contact with the defects and impurities of the insulator and its surfaces. This provides a basis for a hybrid metal–insulator–semiconductor and semiconductor–semiconductor heterojunction technology (Chen *et al.*, 1982; Hotta *et al.*, 1982).

With an increase in carrier density beyond $6-8 \times 10^{11}$ cm^{-2} in GaAs modulation-doped single interfaces, it is possible to populate a second band of two-dimensional electrons. The second subband consists of electrons in the second confined quantum state of the potential well formed by the heterojunction band-edge potential step and the attractive potential of the ionized impurity. The electrons are, of course, free to move parallel to the interface, but occupation of the second subband introduces new scattering mechanisms that reduce mobility. Coincident with populating the second subband, decreases in Hall density and Hall mobility have been seen, the clearest results having been

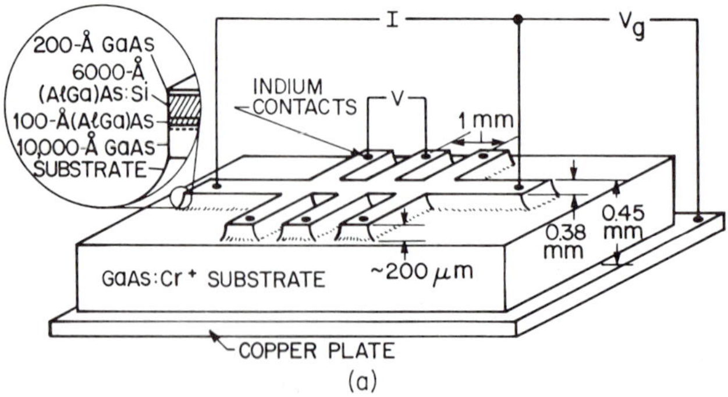

Fig. 8. (a) Sample configuration for application of backside gate voltage V_g and measurement of resistance and Hall effect in selectively doped single-interface structure. (b) Variation in mobility, carrier density, and conductivity at a backside-gated selectively doped GaAs–(Al,Ga)As single-interface structure at 4.2 K (Störmer *et al.*, 1981c).

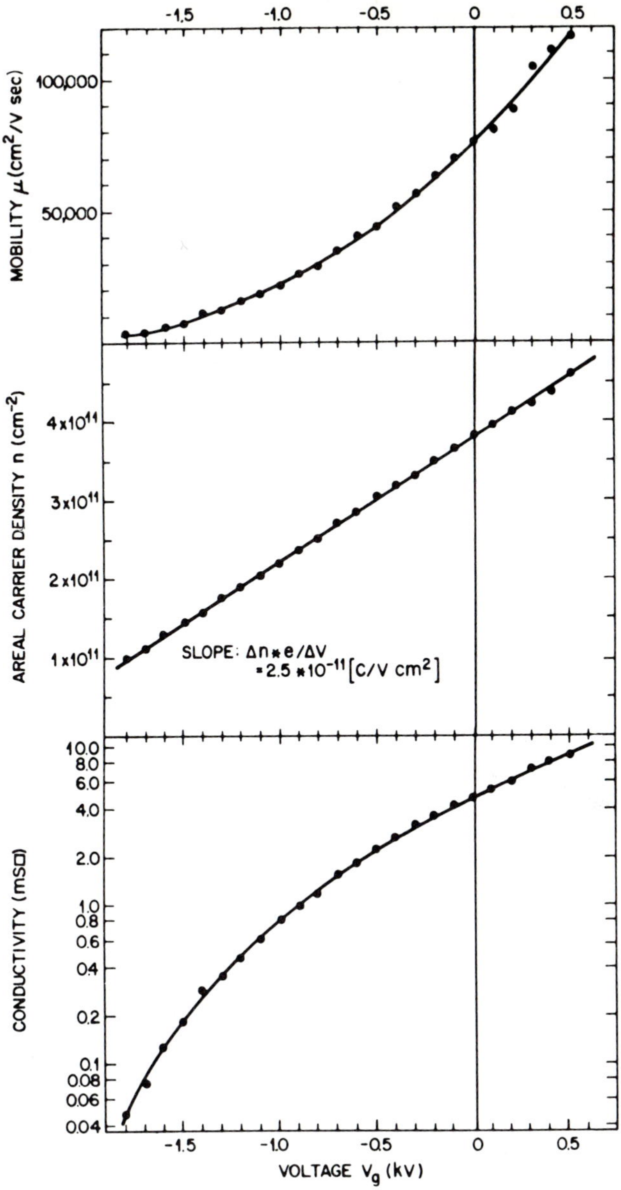

(b)

Fig. 8. *(Continued)*

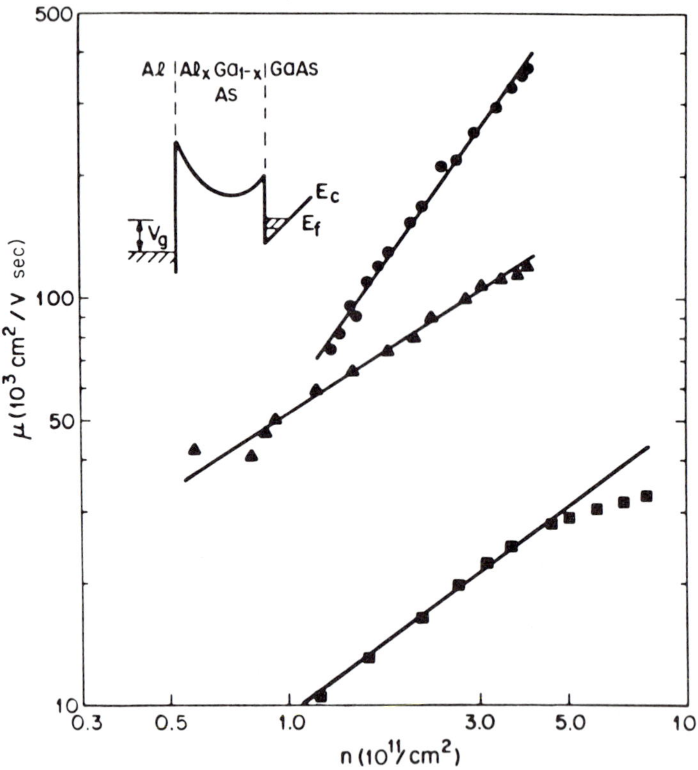

Fig. 9. Dependence of electron mobility on carrier density at a selectively doped GaAs–(Al,Ga)As single interface with carrier density controlled by an Al metal gate on top of (Al,Ga)As layer (Tsui *et al.*, 1981). The lowest curve (■, at 4.2 K) refers to a device with no undoped (Al,Ga)As spacer layer. The upper curves (●, at 4.2 K; ▲, at 78 K) are for a device with a 165-Å-thick undoped spacer layer.

established for the backside gate samples mentioned earlier (Störmer *et al.*, 1982a; Störmer, *et al.*, 1981c). At positive gate voltages, the occupation of the second subband is clearly observed in Shubnikov–de Haas measurements by appearance of a new magnetoresistance oscillation period. The sum of the occupation of the first and second subbands remains linearly dependent on gate voltage. Combination of Hall and Shubnikov–de Haas data shows that on occupying the new subband, mobility of electrons in the first subband drops by about 25% and the new (second) subband has carrier mobility ≈ 50% less than the first subband (Fig. 10). The drop in mobility of the first subband is produced by intersubband scattering, while the lower mobility of the second subband is a result of its lower Fermi wave vector magnitude. Closely related effects occur in the modulation-doped superlattice, as calculated by Mori and Ando (1979).

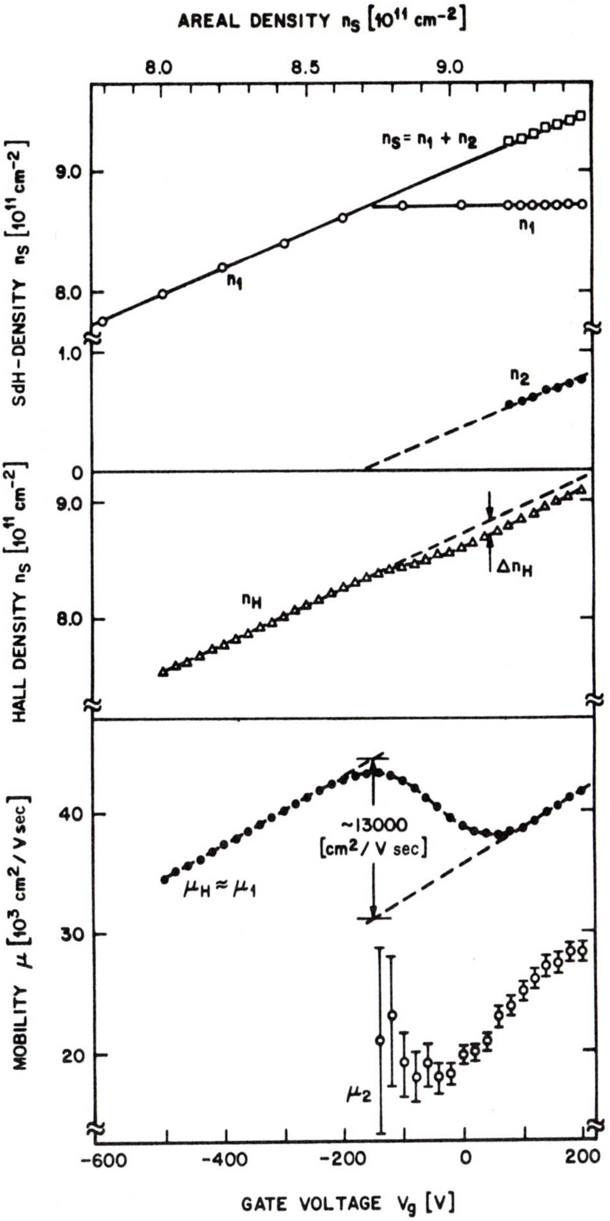

Fig. 10. Effects of occupation of upper quantum subband at selectively doped single interface. Data are shown for electron mobilities, Hall effect electron density, and electron subband densities from Shubnikov–deHaas magnetoresistance oscillations as a function of gate voltage at a back-side-gated single interface, where 1 and 2 refer to first and second quantum subbands (Störmer *et al.*, 1982a).

C. Dependences on Crystal System, Growth Conditions, and Sequences

The electron mobilities that can be obtained with the modulation doping technique are sensitively dependent on the crystal system and compositions examined as well as on the conditions of epitaxial growth and the sequence in which the layers are grown. The dependence on extrinsic parameters is especially noticeable because of the reduced scattering from the introduced dopant atoms. In addition to the widely studied GaAs–$Al_xGa_{1-x}As$ (100)-crystal-face superlattices and single interfaces, enhanced mobilities have been seen in $In_xGa_{1-x}As$–$In_yAl_{1-y}As$ on InP heterostructures (Kastalsky *et al.,* 1982), where higher MD electron mobilities might be expected because of a lower conduction band effective mass but where alloy scattering in the $In_xGa_{1-x}As$ channel might limit mobilities. Enhanced mobilities up to 10^5 cm^2/V s have been produced in this system. In the GaAs–$Al_xGa_{1-x}As$ system, enhanced hole mobilities and two-dimensional hole–gas conduction have also been observed (Störmer and Tsang, 1980; Störmer *et al.,* 1983a), but no reports of modulation doping of other crystal surface orientations have at this writing appeared. Drummond *et al.* have found and determined the effects of substrate temperature, AlAs mole fraction, and background impurity concentration on mobility enhancement in single-period and multiple-period structures (Drummond *et al.,* 1982a).

Another factor that, somewhat surprisingly, produced a strong influence on the properties of modulation-doped heterojunction is the order in which layers are grown. Almost universally, an MBE-grown heterojunction in which an undoped GaAs channel layer is grown *before* a doped $Al_xGa_{1-x}As$ barrier layer will show better behavior, i.e., more carriers and higher mobility, than the same layers grown in the reverse order (Drummond *et al.,* 1981b). This was surprising at first sight because of the initial expectation that layer properties and purity would be equivalent for layers grown with the same parameters but at different places in layered structures. Furthermore, carrier scattering in modulation-doped multilayers is more than twice as strong as at modulation-doped single interfaces on GaAs, as evidenced by the fact that the highest single-interface mobilities are significantly more than two times the highest multilayer mobilities. Subsequent work on the optical and structural properties of single- and multiple-quantum-well structures demonstrated that the properties of an $Al_xGa_{1-x}As$ interface with GaAs grown on top of it are sensitively dependent on the thickness of the $Al_xGa_{1-x}As$ and the composition of underlying layers (Miller *et al.,* 1982; Gossard *et al.,* 1982). The purity and smoothness of interfaces on $Al_xGa_{1-x}As$ under common conditions of MBE growth are poorer than for interfaces on GaAs, and furthermore, the quality of the $Al_xGa_{1-x}As$ interface degrades with increasing thickness of the $Al_xGa_{1-x}As$ layer. Significant improvements in the photoluminescence and the smoothness of MBE-grown GaAs single-quantum-well layers on $Al_xGa_{1-x}As$ were obtained by incorporation of one or more thin (10–100-Å) layers of GaAs in the $Al_xGa_{1-x}As$ just before growth of the single GaAs quantum well (Gossard *et al.*, 1982). Marked improvements in electron mobilities at modulation-doped single interfaces of GaAs on $Al_xGa_{1-x}As$ have been obtained by growth of a series of thin GaAs layers in the $Al_xGa_{1-x}As$ material just before the interface to the undoped GaAs conducting channel (Drummond *et al.*, 1983). This behavior could be produced by enrichment of surface impurity concentrations with associated increases in surface roughness during $Al_xGa_{1-x}As$ growth and by decreased impurity concentrations and decreased roughness during GaAs growth.

D. Mobilities of Hot Carriers

Carrier mobilities in modulation-doped heterojunction structures have been found to depend strongly on the strength of the external electric field. The effect of the field is to heat the carriers and thus to allow new momentum relaxation processes, including the emission of phonons. Drummond *et al.* (1981c, 1982b) and Inoue *et al.* (1981) have reported large decreases in mobility at relatively modest fields. Shah *et al.* (1983) reported the simultaneous determination of the temperature and mobility of electrons in modulation-doped quantum wells as a function of applied field. In the experiments of Shah *et al.*, the electron energy distribution function was determined from the measurements of luminescence spectra. Up to fields of 150 V/cm, the distribution function was found to be well characterized by a carrier temperature T_c higher than the lattice temperature ($T_L \approx 2$ K). Figure 11 shows the results obtained in a sample with a

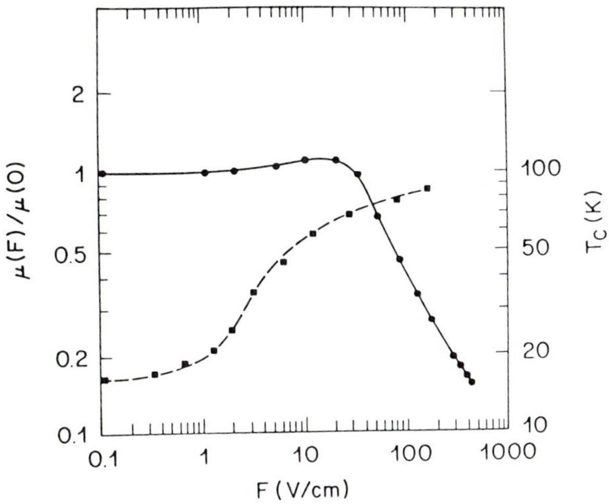

Fig. 11. Variation of T_c and normalized mobility with applied electric field for the GaAs–$Al_{0.2}Ga_{0.8}As$ MQW heterostructure, where $T_L = 2$ K (●, mobility; ■, electron temperature).

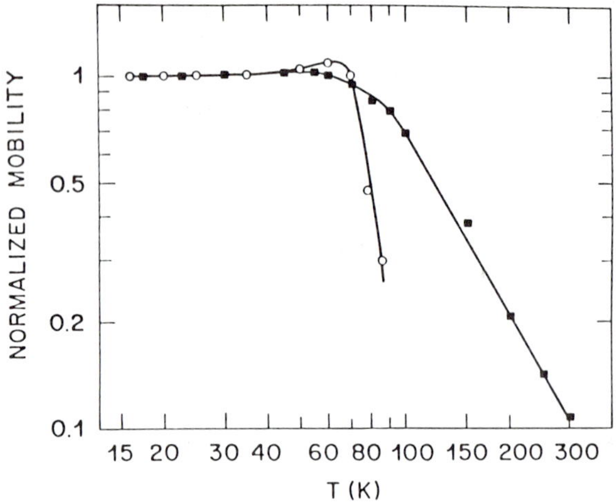

Fig. 12. For the GaAs–$Al_{0.2}Ga_{0.8}As$ MQW heterostructure the variation of normalized high-field mobility $\mu(F)/\mu(0)$ with T_c (indicated by ○, where $T = T_c$, $T_L = 2$ K, and $F \neq 0$) and the variation of normalized low-field mobility $\mu_0(T_L)/\mu_0(2\text{ K})$ with lattice temperature (indicated by ■, where $T = T_L = T_c$ and $F = 0$).

low field mobility of $\mu_0 = 63{,}000\ \text{cm}^2/\text{V sec}$ and $n = 7 \times 10^{11}\ \text{cm}^{-1}$ in each GaAs well. The electron temperature changed measurably in fields as low as 0.3 V/cm, and for 10 V/cm $T_c = 50$ K. On the other hand, the mobility $\mu(F)$ can be seen to decrease rapidly in fields larger than ~40 V/cm and $T_c \gtrsim 70$ K. The variation of $\mu(F)$ with T_c is plotted in Fig. 12, together with the measured variation of μ_0 with T_L. For $T_c \leq 70$ K the two curves are similar, suggesting that the effects can be described as due to electric-field-induced carrier heating. For $T_c \gtrsim 70$ K and $F \gtrsim 30$ V/cm there is a substantial decrease in $\mu(F)$ that is not yet understood. Field-induced ionization of traps/impurities or generation of large nonequilibrium phonon population is the most likely interpretation of this behavior. In results from a different sample Shah *et al.* (1983) have observed that the variation of $\mu(F)$ for fields $F \lesssim 1$ kV/cm is largely due to electric field heating of the electrons. This result suggests that considerable improvement in device performance should be attainable with modulation doping. However, further research in materials fabrication and device operation is certainly required.

E. Radiation Resistance

Another potentially important property related to device application of modulation-doped structures is resistance to radiation damage. Changes in transport properties under gamma-ray and electron-beam radiation have been studied for modulation-doped GaAs single interfaces (Tsui *et al.,* 1983). The MD GaAs structures were found to be highly resistant to radiation damage, particularly in comparison to silicon MOS structures. The apparent reason for this is that the relatively perfect GaAs–(Al,Ga)As interface lacks the defect centers and dangling bonds that act as trapping centers under electron–hole pair creation in the relatively amorphous $Si-SiO_2$ interface. Applications of this feature would include resistance to energetic beam lithography processes and operation in radiation environments.

V. TWO-DIMENSIONAL PHENOMENA IN MODULATION-DOPED STRUCTURES

The confinement of electrons to two-dimensional motion with long scattering lifetimes makes modulation-doped structures fruitful media for study of the physics of two-dimensional electrons. For more than ten years there has been interest in the properties of the two-dimensional electron systems that occur in space-charge layers at semiconductor interfaces (Ando *et al.,* 1982). In these systems, because of potential discontinuities and space-charge electric fields, the electron motion normal to the interfaces is quantized into discrete energy levels. The electrons are free to move in a plane parallel to the interface in a fashion that is similar to that of an idealized two-dimensional electron gas. The classic cases are those of the accumulation and inversion layers at metal–insulator—semiconductor (MIS) and metal–oxide–semiconductor (MOS) structures. There has also been an increasing interest in the 2D electron systems at semiconductor heterostructures made by molecular beam epitaxy.

The carriers confined at semiconductor interfaces constitute some of the most interesting systems for studies of electron behavior under conditions of reduced dimensionality. Extensive experimental studies have been carried out by transport, magnetotransport, and optical measurements. The optical methods include several techniques. Inelastic light scattering has been introduced as a novel tool in the investigation of the elementary excitations of 2D semiconductor plasmas. Among the more exciting developments in the field is the discovery of the fractional quantization of the Hall effect (Tsui *et al.,* 1982b). In this section we review some of these developments. We focus on the inelastic light scattering work and the quantized Hall effect.

A. INELASTIC LIGHT SCATTERING

Burstein *et al.* (1979) pointed out that resonant inelastic light scattering has the sensitivity required for studies of electrons at semiconductor interfaces. The method also yields separate spectra of single-particle and collective excitations, a feature that is important in the determination of electron energy levels and Coulomb interactions. This proposal was soon followed by the observations of light scattering by *intersubband excitations,* between discrete energy levels, of electrons in modulation-doped GaAs–(AlGa)As heterostructures (Abstreiter and Ploog, 1979; Pinczuk *et al.,* 1979a). Although the method had been used in studies of 3D semiconductor plasmas (Mooradian, 1972; Klein, 1975; Pinczuk *et al.,* 1979b) and of semiconductor heterostructures (Manuel *et al.,* 1976; Sai-Halasz *et al.,* 1978; Merlin *et al.,* 1980; Colvard *et al.,* 1980), these were the first observations of light scattering by excitations of electrons confined at semiconductor interfaces.

The 2D electron systems at GaAs–(AlGa)As heterostructures have also been studied extensively. In the case of intersubband spectroscopy the results have revealed energy-level structures and resonant behavior of scattering intensities (Abstreiter and Ploog, 1979; Pinczuk *et al.,* 1979a), collective Coulomb interactions (Pinczuk *et al.,* 1980a), correlations with electron mobilities (Pinczuk *et al.,* 1981a), and the effects of photoexcitation at high intensities. These results confirmed the prediction of Burstein *et al.* (1979) and yielded substantial information on the physical properties of electrons in the heterostructures. There are also two reports on the observation of excitations associated with *in-plane* motion. One involves the transitions between Landau levels in a large magnetic field normal to the interface (Worlock *et al.,* 1981; Tien *et al.,* 1982). The other is the observation and measurement of the dispersion of the plasma oscillation of the layered electron gas in doped multiple heterostructures (Olego *et al.,* 1982).

Resonant light scattering has also been observed at other semiconductor interfaces. Among them are the accumulation layers at MIS structures of InAs (Ching *et al.,* 1980), Si (Abstreiter *et al.,* 1982), and InP (Abstreiter *et al.,* 1983). Photoexcited electron–hole systems have been studied in GaAs doping superlattices (Döhler *et al.,* 1981; Zeller *et al.,* 1982b). Light scattering by 2D electron systems has also been reported in doped Ge–GaAs heterostructures (Merlin *et al.,* 1982).

1. Light-Scattering Mechanisms and Selection Rules

The light-scattering mechanisms and selection rules that apply to 2D electron plasmas in semiconductors have been considered by Burstein *et al.* (1979, 1980). Within the effective-mass approximation, light scattering is anticipated to be similar to that from 3D semiconductor plasmas. Resonant enhancements at this writing are expected for photon energies near optical gaps that involve directly the confined electrons. In the case of GaAs layers these are the fundamental E_0 and the spin-orbit split-off $E_0 + \Delta_0$ gaps near the Γ point of the Brillouin zone. In *n*-type Ge layers the resonance occurs at the L point near the E_1 and $E_1 + \Delta_1$ optical gaps. In *p*-type channels of Si the resonance is at the E_0 energy gap near the Γ point.

The nearly backscattering geometry shown in Fig. 13 is frequently used.

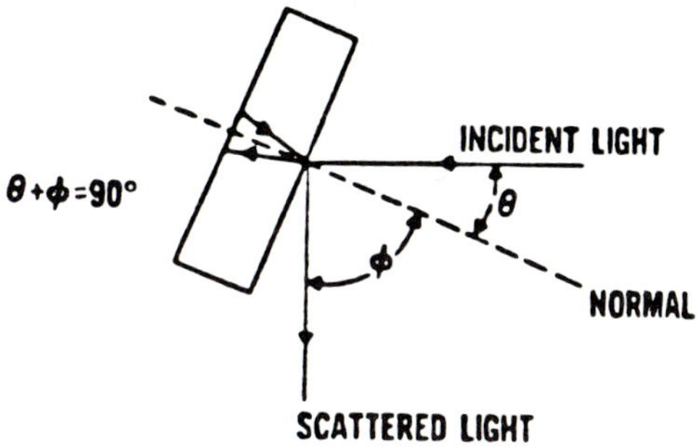

Fig. 13. Nearly backscattering geometry.

Light propagates inside the sample along directions close to the normal to the plane of the 2D system. Under these conditions the in-plane and normal components of the scattering wave vector are given by

$$k_{\parallel} = (2\pi/\lambda)(\sin\theta - \cos\theta) \tag{2}$$

and

$$k_{\perp} = (4\pi/\lambda)\eta(1 - (1/4\eta^2)), \tag{3}$$

where λ is the laser wavelength and η the refractive index. The component $k_{\parallel}$ can be varied from a small value, at $\theta \approx 45°$, up to a maximum of $\sim 10^5\ \text{cm}^{-1}$. In GaAs $\eta \approx 3.6$. Then, $k = 7 \times 10^5\ \text{cm}^{-1}$ near the $E_0 + \Delta_0$ energy gap at ~ 1.9 eV. Near the E_0 gap at ~ 1.5 eV, $k_{\perp} \approx 5 \times 10^5\ \text{cm}^{-1}$. The majority of the light-scattering experiments in GaAs layers have been carried out near the $E_0 + \Delta_0$ resonance. Two types of spectra are measured. The *polarized* spectra, obtained with parallel incident and scattered light polarizations, are assigned to *collective* charge-density excitations of the electrons. *Depolarized* spectra, in which the polarizations are orthogonal, are assigned to *single-particle* spin-density excitations. These assignments are similar to those that explain the spectra of bulk *n*-type GaAs (Burstein *et al.*, 1979, 1980).

2. Intersubband Excitations

The initial light-scattering experiments involved intersubband excitations. Figure 14 illustrates the two possible types of electron transitions between states in two-dimensional subbands. Studies of intersubband excitations are carried out by means of a nearly backscattering geometry with $\theta \approx 45°$, in which $k_{\parallel}$ is small. Therefore, provided there is wave vector conservation ($k_{\parallel} = q$) the intersubband transitions can be considered *vertical* as in Fig. 14a.

In the spectra shown in Fig. 15, z and $\bar{z}$ are the propagation directions of the incident and scattered light, close to (001) and (00$\bar{1}$) crystals axes. The in-plane directions of light polarizations x' and y' are (110) and (1$\bar{1}$0). The depolarized

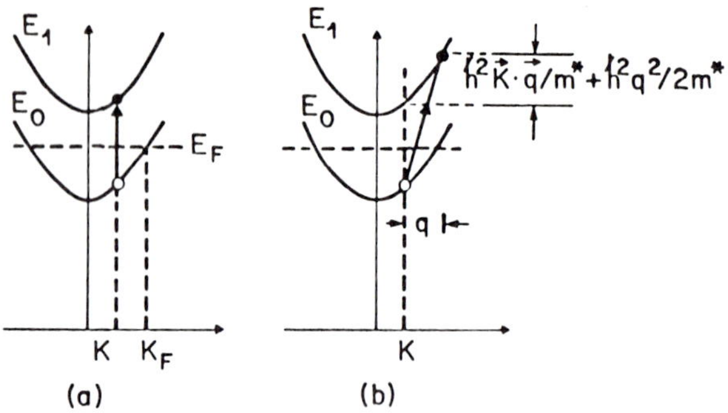

Fig. 14. Schematic representation of intersubband transitions: (a) vertical transitions, (b) non-vertical transitions ($q \neq 0$).

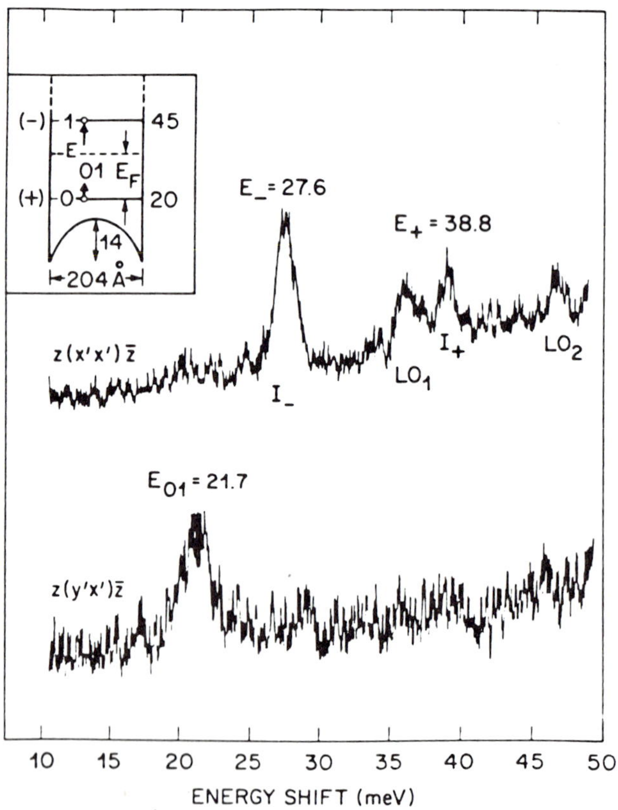

Fig. 15. Light-scattering spectra, in the range of the lowest intersubband transition E_{01}, of GaAs–$Al_{0.18}Ga_{0.82}As$ MQW heterostructure at $n_S = 4.2 \times 10^{11}$ cm^{-2} and $T = 2$ K. The inset shows energy levels, band bending, and the Fermi energy.

$z(y'x')\bar{z}$ spectrum shows a single band that has been assigned the lowest spin-density intersubband excitation. Single-particle intersubband energy close to the spacing between the subbands is represented by E_{01} (Pinczuk *et al.*, 1980). In the polarized $z(x'x')\bar{z}$ spectrum the peaks at $E_+ = 38.8$ meV and $E_- = 27.6$ meV, labeled I_+ and I_-, are assigned (Pinczuk *et al.*, 1980a) to the coupled collective intersubband-LO phonon excitations considered by Burstein *et al.* (1980). Similar coupled modes were also observed by Abstreiter *et al.* (1981).

a. Depolarization Field Effects. The difference between the energies of single-particle and collective excitations are a manifestation of the depolarization electric field (or resonant screening) associated with intersubband excitations. The effect, first predicted by Chen *et al.* (1976), has been invoked in the interpretation of subband optical absorption in Si MOS structures (Allen *et al.*, 1976; Ando, 1977) and in InSb and InAs MIS structures (Chang *et al.*, 1980). Light-scattering spectra like those shown in Fig. 15 have for the first time made possible simple determination of depolarization field effects.

The collective intersubband excitations have been considered by Dahl and Sham (1977). This treatment has been extended to incorporate the specific conditions that exist in multiple quantum well (MQW) heterostructures (Pinczuk *et al.*, 1980a; Pinczuk and Worlock, 1982), including the limitations associated with the mirror symmetry of the quantum wells. In the case of results like those shown in Fig. 15, the coupled mode energies are the solutions of

$$1 - [(E_{\mathrm{T}}^2 - E_{\pm}^2)/(E_{\mathrm{L}}^2 - E_{\pm}^2)][E_{\mathrm{p}}^2/(E_{\pm}^2 - E_{01}^2)] = 0, \tag{4}$$

where $E_{\mathrm{T}} = 33.6$ meV and $E_{\mathrm{L}} = 36.7$ meV are the energies of TO and LO phonons of GaAs. The plasma energy E_{p} describes the depolarization field effects. The values of E_{P} that best fit the data from a number of samples have been interpreted in terms of Coulomb matrix elements for the transition (Pinczuk *et al.*, 1980a; Pinczuk and Worlock, 1982). The values of Coulomb matrix elements determined from the experiments are in good agreement with the ones evaluated by means of subband envelope functions of simple model calculations (Pinczuk *et al.*, 1980a, 1980c; Fishman *et al.*, 1981). Separate light-scattering spectra of single-particle and collective intersubband excitations have also been obtained from single layer 2D electron systems at modulation-doped GaAs–(AlGa)As heterostructures (Pinczuk *et al.*, 1981b; Pinczuk and Worlock, 1982).

b. Correlation with Transport Properties. One of the intriguing questions in the light-scattering spectroscopy of intersubband excitations is the relation between spectral line shapes and the transport properties. One such correlation has been identified (Pinczuk *et al.*, 1981a). It takes the form of an unusual dependence of the widths of spectral bands on incident photon energy.

These studies have been carried out in the four samples whose mobilities are shown in Fig. 6. In Fig. 16 we show $z(x', y')\bar{z}$ spectra obtained from three of the

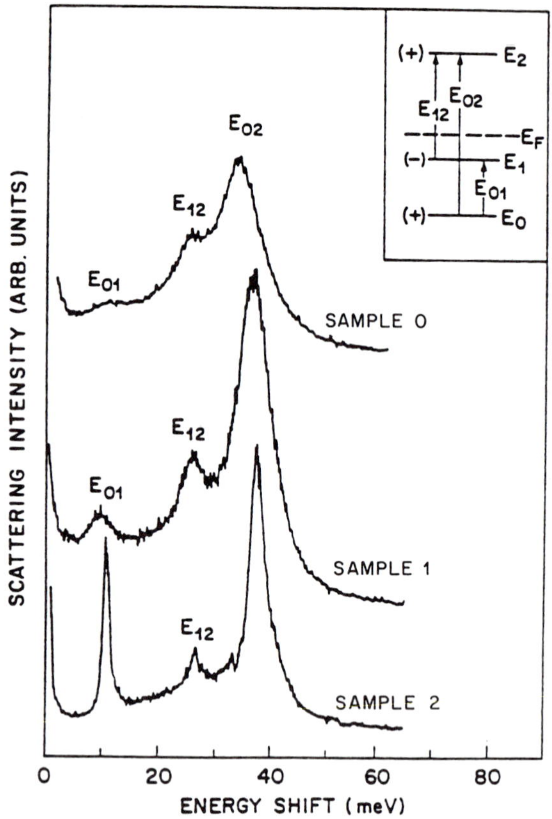

Fig. 16. Depolarized backscattering spectra of three modulation-doped MQW heterostructures [GaAs–($Al_{0.12}Ga_{0.88}$)As at $T = 2$ K; $z(y', x')\bar{z}$, $\hbar\omega_{LASER} = 1.900$ eV] (the mobilities are shown in Fig. 6). Sample 0: $d_1 = 255$ Å, $d_2 = 302$ Å, $d_3 = 0$; Sample 1: $d_1 = 245$ Å, $d_2 = 292$ Å, $d_3 = 50$ Å; Sample 2: $d_1 = 250$ Å, $d_2 = 292$ Å, $d_3 = 151$ Å. The inset shows the assignments of the single-particle intersubband excitations.

samples. The bands are assigned to the single particle intersubband excitations shown in the inset. We can see that the widths of all the bands decrease with increasing Hall mobility, as well as with the thickness of the undoped (AlGa)As spacer.

Even more striking is the way the widths depend on laser photon energy. Figure 17 shows the FWHM of the band of the E_{01} transition as function of incident photon energy for the three samples with the highest mobilities. Although a correlation between FWHM and electron mobilities could be anticipated, the results in Fig. 17 show that in samples with lower mobility there is an unexpected dependence of width on incident photon energy, with a peak at photon energies close to the maximum in resonant enhancement (indicated as E_R in Fig. 17).

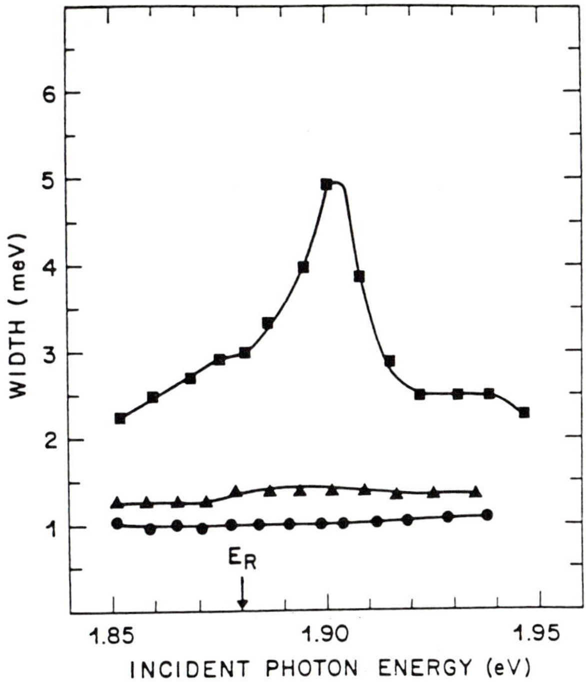

Fig. 17. Full width at half-maximum of the E_{01} transition for GaAs–($Al_{0.12}Ga_{0.88}$)As as a function of incident photon energy. (■, $d_3 = 50$ Å; ▲, $d_3 = 99$Å; ●, $d_3 = 151$ Å.)

In the interpretation of these results it was proposed (Pinczuk *et al.*, 1981a) that wave vector relaxation processes related to those that limit the electron mobilities play a role in resonant light scattering by intersubband excitations. In these cases there is breakdown in the condition of wave vector conservation. The intersubband transitions active in nearly backscattering spectra no longer need to be vertical, as in Fig. 14a, but also occur with an in-plane wave vector transfer $q \neq k_{\parallel}$ as in Fig. 14b. The energies of nonvertical transitions are shifted from the subband spacing as shown in Fig. 14b, where $|\mathbf{K}| \leq K_F$ the Fermi wave vector. The spectral bands of nonvertical transitions are expected to show considerable width or even to appear as a quasi-continuum. As it occurs in *n*-type GaAs (Pinczuk *et al.*, 1979b) and *p*-type GaAs (Olego and Cardona, 1979), light scattering with relaxed wave vector conservation has a strong resonant behavior.

c. Photoexcited Electrons. Resonant light scattering by photoexcited electrons was first reported by Pinczuk *et al.* (1981c) in *undoped* GaAs–(AlGa)As MQW heterostructures. Zeller *et al.* (1982b) studied the effects of photoexcitation in the modulation-*doped* heterostructures. Experiments like these allow light-scattering studies of 2D systems in which the electron density can be varied by changing the photoexcitation intensity.

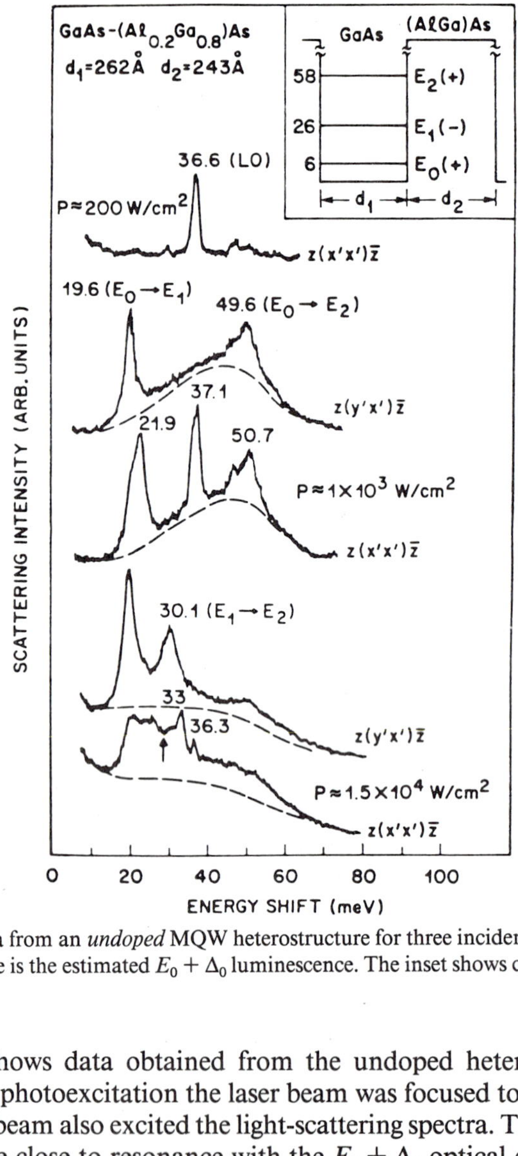

Fig. 18. Spectra from an *undoped* MQW heterostructure for three incident laser power densities. The dotted line is the estimated $E_0 + \Delta_0$ luminescence. The inset shows calculated quantum-well energy levels.

Figure 18 shows data obtained from the undoped heterostructures. To achieve carrier photoexcitation the laser beam was focused to a spot of radius $\sim 10\ \mu$m. This beam also excited the light-scattering spectra. The incident photon energies are close to resonance with the $E_0 + \Delta_0$ optical energy gap. At a relatively low power density of $P \sim 200$ W/cm^2, the spectrum shows only the peak of the LO phonon of GaAs. This is an indication that at these power densities carrier photoexcitation is low. When the power density is increased to $\sim 10^3$ W/cm^2 new bands appear. The ones in $z(y'x')\bar{z}$ spectra are assigned to the spin-density, single-particle intersubband transitions of photoexcited *electrons,* as indicated Fig. 18. The bands in the $z(x'x')\bar{z}$ spectrum are assigned to the corresponding collective excitations coupled to LO phonons. Transitions by the photoexcited holes were not observed. Their scattering cross section is believed to be much smaller because it is not enhanced at the $E_0 + \Delta_0$ energy gap (Burstein *et al.*, 1979).

At a higher power density of $P \sim 1.5 \times 10^4$ W/cm^2 there is a new band, at 30.1 meV, in the $z(x'y')\bar{z}$ spectrum. This has been interpreted as arising from transitions of photoexcited electrons that, at large power densities, populate the next-higher conduction subband. At these power densities the $z(x'x')\bar{z}$ spectra, assigned to collective excitations, are rich in structure related to the occupation of more than one conduction subband. The dip in the scattering intensity just below the $E_0 \rightarrow E_1$ transition, indicated by an arrow, has been explained by the coupling of collective excitations associated with odd parity $E_0 \rightarrow E_1$ and $E_1 \rightarrow E_2$ transitions.

These results show that resonant light scattering by photoexcited electrons is a powerful tool for the investigation of 2D electron systems. Among the more exciting applications are the studies in GaAs doping multilayer structures. This work has revealed the 2D character of the conduction band states in these structures (Döhler *et al.*, 1981) and, at high photoexcitation intensities, has demonstrated the transition to 3D behavior (Zeller *et al.*, 1982a).

3. Plasma Oscillations

Olego *et al.* (1982) have reported light scattering by the *plasma oscillation* in modulation-doped GaAs–(AlGa)As MWQ heterostructures. The oscillations are the collective modes associated with the in-plane, 2D motion of the electrons confined in the GaAs layers. Like the plasmons of an idealized layered electron gas (Fetter, 1974), these modes are expected to show large dispersive effects in the plasma frequency (as a dependence of $k_\parallel$ and $k_\perp$) (Das Sarma and Quinn, 1982; Bloss and Brody, 1982). The light-scattering determinations confirmed these predictions.

The experiments were carried out with photon energies (~ 1.58 eV) close to the fundamental optical energy gap of GaAs (~ 1.52 eV). This allowed larger penetration depths (~ 2 μm) than those near the $E_0 + \Delta_0$ gap (~ 0.5 μm), while retaining the benefits of resonant enhancements and relatively low luminescence backgrounds. Figure 19a shows a spectrum from a GaAs–($Al_{0.2}Ga_{0.8}$)As heterostructure with $d_1 = 262$ Å, $d_2 = 317$ Å, and $n_s = 7.3 \times 10^{11}$ cm^{-2}. The E_{01} and $E_{\bar{0}1}$ bands are associated with single-particle and collective intersubband excitations. In addition, there is a new low energy peak at 3.5 meV. Its most remarkable feature, displayed in Fig. 19b, is that its frequency depends on the angle of incidence. Through Eq. (2) this implies a dependence on $k_\parallel$. Such a dispersive behavior led Olego *et al.* (1982) to assign the peak to the plasma oscillation of the MQW heterostructure.

Figure 20 shows the plasma dispersion measured in two samples. These

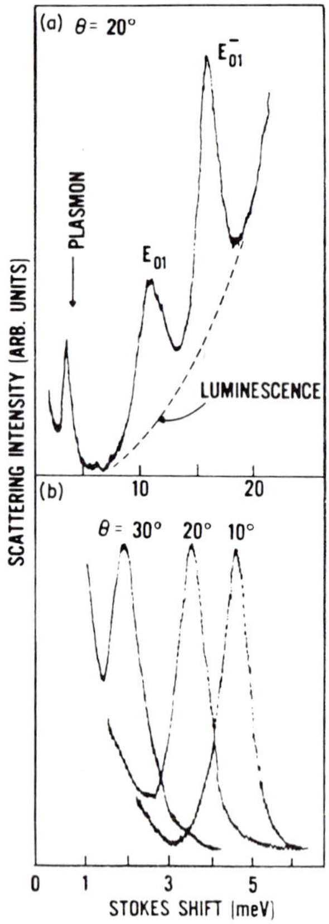

Fig. 19. (a) Nearly backscattering spectrum showing the plasmon line of a modulation-doped GaAs–(AlGa)As MQW heterostructure, (b) plasmon line for different angles of incidence.

results were interpreted by means of a plasma frequency of the form (Fetter, 1974; Das Sarma and Quinn, 1982; Bloss and Brody, 1982)

$$\omega_P = \left[\frac{2\pi ne^2}{\epsilon_M m^*} k_\parallel \frac{\sinh k_\parallel d}{\cosh k_\parallel d - \cos k_\perp d}\right]^{1/2}, \tag{5}$$

where m^* is the effective mass, ϵ_M is the dielectric constant of the multilayers, and $d = d_1 + d_2 + 2d_3$. The solid lines of Fig. 20 are the evaluations of Eq. (5). The dashed lines are the extrapolations of the linear behavior. The agreement between measured and calculated values is very good. The small differences can be explained by the uncertainties in the values of the sample parameters. On exciting the plasmons, the possibility exists for emission of radiation by radiative de-excitation. This was observed originally in silicon MOSFETs (Tsui *et al.*, 1980a) and more recently from single-interface modulation-doped GaAs

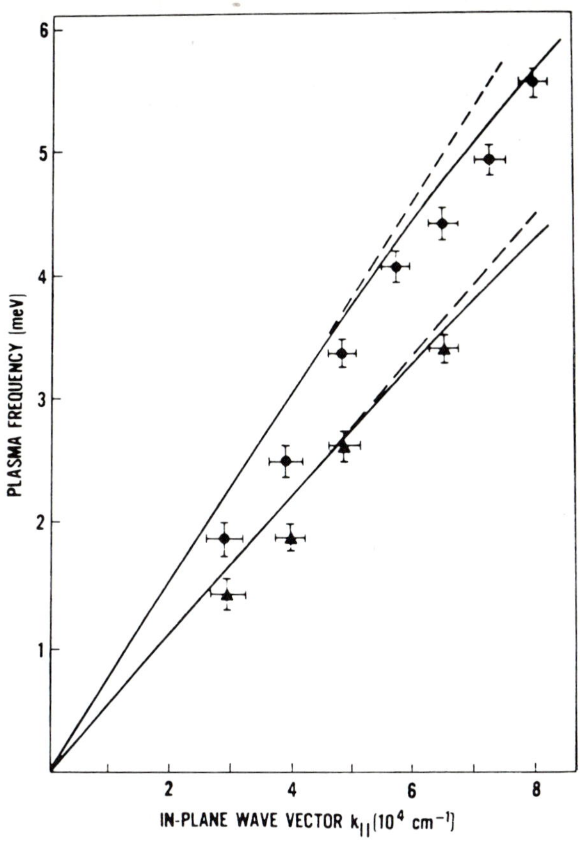

Fig. 20. Dispersions of the plasma frequency measured in two samples. (●: $d = 890$ Å, $n = 7.3 \times 10^{11}$ elec cm^{-2}, $k_{\perp}d = 4.94$; ▲: $d = 827$ Å, $n = 5.5 \times 10^{11}$ elec cm^{-2}, $k_{\perp}d = 4.67$.) The lines are calculations based on Eq. (5).

heterostructures excited by passing a current along the channel (Höpfel *et al.*, 1982). A grating structure in close proximity to the two-dimensional electron gas is necessary to couple out the far-infrared radiation. The radiation intensity is apparently dependent on electron mobility, and high mobility samples with undoped barrier spacer layers show weaker plasmon emission than lower mobility samples. Intensities up to $\approx 3 \times 10^{-8}$ W at 3- to 5-μm wavelengths were measured. The dependence on mobility suggests that impurity scattering is instrumental in exciting the plasmons.

B. Magnetic Field Effects

Application of a magnetic field B perpendicular to a plane of electrons quantizes the two remaining degrees of translation freedom into cyclotron

(Landau) orbits. The spacing between Landau levels $\hbar\omega_c = eB/m^*c$ has been measured directly in modulation-doped structures by cyclotron resonance absorption in MD multilayers (Störmer *et al.*, 1979) and by excitation between Landau levels with Raman scattering (Tien *et al.*, 1982). The Landau-level spectrum has also been probed by magnetophonon scattering, in which resistance singularities are observed in a single-interface structure when the separation between Landau levels is equal to or is a submultiple of an LO phonon energy (Tsui *et al.*, 1980b). The Landau-level separation is also responsible for Shubnikov–de Haas magnetoresistance oscillations that occur when Landau levels pass through the Fermi energy as magnetic field is increased. The period of the oscillations accurately measures the carrier density (Störmer *et al.*, 1981c; Tsui *et al.*, 1981). The cyclotron resonance measures an effective electron mass at the Fermi energy that is slightly larger than the effective mass at the bottom of the conduction band because of nonparabolicity in the conduction band dispersion. The magnetophonon resonance measures a polaron mass that is not measurably altered from the cyclotron mass. The Raman scattering from excitation between Landau levels had been unexpected because no theory predicts scattering from in-plane motion of carriers. The cyclotron mass deduced from the Raman excitation is nearly equal to the GaAs conduction band mass at fields up to 14 T. At higher magnetic fields, where the extreme quantum limit of all electrons in the lowest Landau level is reached, anomalies were observed in the frequency, intensity, and selection rules of the transitions (Tien *et al.*, 1982). Cyclotron resonance observations by far-infrared transmission through modulation-doped GaAs–$Al_xGa_{1-x}As$ single interfaces have shown oscillations of the cyclotron resonance linewidth that correlate with Landau-level filling (Englert *et al.*, 1983). Both single-interface and multilayer MD samples have shown far-infrared emission at the cyclotron resonance frequency when excited by passing current through the layers (Höpfel *et al.*, 1982).

C. Quantum Hall Effect

At fields above ~4 T at temperatures below ~4.2 K, the Shubnikov–de Haas oscillations referred to above in the electrical resistance versus perpendicular magnetic field of high-mobility modulation-doped GaAs–$Al_xGa_{1-x}As$ heterojunctions exhibit remarkable behavior, with broad resistance minima approaching zero resistance (Störmer *et al.*, 1982b). Some samples have shown resistances below 10^{-6} ohms/sq corresponding to a resistivity in the current-carrying volume of the heterostructure of less than 10^{-12} ohm cm, which is less than the resistivity of any known normal-state material. At the same fields as the minima of parallel resistance, the Hall resistance, defined as Hall voltage divided by current along the layer, assumes wide, flat plateaus (Tsui and Gossard, 1981) (Fig. 21). The value of Hall resistance at these plateaus is accurately h/ne^2, where h is Planck's constant and n an integral number of filled Landau

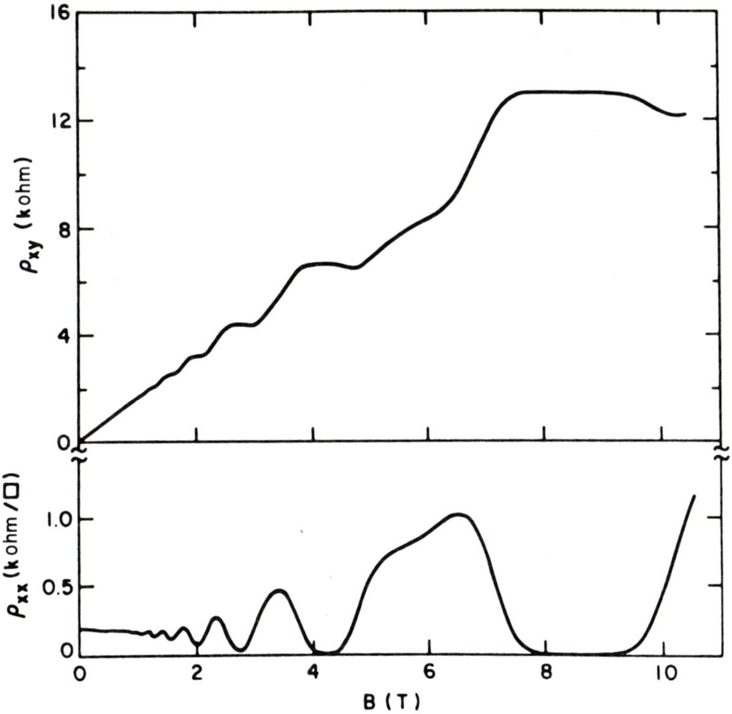

Fig. 21. Quantum Hall effect observed in Hall resistance ρ_{xy} and parallel resistance ρ_{xx} of selectively doped GaAs–(Al,Ga)As single interface (GaAs–$Al_{0.3}Ga_{0.7}As$ at 4.2 K). Plateau in ρ_{xy} between 8 and 9 T has value $h/2e^2$, corresponding to occupation of single, spin-degenerate Landau level (Tsui and Gossard, 1981).

levels, including spin, at the operating field. This effect is the *quantized Hall effect* (QHE), which had been discovered initially in silicon MOSFET structures (von Klitzing *et al.,* 1980). The Hall resistance plateaus and resistance minima occur at fields in which an integral number of Landau levels are filled, although the explanation of the extreme width and flatness of the plateaus is still incomplete. At temperatures below 100 mK, the width of the plateaus has been measured to be more than 90% of the center-to-center field difference between plateaus (Paalanen *et al.,* 1982).

Relative to silicon structures, the GaAs heterostructure has a lower effective mass, requiring a lower magnetic field to produce a given separation between Landau levels. Furthermore, because the density of interface states is less at the GaAs–$Al_xGa_{1-x}As$ junction than at a silicon–oxide interface, the quantized Hall effect can be observed at lower carrier density in GaAs heterojunctions, allowing use of lower magnetic fields to reach a given plateau and allowing access to the extreme quantum limit in which all electrons occupy the lowest Landau level.

The Hall resistance plateaus have been found to be flat to at least one part in 10^6 over field ranges greater than 0.1 T in samples with the flattest plateaus. This enables use of the GaAs QHE for a fundamental standard of resistance and as a means for determination of e^2/h and thus the fine structure constant (Tsui *et al.,* 1982a). The flatness and width of the Hall resistance plateaus are not equal among samples of comparable mobility, however, and the mechanism for occurrence of the flat plateaus is apparently dependent on details of the structure of the junction. A likely source of the plateau width is impurity-produced potential fluctuations that produce electron paths along lines of intersecting Landau and Fermi levels (Kazarinov and Luryi, 1982). These lines would form a connected cluster only for magnetic fields in ranges between the plateaus; the resistance is normal in these ranges. On the plateaus, the current would flow in extended nondissipating states parallel to the bridge and perpendicular to the electric field, whence the vanishing resistivity described earlier.

On increasing the magnetic field acting on a modulation-doped GaAs–$Al_xGa_{1-x}As$ interface into the region of the extreme quantum limit, where all electrons are the lowest Landau level, additional new features have unexpectedly been seen (Tsui *et al.,* 1982b) (Fig. 22). In the vicinity of magnetic fields corresponding to fractional occupation of the lowest Landau level, additional minima are observed in the resistance, and additional plateaus are seen in the Hall resistance. Minima and plateaus have been found at Landau level occupations of $\frac{1}{3}$, $\frac{2}{3}$, $\frac{2}{5}$, $\frac{3}{5}$, $\frac{4}{5}$, and $\frac{2}{7}$ as well as at $\frac{4}{3}$ and $\frac{5}{3}$ (Störmer *et al.,* 1983d) (Fig. 23). At odd-denominator fractional Landau level occupation f, the Hall resistance plateau values tend toward values of h/fe^2. The observability of the new features requires temperatures below 1.6 K and very high magnetic fields. Whereas peaks and gaps in the energy spectrum of Landau levels had been invoked to explain the QHE features at integral occupation of the Landau levels, no such features occur at fractional positions for noninteracting electrons. The new phenomena at fractional occupation require a fundamentally new explanation. Such an explanation has been devised by Laughlin (1983), who has found wave functions corresponding to new forms of interacting condensed electron quantum liquid that would become stable at fractional Landau-level occupation. These are new many-particle states that are separated by energy gaps from the excitations of the system. The excitations are also novel, consisting of fractionally charged quasi-electrons and quasi-holes.

The quantized Hall effect has also been observed in the two-dimensional holes at a p-type modulation-doped (Al,Ga)As : Be – GaAs interface (Störmer *et al.,* 1983a). Accurately quantized plateaus and plateaus at fractional Landau-level occupation are also seen in this case. The hole gas shows several cyclotron resonance peaks.

In principle, the quantized Hall effect could also be observed in multiple-layer structures as well as in the previously studied single-interface MOSFET

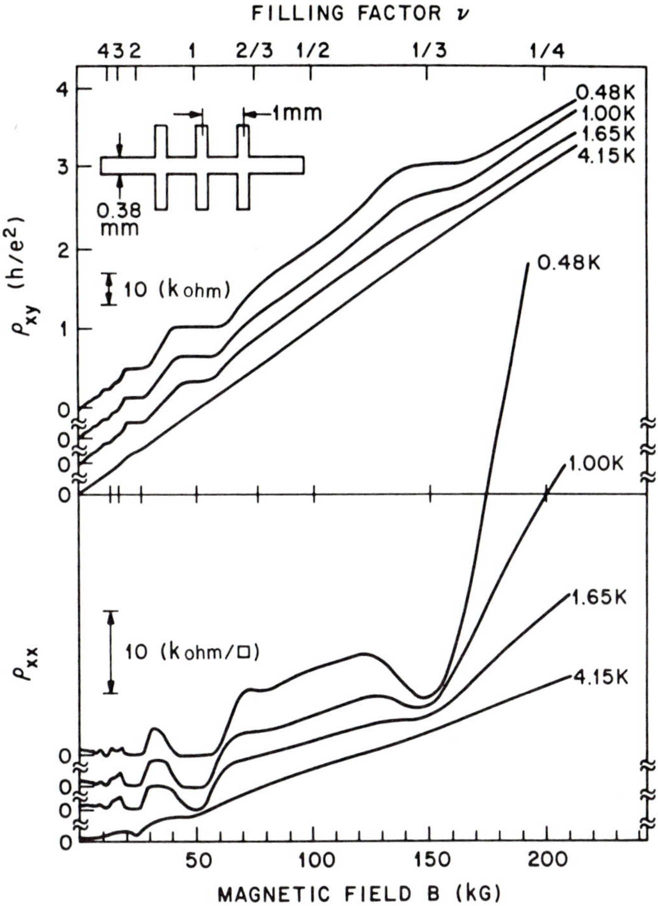

Fig. 22. Magnetoresistance and Hall effect in extreme quantum limit. Plateau in ρ_{xy} at 15 T and 0.48 K has the value $3h/e^2$ at $\nu = \frac{1}{3}$, corresponding to $\frac{1}{3}$ occupation of lowest Landau level (Tsui *et al.*, 1982b).

and GaAs structures. This has been successfully attempted (Störmer *et al.*, 1983b) by growth of a multiple-layer GaAs–$Al_xGa_{1-x}As$ structure. Structures with multiple layers also allow study of extensive properties of interface and two-dimensional systems that might otherwise be too weak to measure. For example, the first measurement of the magnetic susceptibility of the two-dimensional electron gas was accomplished by means of a stack of over 4000 epitaxial multilayers assembled from the modulation-doped structure described. By using a SQUID magnetometer, oscillations of the field-dependent susceptibility (de Haas–van Alphen oscillations) were observed with the same period as the electrical resistance oscillations, giving information on the electronic density of states (Störmer *et al.*, 1983b).

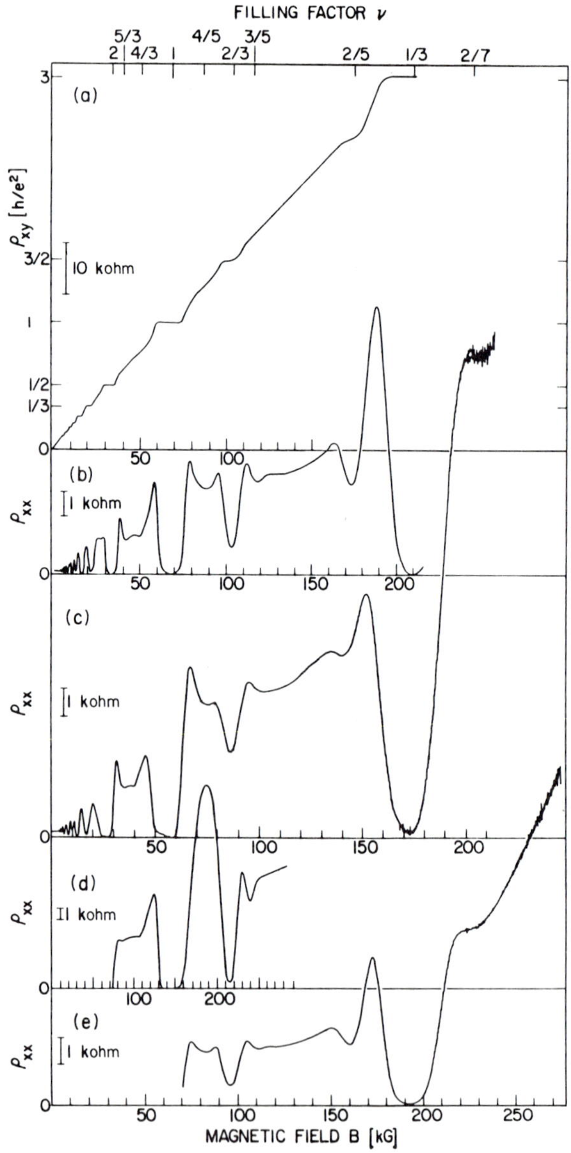

Fig. 23. Magnetoresistance and Hall effect data at high magnetic fields and at temperatures ~0.55 K in four GaAs–(Al,Ga)As selectively doped heterostructures. Resistance minima and development of plateaus and minima in Hall resistance at fractional Landau level filling ($\nu = \frac{1}{3}, \frac{2}{3}, \frac{4}{3}, \frac{5}{3}, \frac{2}{5}, \frac{3}{5}, \frac{4}{5}$, and $\frac{2}{7}$) are evidence of new ordered states. Magnetic fields have been scaled to fit a common filling factor scale (top). Curves (a), (b), (c), and (e) refer to 2D-electron systems and curve (d) to a 2D-hole system (Störmer *et al.*, 1983b).

VI. CONCLUSIONS

The ability to selectively dope semiconductor heterojunctions has greatly expanded the range of performance and phenomena accessible in doped semiconductors. The modulation doping technique permits wide flexibility and innovation in structure design but simultaneously puts strict demands on the crystal growth parameters associated with control, smoothness, stability, and purity. The contemporary molecular beam epitaxy technique is showing success in meeting these challenges, and it is not unreasonable to expect continued progress in development and production of modulation-doped structures.

ACKNOWLEDGMENTS

We wish to acknowledge the collaboration of W. Wiegmann, H. L. Störmer, D. C. Tsui, J. Shah, J. M. Worlock, D. Olego, R. Dingle, M. A. Paalanen, M. E. Cage, R. F. Dziuba, B. F. Field, S. Luryi, R. F. Kazarinov, V. Narayanamurti, T. Haavasoja, R. C. Miller, P. M. Petroff, W. T. Tsang, M. B. Panish, and J. Brooks in the work described here.

REFERENCES

Abstreiter, G., and Ploog, K. (1979). *Phys. Rev. Lett.* **42,** 1308–1311.
Abstreiter, G., Zeller, C., and Ploog, K. (1981). *In* "Gallium Arsenide and Related Compounds 1980" (H. S. Thim, ed.), pp. 741–749. Institute of Physics, Bristol.
Abstreiter, G., Claessen, U., and Traenkle, G. (1982). *Solid State Commun.* **44,** 673–676.
Abstreiter, G., Huber, R., Traenkle, G., and Vinter, B. (1983). *Solid State Commun.* **47,** 651–654.
Allen, S. J., Jr., Tsui, D. C., and Vinter, B. (1976). *Solid State Commun.* **20,** 425–428.
Ando, T. (1977). *Solid State Commun.* **21,** 133–136.
Ando, T., Fowler, A. B., and Stern, F. (1982). *Rev. Mod. Phys.* **54,** 437–672.
Bloss, W. L., and Brody, E. M. (1982). *Solid State Commun.* **43,** 523–528.
Burstein, E., Pinczuk, A., and Buchner, S. (1979). *In* "Physics of Semiconductors" (B. L. H. Wilson, ed.), pp. 1231–1234. Institute of Physics, Bristol.
Burstein, E., Pinczuk, A., and Mills, D. L. (1980). *Surf. Sci.* **98,** 451–468.
Chang, H. R., Reisinger, H., Schaffler, F., Scholz, J., Wiesinger, K., and Koch, F. (1980). *J. Phys. Soc. Jpn., Suppl. A* **49,** 955–958.
Chang, L. L., Esaki, L., and Tsu, K. (1974). *Appl. Phys. Lett.* **24,** 593–595.
Chang, L. L., Sakaki, H., Chang, C-A., and Esaki, L. (1977). *Phys. Rev. Lett.* **38,** 1489–1493.
Chen, C. Y., Cho, A. Y., Gossard, A. C., and Garbinski, P. A. (1982). *Appl. Phys. Lett.* **41,** 360–362.
Chen, W. P., Chen, Y. J., and Burstein, E. (1976). *Surf. Sci.* **58,** 263–265.
Ching, L. Y., Burstein, E., Buchner, S., and Wieder, H. H. (1980). *J. Phys. Soc. Jpn., Suppl. A* **49,** 951–954.
Colvard, C., Merlin, R., Klein, M. V., and Gossard, A. C. (1980). *Phys. Rev. Lett.* **45,** 298–301.
Dahl, D., and Sham, L. J. (1977). *Phys. Rev. B* **16,** 651–661.
Das Sarma, S., and Quinn, J. J. (1982). *Phys. Rev. B* **12,** 7603–7618.
Döhler, G. H., Künzel, H., Olego, D., Ploog, K., Ruden, P., Stolz, H. J., and Abstreiter, G. (1981). *Phys. Rev. Lett.* **47,** 864–867.
Dingle, R. (1975). *In* "Festkörperprobleme (Advances in Solid State Physics)" (H. J. Queisser, ed.), Vol. 15, p. 21. Pergamon, Oxford.
Dingle, R., Wiegmann, W., and Henry, C. H. (1974). *Phys. Rev. Lett.* **33,** 827–829.
Dingle, R., Störmer, H., Gossard, A. C., and Wiegmann, W. (1978). *Appl. Phys. Lett.* **33,** 665–667.

Drummond, T. J., Morkoc, H., and Cho, A. Y. (1981a). *J. Appl. Phys.* **52,** 1380–1386.
Drummond, T. J., Morkoc, H., Su, S. L., Fischer, S., and Cho, A. Y. (1981b). *Electron. Lett.* **17,** 870–871.
Drummond, T. J., Keever, M., Kopp, W., Morkoc, H., Hess, K., Streetman, B. G., and Cho, A. Y. (1981c). *Electron. Lett.* **17,** 545–547.
Drummond, T. J., Fischer, R., Morkoc, H., and Miller, P. (1982a). *Appl. Phys. Lett.* **40,** 430–432.
Drummond, T. J., Kopp, W., Morkoc, H., and Keever, M. (1982b). *Appl. Phys. Lett.* **41,** 277–279.
Drummond, T. J., Klem, J., Arnold, D., Fisher, R., Thorne, R. E., Lyons, W. G., and Morkoc, H. (1983). *Appl. Phys. Lett.* **42,** 615–617.
Englert, T., Maan, J. C., Uihlein, C., Tsui, D. C., and Gossard, A. C. (1983). *Physica* **118B,** 631–633.
Esaki, L., and Tsu, R. (Unpublished.)
Fetter, A. L. (1974). *Ann. Phys. (N.Y.),* **88,** 1–25.
Fishman, G., and Calecki, D. (1983). *Physica* **117*B*** and **118*B*,** 744–746.
Fishman, G., Pinczuk, A. Worlock, J. M., Störmer, H. L., Gossard, A. C., and Wiegmann, W. (1981). *J. de Phys., 12, Suppl. C* **6,** 305–308.
Gossard, A. C. (1979). *Thin Solid Films* **57,** 3–13.
Gossard, A. C., Wiegmann, W., Miller, R. C., Petroff, P. M., and Tsang, W. T. (1982). *Int. Symp. MBE Related Clean Surface Tech., 2nd Tokyo, Japan,* p. 39.
Hayashi, I., Panish, M. B., and Foy, P. W. (1969). *IEEE J. Quantum Elctron.* **QE-5,** 211.
Hersee, S. D., Hirtz, J. P., Baldy, M., and Duchemin, J. P. (1982). *Electron. Lett.* **18,** 1076–1078.
Hiyamizu, S. (1982). *Int. Symp. Molecular Beam Epitaxy Related Clean Surface Tech., 2nd Tokyo, Japan,* p. 113.
Höpfel, R., Lindemann, G., Gornik, E., Stangle, G., Gossard, A. C., and Wiegmann, W. (1982). *Surf. Sci.* **113,** 118–123.
Hotta, T., Sakaki, H., and Ohno, H. (1982). *Jpn. Appl. Phys.* **21,** L122–124.
Hwang, J. C. M., Kastalsky, A., Störmer, H. L., and Keramidas, V. G. (1982). *Int. Symp. MBE Related Clean Surface Tech., 2nd* (Post-deadline paper), *Tokyo, Japan.*
Inoue, M., Hiyamizu, S., Hida, H., Hashimoto, H., and Inmishi, Y. (1981). *J. de Phys. C* **7,** 19–20.
Ishikawa, T., Saito, J., Sasa, S., and Hiyamizu, S. (1982). *Jpn. J. Appl. Phys. 2* **21,** L675–L677.
Kastalsky, A., Dingle, R., Cheng, K. Y., and Cho, A. Y. (1982). *Appl. Phys. Lett.* **41,** 274–277.
Kazarinov, R. F., and Luryi, S. (1982). *Phys. Rev. B* **25,** 7626–7630.
Klein, M. V. (1975). *In* "Light Scattering in Solids" (M. Cardona, ed.), pp. 147–204. Springer-Verlag, Heidelberg.
Laughlin, R. B. (1983). *Phys. Rev. Lett.* **50,** 1395–1397.
Manuel, P., Sai-Halasz, G. A., Chang, L. L., Chang, C. A., and Esaki, L. (1976). *Phys. Rev. Lett.* **37,** 1701–1704.
Merlin, R., Colvard, C., Klein, M. V., Morkoc, H., Cho, A. Y., and Gossard, A. C. (1980). *Appl. Phys. Lett.* **36,** 43–45.
Merlin, R., Pinczuk, A., Beard, W. T., and Wood, C. E. E. (1982). *J. Vac. Sci. Technol.* **21,** 516.
Miller, R. C., Tsang, W. T., and Munteanu, O. (1982). *Appl. Phys. Lett.* **41,** 374–376.
Mimura, T. (1982). *Surf. Sci.* **113,** 454.
Mooradian, A. (1972). *In* "Laser Handbook" (F. T. Arecchi and E. O. Schulz-DuBois, eds.), Vol. 2, pp. 1409–1456. North-Holland, Amsterdam.
Mori, S., and Ando, T. (1979). *Phys. Rev. B* **19,** 6433.
Olego, D., and Cardona, M. (1979). *Solid State Commun.* **32,** 375–378.
Olego, D., Pinczuk, A., Gossard, A. C., and Wiegmann, W. (1982). *Phys. Rev. B* **25,** 7867–7870.
Paalanen, M. A., Tsui, D. C., and Gossard, A. C. (1982). *Phys. Rev. B,* **25,** 5566.
Petroff, P. M., Gossard, A. C., Wiegmann, W., and Savage, A. (1978). *J. Cryst. Growth* **44,** 5.
Pinczuk, A., and Worlock, J. M. (1982). *Surf. Sci.* **113,** 69–84.
Pinczuk, A., Störmer, H. L., Dingle, R., Worlock, J. M., Wiegmann, W., and Gossard, A. C. (1979a). *Solid State Commun.* **32,** 1001–1003.

Pinczuk, A., Abstreiter, G., Trommer, R., and Cardona, M. (1979b). *Solid State Commun.* **30,** 429–432.

Pinczuk, A., Worlock, J. M., Störmer, H. L., Dingle, R., Wiegmann, W., and Gossard, A. C. (1980a). *Solid State Comm.* **36,** 43–46.

Pinczuk, A., Worlock, J. M., Störmer, H. L., Dingle, R., Wiegmann, W., and Gossard, A. C. (1980b). *Surf. Sci.* **98,** 126–133.

Pinczuk, A., Worlock, J. M., Störmer, H. L., Dingle, R., Wiegmann, W., and Gossard, A. C. (1980c). *J. Phys. Soc. Jpn., Suppl. A* **49,** 1025–1028.

Pinczuk, A., Worlock, J. M. Störmer, H. L., Gossard, A. C., and Wiegmann, W. (1981a). *J. Vac. Sci. Technol.* **19,** 561–563.

Pinczuk, A., Worlock, J. M., Störmer, H. L., Gossard, A. C., Wiegmann, W., and Dingle, R. (1981b). *Bull. Am. Phys. Soc.* **26,** 254.

Pinczuk, A., Shah, J., Gossard, A. C., and Wiegmann, W. (1981c). *Phys. Rev. Lett.* **46,** 1341–1344.

Price, P. J. (1982). *Surf. Sci.* **113,** 199–210.

Sai-Halasz, G. A., Tsu, R., and Esaki, L. (1977). *Appl. Phys. Lett.* **30,** 651–653.

Sai-Halasz, G. A., Pinczuk, A., Yu, P. Y., and Esaki, L. (1978). *Solid State Commun.* **25,** 381–384.

Shah, J., Pinczuk, A., Störmer, H. L., Gossard, A. C., and Wiegmann, W. (1983). *Appl. Phys. Lett.* **42,** 55–57.

Störmer, H. L., and Tsang, W. T. (1980). *Appl. Phys. Lett.* **36,** 685–687.

Störmer, H. L., Dingle, R., Gossard, A. C., Wiegmann, W., and Sturge, M. D. (1979). *Solid State Comm.* **29,** 705–709.

Störmer, H. L., Pinczuk, A., Gossard, A. C., and Wiegmann, W. (1981a). *Appl. Phys. Lett.* **38,** 691–693.

Störmer, H. L., Gossard, A. C., Wiegmann, W., and Baldwin, K. (1981b). *Appl. Phys. Lett.* **39,** 912–914.

Störmer, H. L., Gossard, A. C., Wiegmann, W. (1981c). *Appl. Phys. Lett.* **39,** 493–495.

Störmer, H. L., Gossard, A. C., and Wiegmann, W. (1982a). *Solid State Commu.* **41,** 707–709.

Störmer, H. L., Tsui, D. C., and Gossard, A. C. (1982b). *Surf. Sci.* **113,** 32–38.

Störmer, H. L., Chang, A., Schlesinger, Z., Tsui, D. C., Gossard, A. C., and Wiegmann, W. (1983a). *Phys. Rev. Lett.* **50,** 1953–1956.

Störmer, H. L., Haavasoja, T., Narayanamurti, V., Gossard, A. C., and Wiegmann, W. (1983b). *J. Vac. Sci. Tech. B* **1,** 423–426.

Tien, Z. J., Worlock, J. M., Perry, C. H., Pinczuk, A., Aggarwal, R. L., Störmer, H. L., Gossard, A. C., and Wiegmann, W. (1982). *Surf. Sci.* **113,** 89–93.

Tsui, D. C., and Logan, R. A. (1979). *Appl. Phys. Lett.* **35,** 99–101.

Tsui, D. C., and Gossard, A. C. (1981). *Appl. Phys. Lett.* **38,** 550–552.

Tsui, D. C., Gornik, E., and Logan, R. A. (1980a). *Solid State Comm.* **35,** 875–877.

Tsui, D. C., Englert, T., Cho, A. Y., and Gossard, A. C. (1980b). *Phys. Rev. Lett.* **44,** 341–344.

Tsui, D. C., Gossard, A. C., Kaminsky, G., and Wiegmann, W. (1981). *Appl. Phys. lett.* **39,** 712–714.

Tsui, D. C., Gossard, A. C., Field, B. F., Cage, M. E., and Dziuba, R. F. (1982a). *Phys. Rev. Lett.* **48,** 3–6.

Tsui, D. C., Störmer, H. L., and Gossard, A. C. (1982b). *Phys. Rev. Lett* **48,** 1559–1562.

Tsui, D. C., Gossard, A. C., and Dolan, G. J. (1983). *Appl. Phys. Lett.* **42,** 180–182.

von Klitzing, K., Dorda, G., and Pepper, M. (1980). *Phys. Rev. Lett.* **45,** 494–497.

Weisbuch, C., Miller, R. C., Dingle, R., Gossard, A. C., and Wiegmann, W. (1981). *Solid State Comm.* **37,** 219–222.

Wolfe, C. M., Stillman, G. E., and Lindley, W. T. (1970). *J. Appl. Phys.* **41,** 3088–3091.

Worlock, J. M., Pinczuk, A., Tien, Z. J., Perry, C. H., Störmer, H. L., Dingle, R., Gossard, A. C., Wiegmann, W., and Aggarwal, R. L. (1981). *Solid State Commun.* **40,** 876–880.

Zeller, C., Vinter, B., Abstreiter, G., and Ploog, K. (1982a). *Phys. Rev. B* **26,** 2124–2132.

Zeller, C., Abstreiter, G., and Ploog, K. (1982b). *Surf. Sci.* **113,** 85–88.

7 QUANTUM-WELL HETEROSTRUCTURE LASERS

N. Holonyak, Jr., and K. Hess

Electrical Engineering Research Laboratory
and Materials Research Laboratory
University of Illinois at Urbana-Champaign
Urbana, Illinois

I. INTRODUCTION

Since 1962 and the first construction of semiconductor lasers (Hall *et al.*, 1962; Nathan *et al.*, 1962; Holonyak and Bevacqua, 1962; Quist *et al.*, 1962), much effort has been expended in developing increasingly sophisticated versions of this smallest of lasers. [For a comprehensive account of this work see Bogdankevich *et al.* (1976), Kressel and Butler (1977), Casey and Panish (1978), and Thompson (1980).] Of most importance, beyond the initial demonstration of homojunction GaAs and $GaAs_{1-x}P_x$ lasers, has been the double heterostructure (DH) laser first proposed by Kroemer (1963) and by Alfëro v (1963) and initially demonstrated in continuous room temperature [continuous-wave (cw) 300-K] operation by Alfërov *et al.* (1970) and then by Hayashi *et al.* (1970). Since the introduction of the DH laser and its cw 300-K operation, semiconductor laser work has been dominated by this one theme, further and further development of the double heterojunction and its crystal technology.

It is the heterojunction and the sophisticated liquid-phase epitaxial (LPE), vapor-phase epitaxial (VPE) or chemical vapor deposition (CVD), and molecular-beam epitaxial (MBE) crystal growth procedures (devised for heterojunc-

ISBN 0-12-170470-X

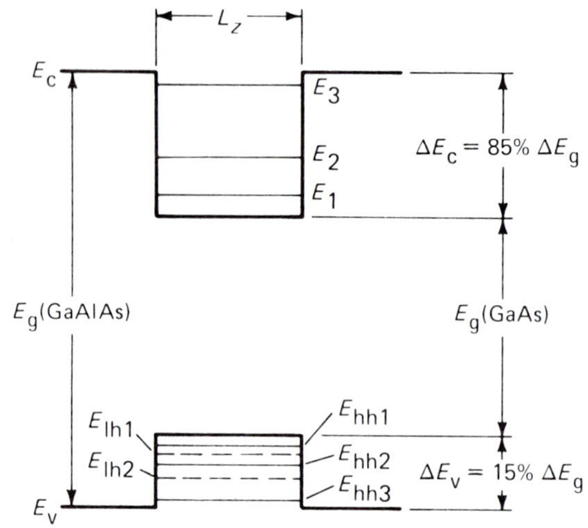

Fig. 1. Single GaAs quantum well of size $L_z(<\lambda \sim h/p)$ confined between two higher-gap ($\Delta E_g = \Delta E_c + \Delta E_v$) $Al_xGa_{1-x}As$ ($0 < x \leq 1$) layers. Confined-particle electron states are indicated by E_1, E_2, E_3; heavy-hole states by E_{hh1}, E_{hh2}, E_{hh3}; and light-hole states by E_{lh1}, E_{lh2}. (From Holonyak *et al.*, 1980a. Reprinted from *IEEE J. Quantum Electron.* **QE-16,** 170–186. © 1980 IEEE.)

tions) that have made possible quantum-well heterostructure (QWH) lasers. For example, the first cw 300-K QWH lasers, photopumped samples (Holonyak *et al.*, 1978a) and p–n diodes (Dupuis *et al.*, 1979b), were simply $Al_xGa_{1-x}As$–GaAs DHs with the GaAs active region, a single quantum well, reduced in thickness from $L_z \gtrsim 500$ Å to $L_z \sim 200$ Å, i.e., to a thickness such that the carrier de Broglie wavelength $\lambda = (h/p) \sim L_z$. For $\lambda \sim L_z$, the excess carriers exhibit confined-particle (particle-in-a-box) behavior and are "lifted" in energy above the band edge to a higher and sharper density of states, in fact, to a staircase density of states (see Figs. 1 and 2). This and other quantum-size-effect (QSE) behavior, which can be realized in single- and multiple- (coupled) semiconductor quantum wells, lead to a number of fundamental and practical effects in carrier collection, recombination, luminescence, and stimulated emission, which are the subject of this chapter.

II. QUANTUM-SIZE EFFECTS

The development of more and more refined methods of growing III–V semiconductor heterojunctions, including single and multiple lattice-matched ultrathin layers, has led basically to a rediscovery of an old idea, quantum-size effects (QSE) in thin layers [e.g., see Lutskii (1970), Elinson *et al.* (1972), Shik (1974), Dingle (1975), Holonyak *et al.* (1980a)]. Over 25 years ago it was

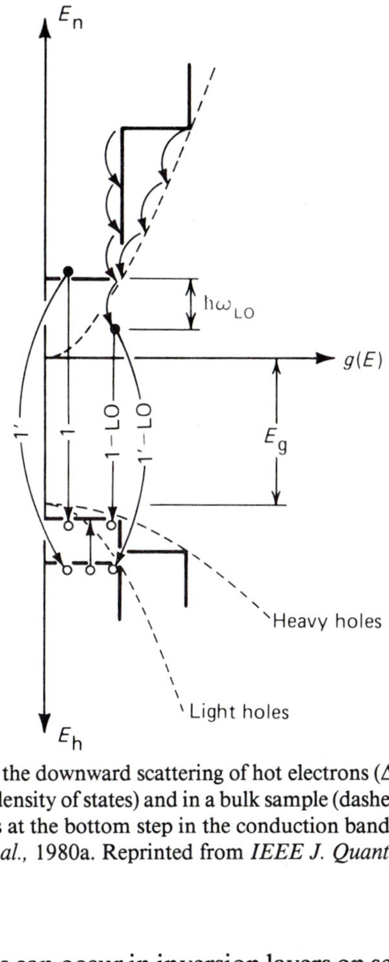

Fig. 2. Comparison of the downward scattering of hot electrons ($\Delta E \sim \hbar\omega_{LO}$) in a quantum-well heterostructure (step density of states) and in a bulk sample (dashed lines, parabolic density of states). The virtual process at the bottom step in the conduction band is identified with an $\hbar\omega_{LO}$ label. (From Holonyak *et al.*, 1980a. Reprinted from *IEEE J. Quantum Electron.* **QE-16,** 170–186. © 1980 IEEE.)

appreciated that QSEs can occur in inversion layers on semiconductor surfaces (Schrieffer, 1955). These effects have long been a part of field-effect studies (Stern and Howard, 1967) and now can more naturally be studied in III–V QWHs, including III–V superlattices (SLs).

A simple case that is fundamental to the problem of light emission in a QWH is that of a narrower-gap III–V semiconductor sandwiched (lattice matched) between two wider-gap confining layers. Typical examples are the AlAs–$Al_xGa_{1-x}As$–GaAs system, the InP–$In_{1-x}Ga_xP_{1-z}As_z$ system, and, at higher gap (visible), the $In_{1-x'}Ga_{x'}P_{1-z'}As_{z'}$–$In_{1-x}Ga_xP_{1-z}As_z$ system. For the sake of illustration we show in Fig. 1 the $Al_xGa_{1-x}As$–GaAs system, which might be a photopumped sample or, with doped confining layers, a p–n diode. (We will regard this as the prototype system for most of our discussion.) Excess, photo-generated or injected, electrons and holes collected in the GaAs active layer of

Fig. 1 are confined in a finite conduction-band well and a finite valence-band well, and the usual band-to-band recombination process, for $L_z \lesssim 500$ Å, must be modified from the case of a bulk (GaAs) layer. The energy spectrum of carriers in the narrower-gap center layer, for $\lambda = (h/p) \sim L_z < 500$ Å, is commonly determined by using the effective mass approximation and assuming that the electron or hole Hamiltonian can be separated into a component z normal to the layer and into the usual (unconfined) Bloch-function components (x, y) in the plane of the layer. The carrier eigenvalues then are

$$E(n,k_x,k_y) = E_n + (\hbar^2/2m^*_{n,p})(k_x^2 + k_y^2), \tag{1}$$

where E_n is the nth confined-particle (finite-well) energy eigenvalues of the z portion of the Hamiltonian, $m^*_{n,p}$ is the electron or hole effective mass, $\hbar$ is Planck's constant $(h/2\pi)$, and k_x and k_y are the usual x and y components of the crystal "momentum." In Fig. 1 the E_n values of Eq. (1) are labeled E_1, E_2, E_3 for electrons, $E_{\mathrm{hh1}}, E_{\mathrm{hh2}}, E_{\mathrm{hh3}}$ for heavy holes, and $E_{l\mathrm{h1}}, E_{l\mathrm{h2}}$, for light holes. Below E_1 there are no allowed electron states. Then as shown in Fig. 2 there is a subband (upward, $E\uparrow$) with a constant density of states (per unit area) (Elliott and Gisbon, 1974)

$$g(E)\, dE = \{m_n^*/(\pi\hbar^2)\}\, dE. \tag{2}$$

At E_2 (and then E_3) another subband appears with another step in the density of states (Fig. 2). The same pattern of subbands (downward, $E\downarrow$) and steps in the density of states is repeated in the valence band for heavy and light holes.

It is worth mentioning that the prototype single-well structure of Fig. 1 is itself a useful practical configuration (Holonyak *et al.*, 1978a; Dupuis *et al.*, 1979b). In many cases, however, a multiple-well QWH is more desirable or even essential, particularly for well sizes $L_z \lesssim 100$ Å (Shichijo *et al.*, 1978). For the case of a multiple-well QWH, the energies E_n of Eq. (1) (and the steps in Fig. 2) can be determined by solving Schrödinger's equation (for electrons, heavy holes, and light holes) in each layer and by, as usual, matching the solutions across the layer boundaries. This procedure is practical for a small number ($\lesssim 6$) of quantum wells, and without much difficulty can be carried out on a small computer (Kolbas, 1979). Beyond ~ 6 wells the allowed confined-particle states begin to form energy bands (minibands) and then a Kronig–Penney analysis, for layer sizes exceeding 5–10 monolayers, is an adequate approximation for calculating the lower electron, heavy-hole, and light-hole bands. We illustrate this procedure on an AlAs–GaAs superlattice (SL) because of its importance as a later experimental example. Six periods of such a SL confined between two $Al_xGa_{1-x}As$ layers are shown in Fig. 3.

A classic Kronig–Penney analysis leads to the familiar expression

$$1 \geq |[(\alpha^2 - \beta^2)/(2\alpha\beta)] \sinh(\alpha L_{\mathrm{B}}) \sin(\beta L_z) + \cosh(\alpha L_{\mathrm{B}}) \cos(\beta L_z)|, \tag{3}$$

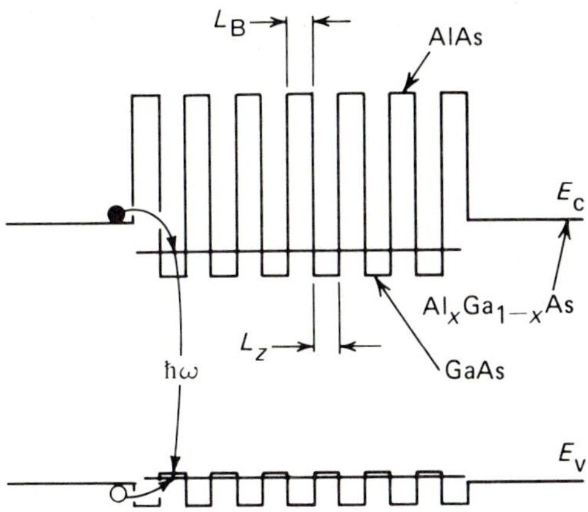

Fig. 3. Energy-band diagram of an $Al_xGa_{1-x}As$–AlAs–GaAs QWH with an all-binary active region that is free of alloy disorder and clustering. [After Vojak *et al.* (1981).]

where

$$\alpha \equiv [2m_1^*(V - E)]^{1/2}/\hbar \tag{4}$$

and

$$\beta \equiv [2m_2^* E]^{1/2}/\hbar. \tag{5}$$

Note that the analysis based on Eq. (3) is not strictly valid for different masses in the quantum wells and barrier layers. This can lead to some error, especially in extreme cases (e.g., as already mentioned, AlAs–GaAs superlattices with periods of only a few lattice constants). Nevertheless, for appropriate mass and size parameters, Eq. (3) can be solved for the allowed energies E of electrons in the conduction band and holes in the valence band (layer sizes exceeding 5–10 monolayers). These energies, with L_B (AlAs) = 20 Å, are shown in Fig. 4 with the 0 boundary used as a reference for the conduction-band edge and the valence-band edge. The electron effective mass m_1^* in the AlAs barriers is taken as $0.15m_0$, where m_0 is the free-electron mass. In the GaAs quantum wells the band nonparabolicity and the electron effective mass m_2^* is approximated (Kolbas, 1979) as

$$m_2^* = (0.0665 + 0.0436E + 0.236E^2 - 0.147E^3)m_0, \tag{6}$$

where E is in electron volts. The conduction-band discontinuity or barrier height V in Eq. (4) is taken as

$$V = 0.85[E_\Gamma(1, T) - E_\Gamma(0, T)], \tag{7}$$

with the energy of $E_g(\Gamma)$ in $Al_xGa_{1-x}As$ as a function of crystal composition x

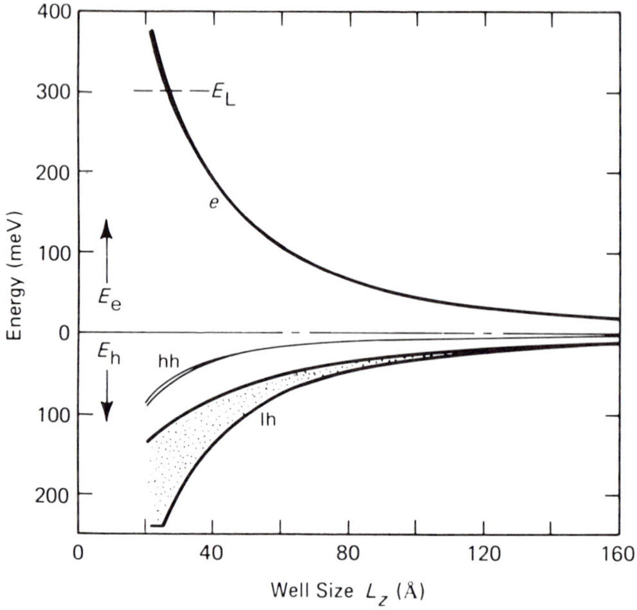

Fig. 4. Lowest ($n = 1$) confined-particle energy bands (minibands) for electrons e, heavy holes hh, and light holes *l*h as a function of well size L_z for GaAs quantum wells coupled by AlAs barriers of size $L_B = 20$ Å. (See Fig. 3.) [After Vojak *et al.* (1981).]

and temperature T given as

$$E_\Gamma(x, T) = \begin{cases} 1.519 + 1.247x - 5.405 \times 10^{-4}T^2/(T + 204), & x < 0.45, \quad (8a) \\ 1.519 + 1.247x + 1.147(x - 0.45)^2 & \\ -5.405 \times 10^{-4}T^2/(T + 204), & x > 0.45. \quad (8b) \end{cases}$$

For heavy holes (hh) the effective masses (Kolbas, 1979) are taken as $m_1^*(\text{hh}) = 0.75m_0$ (AlAs) and $m_2^*(\text{hh}) = 0.45m_0$ (GaAs), and for light holes (*l*h) $m_1^*(l\text{h}) = 0.15m_0$ (AlAs) and $m_2^*(l\text{h}) = 0.087m_0$ (GaAs). In accord with Eq. (7), the valence-band discontinuity is taken as

$$V = 0.15[E_\Gamma(1, T) - E_\Gamma(0, T)] \quad (9)$$

and is determined from the E_Γ values of Eq. (8).

The expressions outlined, with the AlAs barrier widths arbitrarily chosen as L_B (AlAs) = 20 Å, lead to the $n = 1$ electron e, heavy hole hh, and light hole *l*h energies shown in Fig. 4. At smaller and smaller well size L_z, the electron band e and the heavy-hole band hh do not broaden much when compared to the light-hole band *l*h, which is the case of a low mass (*l*h) carrier in a relatively shallow well. For barrier widths L_B smaller than 20 Å, the energy bands tend to broaden equally above and below the band centers, and for $L_B > 20$ Å contract to smaller widths. In the region $L_z \lesssim 25$ Å, the lh band broadens to the "top" of the valence band well, and this small size is an obvious limit. Similarly, the e band extends upward into the region of the L or X indirect minima of the GaAs quantum wells, where decreasing size also "lifts" the L and X minima to higher energy. This behavior (Vojak *et al.*, 1981) leads to a higher-energy, size-induced direct–indirect transition than can be realized in bulk (disordered) $Al_xGa_{1-x}As$ (Camras *et al.*, 1982). Figures 4 and 5 make it clear that at smaller layer size QSE can be used to effect profound changes in the electron and the hole energies in a semiconductor crystal. For example, in Fig. 5 the curve $E_1(e \rightarrow h)$ shows, for an AlAs–GaAs superlattice ($L_B = 100$ Å), how high in energy direct-gap electron–hole recombination can be made by small ($L_z = 15$–20 Å) GaAs well size (Camras *et al.*, 1982).

Not only is the shift of electron and hole states to higher energy an important effect in thin semiconductor layers, the staircase density of states introduced by small layer sizes is of some consequence also. Figure 2 shows that $n = 1$ (or 2, 3, . . .) electrons can recombine with $n = 1$ (or 2, 3, . . .) heavy holes (nth $e \rightarrow hh$ transition, $\Delta n = 0$) or with $n' = 1'$ (or $2'$, $3'$, . . .) light holes (n'th $e \rightarrow lh$ transition, $\Delta n = 0$). The selection rule $\Delta n = 0$ is commonly assumed in the recombination process but is not entirely valid (Shik, 1979) and, in addi-

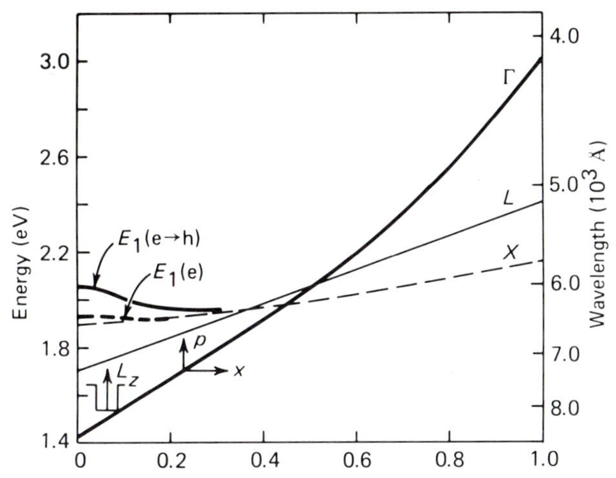

Fig. 5. Total energy (300 K) as a function of crystal composition of the $n = 1$ confined-particle electron state $E_1(e)$ and electron and hole states $E_1(e \rightarrow h)$ at the point of the (size-induced) direct–indirect crossover with the $n = 1$ L or X confined-particle bands of a QWH. The Γ, L, and X band edges of bulk $Al_xGa_{1-x}As$ are shown as references. Note that pressure p, composition x, or size L_z can lead to a direct–indirect transition. [After Camras *et al.* (1982).]

tion, can be further upset by flaws in the QWH (or SL). Figure 2 shows that electron–hole recombination in a QWH proceeds from a conduction-band subband, with a steplike density of states, to a similar group of valence-band states. In contrast, in a *bulk* sample (e.g., GaAs) there are no "steps" in the density of states. Instead, the density of states is parabolic, decreasing monotonically to zero at the band edges. (See the dotted envelopes in Fig. 2.) Thus, the electrons and holes cannot all thermalize into a narrow energy range (as at a "step"), and the recombination (radiation) to some extent must be broadened. This can be avoided in a QWH unless there is much band filling.

Another interesting feature of QWHs is apparent in Fig. 2. Photogenerated or injected excess carriers scatter downward in energy (thermalize) to a lesser density of states by phonon emission (LO phonon, $\hbar\omega_{LO}$). In a bulk sample ($L_z \gtrsim 500$ Å) the phonon emission intensity I_{em} is proportional to the final density of states or $I_{em} \propto \sqrt{E - \hbar\omega_{LO}}$; similarly, phonon absorption I_{abs} is proportional to $\sqrt{E + \hbar\omega_{LO}}$, or clearly $I_{abs} > I_{em}$. Thus, in a bulk sample carrier thermalization by phonon emission has a tendency to be limited by the decreasing density of states. On the other hand, within the constant density-of-states region of a quasi-two-dimensional quantum-well system, no such limitation occurs (e.g., see the two phonon "paths" in Fig. 2). Figure 2 shows also that from the lowest (or last) conduction-band step E_1 in the density of states a virtual transition is possible that makes recombination and laser operation possible $\hbar\omega_{LO}$ below the confined-particle energy gap, i.e., below the fundamental absorption (Coleman *et al.*, 1981a,b). This will be discussed in detail in a later section.

III. THE III–V QWH LASER CRYSTAL

As already mentioned, the key to the existence of QWH and SL lasers has been the high degree of sophistication that has been developed in III–V semiconductor crystal growth. All three methods, MBE (van der Ziel *et al.*, 1975), LPE (Rezek *et al.*, 1977), and VPE [or metalorganic chemical vapor deposition (MOCVD), Kolbas *et al.* (1978)] have been used successfully to grow QWH and SL lasers. Basically, any well-behaved semiconductor epitaxial growth process, one that is capable of growing flat layers on low-index planes and that can be operated relatively slowly or that can be "switched" on or off rapidly, can be used for the growth of QWHs and SLs. Since our main interest here is not the details of various crystal growth processes, we will merely illustrate some of the various QWH and SL structures that have been grown and refer the reader to the literature for details of the various crystal growth processes. [For detailed reviews, see for MBE Cho and Arthur (1975), for LPE Rezek *et al.* (1980), and for MOCVD (a form of VPE) Manasevit (1971), Dupuis *et al.* (1971a), and Coleman *et al.* (1981c).]

Perhaps the method most widely used to construct DH lasers has been LPE, and indeed this process employing the $InP-In_{1-x}Ga_xP_{1-z}As_z$ system has been the first used to fabricate QWH $p-n$ diode lasers (Rezek *et al.*, 1977). An example is that shown in Fig. 6 (Rezek *et al.*, 1980), which is for the case of six $L_z \sim 160$-Å $In_{1-x}Ga_xP_{1-z}As_z$ ($x \sim 0.13$, $z \sim 0.29$) narrower-gap quantum wells [E_g(77 K) ~ 1.16 eV] coupled by five $L_B \sim 250$-Å $In_{1-x'}Ga_{x'}P_{1-z'}As_{z'}$ ($x' \sim 0.02$, $z' \sim 0.04$) wider-gap barriers [E_g(77 K) ~ 1.37 eV]. The region between the arrows is the active region, which is confined on the top by a p-type InP hole-injecting (and contact) layer and on the bottom by electron-injecting n-type InP grown first on the n-type InP substrate. The QWH laser diode employing this structure is assembled in the usual oxide-defined stripe configuration.

The LPE crystal growth method has been used also to grow visible-spectrum $In_{1-x}Ga_xP_{1-z}As_z$ ($x \sim 0.83$, $z \sim 0.35$) QWH lasers on $GaAs_{1-y}P_y$ substrates (Chin *et al.*, 1980c). So far, however, this area of work has suffered because of the poor quality of the $GaAs_{1-y}P_y$ substrates (owing to defects and surface crosshatch) and will not be mentioned further.

The greatest amount of work performed on QWH and SL lasers has been on

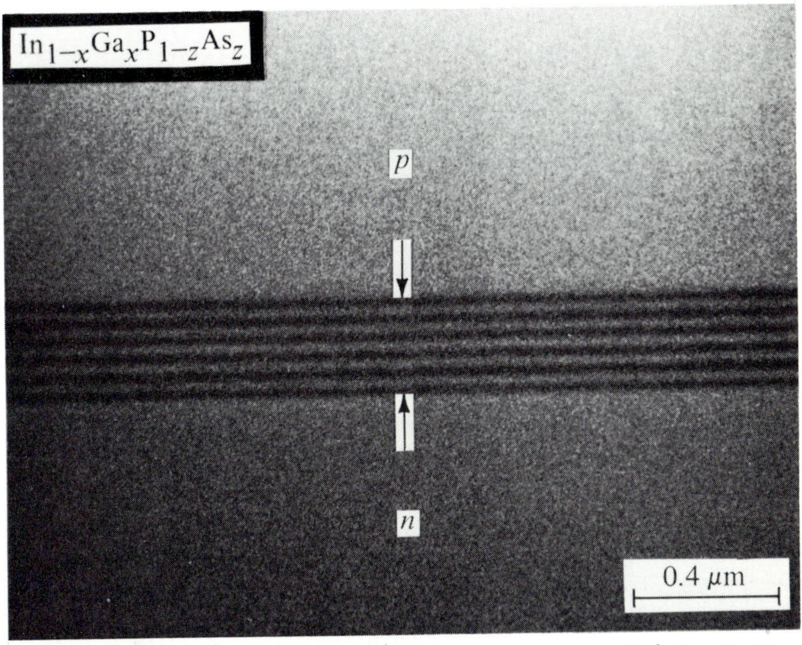

Fig. 6. SEM photomicrograph of a $p-n$ QWH laser with six $L_z \sim 160$-Å $In_{1-x}Ga_xP_{1-z}As_z$ ($x \sim 0.13$, $z \sim 0.29$) quantum wells coupled by five $L_B \sim 250$-Å wider-gap $In_{1-x'}Ga_{x'}P_{1-z'}As_{z'}$ ($x' \sim 0.02$, $z' \sim 0.04$) barriers. [After Rezek *et al.* (1980). Reprinted with permission from *Journal of Electronic Materials,* Vol. 9, No. 1, 1980, a publication of The Metallurgical Society of AIME, Warrendale, Pennsylvania.]

the $Al_xGa_{1-x}As$–GaAs system, and in further discussion we will concentrate on this system. Both MBE and MOCVD crystals have been important in this work. An example of an MBE $Al_xGa_{1-x}As$–GaAs QWH crystal is shown in Fig. 7, which shows in a shallow-angle (0.3°) cross section (Holonyak *et al.*, 1979a) six $L_z \sim 28$-Å GaAs quantum-well layers coupled by five $L_B \sim 100$-Å $Al_xGa_{1-x}As$ ($x \sim 0.4$) barriers (region between the arrows). Similar composition confining layers are located on either side of the quantum-well active region. The waviness of the layers apparent in Fig. 7 is simply an artifact of shallow-angle cross sectioning (i.e., the polishing procedure) and does not indicate that the layers in fact are wavy. An interesting property of the QWH shown in Fig. 7 is that it has been operated (photopumped) as a cw 300-K laser at the relatively short wavelength of 7270 Å (Morkoç *et al.*, 1982). Even shorter wavelengths have been achieved on pulse operated (300-K) MOCVD $Al_xGa_{1-x}As$–AlAs–GaAs QWHs [6785 Å by Vojak *et al.* (1981) and 6500 Å by Camras *et al.* (1982)].

It is clear that a quantum-well laser structure as shown in Fig. 7 can be arranged in a *p–n* diode configuration. Figure 8 shows a shallow-angle (~0.64°) cross section of such a QWH grown by MOCVD. The active region consists of six $L_z \sim 120$-Å GaAs quantum wells coupled by five $L_B \sim 120$-Å

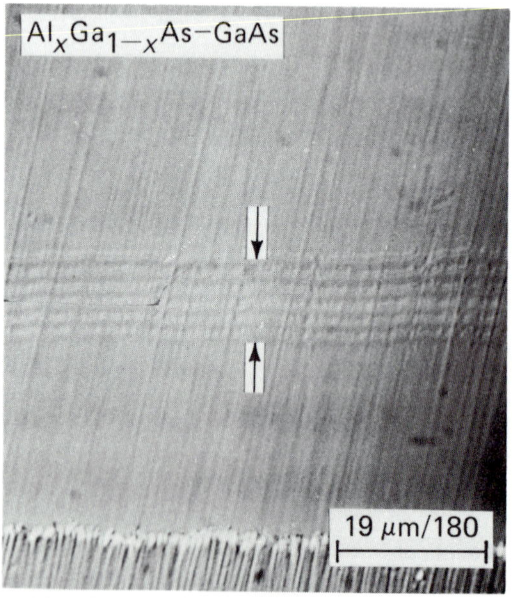

Fig. 7. Shallow-angle cross section (~0.32°) of an $Al_xGa_{1-x}As$–GaAs QWH grown by MBE. The active region (indicated by arrows) consists of six $L_z \sim 28$-Å GaAs quantum wells coupled by five $L_B \sim 100$-Å $Al_xGa_{1-x}As$ ($x \sim 0.4$) barriers. [After Morkoç *et al.* (1982).]

$Al_xGa_{1-x}As$ ($x \sim 0.3$) barriers. The bottom *n*-type confining layer ($\sim$0.8-μm thick) is $x \sim 0.4$ $Al_xGa_{1-x}As$ with a Se doping of $n_d \sim 10^{18}/cm^3$. The top *p*-type confining layer ($\sim$1.1-μm) is $x \sim 0.4$ $Al_xGa_{1-x}As$ with a Zn doping of $n_a \sim 5 \times 10^{17}/cm^3$. The substrate GaAs is obvious below the lowest arrow, and just above the topmost arrow is a *p*-type GaAs contact layer. The MOCVD $Al_xGa_{1-x}As$–GaAs QWH of Fig. 8, assembled in a stripe configuration, has the distinction of being the first *multiple* quantum-well laser diode to be operated cw at 300 K (Dupuis *et al.*, 1979c).

Besides QWHs of one to six coupled quantum wells in the active region, SL configurations are interesting for various studies, including luminescence, absorption, and impurity diffusion or implantation studies. A large $Al_xGa_{1-x}As$–GaAs SL of 121 periods ($L_B \sim 80$ Å, $L_z \sim 85$ Å) grown by MOCVD is shown in Fig. 9. Although there are $\sim$240 interfaces in this SL, when adequately heat sunk and photopumped it is capable of low-threshold cw 300-K laser operation (Coleman *et al.*, 1981b). This indicates that there is a neligible density of defects at the layer interfaces.

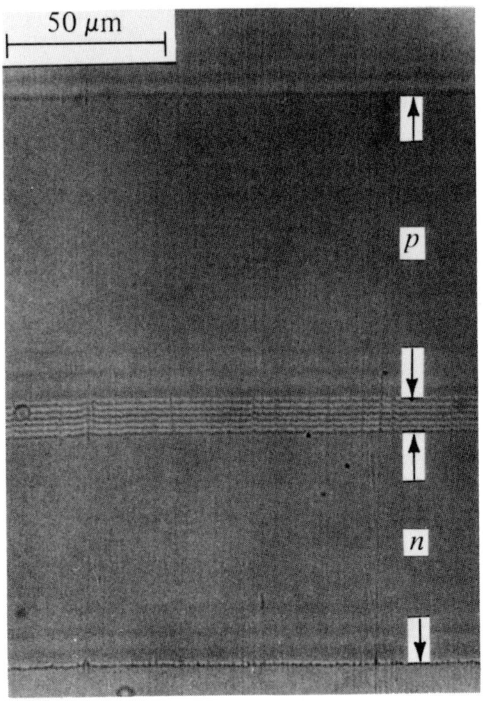

Fig. 8. Shallow-angle ($\sim$0.64°) cross section of an $Al_xGa_{1-x}As$–GaAs QWH laser grown by MOCVD with six $L_z \sim 120$-Å GaAs quantum wells coupled by five $L_B \sim 120$-Å $Al_xGa_{1-x}As$ ($x \sim 0.3$) barriers. [This is the first multiple-well *p*–*n* QWH laser, a diode, to operate cw at 300 K. See Dupuis *et al.* (1979c) and Holonyak *et al.* (1980a).]

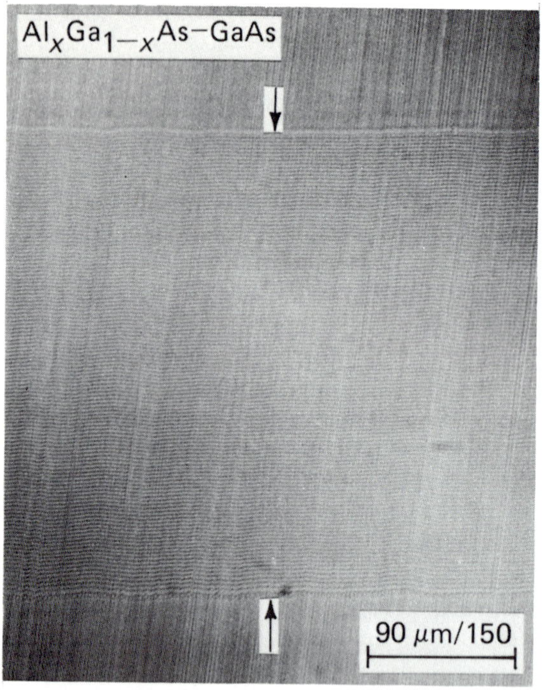

Fig. 9. Shallow-angle (~0.38°) cross section of a 121-period $Al_xGa_{1-x}As-GaAs$ ($x \sim 0.5$) superlattice (region between arrows) that is capable of cw 300-K laser operation. The SL is grown by MOCVD. [After Coleman *et al.* (1981b).]

As already mentioned, an artifact of shallow-angle cross sectioning is the fact that the layers, because of polishing scratches and undulations, appear wavy. Transmission-electron-microscope (TEM) measurements on MOCVD $Al_xGa_{1-x}As-GaAs$ and AlAs–GaAs SLs show that in fact the layers are flat and abrupt within the measuring limits of the instrument (Coleman *et al.*, 1981c). An example is shown in Fig. 10, which is a TEM image of part of a 50-period all-binary AlAs–GaAs SL grown by MOCVD (Coleman *et al.*, 1981c). As shown in the figure, the AlAs barriers are ~105 Å thick and the GaAs wells are ~145 Å.

Another form of SL [grown by means of metalorganic vapor-phase epitaxy (MOCVD) by Ludowise and Dietze (1982)] that is interesting for experimental reasons is shown in Fig. 11. Figure 11 is a shallow-angle (<0.5°) cross section of a 66-period $GaAs-In_xGa_{1-x}As$ ($x \sim 0.2$) *strained* SL consisting of $L_B \sim 75$-Å GaAs barriers and $L_z \sim 75$ Å $In_xGa_{1-x}As$ quantum wells. The strained SL (strain of 7×10^{-3}) is the ~1-μm-thick region shown between the two vertical arrows in the center of Fig. 11. At the top of the SL is an ~800-Å $In_xAl_yGa_{1-x-y}As$ confining layer and at the bottom of the SL is a similar

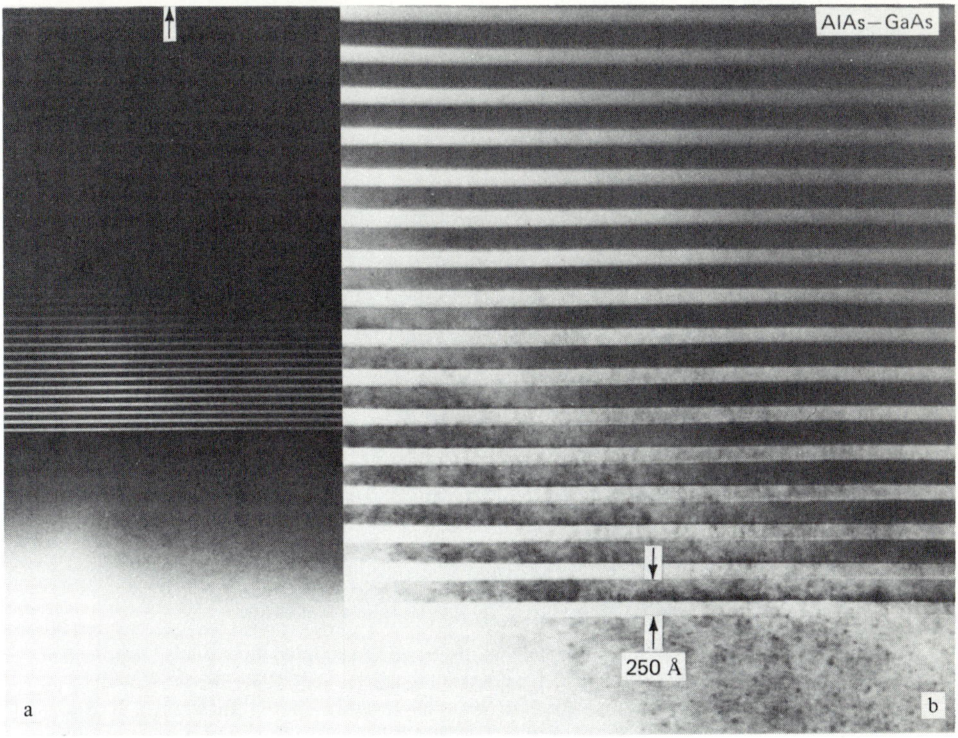

Fig. 10. TEM photomicrograph of a 50-period AlAs–GaAs superlattice grown by MOCVD. The light layers are AlAs ($L_B \sim 105$ Å), the dark layers GaAs ($L_z \sim 145$ Å). Uneven sample thinning causes the layer fading at lower magnification in (a). [SL crystal by Coleman; TEM photomicrograph by Mochel.]

quaternary confining layer (i.e., the region between the two head-to-head arrows at the bottom right in Fig. 11). The SL shown in Fig. 11, a SL with higher-gap confining layers, is perhaps one of the most sophisticated structures of this type that has ever been grown and is shown here because it, too, is capable of 300-K laser operation (Ludowise *et al.,* 1983b), as will be discussed later.

Before concluding this section, we mention that for some measurements on QWH- and SL-laser samples there are certain advantages (and convenience) in using photopumping as the means of excitation. In this method of excitation, extremely small rectangular samples can be employed (see Fig. 12). Then the Fabry–Perot cavity end losses (say for a 10–50-μm-wide sample) are large and make it possible to observe band filling and carrier recombination over a large energy range (Kolbas *et al.,* 1978). Also, a rectangular sample permits laser oscillations to occur across *and* along a sample and aids in identification of phonon-assisted recombination (Holonyak *et al.,* 1979b). A further advantage

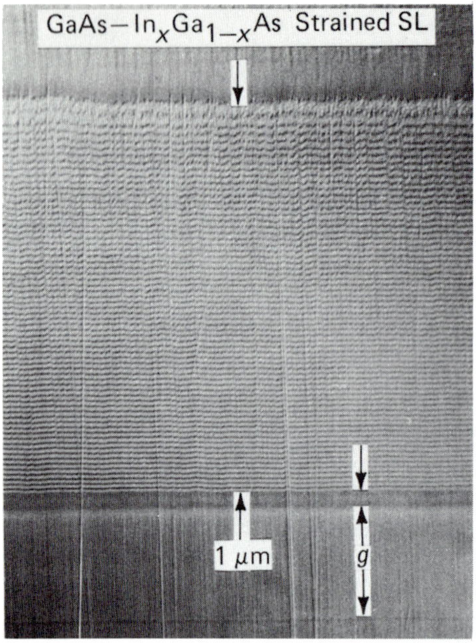

Fig. 11. Shallow-angle (<0.5°) cross section of a 66-period strained SL. A graded region (g) of $In_xGa_{1-x}As$ ($0 \leq x \leq 0.1$) is grown first on a GaAs substrate (by MOCVD). This is followed by an ~800-Å $In_xAl_yGa_{1-x-y}As$ confining layer (arrows at lower right). The ~1-μm strained SL is next (center arrows), with an ~800-Å $In_xAl_yGa_{1-x-y}As$ ($x \sim 0.1$, $y \sim 0.6$) confining layer on top. (The upper layers appear thicker because of rounding of the "corner" formed by the slant cross section and the wafer surface.) [After Ludowise *et al.* (1983b).]

of photoexcitation is that undoped samples can be employed and processes intrinsic to the crystal can be observed. In addition, a thin-QWH (or SL) sample can be compressed under diamond into a metal heat sink and, because of the excellence of the heat sinking, can be driven at least an order of magnitude higher in intensity than a diode.

The QWH or SL sample preparation for photopumping consists in removing the substrate (and where necessary top contact layer) by polishing and selective etching. The resulting thin wafers are then cleaved into 10–50 × 50–250-μm (or sometimes bigger) rectangles (Fig. 12) and are compressed and heat sunk in indium under sapphire or in annealed copper under diamond (Holonyak and Scifres, 1971). The samples are photopumped with the cavity-dumped or cw output of an Ar^+ laser ($\lambda \sim 5145$ Å) as shown in Fig. 12. For a sample as shown in Fig. 12 (where, incidentally, the GaAs layer shown as $L_z \lesssim 200$ Å can be actually a multiple quantum-well active region, e.g., as in Fig. 3), the excess electron–hole pairs are generated predominantly in the thicker, wider-gap $Al_xGa_{1-x}As$ confining layers since the thin GaAs quantum wells contribute

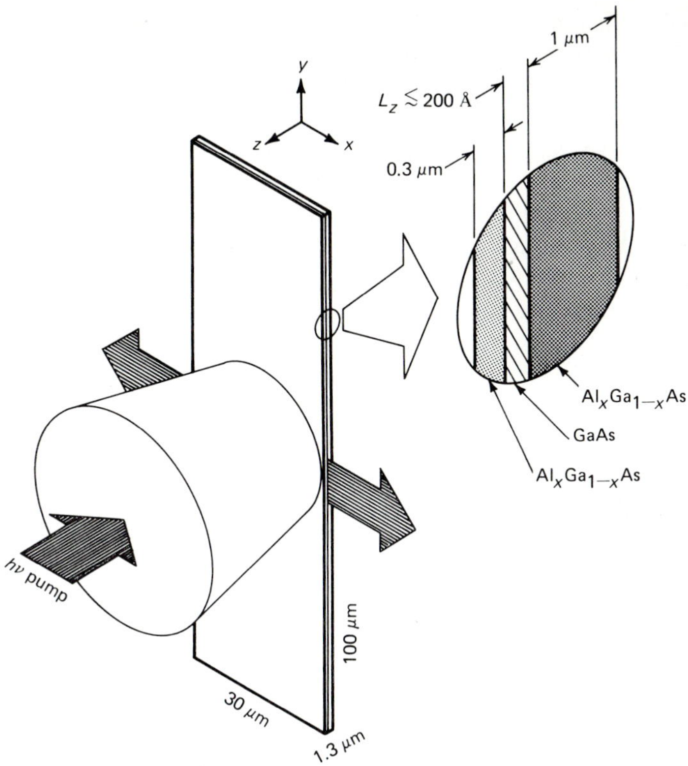

Fig. 12. Schematic diagram of a cleaved rectangular $Al_xGa_{1-x}As$–GaAs QWH sample photopumped (pulsed or cw) with an Ar^+ laser. The GaAs quantum-well active region can actually be a multiple-well QWH or an all-binary active region as shown in Fig. 3. (From Holonyak *et al.*, 1980a. Reprinted from *IEEE J. Quantum Electron.* **QE-16,** 170–186. © IEEE.)

little attenuation (absorption) of the incident photon flux. In other words, for single or multiple (m) quantum-well layers, $L_z\alpha \ll 1$ or $mL_z\alpha \ll 1$. Excess carriers generated in the $Al_xGa_{1-x}As$ confining layers diffuse and are collected by the active region if it is not too thin and it offers sufficient scattering and carrier energy loss for electron and hole thermalization in the GaAs quantum well(s).

IV. ELECTRON–HOLE COLLECTION IN A QUANTUM WELL

A. Carrier Collection in a QWH

It is clear from the preceding discussion that if excess carriers are photogenerated or injected in the confining layers of a single quantum-well heterostructure within a diffusion length ($\lesssim 1$ μm) of the narrower-gap active region, the carriers will diffuse to the quantum well, be collected, and recombine in the well (GaAs) if it is large enough (L_z). This is the case for curve (a) [L_z (GaAs) ~ 200 Å] of Fig. 13, which is evident from the small amount of direct-gap ($x \sim 0.4$) $Al_xGa_{1-x}As$ recombination radiation (77 K) at 6400–6200 Å compared to the large quantum-well (GaAs) emission intensity on the confined-particle transitions $n = 1-4$ (e → hh) and $n' = 1'-3'$ (e → lh) observed in the range 8500–7000 Å. In contrast to sample (a) of Fig. 13, sample (b) [L_z (GaAs) ~ 80 Å], although photopumped also on the side of the 0.3-μm-thick confining layer (or well within a diffusion length of the quantum well), exhibits a very large $Al_xGa_{1-x}As$ confining-layer (band-to-band) recombination-radiation intensity and much weaker confined-particle ($n = 1$, $n' = 1'$) recombina-

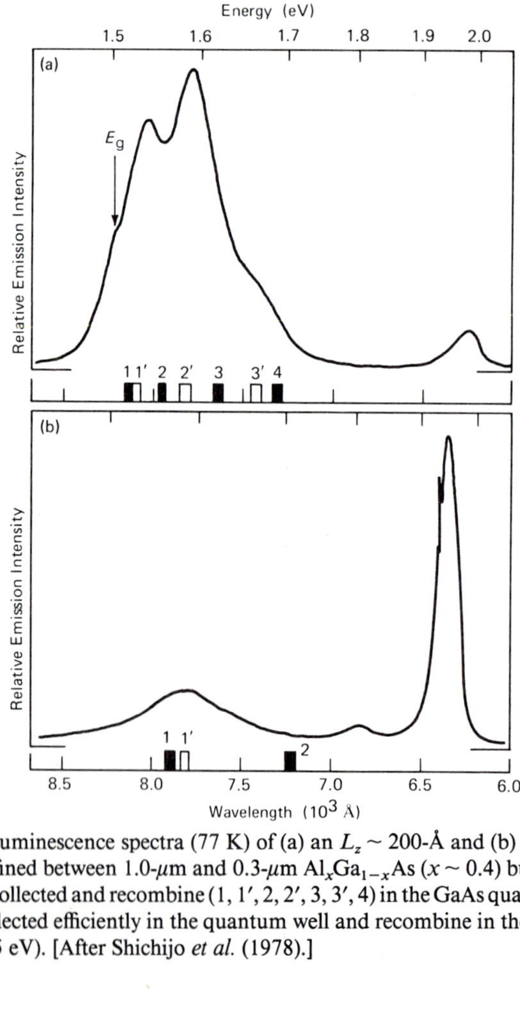

Fig. 13. Photoluminescence spectra (77 K) of (a) an $L_z \sim$ 200-Å and (b) an $L_z \sim$ 80-Å GaAs quantum well confined between 1.0-μm and 0.3-μm $Al_xGa_{1-x}As$ ($x \sim 0.4$) bulk layers. In (a) the excess carriers are collected and recombine (1, 1′, 2, 2′, 3, 3′, 4) in the GaAs quantum well; in (b) the carriers are not collected efficiently in the quantum well and recombine in the confining layers at high energy (~ 1.95 eV). [After Shichijo *et al.* (1978).]

tion radiation. It is evident that a fundamental problem occurs when the GaAs quantum-well thickness is reduced to the range $L_z < 100$ Å. When L_z is made smaller and approaches the electron scattering path length l_p, electrons (which are not confined) are not scattered sufficiently in distance L_z to thermalize in large density from the band edge of the confining layers to the lower confined-particle states of the quantum well. Since holes scatter in a smaller distance and electrons in a larger distance, this effect (quantum-well capture of a hole but not an electron) has been used to measure the valence-band discontinuity ΔE_v of $In_{1-x}Ga_xP_{1-z}As_z$ ($x \sim 0.13$, $z \sim 0.29$) relative to InP (Chin *et al.*, 1979) and of $Al_xGa_{1-x}As$ relative to GaAs (Dupuis *et al.*, 1978b). The basis for this measurement is sketched in the inset of Fig. 14. Figure 14 shows also the laser operation (77 K) that occurs in the process of hot electrons [$(p^2/2m_n) \sim \Delta E_c$] recombining with $n = 1-6$ holes captured by the quaternary quantum well in the range ~80 meV below E_g (InP) ≈ 1.415 eV. A long-luminescence tail involving also bound-electron states extends to the quaternary energy gap [E_g (InGaPAs) ~

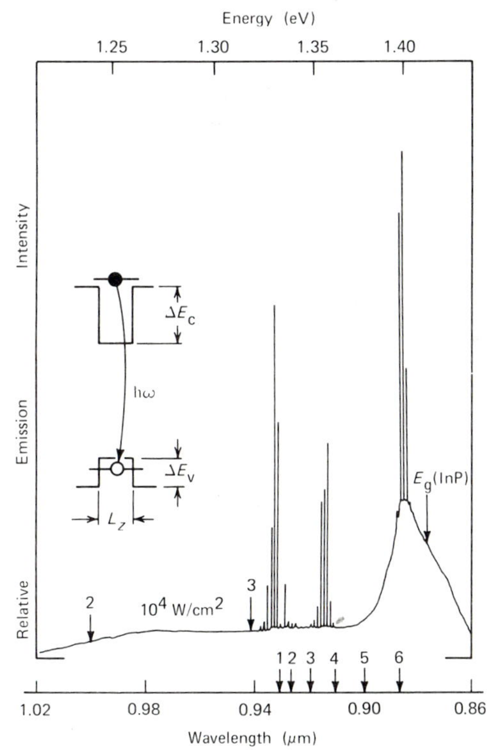

Fig. 14. Photoluminescence spectrum (77 K) of a single $In_{1-x}Ga_xP_{1-z}As_z$ ($x \sim 0.13$, $z \sim 0.29$) quantum well thin enough ($L_z \sim 150$ Å) to restrict downward electron scattering from the thick InP confining layers in which they are generated. [After Rezek *et al.* (1980).]

1.17 eV]. Figure 14 shows that

$$\Delta E_g = E_g\,(\text{InP}) - E_g\,(\text{InGaPAs}) = 0.245$$
$$= \Delta E_v + \Delta E_c = 0.80 + 0.165 \quad \text{eV} \tag{10}$$

or

$$\Delta E_c \sim 160 \quad \text{meV} \sim \tfrac{2}{3}\,\Delta E_g \qquad \text{and} \qquad \Delta E_v \sim \tfrac{1}{3}\,\Delta E_g. \tag{11}$$

The basic asymmetry illustrated above in respect to how electrons and holes behave and are collected in a QWH (Shichijo *et al.,* 1978; Chin *et al.,* 1979; Holonyak *et al.,* 1980a) suggests immediately that this asymmetry can be used to enhance electron multiplication and suppress hole multiplication in an avalanche detector. The detailed features of such a layered avalanche detector have been described by Chin *et al.* (1980b).

The model and problem of how carrier collection proceeds in a quantum well is complicated by several factors: (1) the need to know the quantum-mechanical reflection coefficient for the Bloch functions at the layer interface (the effective mass approximation is not applicable in this case), (2) the need to know the phonon scattering rates, including intersubband and intervalley scattering, and (3) the need to take into account the finite probability of re-emission (by phonon absorption) of a captured electron. These features complicate the problem unless various simplifying assumptions are made. We assume that an electron is not re-emitted once it is captured, and we neglect quantum reflection and resonances. The problem then reduces to a solution of the semiclassical transport equations. After the electron traverses the confining-layer–quantum-well interface, it is high above the GaAs band edge and has a de Broglie wavelength much smaller than the quantum-well size L_z (80 Å in say Fig. 13). Therefore we can neglect also size quantization in the electron–phonon interaction and can use the three-dimensional limit of the scattering equations. We mention that the first treatment (Shichijo *et al.,* 1978) of this problem by a method of this kind involved the use of the Fermi "age theory" and an electron mean free path $l_p \sim 63$ Å. This treatment is in good agreement with experimental results even though it does not distinguish between scattering to the various valleys. It does show, however, that holes (as in Fig. 14) are collected more effectively than the electrons. Below we consider the carrier collection problem (for unconfined carriers) and present data obtained via a Monte Carlo simulation. First, however, we digress and present a treatment of the electron–phonon interaction (and scattering) appropriate to the situation in a QWH.

B. The Electron–Phonon Interaction in a QWH

The consequences of electron confinement in heterolayers are understood in some detail. Less attention has been paid, however, to the confinement of phonons. For electrons the band-edge discontinuity determines the confinement, but for phonons it is determined by different material parameters, e.g., the ionic masses of the layer constituents and the differences in the dielectric functions. Therefore we assume as an approximation bulklike phonons and two-dimensional electrons or under different conditions three-dimensional electrons and Rayleigh modes for the phonons. The validity of the approximation will depend to a large extent on the phonon and electron wavelengths, the material parameters, and the specific physical quantity under consideration. For a single GaAs well sandwiched between $Al_xGa_{1-x}As$ ($x < 0.5$) confining layers, the interaction of low-energy electrons with long-wavelength phonons is probably described sufficiently well by using a bulk-phonon picture and quasi-two-dimensional electrons. On the other hand, high-energy electrons in a SL interacting with a specific phonon mode would require a different approximation. In the latter case, size quantization may be of minor importance for the electrons (at higher energy), and the SL might be an ideal Bragg reflector for the specific phonon mode considered.

We consider first the matrix element for the polar-optical interaction. We assume that the electrons reside in the lowest subband of an infinite square well. Using effective mass theory, we have for the wave function ψ_{el}

$$\psi_{\text{el}} = (2/L_z A)^{1/2} \sin(\pi z/L_z) \exp(i\mathbf{kr}), \tag{12}$$

where A is the area and $\mathbf{k}$ and $\mathbf{r}$ are two-dimensional vectors. If the electron interacts with bulk phonons, we can use the Fröhlich Hamiltonian $H_{\text{el-ph}}$

$$H_{\text{el-ph}} = \sum_{\mathbf{q}_{3\text{d}}} (C/q_{3\text{d}})[a_{\mathbf{q}_{3\text{d}}} \exp(i\mathbf{q}_{3\text{d}}\mathbf{r}_{3\text{d}}) - a^{\dagger}_{\mathbf{q}_{3\text{d}}} \exp(-\mathbf{q}_{3\text{d}}\mathbf{r}_{3\text{d}})], \tag{13}$$

which gives for phonon emission

$$M = \langle \mathbf{k}',(N_q + 1)|H_{\text{el-ph}}|\mathbf{k}, N_q\rangle$$

or (14)

$$M = (-C/\mathbf{q}_{3\text{d}})\, \delta_{\mathbf{k}-\mathbf{k},\mathbf{q}}(N_q + 1)^{1/2} B$$

with B given as

$$B \equiv (2/L_z) \int_0^{L_z} \exp(-iq_z z) \sin^2(\pi z/L_z)\, dz. \tag{15}$$

It is important to notice that the calculation for an interface mode would proceed in a similar manner and give similar results except that

$$M^2 \propto 1/q \tag{16}$$

instead of

$$M^2 \propto 1/q_{3\text{d}}^2. \tag{17}$$

If the various modes are not handled appropriately, divergences arise since

$$\sum_{q_{3d}} \to \int q_{3d}^2 \, dq_{3d} \sin\theta \, d\theta \, d\phi \tag{18}$$

and the $(1/q_{3d}^2)$ term in M^2 is canceled. Similarly,

$$\sum_{q} \to \int q \, dq \, d\phi, \tag{19}$$

which is appropriate for interface modes, leads to cancellation of the $1/q$ term.

For $q_z L_z \lesssim 1$ Eqs. (14) and (15) give for the total scattering rate involving phonon emission

$$1/\tau_{\rm em} = [|C|^2 V/(4\pi\hbar)](N_q + 1) \int \delta(E_{k-q} - E_k + \hbar\omega_{\rm LO}) \, dq \, d\phi. \tag{20}$$

This scattering rate is independent of the layer width L_z. The reason for this is the form of the matrix element, which favors small momentum exchange. Therefore, the higher Fourier components of the quantum well do not add appreciably to scattering, and the result is constant in the limit of small $q_z L_z$. Overall, however, the scattering rate is enhanced slightly over the bulk rate because phonons with practically any q_z value can be emitted or absorbed (Hess, 1979).

For problems of electronic transport the *momentum* scattering time is most important. For optical problems involving QWHs, it is the *energy* relaxation that matters most since this time constant determines the capability of a quantum well to collect carriers. The energy loss $(\partial E/\partial t)|_{\rm c}$ for scattering by optical phonons is given (Conwell, 1967) as

$$(\partial E/\partial t)|_{\rm c} = \hbar\omega_{\rm LO}(1/\tau_{\rm abs} - 1/\tau_{\rm em}). \tag{21}$$

For scattering of quasi-two-dimensional electrons by polar-optical phonons, $1/\tau_{\rm em}$ is given by Eq. (20). The integral in Eq. (20) is elliptic and can be approximated in the following way (with a similar treatment for $1/\tau_{\rm abs}$):

$$1/\tau_{\rm abs} = [F_0 2^{1/2} e/(m^* E)^{1/2}] N_q I_{\rm abs}^{\rm 2d}, \tag{22a}$$

$$I_{\rm abs}^{\rm 2d} \approx 1.25(E/\hbar\omega_{\rm LO})^{0.42}; \tag{22b}$$

$$1/\tau_{\rm em} = [F_0 2^{1/2} e/(m^* E)^{1/2}](N_q + 1) I_{\rm em}^{\rm 2d}, \tag{22c}$$

$$I_{\rm em}^{\rm 2d} \approx 1.51(E/\hbar\omega_{\rm LO})^{0.17} H(E - \hbar\omega_{\rm LO}). \tag{22d}$$

Here F_0 is the polar coupling field (~6000 V/cm for GaAs). Figure 15 shows the terms $I_{\rm abs}^{\rm 2d}$ and $I_{\rm em}^{\rm 2d}$ together with the corresponding terms in three dimensions. It is important to note that, in contrast to the three-dimensional case, the emission integral is always larger than the absorption integral. This means that for higher-energy electrons collected in the active region of a QWH-stimulated phonon emission (the term proportional to N_q) dominates in comparison to absorption while in three-dimensional systems absorption is larger or equal.

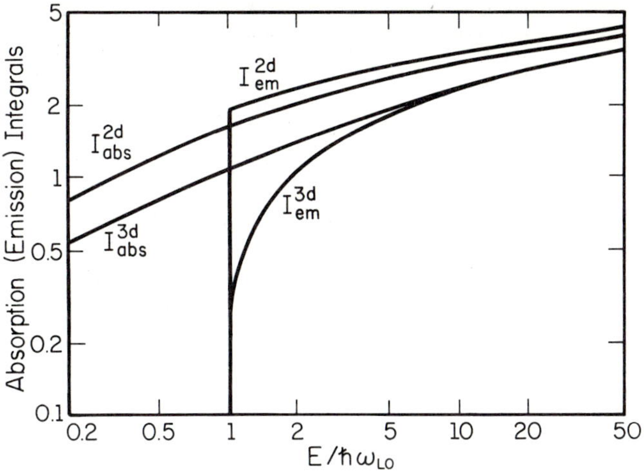

Fig. 15. Numerical solutions and curves for phonon absorption (abs) and phonon emission (em) integrals (I) as a function of electron energy (E) in units of $\hbar\omega_{\mathrm{LO}}$. In the three-dimensional case $I^{3d}_{abs} \geq I^{3d}_{em}$. In the two-dimensional case, however, $I^{2d}_{abs} < I^{2d}_{em}$ for $E > \hbar\omega_{\mathrm{LO}}$. [After Hess *et al.* (1980a). Reprinted with permission from *Solid State Commun.* **34.** Copyright 1980, Pergamon Press.]

Besides the importance of the polar-optical interaction in carrier scattering, it is important to consider also the matrix element and scattering for the case of the deformation-potential interaction. The electron–phonon interaction via the deformation potential is not restricted to small momentum exchange (short-range forces). For scattering by acoustic-phonon emission we have

$$M^2 = (N_q + 1)[\hbar q_{3d}/(2u_l\rho V)]Z_A^2\, \delta_{\mathbf{k}-\mathbf{k}',\mathbf{q}}. \tag{23}$$

Here ρ is the mass density, V the crystal volume, u_l the longitudinal sound velocity, and Z_A the deformation potential. For small values of the phonon energy $\hbar\omega_q$, M^2 is independent of q. Therefore the quantum well is expected to open up scattering channels in this case, and for small L_z the scattering is strongly enhanced. Using the golden rule, we obtain for the scattering rate for emission

$$1/\tau_{em} = 2\pi \sum_{q_{3d}} (N_q + 1) \times [\omega_q/(2\rho u_l^2 V)]Z_A^2\, \delta_{\mathbf{k}-\mathbf{k}',\mathbf{q}}\, \delta(E_{k+q} - E_k + \hbar\omega_q). \tag{24}$$

Then for $N_q \approx kT/\hbar\omega_q$ the total scattering rate becomes

$$1/\tau = 1/\tau_{em} + 1/\tau_{abs} \approx (\tfrac{3}{2}kTm^* Z_A^2)/(\hbar^3 u_l^2 \rho L_z). \tag{25}$$

This equation is valid for the lowest confined-particle state. Note the independence of electron energy.

For highly energetic electrons cascading down the quantum well, optical intervalley-deformation-potential scattering is more important. Its functional dependence on the various parameters is similar to that of acoustic scattering. For very high energies, however, size quantization is of minor importance and a three-dimensional treatment of the problem may be more appropriate [see Conwell (1967)]. For this case $1/\tau_{em}$ is given by

$$1/\tau_{em} = D^2(m^*)^{3/2}(N_q + 1)(E - \Delta E - \hbar\omega_q)^{1/2}/\sqrt{2}\pi\hbar^3\rho\omega_q. \qquad (26)$$

Here D is the coupling constant, m^* the density-of-states effective mass, ΔE the valley separation, and $\hbar\omega_q$ the energy of the intervalley phonons. An analogous equation holds for absorption. *A priori* it is not clear that Z_A and D are the same in bulk material and layered structures. It has been shown by Mon *et al.* (1981), however, that for small band-edge discontinuities the deviations of these constants from their bulk values are small.

We mention that the scattering rates as a function of energy vary widely, ranging from $\sim 10^{14}$ sec^{-1} at higher electron energies to $\sim 10^{12}$ sec^{-1} at lower energies. Overall, the scattering rates in SLs are enhanced over the bulk values. For layer widths of the order of 100 Å, this enhancement is weak ($\lesssim 30\%$). For very small layer widths, however, the enhancement can be appreciable. We will see later that the generation of high phonon densities can enhance the electron–phonon interaction still further. This effect is connected with the very high power densities dissipated in QWH lasers.

C. Monte Carlo Simulation of Carrier Collection

A Monte Carlo stimulation procedure, which can be applied to the problem of carrier collection, has been described by Shichijo and Hess (1981). This method, which includes the pseudopotential band structure (Cohen and Bergstresser, 1966), has been used to obtain the results described in this subsection (Tang *et al.*, 1982). The physical basis for this procedure is the same as that described in Subsection IV.A; however, now the details of the scattering mechanisms (intervalley and polar-optical scattering) and, of course, the pseudopotential band structure are included. The results exhibit a significant difference between the case of electrons starting at the beginning of the collection process in the Γ valley and the circumstance of electrons starting to cascade down the quantum well in the X valleys. Therefore the collection efficiency of a quantum well will depend sensitively on whether the confining layers are direct or indirect gap.

Electrons starting at X in the $Al_xGa_{1-x}As$ confining layers are transmitted with higher probability to the X valleys of the GaAs layer than to the Γ or L valleys because of the conservation of crystal momentum. Figure 16 shows the percentage of electrons having emitted at least one phonon and thus having descended (at least partially) into the GaAs quantum well. According to our previous assumptions, these electrons are collected. The lower curve is for

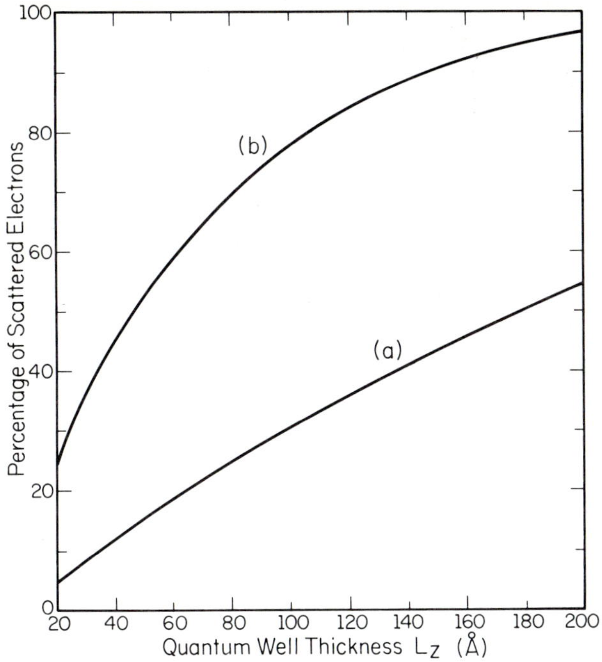

Fig. 16. Percentage of electrons scattered into a *single* GaAs quantum well from two $Al_xGa_{1-x}As$ confining layers by phonon emission (see Fig. 12). The curve (a) corresponds to electrons at Γ and curve (b) to electrons at X in both the $Al_xGa_{1-x}As$ confining layers *and* in the GaAs quantum well.

electrons starting at Γ. We see that only few Γ electrons are collected, especially for quantum-well sizes below $L_z = 100$ Å. The reason for this is the high-speed ($\sim 5 \times 10^7$ cm/sec) and the low scattering rates of the electrons in the Γ valley. Electrons in the X valleys are collected with much higher probability because of their low velocities and high phonon scattering rate. The same consideration (particle velocity) is true also for holes, as discussed earlier. For X electrons, as for Γ electrons, we observe a critical decrease of the percentage of scattered electrons for well sizes below $L_z \sim 100$ Å.

The carrier distribution functions at distance $L_z = 80$ and 200 Å from the $Al_xGa_{1-x}As$ confining layer (the source) for X electrons transferring into a GaAs well are shown in Fig. 17. These distributions exhibit similar features, with a considerable (energy) tail for the case of a 200-Å distance and only a short tail for the smaller size of 80 Å. Of particular interest, the Monte Carlo calculation shows considerable structuring of the carrier distribution with respect to the phonon energy. The distribution obtained as described here is, of course, not for steady-state conditions. The steady-state distribution would, however, reflect similar features if electrons are depleted at lower energies because of high recombination rates. Otherwise, electron accumulation at lower energies, or

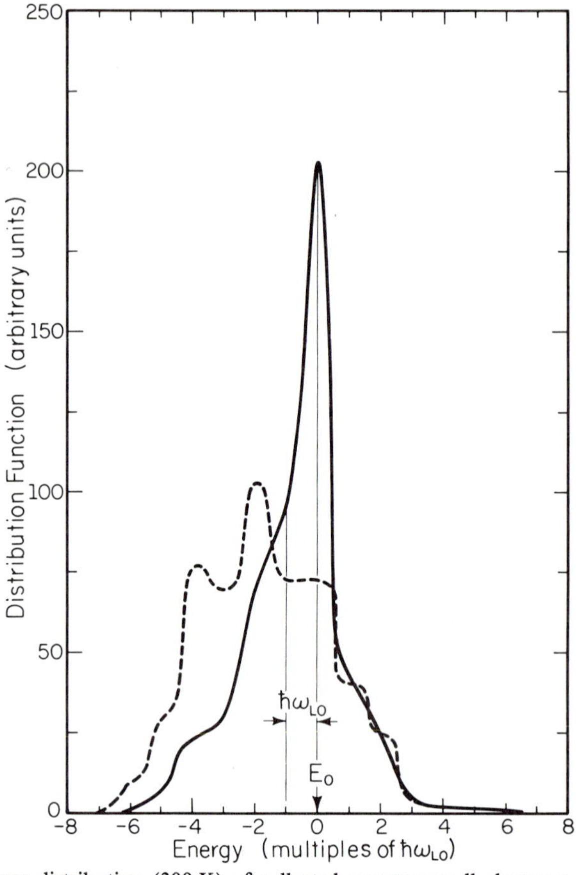

Fig. 17. Energy distribution (300 K) of collected quantum-well electrons (in multiples of $\hbar\omega_{LO}$) for well sizes L_z (GaAs) = 80 Å (solid line) and $L_z = 200$ Å (dashed line). The electron "injection" energy is $\Delta E_c \sim 0.53$ eV.

even the filling of the quantum well, can be expected. Some features of steady-state distributions are described in the next subsection.

D. Electron and Phonon Distributions

After the initial phonon emission, electrons are essentially captured in the quantum well (except for re-emission by phonon absorption) and are bound to "percolate" down in energy (Fig. 2) and thermalize, either by phonon emission or electron–electron interaction. The initial energy of an electron arriving at the active region of a QWH laser is extremely high, of the order of the band-edge discontinuity. To achieve comparable energies of, for example, 0.5 eV in GaAs by means of external electric fields F, it is necessary that $F \gtrsim 10^5$ V/cm. Such fields can be achieved in reverse-biased $p-n$ junctions if the electron concentration is low. In heterojunction lasers the electron density can be as high as $\sim 10^{18}$ cm^{-3}, and the energy of the injected electrons above the GaAs band edge is ~ 0.5 eV. These values correspond to a power loss of

$$en\mu F^2 \sim 10^{12} \quad \text{W/cm}^3 \tag{27}$$

if μ is around 10^3 cm^2/V sec. This means that enormous power densities are transferred to the lattice vibrations as the carriers thermalize, and neither the excess electrons nor the phonons can be close to thermal equilibrium under such circumstances. The actual form of the distribution functions is, of course, difficult to calculate. A qualitative discussion of some important features is given in this subsection.

We consider two limiting cases. In the case of extremely strong electron–electron interaction (which probably is not present in practice) a Fermi–Dirac distribution is established, i.e.,

$$f_c(E) = 1/[1 + \exp(E - F_c)/kT_c]. \tag{28}$$

In the equation above E is the electron energy, F_c is the quasi-Fermi level, and T_c the electron temperature, which is always higher than the temperature of the crystal lattice. A method to calculate T_c has been described earlier (Shah, 1978). If this method is applied to a QWH, the electron temperature is $T_c \sim 300$ K, even for rather low lattice temperatures. Even though this method (Shah, 1978) is based on the assumption that the phonon system is in equilibrium, the temperature of the electrons will be substantially larger than the lattice temperature under lasing conditions, including when the phonon system is highly perturbed.

In the limit of weak electron–electron interaction, a distribution function composed of "spikes" spaced $\hbar\omega_{\mathrm{LO}} \sim 36$ meV apart is expected because of the strong polar-optical phonon scattering. This can be easily seen from the Boltzmann equation if scattering by optical phonons alone is important. These spikes should occur at the locations of the arrowheads shown in Fig. 2. The curved arrows represent the process of electrons cascading downward by the emission of LO phonons. Modeling these spikes as delta functions results in the following form for the carrier distribution function

$$f_c(E) = \sum_{m=0}^{r} C_m\, \delta[E - (\Delta E_c - m\hbar\omega_{\mathrm{LO}})/kT_c]. \tag{29}$$

The constants C_m above are determined by the scattering rates and normalization, with $r = \mathrm{int}(\Delta E_c/\hbar\omega_{\mathrm{LO}})$.

Experiments showing emission peaks with spacing of $\hbar\omega_{\mathrm{LO}}$ are strongly indicative of the validity of this model. In most practical cases, however, the distribution function will be a mixture of the functions deduced for the two extreme cases, and, of course, most experimental situations are not ideal enough to confirm the preceding equation.

As a consequence of the carrier thermalization, the phonon occupation number increases. The nonthermal phonon occupation N_q^{nt} can be calculated from the equation

$$\partial N_q^{nt}/\partial t = \partial N_q^{nt}/\partial t|_{\text{ph-el}} + \partial N_q^{nt}/\partial t|_{\text{nonel}}. \tag{30}$$

The gain rate from the electron–phonon interaction $(\partial N_q^{nt}/\partial t)_{\text{ph-el}}$ can be calculated from the formulas for $1/\tau_{\text{em}}$ and $1/\tau_{\text{abs}}$ given earlier. Because of the great complexity of the phonon–phonon interaction, an evaluation of the nonelectronic term $(\partial N_q^{nt}/\partial t)_{\text{nonel}}$ has never been achieved. Under certain conditions, however, a single-mode relaxation time $\tau(q, T_L)$ can be defined (Hess *et al.*, 1980a). We adopt this idea here and insert below the experimental value of $\tau_{\text{p}} \sim 5 \times 10^{-12}$ sec. A lower limit on the phonon generation can be calculated by using the emission probabilities for the lowest subband and polar-optical scattering only as given by Eqs. (22). The "percolation phonons" (Kroemer, 1981) are emitted by an electron with energy E_r and a probability per unit time $S \equiv 1/\tau_{\text{em}}(\mathbf{q}, E_r)$. Using $\partial N_q^{nt}/\partial t = 0$ and an electron distribution as given by Eq. (29) with spikes at energies E_r, we obtain for the occupation number N_q^{nt} of phonons with wave vector of absolute value $q_{2\text{d}}$

$$N_q^{nt} = \sum_r \left[N_r \int_0^{2\pi} S(E_r, q)\, d\phi \right] \left[\sum_{q_{3\text{d}}, r} S(E_r, q_{3\text{d}}) \right]^{-1} (\tau_{\text{p}}/\tau_{\text{e}}). \tag{31}$$

Here N_r is the number of electrons with lifetime $\tau_{\text{e}} \sim 10^{-9}$–$10^{-10}$ sec (spontaneous or stimulated). As shown by Hess *et al.* (1980a) N_q^{nt} exhibits two sharp peaks. The peak at very low values of q is probably screened. The other peak is found at

$$q_r \sim 1.8(2r\hbar\omega_{\text{LO}} m^*)^{1/2}/\hbar. \tag{32}$$

There are two important consequences of these results: (1) If no phonon reflections are assumed (by layer boundaries, etc.) and if reasonable numbers are inserted in Eq. (31), then N_q can assume values greater than unity. We note that this is a lower limit since all phonon emission processes have been neglected except that of the polar optical in one subband. (2) Since the phonons are emitted preferentially around certain wavelengths, the construction of Bragg reflectors and further enhancement of N_q^{th} may be possible (Holonyak *et al.*, 1980b). In this connection recall also that for large N_q^{th} emission is always larger than absorption for higher electron energies. These results are important for the understanding of the phonon side bands observed in QWH lasers (see Section VI).

V. QUANTUM-WELL BAND FILLING

Besides the importance of carrier collection and thermalization in a QWH or SL laser, the constraint of size makes it possible to observe a large amount of carrier band filling, particularly in a single-well QWH or one with just a few quantum wells. Simple charge analysis shows that the excess carrier density in a QWH is given as

$$\delta n = (J/e)(\tau/mL_z), \tag{33}$$

where J is the actual or effective current density ($J \sim W/\hbar\omega_{\text{pump}}$ for a photo-pumped sample), e the magnitude of the charge of an electron, $\tau \approx 10^{-8}$–10^{-9} sec the carrier lifetime (which decreases with stimulated emission), and m the number of quantum wells in the sample active region. Figure 2 indicates that the density-of-states of a QWH increases in a steplike manner, but for convenience the usual parabolic dependence (dashed lines in Fig. 2) can be employed for approximate analysis. Thus the height to which electrons fill the well (a GaAs well) above the band edge (or above the first confined-particle state) can be estimated as

$$\delta E \sim 3.65 \times 10^{-15}(1/m_n^*)(\delta n)^{2/3} \quad \text{eV}, \tag{34}$$

where $m_n^* \sim 0.067$ and δn (cm^{-3}) is given by Eq. (33).

If photopumping is employed on an AlAs–$Al_xGa_{1-x}As$–GaAs QWH or SL sample that is heat sunk in Cu under diamond (Holonyak *et al.*, 1978a), an excitation level exceeding 10^4 W/cm^2 (or 4×10^3 A/cm^2) can easily be achieved before heating becomes a problem. At this excitation level (and with all the carriers collected from the confining layers), Eqs. (33) and (34) show that δE for a 200-Å GaAs quantum well can be as high as 300 meV, and an unusually wide emission spectrum results. This is evident in Fig. 18, which shows (77-K) laser data on three single-well $L_z \sim 200$-Å $Al_xGa_{1-x}As$ ($x \sim 0.6$) QWH samples of width (a) 15 μm, (b) 31 μm, and (c) 26 μm. The sample confining layers are 0.1 μm thick on one side and 1.0 μm on the other and are indirect gap and not the source of the laser emission evident in Fig. 18.

For the single-well QWH of Fig. 18, it is clear that the active-region size is adequate to collect efficiently the carriers photogenerated in the bulk-size confining layers. The "hot" carriers then relax in the well, as well as recombine, and support laser operation in the unusually large range from the infrared to the visible red ($\lambda \lesssim 7000$ Å). It is evident that the carrier-thermalization process is not complete inasmuch as laser operation is observed at such short wavelengths. For curve (a) laser operation occurs on the lowest confined-particle transitions [$n = 1$ (e $\rightarrow$ hh) and $n' = 1'$ (e $\rightarrow$ lh)], making it obvious that carrier scattering in an $L_z \sim 200$-Å well can proceed all the way from the top of the conduction-band (ΔE_c) and the valence-band (ΔE_v) wells to the lowest states.

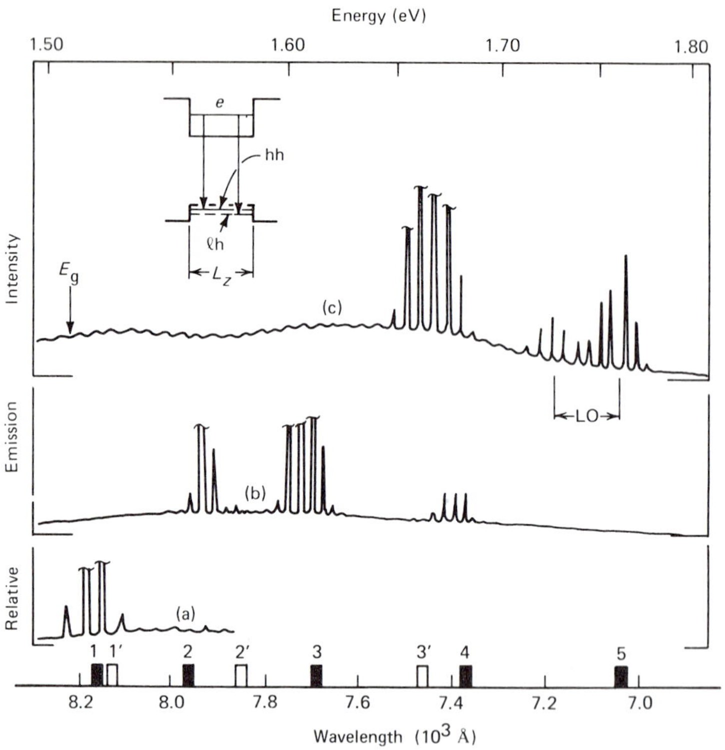

Fig. 18. Stimulated emission (77 K) on the confined-particle transitions (1, 1′, 2, 2′, 3, 3′, 4, 5) of a single-well ($L_z \sim 210$ Å) $Al_xGa_{1-x}As$–GaAs ($x \sim 0.6$) QWH grown by MOCVD. The laser operation (c) at energy $E_5 - \hbar\omega_{LO}$ agrees with the results of Fig. 17. (From Holonyak *et al.*, 1980a. Reprinted from *IEEE J. Quantum Electron.* **QE-16,** 170–186. © IEEE.)

Curves (b) and (c) show that laser operation can occur also on the higher confined-particle transitions of the $L_z \sim$ 200-Å GaAs quantum well, in fact, up to ~300 meV higher than the GaAs band edge or up to the L indirect minima. This form of laser operation, i.e., a single quantum well filled with excess carriers, has been used, with its broad emission spectrum and high-energy cutoff where the well becomes indirect, to measure the L band edge in GaAs (Dupuis *et al.*, 1978a).

If the quantum-well size is doubled and is in the range $L_z = 400-500$ Å (i.e., approaching the size of an ordinary double heterostructure), the carrier collection and scattering are complete down to the lowest states, and all of the recombination occurs near the bottom of the well. If, as already discussed in connection with Fig. 13, the quantum-well size is reduced much under $L_z \sim$ 100 Å, the carrier collection becomes very inefficient, and small size becomes a constraint on single-well laser operation. Small well sizes (< 100 Å) can be used,

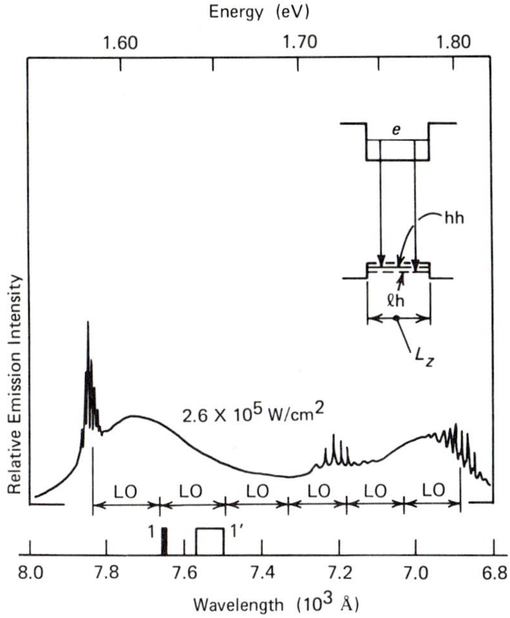

Fig. 19. Laser spectra (77 K, square wave) of an MOCVD $Al_xGa_{1-x}As$ QWH with two $L_z \sim$ 50-Å GaAs quantum wells coupled by a single $L_B \sim$ 50-Å $Al_xGa_{1-x}As$ ($x \sim 0.3$) barrier. The confining layers are $x \sim 0.40$ $Al_xGa_{1-x}As$. The coupling of the wells enhances the carrier scattering and the collection efficiency compared to a single well (see Fig. 12). The $\Delta E \sim 2\hbar\omega_{LO}$ separation of the two sets of high-energy modes agrees with the results of Fig. 17. [After Tang *et al.* (1982).]

however, if multiple wells are employed or if the carriers are confined near the well (Burnham *et al.*, 1982b). An example is shown in Fig. 19, which is for the case of two $L_z \sim$ 50-Å GaAs wells coupled by an $L_B \sim$ 50-Å $Al_xGa_{1-x}As$ ($x \sim 0.3$) barrier. (The 1- and 0.3-μm-thick confining layers are $x \sim 0.4$ $Al_xGa_{1-x}As$.) This QWH has an overall active-region size of 150 Å, or a size comparable to the 200-Å single-well QWH of Fig. 18. This small total active-layer size (not too small, $\lesssim$ 100 Å; not too big, $\gtrsim$ 400 Å) permits band filling to as much as ~300 meV above E_g (GaAs). In accord with Fig. 17 and the distribution function given by Eq. (29), the laser operation at high energy in Fig. 19 occurs on two sets of modes ($\lambda \sim$ 7200, 6900 Å) separated by $2\hbar\omega_{LO}$. This behavior is characteristic of the carrier scattering process as has been shown by a Monte Carlo analysis [see Section IV and Fig. 16; also Tang *et al.* (1982)].

VI. PHONON-ASSISTED LASER OPERATION

Besides the higher-energy laser operation evident at $\lambda \sim$ 7200 and 6900 Å ($\Delta E \sim 2\hbar\omega_{LO}$) in Fig. 19, a more interesting form of laser operation is obvious at lower energy. This laser operation (at $\lambda \sim 7830$ Å) occurs $\hbar\omega_{LO}$ below the lowest confined-particle transitions and on more closely spaced modes than does the laser operation at higher energy. A clearer example is shown in Fig. 20, which is for a photopumped rectangular QWH sample (50×150 μm) with four 50-Å GaAs wells coupled by three 70-Å $Al_xGa_{1-x}As$ ($x \sim 0.35$) barriers (Holonyak *et al.*, 1979b). In addition to the widely spaced edge-to-edge modes (50-μm sample width) supported by recombination on the confined-particle transitions $n = 1$ and $n' = 1'$ (a) (photoexcitation), more closely spaced end-to-end laser modes (150-μm sample length) shifted downward in energy by $\hbar\omega_{LO} \approx 36$ meV are observed to turn-on abruptly (b) and then predominate at still higher excitation (c). In effect, the small rectangular sample with its two sets of spatially orthogonal modes acts as a phonon spectrometer.

An interesting aspect of the phonon-assisted laser operation of Fig. 19 or 20 is that the lowest energy transitions can involve cluster-induced states (Holonyak *et al.*, 1980c) and not just the virtual transitions sketched in Fig. 2. For various (crystal-growth) reasons it is possible that the $Al_xGa_{1-x}As$ coupling barriers of

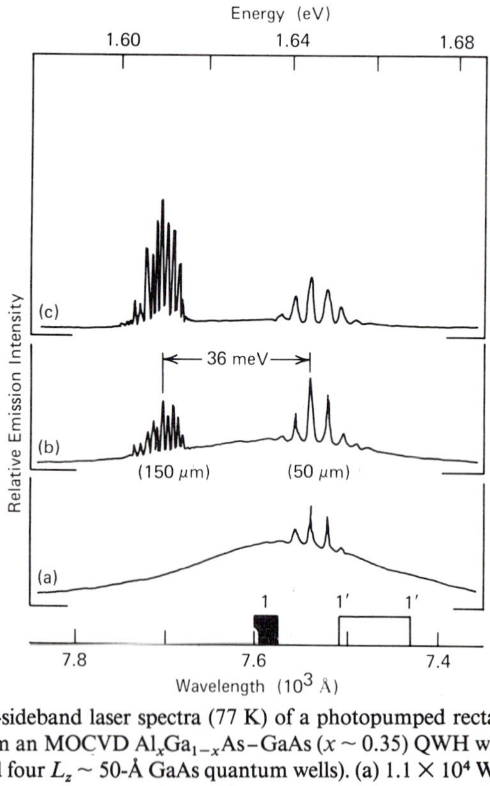

Fig. 20. Phonon-sideband laser spectra (77 K) of a photopumped rectangular sample (50×150 μm) cleaved from an MOCVD $Al_xGa_{1-x}As$–GaAs ($x \sim 0.35$) QWH wafer (three $L_B \sim 70$-Å coupling barriers and four $L_z \sim 50$-Å GaAs quantum wells). (a) 1.1×10^4 W/cm², pulse-excited, (b) 2.5×10^4 W/cm², and (c) 3.7×10^4 W/cm². [After Holonyak *et al.* (1979b, 1980a).]

the samples of Fig. 19 or 20 are not ideal and that GaAs clusters within the barriers lead to increased coupling between the GaAs quantum wells. For example, consider the two-well QWH sample of Fig. 19 and for this case the model diagram in the inset of Fig. 21. The single coupling barrier is $x \sim 0.3$ $Al_xGa_{1-x}As$ and the confining layers are higher in composition ($x \sim 0.4$) and thus energy gap. A small satellite or secondary quantum well representing a GaAs cluster of size L_c is shown in the center of the coupling barrier, which itself is of size $L_B = L_z \sim 50$ Å. Of course, the pancakelike clusters could be anywhere within the barrier, with a certain density per unit area, including at one edge or the other and then would simply cause an increase in the size of either one of the primary wells. The solid curves of Fig. 21 show how the lowest confined-particle electron state (e) of the overall system decreases in energy for an edge cluster or a center cluster as L_c varies in size from 0 to 50 Å; at $L_c = 50$ Å there is no barrier, just one big 150-Å well. The dashed curves are for the combined case of electron and hole (e + h). An edge cluster of size $L_c \sim 15$ Å is sufficient to lead to a reduction in energy of $\hbar\omega_{LO} \sim 36$ meV for the electron–hole recombination radiation.

As will be clear later, phonon-assisted laser operation of a QWH does not necessarily require actual final states, e.g., cluster-induced states. More important is high excitation level (10^3 W/cm^2). This can be illustrated by considering the case of a large "detector" quantum well coupled to a phonon-generating array of much smaller quantum wells as illustrated by the inset of Fig. 22 (Coleman *et al.*, 1980). The active region shown is confined on one side by 1 μm

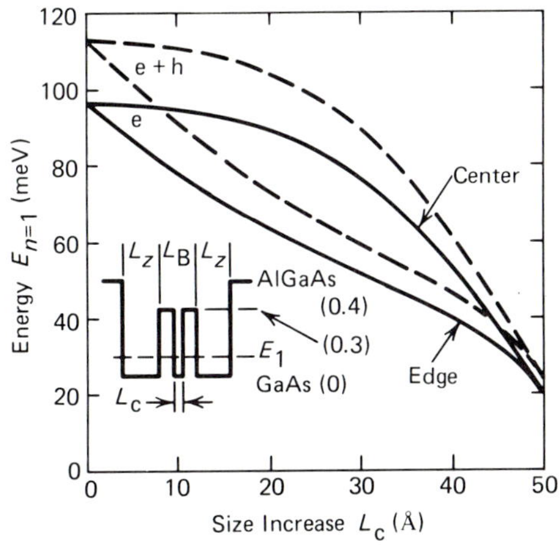

Fig. 21. Diagram showing how a "pancakelike" GaAs cluster L_c at a barrier edge or in its center lowers the energy of the first electron e or electron–hole (e + h) confined-particle state as the cluster size increases. The parameters in the inset (where $L_z = L_B \sim 50$ Å) correspond to the QWH sample of Fig. 19.

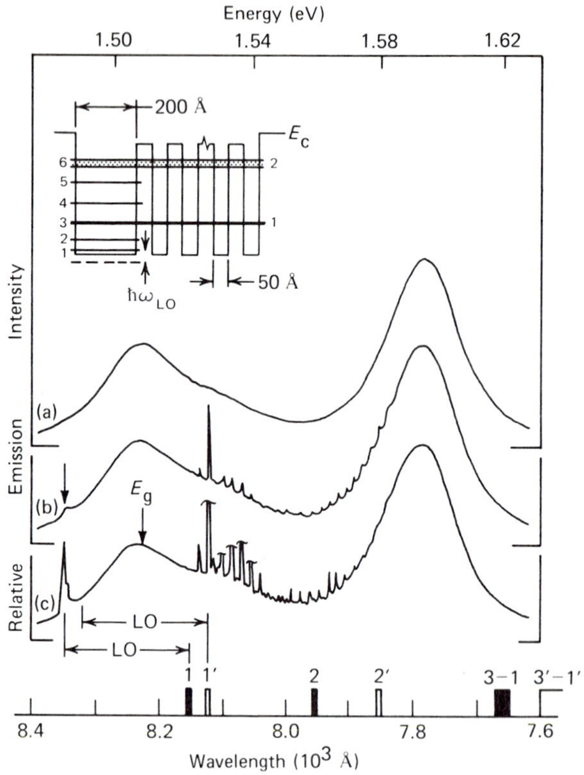

Fig. 22. Emission spectra (77 K) of a photopumped MOCVD $Al_xGa_{1-x}As$–GaAs ($x \sim 0.32$) QWH with a 200-Å GaAs "detector" quantum well coupled ($L_B \sim$ 50-Å barriers) to seven 50-Å GaAs wells as in the inset. With increase in the photopumping from (a) 6.2×10^4 W/cm^2 to (b) 7×10^4 W/cm^2 the $n' = 1'$, e → *l*h transition of the 200-Å well begins to lase while simultaneously a spontaneous bump [(b) ↓] appears one phonon below the $n = 1$, e → hh transition. At still higher excitation [(c) 10^5 W/cm^2] the induced phonon–sideband transition (1-LO, ↓) also lases. [After Coleman *et al.* (1980).]

and on the other by 0.3 μm of $x \sim 0.35$ $Al_xGa_{1-x}As$. For the sample of Fig. 22 the large "detector" well is 200 Å of GaAs coupled via 50-Å $x \sim 0.32$ $Al_xGa_{1-x}As$ barriers to seven 50-Å GaAs wells. The confined-particle (electron) energy levels of the 200-Å GaAs well (1–6) and of the 50-Å multiwell section (1–2) are shown on the left-hand and right-hand sides, respectively, of the inset.

Carriers are photogenerated at higher energy in the thick confining layers and are collected in the active region, mainly in the multiple 50-Å well section. The carriers scatter downward generating phonons in a layered structure (the seven coupled 50-Å GaAs wells) that can act as a phonon cavity. At lower energy the carriers can recombine (the peak at $\lambda \sim 7780$ Å) or at low energy tunnel into the bigger well for further scattering and recombination. The phonons generated in the 50-Å multiple-well section assist the further downward scattering in the single large "detector" well and lead to the unique form of *below-gap* laser operation (Coleman *et al.*, 1980) shown at $\lambda \approx 8343$ Å (1.486 eV < $E_g = 1.508$ eV) in Fig. 22 and at $\lambda \approx 8400$ Å (1.476 eV or $E_g - \hbar\omega \sim 32$ meV) in Fig. 23 for a similar QWH with a still larger "detector" well (500 Å).

The photoluminescence data of Fig. 22 are unique. The high-level spontaneous emission intensity obtained by photopumping across part of the width of the sample is shown by curve (a). When the excitation level is increased from 6.2×10^4 to 7.0×10^4 W/cm², a *single* laser mode emerges abruptly on the first electron-to-light-hole ($e \rightarrow lh$, $n' = 1'$) transition [curve (b)], and *simulta-*

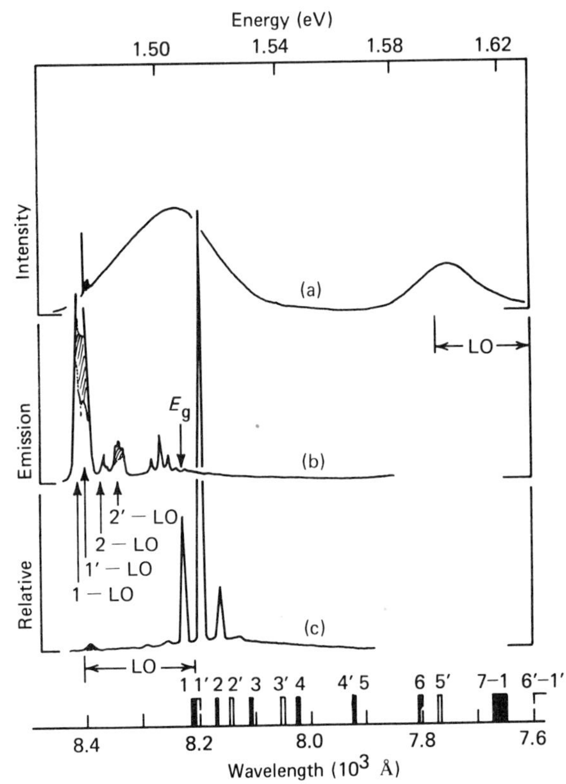

Fig. 23. Emission spectra (77 K) of a photopumped MOCVD $Al_xGa_{1-x}As$–GaAs ($x \sim 0.32$) QWH similar to that of Fig. 21 (see inset) but with a L_{z1} 500-Å GaAs "detector" quantum well coupled ($L_B \sim 50$-Å barriers) to seven $L_{z2} \sim 50$-Å GaAs wells. Induced phonon–sideband laser operation occurs on multiple phonon sidebands below the 1, 1′, 2, and 2′ confined-particle transitions of the large (500-Å) "detector" GaAs quantum well. (a) 1.2×10^3 W/cm³ (cw), (b) 5×10^4 W/cm³ (pulse-excited), and (c) 8×10^4 W/cm³ (pulse-excited). [After Coleman *et al.* (1980).]

neously a prominent bump appears one phonon below the first electron-to-heavy-hole (e → hh, $n = 1$) transition or at $E_1 - \hbar\omega_{LO}$. At still higher pumping level [curve (c)] the $n' = 1'$ mode continues to dominate, and the transition at $E_1 - \hbar\omega_{LO}$ narrows into stimulated emission. The excitation directed on the sample can be adjusted slightly, giving the data of Fig. 24. The reference $n' = 1'$, e → *l*h mode can be tracked and is still identifiable, and at $\hbar\omega_{LO}$ below $E_{1'}$ ($n' = 1'$) and also $\hbar\omega_{LO}$ below E_1 ($n = 1$) *two* bumps (small peaks) appear because of rapid light-hole-to-heavy-hole relaxation. This portion of the spectrum can be expanded (inset of Fig. 24) and gives an accurate measurement of the 1′–1 spacing ($\Delta E = E_{1'} - E_1 = 4.9$ meV) and in turn gives an accurate value for the LO phonon energy of $\hbar\omega_{LO} = 41.0 - 4.9 = 36.1$ meV (see Figs. 22 and 24).

The accuracy and consistency of these results are demonstrated by the data

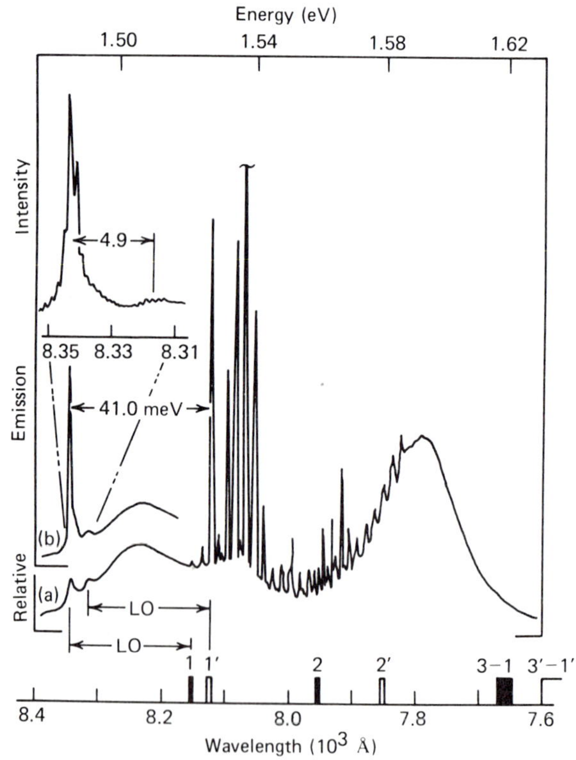

Fig. 24. Induced phonon–sideband laser operation of the sample of Fig. 21 at high enough excitation level to show double peaks at $E_1 - \hbar\omega_{LO}$ and at $E_{1'} - \hbar\omega_{LO}$ or 4.9 meV apart. Because of the accurate measurement of the $n' = 1'$ mode and the $E_1 - \hbar\omega_{LO}$ mode 41.0 meV lower, an accurate measurement of $\hbar\omega_{LO} = 41.0 - 4.9 = 36.1$ meV is possible. (a) 6.9×10^4 W/cm² (pulse-excited) and (b) 1.7×10^5 W/cm². [After Coleman *et al.* (1980).]

of Fig. 23, which shows the induced phonon-sideband laser operation (77 K) of a large, 500-Å GaAs well coupled ($L_B \sim 50$ Å) to a phonon-generating multiple-quantum-well heterostructure consisting of seven GaAs wells coupled with $L_B \sim 50$-Å $Al_xGa_{1-x}As$ ($x \sim 0.32$) barriers. With cw excitation [curve (a)] the phonon sideband of the $n' = 1'$, e $\rightarrow$ lh transition lases from end-to-end along the rectangular sample (254×59 μm). The emission peak at $\lambda \sim 7750$ Å [curve (a)] occurs on the phonon sideband of the $n' = 1'$ transition of the smaller-well (50-Å) section, which is degenerate with the larger-well $n' = 6'$, e $\rightarrow$ lh transition. Note that because of the large size of the phonon "detector" well, $L_{z1} \sim 500$ Å, the lowest confined-particle transitions of this well approach within a few milli-electron-volts the bulk energy gap E_g of GaAs. This removes any ambiguity in the assignment of the lower transitions. At the rather large pulsed excitation level of 5×10^4 W/cm^2 [curve (b)], the sample of curve (a) lases along the sample length at 1-LO ($E_1 - \hbar\omega_{LO}$), 1′-LO, 2-LO and 2′-LO, and edge-to-edge in the range 3-LO to 3′-LO. A narrower sample (23 μm), with large-cavity edge-to-edge losses, shows that laser operation [curve (c)] can be forced upward in energy (for more gain) into the range of the confined-carrier transitions (1, 1′, 2, 2′), while weak phonon-sideband emission is still observed in the range 1-LO to 1′-LO. This laser operation is interpreted as having its origin in an induced process requiring the assistance of "percolation phonons" (Kroemer, 1981) generated in the small-well array.

Figure 25 presents two more examples showing the importance of high-level excitation and its effect on phonon-assisted laser operation. Curve (a) of Fig. 25 shows the case of a cw 300-K QWH laser diode (six $L_z \sim 80$-Å GaAs quantum wells coupled by five $L_B \sim 80$-Å $Al_xGa_{1-x}As$, $x \sim 0.30$, barriers with all of these layers sandwiched between p- and n-type $Al_xGa_{1-x}As$, $x \sim 0.40$, confining and injecting layers) that is driven at $J \sim 2.4 \times 10^3$ A/cm^2 into stimulated emission $\hbar\omega_{LO} \sim 36$ meV below the $n' = 1'$, e $\rightarrow$ lh transition. We mention that cluster states may play a role in this laser operation. When this particular QWH crystal has its GaAs substrate and top GaAs contact layer removed by selective etching, it can be cleaved into thin (~ 1-μm) rectangles that can be more effectively heat sunk (in Cu under diamond) for higher-level photoexcitation than for current-driven diode operation. Curve (b) shows the sample operation at $J \sim 7.9 \times 10^3$ A/cm^2 or at an approximately threefold higher level, and more than threefold higher phonon generation, than the diode of curve (a). In this case the laser operation (cw 300 K) occurs $2\hbar\omega_{LO}$ below the $n =$, e $\rightarrow$ hh transition and, of special note, 13 meV below E_g (GaAs) where cluster-formed states do not exist. This laser operation, just as that of Fig. 22 or 23 [$\hbar\omega < E_g$ (GaAs)], cannot be attributed to cluster-induced states and furthermore requires the involvement of virtual processes.

The experimental results described show that the phonon-assisted laser operation as demonstrated in Figs. 18–20 and 22–25 cannot be explained in conventional terms, and the results are peculiar to QWH lasers. One might

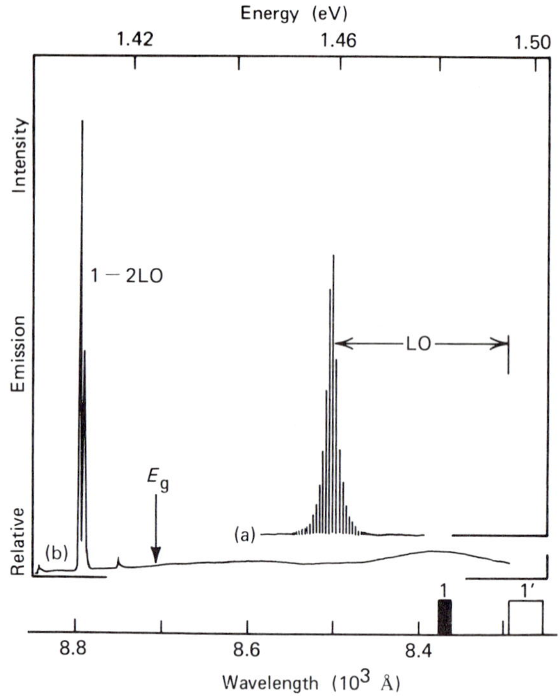

Fig. 25. Comparison of the cw 300-K laser operation of a diode (a) 2.4×10^3 A/cm^2 ($E_{1'} - \hbar\omega_{LO}$) and a photoexcited sample (b) 7.9×10^3 A/cm^2 ($E_1 - 2\hbar\omega_{LO}$) from the same six-well ($L_z \sim 80$-Å), five-barrier ($L_B \sim 80$ Å) MOCVD $Al_xGa_{1-x}As$–GaAs ($x \sim 0.30$) QWH crystal. Note the spontaneous "bump" on (b) near the $n = 1$, e → hh transition.

guess that the electron–phonon interaction is enhanced in a QWH because the electron confinement and phonon zone folding opens up scattering channels. A detailed investigation of this effect (Holonyak *et al.*, 1980b) shows, however, that the coupling constant is enhanced by at most 30–50% in a QWH. The deformation-potential interaction is strongly enhanced only at quantum-well sizes below 10 Å, and the polar-optical interaction has an upper limit for the enhancement (because the matrix element is proportional to $1/q^2$, where $\mathbf{q}$ is the phonon wave vector). This enhancement, although important, does not seem to be sufficient to explain the experiments. However, the matrix elements of the electron–phonon interaction contain as well a factor of $(N_q^{nt} + 1)$ [see Eq. (14)] that can be strongly increased in QWH for the following reasons. The average phonon occupation number N_q is usually assumed to be equal to the thermal equilibrium value

$$N_q = 1/[\exp(\hbar\omega_{LO}/kT) - 1]. \tag{35}$$

We have shown in Eq. (31), however, that a large number of "percolation phonons" (Kroemer, 1981) are created during laser operation of a QWH. As a consequence, the nonequilibrium value N_q^{nt} can be greater than unity even if no phonon reflections at layer boundaries are assumed. If we use a structure (Fig. 3 or the inset of Fig. 22) consisting of a superlattice with layer spacing close to a multiple of the phonon wavelength [Eq. (32)], additional enhancement of N_q^{nt} must be expected since the phonons will be Bragg reflected and therefore will be confined. The actual value of N_q^{nt} under these conditions is difficult to calculate. Notice, however, that $N_q^{nt} = 2$ means an enhancement of the electron–phonon interaction by a factor of 3. Together with the increased coupling constant, the effective enhancement amounts to about a factor of 4 to 5. Therefore, the effective polar coupling constant becomes comparable to the coupling in II–VI compounds, which are known to exhibit phonon sidebands. The interpretation involving enhanced N_q^{nt} is consistent with all the experimental observations as is clearly demonstrated by the switching on of the sidebands at high pumping intensities as in Fig. 22. It is also clear that this effect is mainly observed during laser operation, which further increases N_q^{nt} because of the shorter electron lifetime [Eq. (31)]. It remains to discuss the actual mechanism of the phonon involvement. The experiments indicate two possibilities.

(1) If cluster-induced states are present at a relatively high density, the electrons can scatter into these states with a probability as described in Eqs. (20) and (25). The distribution function for electrons will depend sensitively on the carrier lifetime, density of states, and scattering rate and, for not too large electron–electron interaction rates, will assume the form given by Eq. (29). The appearance of phonon sidebands is then a natural consequence of the higher rates with which electrons scatter into the cluster-induced states. This mechanism can explain the phonon sidebands that are observed above the band gap energy of GaAs.

(2) Below the band gap energy of GaAs, virtual processes must be invoked. It has been proposed by Holonyak *et al.* (1980b) that indirect transitions, i.e., transitions involving both phonon and photon emission, are responsible for these phonon sidebands. A common objection to this explanation is that indirect processes are unimportant in direct semiconductors. However, this is not necessarily true for highly excited systems. A large increase in N_q^{nt} makes the second-order process equally strong as a first-order process because of the proportionality to $N_q^{nt} + 1$.

A typical term appearing in the indirect emission rate is (Hall *et al.*, 1954)

$$\left| \sum_m \frac{\langle f|H_{\text{opt}}|m\rangle\langle m|H_{\text{el-ph}}|i\rangle}{E_m - E_i + \hbar\omega_{\text{LO}}} \right|^2. \tag{36}$$

Here H_{opt} is the Hamiltonian for the electron–photon interaction and as usual the indices i, m, and f stand for initial, intermediate, and final, respectively. The square of the electron–phonon matrix element is, of course, proportional to

$N_q^{nt} + 1$ [Eq. (14)]. There is a slight difficulty in the evaluation of the sum in Eq. (36) because of the possibility of real phonon transitions (the energy denominator vanishes). Therefore it is necessary to proceed in a manner similar to Segall and Mahan (1968) and to introduce a damping term (Laidig *et al.*, 1981a). For moderate values of N_q^{nt} the damping is small, and the value of $|\langle m|H_{\text{el-ph}}|i\rangle|^2$ becomes large. This makes the indirect transition competitive with the direct transition.

VII. CONTINUOUS 300-K QWH LASER OPERATION

Beyond the fact that QWHs and SLs are of major interest for fundamental studies, they are, of course, of practical concern because they permit redesign and improvement of semiconductor lasers. This follows in part because of the steplike density of states of a QWH that replaces, particularly at a band edge, the usual parabolic density of states of a bulk sample (see Fig. 2).

Of first importance, a QWH can be operated continuously (cw) at room temperature (300 K) in stimulated emission. This was first demonstrated (Holonyak *et al.*, 1978a) by photopumping of a single-well $Al_xGa_{1-x}As$–GaAs–$Al_xGa_{1-x}As$ ($x \sim 0.6$) sample ($L_z \sim 200$ Å), followed then by photopumping of a multiple-well MOCVD $Al_xGa_{1-x}As$–GaAs QWH (Holonyak *et al.*, 1978b). Because of the greater flexibility of the latter in the design of the active region of a semiconductor laser, it obviously is of greater importance than the single-well case. Even in rather primitive (1978) versions, multiple-well QWH samples have operated as photopumped cw 300-K lasers at very low thresholds. This is shown by the spectral data of Fig. 26, which are for an MOCVD QWH with four $L_z \sim 80$-Å GaAs wells coupled by three $L_B \sim 80$-Å $Al_xGa_{1-x}As$ ($x \sim 0.35$) barriers (Holonyak *et al.*, 1978b). The QWH sample, with $x \sim 0.4$ $Al_xGa_{1-x}As$ confining layers, is heat sunk in copper under a diamond window, and laser threshold occurs at a photopumping level of $\sim$900 W/cm^2 or an equivalent current density of $J_{\text{eq}} \sim 370$ A/cm^2. The laser operation shown in Fig. 26 is, in fact, achieved with only 7–8 mW of cw power (or $I_{\text{eq}} = 2.9$–3.3 mA) from the pump laser (Ar^+ laser, 5145 Å). It is clear from the large spacing $\Delta\lambda \sim 65$ Å of the small modes on either side of the main mode of curve (b) that the QWH sample is very small ($\lesssim 20$ μm), its edge-to-edge cavity loss large, and its gain very high for it to operate as a cw 300-K laser at so low of a pumping level.

Although the data of Fig. 26 (1978) indicate that very low-threshold photopumped laser operation is possible for a multiple-well QWH, it is only natural that even lower thresholds are possible. This is shown by the data of Fig. 27, which are for a modified single-well QWH in which the excess carriers (photogenerated or injected) are confined near the quantum well in a combination reservoir and waveguide and are "thermionically" emitted into the well (Burnham *et al.*, 1982b, 1983). The inset of Fig. 27 shows schematically the conduc-

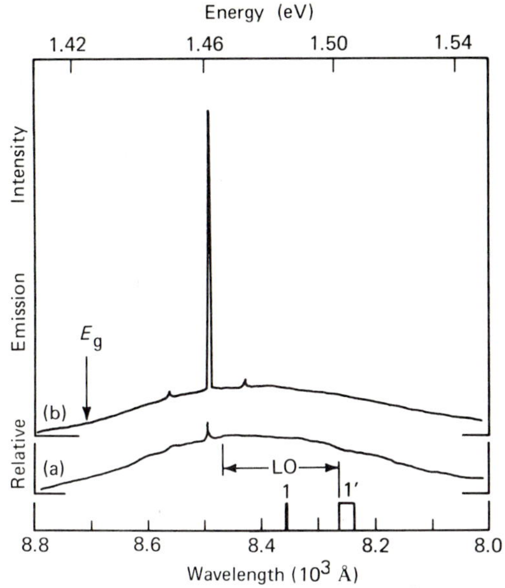

Fig. 26. Low-threshold [(a) 900-W/cm² or J_{eq} ~ 370-A/cm², (b) 1.2×10^3-W/cm²], cw, 300-K photopumped laser operation one phonon below the $n' = 1'$, e → *l*h confined-particle transition of a four-well (L_z ~ 80 Å), three-barrier (L_B ~ 80 Å) MOCVD $Al_xGa_{1-x}As$–GaAs ($x \sim 0.35$) QWH. (From Holonyak *et al.,* 1980a. Reprinted from *IEEE J. Quantum Electron.* **QE-16,** 170–186. © 1980 IEEE.)

tion band of the QWH. An $x \sim 0.05$ $Al_xGa_{1-x}As$ quantum well of size $L_z \sim 60$ Å is "sandwiched" between $x' \sim 0.30$ $Al_{x'}Ga_{1-x'}As$ carrier–capture or confining layers of total thickness $L_{z'} \sim 0.12$ μm. This is the carrier reservoir and waveguide region and is clad on either side with $x'' \sim 0.75$ $Al_{x''}Ga_{1-x''}As$. With only 5 mW of measured power from the Ar^+ laser pump source ($\lambda =$ 5145 Å), and only half of this power focused through windows, lenses, and diamond on the sample, cw 300-K laser threshold is observed. This corresponds to an equivalent current excitation of only $I_{eq} \sim 1$ mA or $J_{eq} \sim 160$ A/cm², which quite likely is as low a threshold as has ever been observed on a semiconductor laser (Burnham *et al.,* 1983).

A further development in photopumped QWH and SL lasers is the demonstration that a strained SL is capable of pulsed *and cw 300-K* laser operation (Ludowise *et al.,* 1983a,b). This is an unusual development because mismatched layers and the resulting strained layers and heterointerfaces are widely regarded to be densely populated with defects and nonradiative recombination centers, enough to preclude stimulated emission at low temperature not alone at 300 K. Contradicting this view, we show in Fig. 28 the cw 300-K photopumped laser operation of the strained GaAs–$In_xGa_{1-x}As$ ($x \sim 0.2$) SL of Fig. 11 ($L_B \sim 75$ Å, $L_z \sim 75$ Å, 66-period SL). The cw 300-K laser operation of Fig.

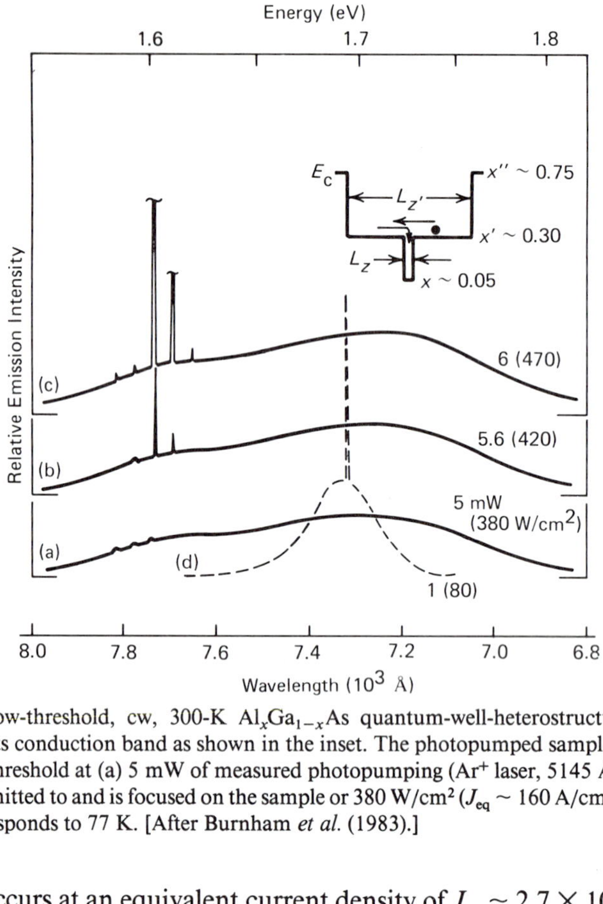

Fig. 27. Low-threshold, cw, 300-K $Al_xGa_{1-x}As$ quantum-well-heterostructure laser ($\lambda \sim$ 7730 Å) with its conduction band as shown in the inset. The photopumped sample (20×40 μm area) reaches threshold at (a) 5 mW of measured photopumping (Ar^+ laser, 5145 Å). Half of this power is transmitted to and is focused on the sample or 380 W/cm^2 ($J_{eq} \sim$ 160 A/cm^2). The dashed curve (d) corresponds to 77 K. [After Burnham *et al.* (1983).]

28, which occurs at an equivalent current density of $J_{eq} \sim 2.7 \times 10^3$ A/cm^2, is not stable. This is not unexpected on account of the large built-in strain (7×10^{-3}) and the obviously high excitation conditions. For example, a less heavily strained (4.5×10^{-3}) $GaAs_{1-x}P_x$–GaAs ($x \sim 0.25$) SL operates at lower threshold and is more stable (Ludowise *et al.*, 1983b). Because of the early nature of these experiments, it is not possible to give a very complete account of work on strained-layer QWHs and SLs or what the ultimate value might be of this class of heterostructures. Nevertheless, it is sufficient at this point to know that a strained-layer QWH or SL can be grown (e.g., by MOCVD) with highly "perfect" heterointerfaces.

Just as for photopumped samples, cw 300-K laser operation of $p-n$ (injection) QWH lasers occurred first for the case of single-well MOCVD $Al_xGa_{1-x}As$–GaAs diodes (Dupuis *et al.*, 1979b) and later in multiple-well configurations (Dupuis *et al.*, 1979c). The first cw 300-K single-well $p-n$ QWH (1978) has the distinction thus far of exhibiting the lowest temperature sensitiv-

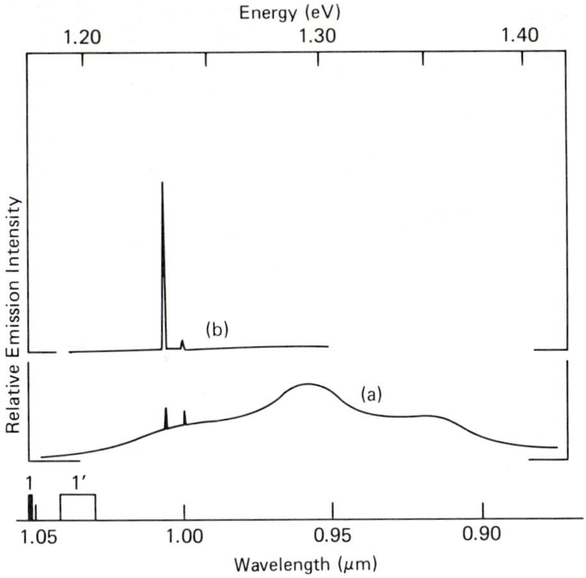

Fig. 28. Continuous 300-K laser operation of the GaAs–$In_xGa_{1-x}As$ ($x \sim 0.2$) strained SL shown in Fig. 11. (a) 6.5×10^3 W/cm², (b) 7.5×10^3 W/cm. At excitation levels below and near threshold (a), band filling is observed (and higher energy peaks) up to $\hbar\omega \sim E_g$ (GaAs). Above threshold the laser peaks predominate but decay as the SL degrades. (The calculated $n = 1$ electron-to-heavy-hole (e → hh) and $n' = 1'$ electron-to-light-hole (e → *l*h) transitions are shown for reference but are not considered accurate because of the built-in lattice deformation.) [After Ludowise *et al.* (1983b).]

ity of all semiconductor lasers. (We return to this later.) The first cw 300-K multiple-well QWH laser diodes (1979) have, as expected, been superseded in performance (Tsang, 1981; Scifres *et al.*, 1982; Burnham *et al.*, 1982a,b) but are still interesting because of their unique tunneling behavior. A slant cross section of this MOCVD QWH wafer is shown in Fig. 8, and a typical diode $I-V$ characteristic (77 K) is shown in Fig. 29 (0.2 mA/div, 0.2 V/div). When the *n*-type-side bias relative to the *p*-type-side approaches ~1.5 V or almost to the region of significant carrier injection [$V \sim E_g$ (GaAs)/e], several local negative resistance regions appear, with the light emission following the $I-V$ characteristic (Vojak *et al.*, 1979). This behavior is caused by tunnel injection of electrons when the *n*-type-side quasi-Fermi level approaches and becomes aligned with the lower confined-particle electron states of the quantum wells adjacent to the *p*-type side. Resonant tunneling as the electron quasi-Fermi level crosses and uncrosses with the lower confined-particle electron states of the *p*-type-side quantum wells permits, in effect, electrons to be "injected" and to recombine with holes entering these wells.

The type of tunneling shown in Fig. 29, which occurs far from $V = 0$ and as

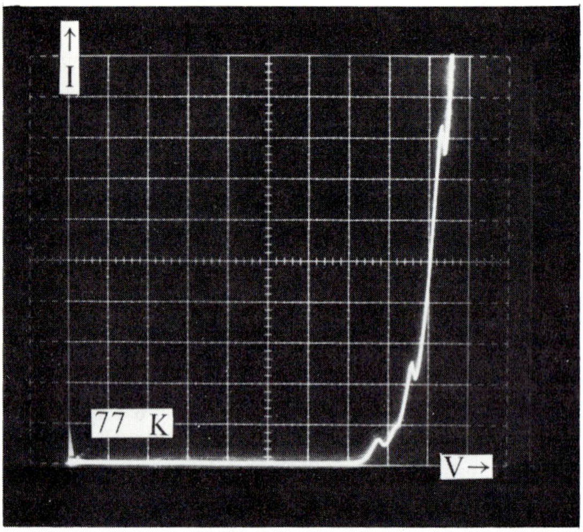

Fig. 29. Tunnel injection current–voltage characteristic (77 K, 0.2 mA/div, 0.2 V/div) of a six-well ($L_z \sim 120$ Å), five-barrier ($L_B \sim 120$ Å) MOCVD $Al_xGa_{1-x}As$–GaAs QWH laser diode fabricated on the wafer of Fig. 8. (From Holonyak *et al.,* 1980a. Reprinted from *IEEE J. Quantum Electron.* **QE-16,** 170–186. © 1980 IEEE.)

usual is very sensitive to layer size (thickness), is remarkable because of, among other reasons, its unusual uniformity across a large (> 3-cm) wafer. This demonstration of the high degree of uniformity of QWHs grown by MOCVD is of special importance when it comes to the fabrication of, for example, the high-power, coupled, multiple-stripe QWH injection lasers of the type described by Scifres *et al.* (1982). This type of laser is shown in Fig. 30 and as is clear consists of ten 3-μm-wide stripe QWH lasers spaced on 10-μm-wide centers. The QWH active region consists of four $L_z \sim 130$-Å $Al_xGa_{1-x}As$ ($x \sim 0.06$) quantum wells coupled by three $L_B \sim 40$-Å $Al_xGa_{1-x}As$ ($x \sim 0.2$) barriers. The bottom n-type $Al_xGa_{1-x}As$ ($x \sim 0.4$) confining layer is ~ 2.5 μm thick and the top p-type $Al_xGa_{1-x}As$ ($x \sim 0.4$) confining layer is ~ 2.0 μm thick. The ten QWH laser stripes are electrically uniform, as expected, and divide the excitation current essentially equally. Hence, the ten QWH laser stripes, which are optically coupled, emit a coherent signal. The QWH laser diode of Fig. 30, *a small diode,* is capable of cw 300-K operation at the remarkable output power of 400 mW (Scifres *et al.,* 1982). In fact, by increasing the number of "stripes" in the QWH laser of Fig. 30 or by just simply making a large area diode version of the QWH of Fig. 27, Burnham and co-workers (Burnham 1982) have been able to demonstrate 1-W semiconductor lasers.

Besides their high power capability, QWH laser diodes give promise of being capable of very-short-wavelength (visible spectrum) operation. For example, Burnham *et al.* (1982a) have reported 7000–7100-Å, cw, 300-K laser operation

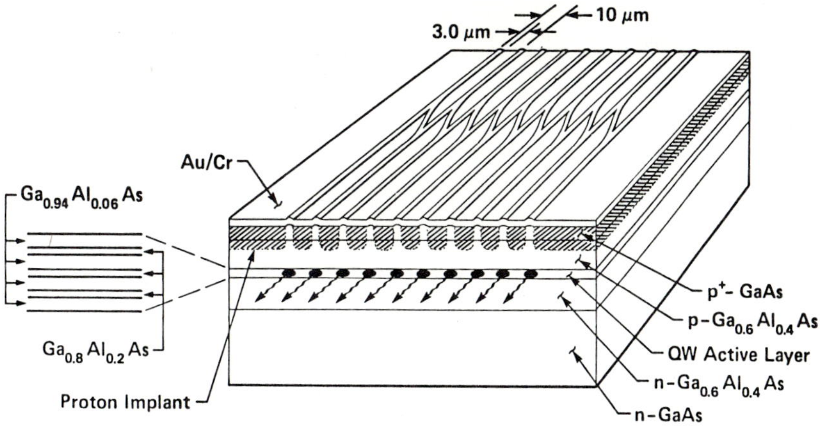

Fig. 30. High-power ($\geq$400-mW), cw, 300-K multiple coupled stripe four-well ($L_z \sim 130$ Å), three-barrier ($L_B \sim 40$ Å) MOCVD $Al_xGa_{1-x}As-Al_{x'}Ga_{1-x'}As$ QWH laser diode. The alloy composition of the quantum wells is $x' \sim 0.06$, the barriers $x \sim 0.2$, and the confining layers $x \sim 0.4$. [After Scifres *et al.* (1982).]

of an $Al_yGa_{1-y}As-Al_xGa_{1-x}As$ ($y > 0.6$, $x \sim 0.27$) QWH with a single-quantum-well active region of size near the bulk limit ($L_z = 400-600$ Å). High-pressure measurements on these QWH lasers (Kirchoefer *et al.*, 1982c) indicate that single-quantum-well laser operation (300 K) in the $Al_xGa_{1-x}As$ system is possible at least to 6700 Å. More sophisticated QWHs of small GaAs well size, as in Figs. 3 and 5 [$E_1(e \rightarrow h)$], indicate that the $AlAs-Al_xGa_{1-x}As-GaAs$ system is capable of 300-K laser operation to wavelengths as short as 6500–6400 Å (Camras *et al.*, 1982). In the case of these more sophisticated QWHs, in which the effects of alloy disorder are minimized, small well size is carried to the point of a size-induced direct–indirect transition. This is indicated in Fig. 5 by the well symbol on Γ that "pulls," with small enough size L_z, Γ confined-particle states up to the point of crossover with L or X confined-particle states, or up to the curve $E_1(e \rightarrow h)$.

In addition, to the features of QWH lasers already mentioned, various measurements indicate that their transient and noise characteristics are improved in comparision with otherwise similar conventional double heterojunctions (DHs). Some of these measurements (Anderson *et al.*, 1981) have been carried out on a multiple-quantum-well $Al_xGa_{1-x}As-GaAs$ heterostructure with six $L_z \sim 80$-Å GaAs quantum wells coupled by five $L_B \sim 80$-Å $Al_xGa_{1-x}As$ ($x \sim 0.3$) barrier layers. The QWH diodes (with $x \sim 0.4$ $Al_xGa_{1-x}As$ confining layers) are typically 200–300 μm long with ~8-μm-wide *p*-type GaAs shallow mesa contact stripes. The cw 300-K stimulated-emission spectra of these QWH lasers are shown in Fig. 31. As shown in Fig. 31 the emission spectrum and large number of modes at 53 mA narrow to a single mode at 65 mA. The single-

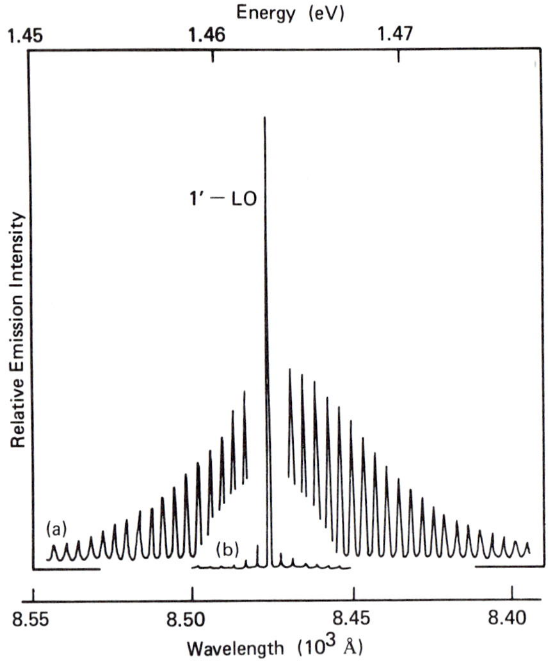

Fig. 31. Continuous 300-K emission spectra of a six-well ($L_z \sim 80$ Å), five-barrier ($L_B \sim 80$ Å) $Al_xGa_{1-x}As$–GaAs QWH laser diode (a) 53 mA and (b) 65 mA. The cw, 300-K QWH laser operates on a single longitudinal mode at energy $\hbar\omega \sim E_{1'} - \hbar\omega_{LO}$. [After Anderson *et al.* (1981).]

mode operation shown in typical of all the diodes fabricated on this QWH wafer.

The intensity fluctuations typical of these QWH lasers are shown (Anderson *et al.*, 1981) in Fig. 32 at the current ratios I/I_{th}: (a) 1.02, (b) 1.07, (c) 1.11, and (d) 1.15. The noise spectra are small and broad without any sharp intrinsic resonance as is usually observed in ordinary DHs of similar stripe geometry. As the excitation current of these QWH lasers is increased, the noise spectrum remains broad and, as expected of shot-noise resonance, shifts to higher frequency as $(I/I_{th} - 1)^{1/2}$.

Earlier work on conventional DH diodes (Petermann, 1979) indicates that the dynamic properties of semiconductor lasers are improved with increased coefficient α giving the spontaneous emission coupled into lasing modes. The coefficient α is given as (Petermann, 1979)

$$\alpha = \lambda^4\xi/2\pi n^3 V\Delta\lambda, \tag{37}$$

where λ is the lasing wavelength, n the refractive index, V the volume of the active region, $\Delta\lambda$ the half-width of the spontaneous emission, and ξ the power-

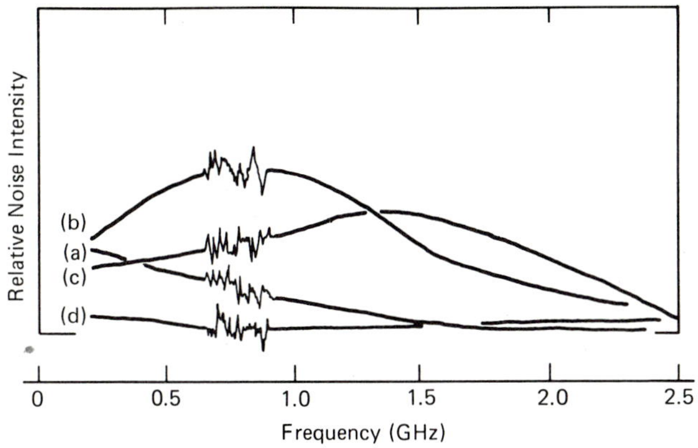

Fig. 32. Noise spectrum of the multiple well (L_z, $L_B \sim 80$ Å, $x \sim 0.30$) $Al_xGa_{1-x}As$–GaAs QWH (300-K, cw) laser diode of Fig. 31 at the normalized currents I/I_{th}: (a) 1.02, (b) 1.07, (c), 1.11, and (d) 1.15. For convenience the noise fluctuations are shown only between 0.7 and 0.9 GHz. [After Anderson *et al.* (1981).]

confinement factor. Perhaps the most important part of Eq. (37) is the inverse dependence on the active region volume V. For a QWH this term can be made small, and α large, thus making possible improved dynamic properties of a semiconductor laser.

Finally, a most important property of a semiconductor laser is its sensitivity to temperature. Usually the current threshold is expressed as

$$J = J_0 \exp(T/T_0), \tag{38}$$

where a high value of T_0 signifies a low sensitivity to increases in temperature. For a single-well QWH a T_0 value as high as 437°C has been measured (Chin *et al.*, 1980a). In more sophisticated QWH diode configurations, Burnham and co-workers have achieved T_0 values as high as 380°C (Burnham, 1982).

To take into account its temperature sensitivity, a theory of the laser threshold current of QWHs has been developed (Hess *et al.*, 1980b) based on the relation

$$J = eL_zR_{sp}/\eta. \tag{39}$$

Here η is the quantum efficiency and L_z is assumed to be equal to the active layer thickness. The total spontaneous emission just below threshold R_{sp} depends sensitively on the joint conduction- and valence-band density of states. Part of the temperature dependence of R_{sp}, and hence of the threshold current, arises from the energy dependence of the density of states. It is therefore clear that a constant density of states as given by Eq. (2) leads to a weaker temperature dependence than for the three-dimensional density of states (proportional to $\sqrt{E}$), or the related case of exponential band tails. This has been demonstrated in detail by Hess *et al.* (1980b), who used the "no-**k**-selection-rule" model of Lasher and Stern (1964). The calculation has been repeated by assuming **k** selection in the recombination process; similar results have been obtained (Dutta, 1982).

While the preceding analyses give a clear indication of a weaker temperature dependence of the threshold current (large T_0) due to the constant density of states of a QWH, the form of the electron–hole distribution function can have similar consequences. The electron (hole) temperature T_c can be much higher than the temperature of the crystal lattice (see Subsection IV.D). For example, T_c can be as large as 200–250 K even if the lattice temperature varies between 4.2 and 100 K. Since T_c enters R_{sp} in Eq. (39) and not the lattice temperature, the threshold current will be less sensitive to the lattice temperature if conditions are such that T_c is raised substantially. However, using a high magnetic field perpendicular to the QWH layers to achieve complete quantization, Arakawa and Sakaki (1982) have confirmed the dominant role of the form of the density of states. Their results clearly indicate that the reduced dimensionality of a QWH causes most of the increase in T_0, as had been pointed out previously by Chin *et al.* (1980a).

VIII. FUNDAMENTAL EFFECTS

Although it is clear that a semiconductor laser can be redesigned with quantum-well layers and be improved in performance, in the preceding presentation we have not concentrated on practical issues as much as on fundamental effects and properties, mainly those, of course, that relate to luminescence behavior and stimulated emission. However, not all that is of interest about QWHs and SLs, and their unique quasi-two-dimensional form, concerns lasers or, for that matter, various other electronic devices. The type of sophisticated layering of a semiconductor that is present in a QWH or SL makes possible other fundamental studies and measurements.

One of the classic experiments performed routinely on a semiconductor sample is to examine its properties, particularly near an energy gap or band edge, by an aborption measurement. Because of the extreme thinness of the active region of a QWH sample such as shown in Fig. 1 (or 6, 7, or 8), it is not at all convenient in this case to make an absorption measurement (generally a low-level measurement) with the goal of revealing the behavior of the confined-particle transitions. A SL sample as shown in Fig. 9, which is high enough in quality to operate as a cw, 300-K photopumped laser, is another matter. Its substrate can be removed, and in the resulting bare $Al_xGa_{1-x}As$–GaAs SL over 1 μm of GaAs is present to establish a good absorption signal. This is a special and an interesting case because a SL provides a basis to partition a GaAs energy band, say the Γ conduction band, with confined-particle states or minibands (see Fig. 4) that can serve as convenient "labels" within a band. Thus a mechanism is provided that makes it possible to examine the validity of band-edge properties high into a band, including at high pressure.

The change in the absorption with pressure of the $Al_xGa_{1-x}As$–GaAs ($x \sim 0.5$) SL of Fig. 9 is shown in Fig. 33 (Kirchoefer *et al.*, 1982a). Markers (↓) indicating the position of the L and X minima are shown for reference. As shown in Fig. 33 the confined-particle transitions (1, 1′, 2, 2′, 3) of the GaAs quantum wells *all* shift (rigidly) with a coefficient of ~11.5 meV/kbar. This behavior is typical of AlAs–$Al_xGa_{1-x}As$–GaAs SLs. These data (Fig. 33) show that the Γ energy band extends high in energy above the L band edge and "moves" with pressure in the same manner high in the band (above L and possibly X) as it does near the fundamental band edge. Also, it is evident that the L indirect minima, and their lesser density of states, have a much weaker effect than do the X minima, which, shifting downward with pressure, interfere with and round off the $n = 3$ e → hh absorption peak.

From the data of Fig. 33, and other similar SL data, it is clear that the Γ

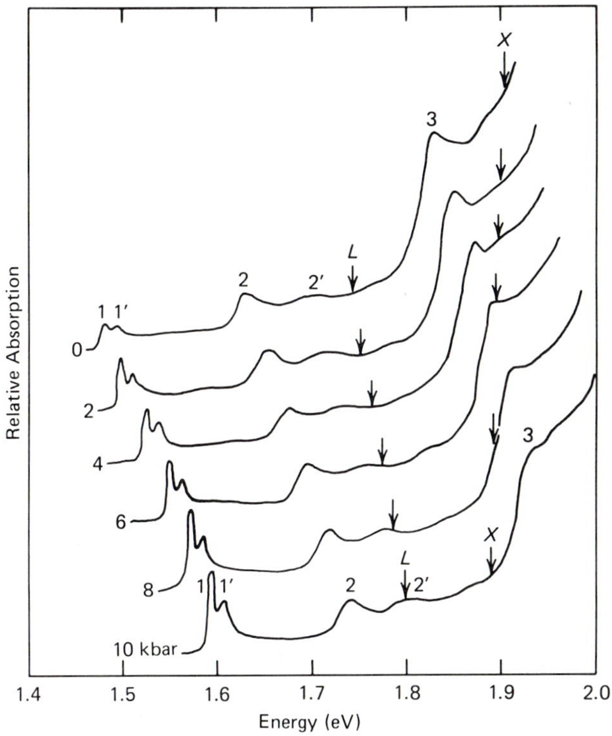

Fig. 33. Relative absorption as a function of photon energy, at hydrostatic pressures from 0 to 10 kbar, of the 121-period $Al_xGa_{1-x}As$ ($x \sim 0.5$, $L_B \sim 80$ Å, $L_z \sim 85$ Å, 300 K) superlattice of Fig. 9 (substrate removed). At higher pressures ($\gtrsim 6$ kbar) the X indirect band edge interferes with the $n = 3$, e → hh transition, which smears into a shoulder. Note that the other peaks sharpen in relative intensity with higher pressures. [After Kirchoefer *et al.* (1982a).]

energy band retains its basic shape and mass properties, with the lattice deformation associated with hydrostatic pressure. The smaller pressure coefficient of 11.5 meV/kbar of an AlAs–$Al_xGa_{1-x}As$–GaAs SL as compared to the 12.5 meV/kbar of bulk GaAs appears to be a true SL effect. In other words, in this lattice-matched system, strain or defects at the heterointerfaces are not a fundamental issue. In contrast to this behavior, a DH or QWH laser diode, in spite of its lesser number of lattice-matched heterointerfaces, exhibits more complicated behavior (Kirchoefer *et al.*, 1982b) and a still smaller pressure coefficient of only ~8.5 meV/kbar (at higher pressures). This significant difference between the pressure coefficients of a bare SL and a QWH or DH diode indicates that built-in (biaxial) strains exist in the active regions of diodes. These probably owe their existence to doping mismatches between the bulk p and n regions of an $Al_xGa_{1-x}As$–GaAs diode (Kirchoefer *et al.*, 1982b) and not to any defects or faults, as such, in the active region.

Superlattices such as shown in Figs. 9, 10, and 11 make it possible to observe another remarkable property of the AlAs–$Al_xGa_{1-x}As$–GaAs system: If an impurity such as Zn is diffused into these SLs or QWHs, impurity-induced layer interdiffusion, or Ga and Al disordering across a heterointerface, occurs at much lower temperatures [<600°C (Laidig *et al.*, 1981b)] than the crystal growth temperature [~750°C (Coleman *et al.*, 1981c)] or the temperature [≳900°C (Chang and Koma, 1976)] for ordinary thermal interdiffusion. In other words, an AlAs–GaAs or $Al_xGa_{1-x}As$–GaAs heterointerface is unstable against Zn diffusion, including at unusually low temperatures (500–600°C). This is true also of strained SLs, e.g., a GaP–$GaAs_{1-x}P_x$ SL (Camras *et al.*, 1983).

An all-binary AlAs–GaAs SL sample into which Zn has been diffused (575°C, 10 min) through a stripe opening in a masking layer of Si_3N_4 on the SL surface is shown in Fig. 34. The SL cross sections of Fig. 34 are prepared on slant angles of ~0.4° and are accidentally skewed somewhat relative to the diffusion stripe openings. Under the Si_3N_4 where the SL is masked during the impurity diffusion, the binary layers [40 periods, L_B (AlAs) ~ 150 Å, L_z (GaAs) ~ 45 Å] remain intact and are clearly distinguishable. In the region under the stripe opening the previously red-gap (1.61-eV) AlAs–GaAs SL becomes disordered *single-crystal* yellow-gap (2.08-eV) $Al_xGa_{1-x}As$ of alloy composition $x = L_B/(L_B + L_z) \approx 0.77$ (see Fig. 5). It is clear from the region under the stripe opening that the high-gap, low-gap heterobarriers of the SL (or a QWH) are unstable against Zn diffusion. That is, the binary layers smear into homogeneous alloy as the Al and Ga interdiffuse under the influence of the (Zn) impurity diffusion.

The fact that diffusion of Zn greatly enhances the interdiffusion of Ga and Al in a SL heterostructure indicates that a modification of the well-known interstitial–substitutional mechanism of Zn diffusion (in a III–V such as GaAs) is required. The older model, of itself, cannot be regarded as complete since it does not suggest or make allowance for any significant motion or

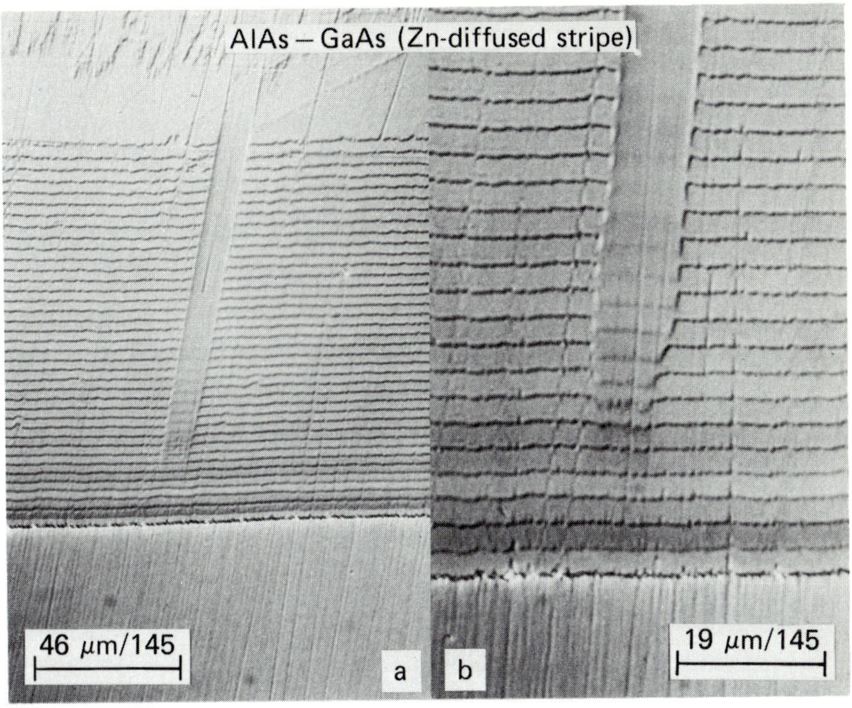

Fig. 34. Shallow-angle (~0.4°) cross section of an AlAs–GaAs SL that, except for a 10-μm stripe, has been masked with Si_3N_4 and has been Zn diffused ($ZnAs_2$) for 10 min at 575°C. Shown are (a) low and (b) high magnifications. [After Laidig *et al.* (1981b).]

instability of lattice constituents; indeed, in a binary such as GaAs it would not be possible to observe easily Ga rearrangement, which, of course, can occur. The data of Fig. 34 (Laidig *et al.*, 1981b) show that the usual interstitial–substitutional mechanism of Zn diffusion in GaAs or $Al_xGa_{1-x}As$ must be modified to include formation of closely associated Zn-vacancy pairs or "molecules" (Zn_i, V) as described by the reaction

$$Zn_i^+ + V^0 \rightleftarrows (Zn_i, V)^+ \rightleftarrows Zn_s^- + 2h^+. \tag{40}$$

In (40) neutral vacancies are represented by V^0, holes by h^+, interstitial Zn by Zn_i, and substitutional Zn by Zn_s. During the process of Zn diffusion the total vacancy concentration c_v is the sum of the neutral-vacancy concentration c_v° and the concentration of Zn-vacancy pairs or "molecules" c_m. The interdiffusion of Al and Ga can be enhanced if

$$c_v = c_v^\circ + c_m \gg c_v^\circ. \tag{41}$$

In this case Al–Ga interdiffusion can proceed via the increased number of column III "vacancies," i.e., via the incipient or pseudovacancies associated with Zn-vacancy pairs or "molecules" having a density c_s approaching that of substitutional Zn. In the interstitial–substitutional Zn diffusion process, which occurs at relatively high Zn concentrations, the substitutional site is partially available for column-III-atom occupancy and hence for Al–Ga interchange. The Al–Ga interchange is particularly effective at a heterobarrier or any region of large Al–Ga gradient. In fact, the model for impurity-induced disordering of a QWH or SL may be quite a bit more complicated than outlined. For example, Van Vechten (1982) has proposed that impurity-induced disordering also involves column V atoms (B atoms, anions) and antisite defects and not just A atoms (cations).

Because of the recent nature of experiments on impurity-induced disordering of a SL or QWH, not all of the ramifications of this work are evident. It is already clear that via Zn diffusion, and photolithography, a SL or QWH can be disordered in an arbitrary lower-gap/higher-gap pattern (Holonyak *et al.*, 1981). This is of obvious interest in various integrated-circuit, integrated-optics applications. On a fundamental level, impurity-induced disordering of a SL or QWH affords a new approach to the study of impurity diffusion and impurity behavior in a III–V semiconductor. Besides the data previously described, it has already been established that Si can be ion implanted into and will disorder an AlAs–GaAs or $Al_xGa_{1-x}As$–GaAs SL (Coleman *et al.*, 1982). No doubt much more will be learned in this area of work, which includes lattice-matched as well as strained-layer QWHs and SLs.

Although no great issue has been raised in this chapter concerning the importance of alloy clustering in $Al_xGa_{1-x}As$–GaAs SLs and QWHs, this is nevertheless a fundamental effect (Holonyak *et al.*, 1980c) of some significance and in many kinds of heterostructures preferably should be eliminated or minimized. This leads to the important idea and the practical embodiment of an alloy-free, cluster-free all-binary AlAs–GaAs SL or QWH [see Coleman *et al.* (1981c) and references therein]. At first sight it might appear that such a heterostructure, for example, as shown in Fig. 3, is of little practical use. The binary AlAs, however, can easily be incorporated into the inside of a SL or QWH without the problem of it being exposed to air and decomposing. The energy-gap variation that typically is accomplished via change of the alloy composition x in $Al_xGa_{1-x}As$ is effected in an all-binary AlAs–GaAs active region by change of the GaAs quantum-well size L_z. This is shown in Fig. 35. By simply reducing the GaAs quantum-well size over a few or gradually over many layers, e.g., $L_{z3} > L_{z2} > L_{z1}$ in Fig. 35, we can achieve the effect of a graded band gap. Also, it is possible to achieve p or n conductivity by simply doping, as desired, the GaAs quantum wells (Holonyak *et al.*, 1982). Thus, it is possible to build a variety of sophisticated semiconductor devices in all-binary artificial material, i.e., in an AlAs–GaAs SL or QWH, that is free of alloy disorder and clustering.

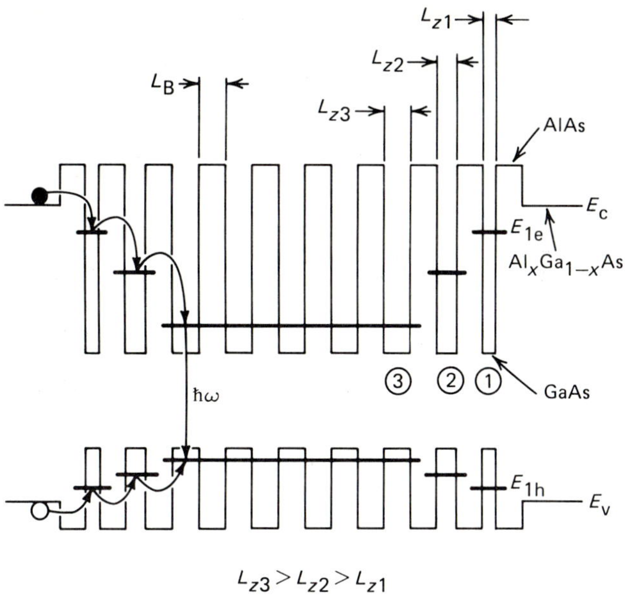

Fig. 35. An all-binary-disorder- and cluster-free AlAs–GaAs quantum-well heterostructure or superlattice showing how quantum-well size can be changed from L_{z3} to L_{z2} to L_{z1} ($L_{z3} > L_{z2} > L_{z1}$) to give the effect of an energy-gap ($E_{1e} - E_{1h}$) change. (Doping on either end, as desired, can be effected in the quantum wells.)

ACKNOWLEDGMENTS

We are grateful to a number of colleagues and former graduate students for their contributions and help in the study of QWHs and SLs. We particularly wish to thank J. Bardeen for his interest, and R. D. Burnham (Xerox), J. J. Coleman (Rockwell), P. D. Dapkus (Rockwell), R. D. Dupuis (Rockwell), and M. J. Ludowise (Varian) for their collaboration in various experiments. For technical support we thank Yuri S. Moroz, B. L. Marshall, and B. L. Payne. This work has been supported by the National Science Foundation and the U.S. Army Research Office.

REFERENCES

Alfërov, Z. I., and Kazarinov, R. F. (1963). Author's certificate 1032155/26-25, U.S.S.R.

Alfërov, Z. I., Andreev, V. M., Garbuzov, D. Z., Zhilyaev, Y. V., Morozov, E. P., Portnoi, E. L., and Trofim, V. G. (1970). *Fiz. Tekh. Poluprovodn.* **4,** 1826–1829. [Engl. Transl. (1971) *Sov. Phys.-Semicond.* **4,** 1573–1575.]

Anderson, E. R., Vojak, B. A., Holonyak, N., Jr., Stillman, G. E., Coleman, J. J., and Dapkus, P. D. (1981). *Appl. Phys. Lett.* **38,** 585–587.

Arakawa, Y., and Sakaki, H. (1982). *Appl. Phys. Lett.* **40,** 939–941.

Bogdankevich, O. V., Darznek, S. A., and Eliseev, P. G. (1976). "Poluprovodnikovie Laseri" ("Semiconductor Lasers"). Nauka, Moscow.

Burnham, R. D. (1982). Private communication.

Burnham, R. D., Scifres, D. R., and Streifer, W. (1982a). *Electron. Lett.* **18,** 507–509.

Burnham, R. D., Streifer, W., Scifres, D. R., Lindstrom, C., Paoli, T. L., and Holonyak, N., Jr. (1982b). *Electron. Lett.* **19,** 1095–1097.

Burnham, R. D., Streifer, W., Scifres, D. R., Holonyak, N., Jr., Hess, K., and Camras, M. D. (1983). *J. Appl. Phys.* **54,** 2618–2622.

Camras, M. D., Holonyak, N., Jr., Hess, K., Coleman, J. J., Burnham, R. D., and Scifres, D. R. (1982). *Appl. Phys. Lett.* **41,** 317–319.

Camras, M. D., Holonyak, N., Jr., Hess, K., Ludowise, M.J., Dietze, W. T., and Lewis, C. R. (1983). *Appl. Phys. Lett.* **42,** 185–187.

Casey, H. C., Jr., and Panish, M. B. (1978). "Heterostructure Lasers." Academic Press, New York.

Chang, L. L., and Koma, A. (1976). *Appl. Phys. Lett.* **29,** 138–141.

Chin, R., Holonyak, N., Jr., Kirchoefer, S. W., Kolbas, R. M., and Rezek, E.A. (1979). *Appl. Phys. Lett.* **34,** 862–864.

Chin, R., Holonyak, N., Jr., Vojak, B. A., Hess, K., Dupuis, R. D., and Dapkus, P. D. (1980a). *Appl. Phys. Lett.* **36,** 19–21.

Chin, R., Holonyak, N., Jr., Stillman, G. E., Tang, J. Y., and Hess, K., (1980b). *Electron. Lett.* **16,** 467.

Chin, R., Holonyak, N., Jr., and Vojak, B. A. (1980c). *J. Appl. Phys.* **51,** 4017–4021.

Cho, A. Y., and Arthur, J. R., Jr. (1975). *In* "Progress in Solid State Chemistry" (G. Somorjai and J. O. McCaldin, eds.), Vol. 10, pp. 157–191. Pergamon, New York.

Cohen, M. L., and Bergstresser, T. K. (1966). *Phys. Rev.* **141,** 789–796.

Coleman, J. J., Dapkus, P. D., Vojak, B. A., Laidig, W. D., Holonyak, N., Jr., and Hess, K. (1980). *Appl. Phys. Lett.* **37,** 15–17.

Coleman, J. J., Dapkus, P. D., Clarke, D. R., Camras, M. D., and Holonyak, N., Jr. (1981a). *Appl. Phys. Lett.* **39,** 864–866.

Coleman, J. J., Dapkus, P. D., Camras, M. D., Holonyak, N., Jr., Laidig, W. D., Low, T. S., Burroughs, M. S., and Hess, K. (1981b). *J. Appl. Phys.* **52,** 7291–7295.

Coleman, J. J., Dapkus, P. D., Thompson, D. E., and Clarke, D. R. (1981c). *J. Cryst. Growth* **55,** 207–212.

Coleman, J. J., Dapkus, P. D., Kirkpatrick, C. G., Camras, M. D., and Holonyak, N., Jr. (1982). *Appl. Phys. Lett.* **40,** 904–906.

Conwell, E. M. (1967). "High Field Transport in Semiconductors" (*Solid State Physics,* Suppl. 9). Academic Press, New York.

Dingle, R. (1975). *In* "Festkörper Probleme" ("Advances in Solid State Physics") (H. J. Queisser, ed.), Vol. 15, pp. 21–48. Pergamon, New York.

Dupuis, R. D., Dapkus, P. D., Kolbas, R. M., and Holonyak, N., Jr., (1978a). *Solid State Commun.* **27,** 531–533.

Dupuis, R. D., Dapkus, P. D., Kolbas, R. M., Holonyak, N., Jr., and Shichijo, H. (1978b). *Appl. Phys. Lett.* **33,** 596–598.

Dupuis, R. D., Moudy, L. A., and Dapkus, P. D. (1979a). *Proc. Int. Symp. GaAs Related Comp. 7th, St. Louis, Missouri, 1978* (C. M. Wolfe, ed.), pp. 1–9. Inst. Phys., London.

Dupuis, R. D., Dapkus, P. D., Chin, R., Holonyak, N., Jr., and Kirchoefer, S. W. (1979b). *Appl. Phys. Lett.* **34,** 265–267.

Dupuis, R. D., Dapkus, P. D., Holonyak, N., Jr., and Kolbas, R. M. (1979c). *Appl. Phys. Lett.* **35,** 487–489.

Dutta, N. K. (1982). *Electronics Lett.* **18,** 451–453.

Elinson, M. I., Volkov, V. A., Lutskii, V. N., and Pinsker, T. N. (1972). *Thin Solid Films* **12,** 383–397.

Elliott, R. J., and Gibson, A. F. (1974). "Solid State Physics and Its Applications." Harper and Row, New York.

Hall, L. H., and Bardeen, J., and Blatt, F. J. (1954). *Phys. Rev.* **95,** 559–560.

Hall, R. N., Fenner, G. E., Kingsley, J. D., Soltys, T. J., and Carlson, R. O. (1962). *Phys. Rev. Lett.* **9,** 366–368.

Hayashi, I., Panish, M. B., Foy, P. W., and Sumski, S. (1970). *Appl. Phys. Lett.* **17,** 109–111.
Hess, K. (1979). *Appl. Phys. Lett.* **35,** 484–486.
Hess, K., Holonyak, N., Jr., Laidig, W. D., Vojak, B. A., Coleman, J. J., and Dapkus, P. D. (1980a). *Solid State Commun.* **34,** 749–752.
Hess, K., Vojak, B. A., Holonyak, N., Jr., Chin, R., and Dapkus, P. D. (1980b). *Solid-State Electron.* **23,** 585–589.
Holonyak, N., Jr., and Bevacqua, S. F. (1962). *Appl. Phys. Lett.* **1,** 82–83.
Holonyak, N., Jr., and Scifres, D. R. (1971). *Rev. Sci. Instrum.* **42,** 1885–1886.
Holonyak, N., Jr., Kolbas, R. M., Dupuis, R. D., and Dapkus, P. D. (1978a). *Appl. Phys. Lett.* **33,** 73–75.
Holonyak, N., Jr., Kolbas, R. M., Laidig, W. D., Vojak, B. A., Dupuis, R. D., and Dapkus, P. D. (1978b). *Appl. Phys. Lett.* **33,** 737–739.
Holonyak, N., Jr., Vojak, B. A., Kolbas, R. M., Dupuis, R. D., and Dapkus, P. D. (1979a). *Solid-State Electron.* **22,** 431–433.
Holonyak, N., Jr., Kolbas, R. M., Laidig, W. D., Altarelli, M., Dupuis, R. D., and Dapkus, P. D. (1979b). *Appl. Phys. Lett.* **34,** 502–505.
Holonyak, N., Jr., Kolbas, R. M., Dupuis, R. D., and Dapkus, P. D. (1980a). *IEEE J. Quantum Electron.* **QE-16,** 170–186.
Holonyak, N., Jr., Kolbas, R. M., Laidig, W. D., Vojak, B. A., Hess, K., Dupuis, R. D., and Dapkus, P. D. (1980b). *J. Appl. Phys.* **51,** 1328–1337.
Holonyak, N., Jr., Laidig, W. D., Vojak, B. A., Hess, K., Coleman, J. J., Dapkus, P. D., and Bardeen, J. (1980c). *Phys. Rev. Lett.* **45,** 1703–1706.
Holonyak, N., Jr., Laidig, W. D., Camras, M. D., Coleman, J. J., and Dapkus, P. D. (1981). *Appl. Phys. Lett.* **39,** 102–104.
Holonyak, N., Jr., Vojak, B. A., Morkoç, H., Drummond, T. J., and Hess, K. (1982). *Appl. Phys. Lett.* **40,** 658–660.
Kirchoefer, S. W., Holonyak, N., Jr., Hess, K., Gulino, D. A., Drickamer, H. G., Coleman, J. J., and Dapkus, P. D. (1982a). *Appl. Phys. Lett.* **40,** 821–824.
Kirchoefer, S. W., Holonyak, N., Jr., Hess, K., Gulino, D. A., Drickamer, H. G., Coleman, J. J., and Dapkus, P. D. (1982b). *Solid State Commun.* **42,** 633–636.
Kirchoefer, S. W., Meehan, K., Holonyak, N., Jr., Gulino, D. A., Drickamer, H. G., Burnham, R. D., and Scifres, D. R. (1982c). *Appl. Phys. Lett.* **41,** 406–407.
Kolbas, R. M. (1979). Ph.D. dissertation, Univ. of Illinois, Urbana, Illinois.
Kolbas, R. M., Holonyak, N., Jr., Dupuis, R. D., and Dapkus, P. D. (1978). *Pis'ma Zh. Tekh. Fiz.* **4,** 69–73. [Engl. transl. (1978) *Sov. Tech. Phys. Lett.* **4,** 28–30.]
Kressel, H., and Butler, J. K. (1977). "Semiconductor Lasers and Heterojunction LEDs." Academic Press, New York.
Kroemer, H. (1963). *Proc. IEEE* **51,** 1782–1783.
Kroemer, H. (1981). *Appl. Phys. Lett.* **38,** 959–961.
Laidig, W. D., Holonyak, N., Jr., Camras, M. D., Vojak, B. A., Hess, K., Coleman, J. J., and Dapkus, P. D. (1981a). *Solid State Commun.* **38,** 301–304.
Laidig, W. D., Holonyak, N., Jr., Camras, M. D., Hess, K., Coleman, J. J., Dapkus, P. D., and Bardeen, J. (1981b). *Appl. Phys. Lett.* **38,** 776–778.
Lasher, G., and Stern, F. (1964). *Phys. Rev.* **133,** A553–A563.
Ludowise, M. J., and Dietze, W. T. (1982). *J. Electron. Mater.* **11,** 59–73.
Ludowise, M. J., Dietze, W. T., Lewis, C. R., Holonyak, N., Jr., Hess, K., Camras, M. D., and Nixon, M. A. (1983a). *Appl. Phys. Lett.* **42,** 257–259.
Ludowise, M. J., Dietze, W. T., Lewis, C. R., Camras, M. D., Holonyak, N., Jr., Fuller, B. K., and Nixon, M. A. (1983b). *Appl. Phys. Lett.* **42,** 487–489.
Lutskii, V. N. (1970). *Phys. Status Solidi A* **1,** 199–220.
Manasevit, H. M. (1971). *J. Electrochem. Soc.* **118,** 647–650.
Mon, K. K., Hess, K., and Dow, J. D. (1981). *J. Vac. Sci. Technol.* **19(3),** 564–566.

Morkoç, H., Drummond, T. J., Camras, M. D., and Holonyak, N., Jr. (1982). *Appl. Phys. Lett.* **40,** 18–19.

Nathan, M. I., Dumke, W. P., Burns, G., Dill, F. H., and Lasher, G. J. (1962). *Appl. Phys. Lett.* **1,** 62–64.

Petermann, K. (1979). *IEEE J. Quantum Electron.* **QE-15,** 566–570.

Quist, T. M., Rediker, R. H., Keyes, R. J., Krag, W. E., Lax, B., McWhorter, A. L., and Zeiger, H. J. (1962). *Appl. Phys. Lett.* **1,** 91–92.

Rezek, E. A., Holonyak, N., Jr., Vojak, B. A., Stillman, G. E., Rossi, J. A., Keune, D. L., Fairing, J. D. (1977). *Appl. Phys. Lett.* **31,** 288–290.

Rezek, E. A., Chin, R., Holonyak, N., Jr., Kirchoefer, S. W., and Kolbas, R. M. (1980). *J. Electron. Mater.* **9,** 1–27.

Segall, B., and Mahan, G. D. (1968). *Phys. Rev.* **171,** 935–948.

Schrieffer, J. R. (1955). *Phys. Rev.* **97,** 641–646.

Scifres, D. R., Burnham, R. D., and Streifer, W. (1982). *Appl. Phys. Lett.* **41,** 118–120.

Shah, J. (1978). *Solid-State Electron.* **21,** 43–50.

Shichijo, H., and Hess, K. (1981). *Phys. Rev. B* **23,** 4197–4207.

Shichijo, H., Kolbas, R. M., Holonyak, N., Jr., Dupuis, R. D., and Dapkus, P. D. (1978). *Solid State Commun.* **27,** 1029–1032.

Shik, A. Y. (1974). *Fiz. Tekh. Poluprovodn.* **8,** 1841–1864. [Engl. transl. (1975) *Sov. Phys.-Semicond.* **8,** 1195–1209.]

Shik, A. Y. (1979). *Pis'ma Zh. Tekh. Fiz.* **5,** 869–871. [Engl. transl. (1979) *Sov. Tech. Phys. Lett.* **5,** 358–359.]

Stern, F., and Howard, W. E. (1967). *Phys. Rev.* **163,** 816–835.

Tang, J. Y., Hess, K., Holonyak, N., Jr., Coleman, J. J., and Dapkus, P. D. (1982). *J. Appl. Phys.* **53,** 6043–6046.

Thompson, G. H. B. (1980). "Physics of Semiconductor Laser Devices." Wiley, New York.

Tsang, W. T. (1981). *Appl. Phys. Lett.* **39,** 786–788.

van der Ziel, J. P., Dingle, R., Miller, R. C., Wiegmann, W., and Nordland, W. A., Jr. (1975). *Appl. Phys. Lett.* **26,** 463–465.

Van Vechten, J. A. (1982). *J. Appl. Phys.* **53,** 7082–7084.

Vojak, B. A., Kirchoefer, S. W., Holonyak, N., Jr., Chin, R., Dupuis, R. D., and Dapkus, P. D. (1979). *J. Appl. Phys.* **50,** 5830–5834.

Vojak, B. A., Laidig, W. D., Holonyak, N., Jr., Camras, M. D., Coleman, J. J., and Dapkus, P. D. (1981). *J. Appl. Phys.* **52,** 621–626.

PART III

METALS

8 SYNTHESIS OF MULTILAYER STRUCTURES BY PHYSICAL VAPOR DEPOSITION TECHNIQUES

Troy W. Barbee, Jr.

Department of Materials Science and Engineering
Stanford University
Stanford, California

I. INTRODUCTION

Since the early 1960s advances in materials synthesis techniques by vacuum vapor condensation have made possible the formation of both highly perfect thin films of equilibrium phases and, more important, nonequilibrium phases of dramatic diversity. Such diversity presents a broad ranging set of opportunities for the materials synthesis practitioner as well as for the general scientific community involved in all aspects of the science and technology of solids. This advance has resulted in the now apparent capability to perform atomic engineering in a true engineering sense. One may, following a set of developing guidelines, use vapor condensation in a controllable and predictable manner to form solid films having nonequilibrium structures and unique nonequilibrium compositions. It is the purpose of this chapter to examine these guidelines, providing, where possible, scientific basis for their use and relevance. Though much of what will be presented has general application to material synthesis by vapor condensation, emphasis will be placed on systems from which multilayer structures may be synthesized.

A series of definitions pertaining to the relationships between the deposited layers and the substrate are presented first. These definitions provide a framework for consideration of the structure–synthesis-process relationship that will

ISBN 0-12-170470-X

be discussed later in the chapter. Fundamental characteristics of the deposition process termed intrinsic here are then discussed. Extrinsic factors specific to the type of vapor source and deposition approaches will then be considered. Examples of a general nature demonstrating important process characteristics will be presented. In the next section an application of multilayers will be discussed and the x-ray diffraction characteristics reviewed. This example will be used to demonstrate important effects relative to synthesis and research opportunities. A short summary statement is then presented.

II. DEFINITIONS OF STRUCTURAL RELATIONSHIPS: SUBSTRATE-DEPOSITED FILM

Metallic multilayer films synthesized by vacuum vapor condensation are nonequilibrium, representing either kinetically or truly metastable materials. Ideally, multilayer films can be described in terms of the number of atomic layers within each component layer of the multilayer structure and by the specific composition of each component layer. In the following a film containing alternating layers of materials A and B, each *a* and *b* atomic layers thick, will be referred to as $A_a B_b$ if the layers are crystalline. If the layers are noncrystalline, the nomenclature will be again $A_a B_b$ with *a* and *b* now the actual layer thicknesses of A and B divided by an appropriate measure of the atomic size of the elements comprising such layers. These definitions point out the "atomic" scale of multilayer structures. If the number of atomic planes in layers A and B are integer, the structure will be termed interfacially commensurate. If either is not integer, the structure will be termed interfacially incommensurate.

Definitions pertaining to the orientation relationships between the component layers A and B and the continuity of atomic position across the interfaces are also important. Epitaxy (arrangement on) refers to there being a specific orientation relationship between the deposit and the substrate. Such relationships are described by the orientation of the normal to the deposit surface and one other specific crystallographic direction in both the substrate and the deposited film. We note here that multilayer films are unique in that both component layers act as substrates during the synthesis process. It is important to recognize that epitaxy does not define any specific relationship between atoms of the constituent layers at their interfaces. It is only the crystallographic relationship between the substrate and deposit that is specified.

The nature of the arrangement of atoms at the interfaces between layers is described (Matthews, 1975; van der Werwe and Ball, 1975) in terms of the coherency of the interface. If the atoms of the overlayer assume the same spatial distribution as the atoms in the substrate (and hence the same atomic structure), the interface is said to be coherent. This is almost always the case during the deposition of the first few monolayers of two isostructural materials. If there is a large difference in the lattice parameters of the two materials, then coherence is lost after several monolayers by the introduction of dislocations at the free surface that move to the interface decreasing the strain energy of the deposit–substrate system.

Other factors are important if the equilibrium structures of the depositing material and the substrate are not the same. Very often the depositing material will assume the crystalline structure of the substrate even though this structure is not observed in the equilibrium phase field of the deposit at the deposition temperature. This effect is termed pseudomorphism (Matthews, 1973; van der Merwe and Ball, 1975; Venables and Price, 1975) and, at this writing, has only been observed for high-symmetry crystalline structures. It is likely that the formation criterion for pseudomorphic phases is that the free energy of the substrate deposit system be minimum. We note here that such pseudomorphic deposits are typically only one to five atomic layers thick when equilibrium phases appear. Therefore metallic multilayers are an optimum class of materials for study of such nonequilibrium pseudomorphic layers.

III. INTRINSIC DEPOSITION PROCESS CHARACTERISTICS

Vapor-condensation-synthesis processes for the formation of metallic multilayers have differed primarily only in the types of vapor sources and the methods used in the control of individual layer thicknesses. In describing these various approaches, it is important to have a general grasp of the important physical characteristics of the vapor condensation process and, additionally, of what experimental parameters, characteristic of specific synthesis process conditions, are of significance. Since these structures are all essentially nonequilibrium, those process characteristics facilitating the formation of nonequilibrium deposits are outlined first.

The vapor deposition process is very effective in the formation of nonequilibrium structures during film synthesis as a result of two intrinsic characteristics. The first is the source of the term vapor quench (VQ) used to describe synthesis of noncrystalline alloy films and is derived from the rate of energy transfer of an incident adatom to the substrate material at the deposition surface. The second is the maximum free energy difference between the metastable products and equilibrium forms accessible by vapor deposition processes.

Several calculations (Lennard-Jones, 1937; Lennard-Jones and Strachan, 1935; Zwanzig, 1960; Goodman, 1962; McCarrol and Ehrlich, 1963) of the kinetics of energy transfer for one-dimensional systems during the condensation of vapor atoms incident end-on a linear array of atoms are available. Early work, based on quantum mechanical models (Lennard-Jones, 1937; Lennard-Jones and Strachen, 1935), predicted quench times of the order of $2/\nu_D$ where ν_D is the Debye frequency of the lattice, if the energy of the incident atom is less than the Debye energy, $h\nu_D$. This energy cutoff, $h\nu_D$ for condensation, is not experimentally observed, and subsequent calculations (Zwanzig, 1960; Goodman, 1962; McCarroll and Ehrlich, 1963) have shown that if the energy of the incident ion is less than 25 times the adatom desorption energy, then a high condensation probability is expected. More importantly, all these calculations have predicted that the quench times are on the order of no more than five atomic vibration periods in agreement with the earlier results. Therefore it appears that a reasonable estimate of quench time is 10^{-12} sec. These calculated quench times are for one-dimensional systems and may not be applicable to a three-dimensional lattice or to a two-dimensional surface as experimentally experienced.

An approximate approach to this many-body problem consists of using the properties of mechanical energy transfer in a three-dimensional crystal lattice to estimate the relaxation time for transfer of the kinetic energy of the incident vapor atom. Elastic stress waves are a mode of transmitting mechanical energy in a three-dimensional lattice. The rate at which a stress wave moves reflects the time necessary to transfer energy from an atom in the wave to an adjacent atom just ahead in the direction of wave propagation. Typical elastic wave velocities are of the order of 5×10^5 cm sec^{-1}, and the thickness of the elastic wave front (Tsai and McDonald, 1977) is only 5 to 10 atomic distances. Crystalline interplanar spacings are typically 2 Å. Therefore, energy transfer occurs over a time period of 10^{-12} to 10^{-11} sec, in agreement with estimates based on one-dimensional lattice models. Although this argument is approximate, it demonstrates that analysis of the elastic behavior of bulk materials yields results that support the conclusions of the more detailed one-dimensional calculations.

It is therefore concluded that the effective quench rates for individual adatoms are of the order of $\Delta T \times 10^{12}$ K sec^{-1}, where ΔT is the absolute temperature decrease experienced by the absorbed atom. The absorbed energy must be transported away from the surface by electron and phonon processes. If the rate of energy deposition is high, a considerable temperature gradient is established near the deposition surface, resulting in an effective surface temperature that is significantly higher than the bulk substrate temperature. Qualitatively, it is expected that the increased surface temperature will result in the metastable state being observed only for macroscopic substrate temperatures considerably lower than the observed bulk transformation temperature for the material in this metastable state.

Thermodynamic limits can also be estimated for the maximum excess free energy (relative to the equilibrium phase) of a metastable phase synthesized by vapor deposition. Baker and Cahn (1971), treating evaporation–condensation as a Carnot cycle, estimated the allowable maximum free energy excess, ΔF_{ex}. Their analysis for deposition onto substrates at low substrate temperatures indicates that $\Delta F_{ex} \lesssim \Delta H_v$ where ΔH_v is the heat of vaporization. Heats of vaporization are in almost all cases quite large. For the elements Ag, Au, Ge, Si,

Mo, and V, they are approximately 2.81, 3.81, 3.8, 5.8, and 5.2 eV/mole, respectively. Free energies of metastable polymorphs of common elemental metals relative to stable phases calculated by Kaufman (1967) are presented in Table I. In all cases the free energy differences shown in Table I are a factor of 50 to 100 times smaller than the heats of vaporization indicating that vapor condensation is, in principle, a powerful process for synthesis of metastable phases such as represented by pseudomorphic structure films and metal multilayer structures. Therefore, since the thermodynamic limits are far in excess of the expected free energy differences, it is very likely that nonequilibrium structures may be synthesized in a routine fashion by physical vapor condensation techniques if the kinetics of the transformation process (metastable ⇒ equilibrium) are sufficiently slow.

Other important intrinsic factors include deposition rate R, surface diffusivity D_s, deposition surface temperature T_s, adatom–substrate interaction energy E_a, adatom–adatom bond energy E_2, and the bulk substrate temperature T_B. The kinetic energy of the incident adatoms may also be considered intrinsic but as this is largely vapor source specific it will be discussed more fully in later sections. These intrinsic factors are determinative in the nucleation and growth of thin films. Since, with metal multilayers the nucleation phase is critical to the formation of the multilayer structure this aspect of deposition will be emphasized. The formalism of Venables and Price (1975) and Voorhoeve (1976) will be followed. It is assumed here that the majority of the atoms incident onto the substrate are absorbed so that nucleation may be considered relative to a flux of atoms, almost all of which remain on the surface. A quasichemical approach is

TABLE I

Relative Free Energies of Various Polymorphs of Some Common Metals (meV/mole)

	Free energy relative to that of stable meV (at 300 K)		
Metal	bcc	fcc	hcp
Ti/Zr	33.0	34.7	0.0
Hf	67.7	34.7	0.0
V/Nb/Ta	0.0	104.3	75.5
Cr/Mo/W	0.0	110.4	86.8
Tc/Re	22.6	14.8	0.0
Ru/OS	59.0	15.6	0.0
Rh/Ir	68.3	0.0	8.5
Ni/Pd/Pt	39.5	0.0	14.7
Cu/Ag/Au	34.3	0.0	10.4
Zn/Cd	22.6	14.3	0.0

taken, i.e., all adatom cluster energies are assumed to be represented by summing nearest-neighbor bond energies.

Kinetic effects are included through consideration of the limits posed by the surface diffusivity D_s and mean absorption stay time τ_a. Substrates are considered ideal with N_0 (cm^{-2}) sites for absorption having equal absorption energy E_a. The absorption energy E_a, includes substrate coordination effects. Re-evaporation of single atoms occurs after a mean stay time τ_a given by

$$\tau_a = \nu \exp(-E_a/kT), \tag{1}$$

where ν is an atomic vibration frequency ($\simeq 10^{12}$ to 10^{13} sec^{-1}), k Boltzmann's constant, and T the absolute temperature. The surface diffusivity is given by

$$D_s = (\alpha\nu/N_0) \exp[-E_D/kT], \tag{2}$$

where E_D is the activation energy for diffusion and $\alpha(\simeq 0.25)$ is a geometrical constant.

The absorption energy E_a is determined by the sites available for absorption. In Fig. 1 absorption sites available on a substrate with a close-packed plane exposed onto which close-packed islands of adatoms have deposited is shown [following Venables and Price (1975)]. Absorption of a single atom onto the substrate (a) has an absorption energy E_a, while absorption onto the adatom layer (c) has an absorption energy $3E_2$. Adatoms bound to a kink site (b) in the first layer have an absorption energy of $E_a + 3E_2$ and (d) at an adatom layer kink site $6E_2$. Atomic motion on the substrate and on the adatom layers

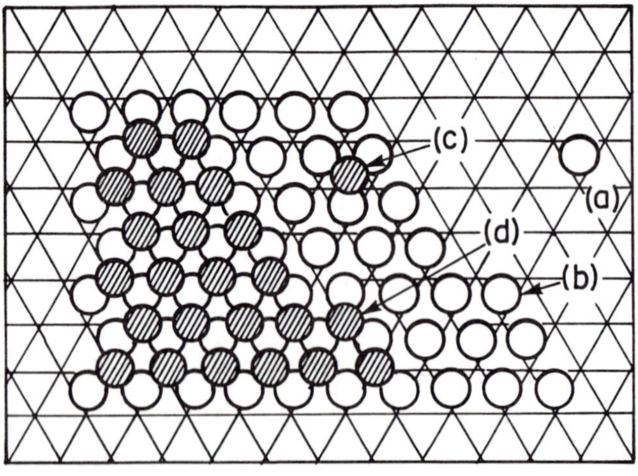

Fig. 1. Binding energies of atoms on a nearest-neighbor bond model for nucleation on a close-packed triangular lattice. Single atoms: (a) E_a, (b) $E_a + 3E_2$, (c) $3E_2$, and (d) $6E_2$ (Venables and Price, 1975).

determine the rate at which adatoms are captured at such sites. Therefore, it is the deposition surface temperature T_s, the arrival rate of atoms R, the bond energies E_a and E_2, and the activation energy for diffusion E_D that are most important and provide a framework for qualitatively describing nucleation (and subsequent growth) as schematically shown in Fig. 2.

The simplest case is one in which adatoms are incident at a rate R onto a substrate held at temperatures sufficiently low such that an atomic jump by surface atoms is unlikely before subsequent atomic layers are deposited. This condition is met when $R/N_0^2D > 1$ and represents the experimental conditions for which noncrystalline deposits may be expected (i.e., Fig. 2a). In the case of directionally bonded solids, noncrystalline films are routinely formed by vapor condensation under such conditions. If noncrystalline films are deposited, the uniformity of atomically thin layers a few atomic diameters in thickness is primarily defined by the spatial uniformity of the incident adatom flux. Therefore, routine synthesis of multilayer films of high quality from two materials forming noncrystalline deposits is possible. Since many alloys can be formed in the noncrystalline state by vapor quenching, an enormously wide range of components for multilayer structures is available.

Note that the effects of the kinetic energy of the adatoms and their energy of condensation have not been considered. This energy source will, at high deposition rates, substantially raise the surface temperature of the deposit so that higher substrate temperatures may not be accessible by increases in the deposition rate R as implied by the condition $R/N_0^2D > 1$. The absorption energy E_a is also relatively unimportant since conditions for all atoms condensing are en-

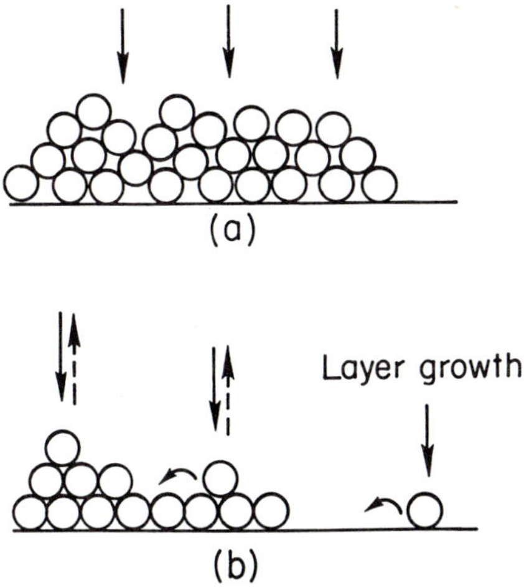

Fig. 2. Schematic diagram showing two regimes of nucleation and growth: (a) $R/N_0^2D > 1$, amorphous, and (b) $E_a \gtrsim 3E_2$, $(R/N_0\nu)\exp(3E_2/kT) \gtrsim 1$, $(R/N_0\nu)\exp(6E_2/kT) \gtrsim 1$, layer growth (Venables and Price, 1975).

forced by the low substrate temperature. Therefore, for the first layer formed it is important the E_a be high, as in the early stages of deposition the first monolayer is strongly affected by E_a, while later stages will be affected by the adatom–adatom bond energy E_2. Also, there is little effect due to the relative interplanar spacing of the substrate material and the deposited film.

It is also possible in this deposition regime to form thin, fine-grain polycrystalline films. Such films typically have the close-packed plane of the crystal structure formed in the plane of deposition with a lateral grain size related to the thickness of the layer. Again, it is clearly possible to form multilayer films due to the lack of lateral mobility of the adatoms on the substrate. Unwanted effects such as alloying and formation of compounds at the interfaces between layers in multilayer deposits are also minimized by the low substrate temperatures.

The second regime is typically that which is termed layer by layer growth (shown in Fig. 2b). This is characterized by $E_a \gtrsim 3E_2$; i.e., the absorption energy on the substrate and the deposit are of the same order and $R/N_0^2D \ll 1$, which allows substantial lateral mobility of the adatoms so that ledge–kink attachment kinetics are of increased importance. This requires that the temperature be low enough so that two-dimensional layers may form even though single atoms may desorb. The condition for growth of two-dimensional layers is $(R/N_0\nu)\exp(6E_2/kT) > 1$ and the condition for re-evaporation of single atoms is $(R/N_0\nu)\exp(3E_2/kT) \gtrless 1$. As pointed out by Voorhoeve (1976), a low critical supersaturation is expected, and if E_a is large, contrary to classical nucleation theory no nucleation barrier is predicted.

If the lattice parameter of the substrate and the deposit are closely matched, coherent interfaces can be formed and the deposit will grow with interfacial dislocations introduced as the strain energy becomes sufficiently large. When the equilibrium structures of the adatom layer and the substrate are dissimilar and the atomic sizes nearly equal, pseudomorphic structures may be formed. Such pseudomorphic structures will grow until interfacial dislocations are formed (Matthews, 1975; van der Merwe and Ball, 1975), reducing the deposit–substrate system energy. It is also possible to achieve layer-by-layer growth when the absorption energy is large but the potential energy of the substrate surface is smooth. This results once again in layer-by-layer growth of the equilibrium phase of the material deposited. It is important to point out that it is this layer-by-layer growth regime that is potentially important in the synthesis of highly perfect metal multilayer films at moderately elevated substrate temperatures.

A variation on the above case, $E_a \gtrsim 3E_2$, occurs when the absorption sites have a spacing that differs widely from the characteristic spacing of the adatoms in their equilibrium structures. This leads to the formation of submonolayer surface phases upon which nucleation of B layers will subsequently occur. In this case the structure of the initial ad-layer is difficult to specify and classical hemispherical cap nucleation behavior may be observed after the initial surface phase formed.

The final case considered here is one where $E_a \ll E_2$; i.e., the interaction of the substrate with the incident adatoms is much smaller than their self-interaction. Classical nucleation theory is appropriate here and adatom cluster nuclei that grow to form hemispherical cap clusters occur at somewhat elevated temperatures. When E_2/kT and therefore E_a/kT are small, nucleation is difficult due to the small population of adatoms on the substrate. Under these conditions, heterogeneous nucleation is controlling and substrate defects have a major influence on nucleation and subsequent growth. Though this particular case has received significant experimental and theoretical attention, its impact on the synthesis of multilayer structures is limited to definition of material combinations that are not suitable except under the condition $R/N_0^2D \gtrsim 1$, as described earlier. Additional guidance as to the macroscopic substrate upon which the initial multilayer component layer is deposited is also given since clustering and hemispherical cap nucleation–growth at this stage will degrade the subsequent layers and the quality of any multilayer structure synthesized with such a substrate–adatom combination unless thick films on which surface morphology is controlled by the deposit characteristics are grown.

In this section the intrinsic characteristics of the vapor deposition process pertinent to the synthesis of multilayer films have been qualitatively discussed. In the following section a specific review of the technological aspects of synthesis techniques is given.

IV. MULTILAYER SYNTHESIS TECHNOLOGY

Multilayer structures can be synthesized, in principle, by using any technique in which the product is formed by means of atom-by-atom processes. Such techniques would include physical vapor condensation (Vossen and Kern, 1978; Maissel and Glang, 1970), chemical vapor deposition (Shaw, 1975; Stringfellow, 1982), and electrochemical deposition (Wright, 1975). Mechanical processing of macroscopic laminates might also yield multilayer structures of specific materials since massive drawing of copper–niobium composites (Bevk *et al.*, 1978) has been demonstrated to result in niobium wires, approximately 100 Å in diameter, contained in a copper matrix. In this section only atom-by-atom processes will be considered, and then only those in which physical vapor condensation technology is used. This choise is based on the fact that these processes dominate reported experimental work.

The earliest reported synthesis work is that of Koeppe (1929) who unsuccessfully attempted the synthesis of multilayers by electrochemical means. Subsequently, Deubner (1930), also using electrochemical techniques, successfully formed multilayer structures of unknown quality. Following this work DuMond and Youtz (1935,1940) produced multilayer structures of copper and gold by using physical vapor condensation from thermal sources. Their objective was to fabricate large-period multilayers for applications as dispersion elements for soft x rays. Although, for metallurgical reasons, the lifetime of these samples was limited, they provided a means for investigation of the interdiffusion of the two components in this binary system at low homologous temperatures T/T_{m}, where T_{m} is the absolute melting temperature.

Since the report by DuMond and Youtz (1935, 1940), many investigators have undertaken the synthesis of multilayer structures by means of physical vapor deposition for a wide variety of both scientific and technological purposes. Both thermal and sputter deposition sources have been applied and substantial success achieved. In the following subsection, first thermal and then sputter technology will be considered.

A. Thermal Source Systems

Multi-vapor-source configurations are used in the synthesis of metal multilayers with thermal sources. These are directly analogous to molecular-beam-epitaxy systems described elsewhere in this volume, except that the sources are not of the Knudsen cell type. In these systems the sources and samples are stationary, the layering introduced through interruption of the vapor streams by means of rotating pin wheel or reciprocating shutters. A schematic of a typical system (Hilliard, 1979) is shown in Fig. 3. Substrates can be held at temperatures from 4 K to above 1300 K. Heating mechanisms include elec-

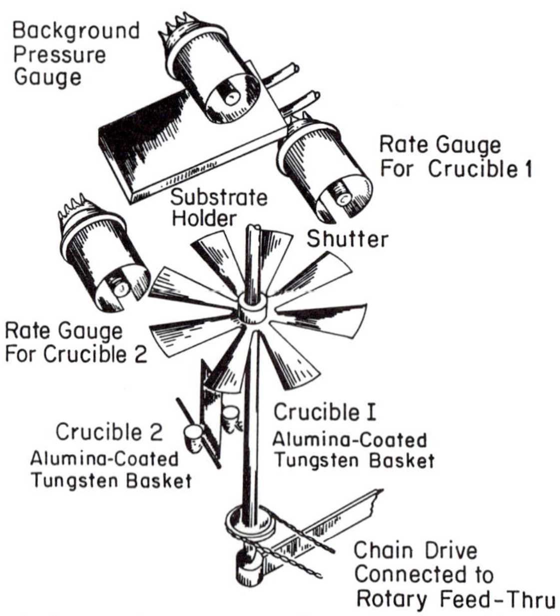

Fig. 3. Schematic diagram of an apparatus used to produce compositionally modulated alloys films using thermal sources (Hilliard, 1979).

tron-beam bombardment and resistive and optical heating. Sample sizes are usually less than 25 cm^2 and are dictated by specific system geometries and heating requirements.

In the application of thermal sources (Maissel and Glang, 1970), the vapor pressure of the evaporant over itself is used as the vapor source. Energies of the evaporant are of the order of the source temperature, being almost always less than ~4000 K (~0.36 eV). Since the vapor pressure is strongly temperature dependent, tight control of the power applied to the source is necessary if constant deposition rates are to be maintained. Deposition rate control is accomplished by using quartz-crystal or ionization-gauge rate monitors in a feedback control loop. These sensors continuously monitor the evaporating species and are arranged in a fixed geometry relative to the substrate surface and the vapor in a fixed geometry relative to the substrate surface and the vapor source. Deposition rates of from 0.2 to 40 nm/sec are typical, necessitating feedback control time constants of fractions of a second. Additionally, the stability of the rate monitors must also be within the range sought: approximately ±1%. Such stability can be achieved with quartz-crystal monitors, although special care to avoid thermally induced drift caused by radiation heating from hot evaporation sources is necessary, particularly with refractory metals.

The importance of deposition rate control other than for the obvious control of individual layer thickness is demonstrated in the work of Flevaris *et al.* (1981) in which commensurate and incommensurate multilayer films of Cu–Ni were synthesized and the character of the x-ray satellite peaks about the (111) reflection determined as shown in Fig. 4a,b. The number of satellite peaks was observed to be a strong function of the commensuration of the multilayer structures. Structures that were nearly commensurate (i.e., integral numbers of atomic planes in both the nickel and copper component layers) exhibited up to six orders of x-ray satellite peaks. A comparison of the powder diffraction scan spectra obtained by using Cu K_α radiation for two samples of essentially the same modulation period but varying in the degree of commensuration demonstrates this result. Therefore, high-quality interfaces, in an x-ray sense, require that deposition rates be both well controlled (±1%) and well known in an absolute sense for high-quality commensurate structures to be formed from dissimilar materials.

A unique monitoring method has been developed (Spiller, 1981; Spiller *et al.*, 1980) and used to fabricate multilayer, soft x-ray mirrors. The procedure developed uses an *in situ,* soft x-ray monitoring system. A soft x-ray reflectometer using characteristic line x-ray sources ($\lambda = 31, 44.8$ on 67.6 Å) is installed in the evaporation system and continually monitors the reflectivity of a multilayer mirror during deposition. This allows accurate control of the individual layer thicknesses (on the basis of property to be used), and additionally gives a experimental record of the x-ray optical thickness of each layer. The ability to correct for any errors introduced during a given synthesis experimental is quite important in x-ray optics work.

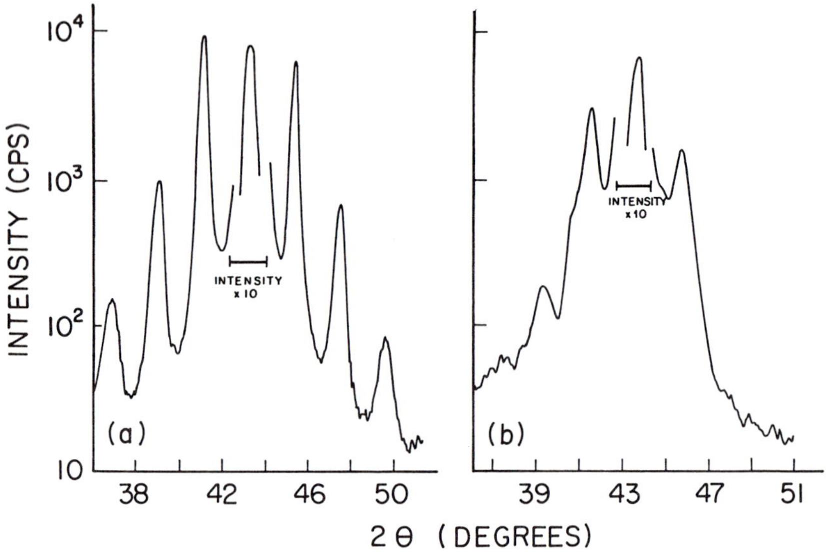

Fig. 4. Two powder diffraction scans (Flevaris *et al.*, 1981) from samples having (a) commensurate and (b) incommensurate layered Cu–Ni structures. The modulation periods are 41.3 Å [curve (a)] and 43.1 Å [curve (b)]. Note the well-defined-composition modulation-induced sidebands of the commensurate sample [curve (a)].

Thermal sources may also be classified as to whether crucibles are required. Crucibles are most often needed when resistive heating is used. Due to the thermal mass typical of such resistively heated sources, the time constant of the feedback control loop is determined by the thermal time constant of the crucible–molten evaporant body. It is likely that the deposition rate control level will not exceed $\pm 5\%$ with such systems. Crucibleless sources typically depend on skull melting (i.e., containment of the molten bath in a solid shell of the same composition) as is characteristic of electron-beam-heated sources and pulse laser vapor sources. Heating with electron-beam sources allows time constants of the order of 1 sec to be attained. In the case of pulse laser sources, a solid surface is illuminated with a short, high-energy-density laser pulse. Control of the deposited energy results in control of the volume of material evaporated. Therefore, in principle, if the deposit is monitored, a very short time constant can be approximated. Crucibleless sources are preferred as a result of the enhance control as well as the elimination of the possible contamination of the evaporant by the crucible material.

Thermal sources also exhibit characteristic deposition profiles; i.e., substrates normal to a vector normal to the bath surface are not uniformly deposited upon at all positions. The effect of the nonuniformity can usually be minimized by placing the substrates sufficiently far from the evaporation source or by moving the substrates so that all regions of experimental interest intercept the same time-integrated flux of evaporant for a given layer. Substrate motion is only necessary when large areas are needed or when specially formed substrates are used, as might be required by x-ray optic applications.

The energy of the evaporant atoms is generally less than ~0.4 eV and any gas scattering may markedly affect the deposition process. Therefore, vacuum requirements are for pressures allowing "line-of-sight" paths for the evaporant atoms, source to substrate; required pressures are always smaller than $\sim 10^{-4}$ torr. A more restrictive boundary condition is that necessary to minimize contamination of the depositing material by reactive components in the ambient atmosphere of the vacuum chamber. The kinetic theory of gases indicates that at 300 K, a pressure of $\sim 5 \times 10^{-7}$ torr results in a number of atoms (if condensed) sufficient to form a monolayer impinge on the substrate per second. Therefore, if the constituents of the evaporator ambient are reactive with the depositing materials, significant contamination can occur that may degrade the structural and physical characteristics of the deposited multilayer. This is the experience in molecular-beam epitaxy, as discussed elsewhere in this volume. It is important to note here that metal, multilayer, single-crystal structures have only been synthesized under ultrahigh-vacuum conditions (Durbin *et al.*, 1982; Durbin, 1983) ($P \lesssim 10^{-8}$ torr) from niobium and tantalum by using a system schematically shown in Fig. 5. These materials are well suited for the formation of single-crystal multilayers, since their lattice mismatch is very small (~0.2%). Also, niobium–titanium multilayer samples deposited in vacuums of $\sim 10^{-6}$ torr (Zheng *et al.*, 1981) had degraded superconducting transition temperatures T_c relative to sputter-deposited samples. This degradation of

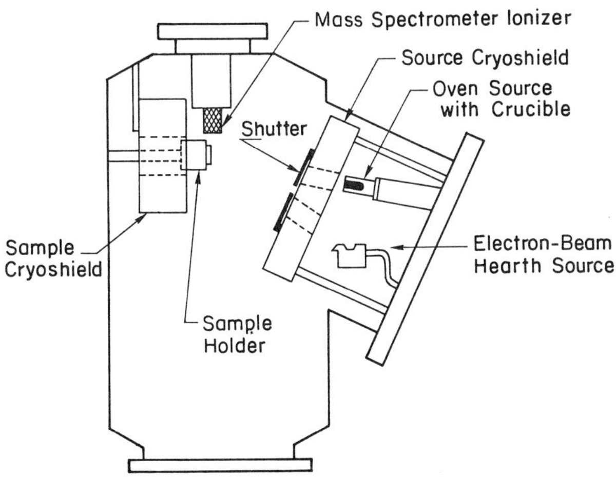

Fig. 5. The cross section of the MBE chamber used by Durbin (1983) to synthesize Nb–Ta single-crystal-composition-modulated films.

T_c was presumably due to contamination of the depositing film by the constituents of the ambient vacuum atmosphere during the evaporation synthesis process.

Process limitations inherent to thermal source synthesis of metal multilayer structures include deposition rate control, contamination of the depositing film by ambient vacuum constituents, and limited sample area. Other limitations result from radiation heating of the substrates during deposition and from fractionation of alloy charges during evaporation. Radiation heating, particularly when high-melting-temperature elements are being deposited, makes substrate temperature control difficult, but with feedback control and a low thermal mass for the heater–substrate body, engineerable. Fractionation of alloy melts as a result of widely differing vapor pressures precludes single-source deposition of many alloys and compounds. In more general deposition experiments this is overcome by use of a single vapor source for each component of the alloy that requires accurate deposition rate control to maintain stoichiometry and many sources if alloys of order higher than binary are to be condensed. Hence, quite complex systems can be envisioned for synthesis of multicomponent, metal multilayer samples.

Irrespective of these limitations, very-high-quality multilayer films can be routinely synthesized by using thermal source systems constructed from commercially available components. If very high purity is required, ultrahigh vacuum is available and, at this time, is the only demonstrated procedure for synthesis of metal, single-crystal multilayer films. It is clear that this is not likely to be the situation for an extended time since other technologies are rapidly developing.

B. Sputter Source Systems

Multisource configurations are also used in sputter deposition systems (Barbee and Keith, 1980; Huggins and Gurvitch, 1983) designed for multilayer synthesis. In these systems the sputter sources are widely separated and the substrates moved past the sources, a single layer being deposited on each pass by a source. A schematic of such a system is shown in Fig. 6. In work using such systems, traditional glow discharge sources and more advanced sources such as magnetron and triode have been applied. Sputter sources use solid targets and can be operated in the deposition up, deposition down, or horizontal deposition modes facilitating coating of nonuniformly configured surfaces as well as very large surface areas.

Sputter sources are solid materials, atoms or atom clusters being ejected from the solid target into the vapor by bombardment of the target surface with energetic particles. The ejected atoms impinge on a substrate and condense to form a film. In most cases noble gases are used as the sputter gas, their ions being thus positively charged. The process is called cathodic sputtering. Ions are

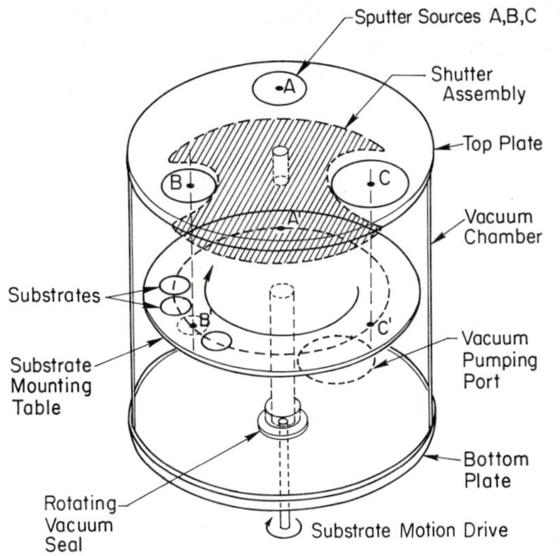

Fig. 6. Schematic diagram of a typical multisource sputtering system used to deposit composition-modulated films. A, B, and C are sputter source positions in which traditional glow discharge, magnetron, or triode sources can be mounted. Substrates are mounted on a rotating table, a layer being deposited as the substrates pass by each source. The individual layer thicknesses are determined by the deposition rate of the sputtered material, the rotation speed of the rotating table, and the source to substrate spacing.

formed by establishing a plasma in much the same manner as a glow discharge is formed in a low-pressure gas by an electric field between two electrodes. In fact, traditional sputter source diodes are termed glow discharge diodes. Since the early 1960s several advances in sputter source technology have occurred. In these new approaches, increases in the efficiency of secondary electrons in ionizing the inert gases by means of magnetic field structures (magnetron sources), increasing the availability of electrons for ionization (triode sputter sources), and independent ion-beam sources have been developed. The advent of these new sources has allowed the technology of sputter deposition to advance in a rapid manner so that significant increases in deposition rate and deposit purity have been attained. In the following, a short discussion of the important features of sputter deposition sources (Vossen and Kern, 1978; Barbee and Keith, 1980) relative to multilayer synthesis is presented. The factors to be considered include sputter source–deposition-surface coupling, energy distribution of the sputtered atoms, and the geometry of the vapor source–substrate configuration.

The sputtering process entails establishing a plasma discharge and imposing a potential of the correct polarity so that ionized gas atoms are accelerated to the cathode surface, where, if of sufficient energy, they dislodge other atoms. These secondary atoms travel from the cathode surface to the deposition surface, being adsorbed to form a deposit. Other types of primary sources exist in which a plasma of the species is generated and accelerated to the deposition surface. Therefore, all such sources are plasma based and source-deposition coupling can be characterized by the degree of plasma-deposition surface interaction.

Since the early 1970s new sputter deposition sources have become available in addition to the conventional diodes. All these can be operated in both dc and rf modes, the nature of the particular application and material to be deposited dictating the mode of excitation. The new sources (Vossen and Kern, 1978; Maissel and Glang, 1970) include triode-supported plasma sources, magnetron sources, and ion-beam sources. A common characteristic of these new approaches is that high deposition rates (approximately an order of magnitude) larger than possible with conventional diodes are observed. In addition, control of the plasma characteristics and the deposition surface environment are possible with these new devices.

The mechanisms of plasma coupling include (Vossen and Kern, 1978) secondary electron bombardment, inert gas ion bombardment, neutral inert gas atom bombardment, and vacuum ultraviolet irradiation. In conventional diode sputtering it is estimated that 10% of the cathode power is deposited at the substrate in rf deposition and 30% for dc sputtering. In both cases it is clear that more than half of this is due to secondary electron bombardment. The distribution of energy among the other factors is less well defined. The primary result of secondary electron bombardment is substrate heating and, in the case of semiconductors or insulators, electronic defects.

Bombardment by either neutral or ionized inert gas ions results in heating of the substrate and enhanced atomic mobility for depositing atoms at the sample surface. It is also possible to induce ion implantation resulting in a high concentration of inert gas being entrapped in the deposit. In addition, the ion bombardment will result in mixing at the deposit surface that may be 1 or 2 nm in extent, degrading interface character.

Magnetron sources, in which a magnetic structure behind the target surface provides a magnetic trap for the secondary electrons active in the plasma ionization process, exhibit deposition rates a factor of 10 larger than those characteristic of conventional diodes. The magnetic structure provides a field normal to the electric field of the source diode that results in a closed toroidal trajectory for the electrons. Such secondary electron trapping allows isolation of the sample from the secondary electrons. In addition, the magnetic structure clamps the plasma at the cathode surface so that the sample can be isolated from bombardment by charged particles. Control of neutral particle bombardment is not typically considered significant, but is is not well understood. It may be quite large since reflection coefficients are considered large for backscattering of low-energy atoms characteristic of plasmas in magnetron systems. In addition, bombardment of the cathode target surface can result in implantation of inert sputtering gas ions so that the target material becomes an alloy, one component being the inert gas.

It is extremely difficult to decouple the sample from the plasma with conventional diode sources. By way of contrast, it is in principle possible to fully decouple (Vossen and Kern, 1978) (except for UV radiation) the plasma from the substrate when magnetron sources are used. This results in a degree of synthesis process control and of purity heretofore not possible with sputtering.

The effects of this sputter plasma-deposition surface coupling (Eltoukhy and Greene, 1979) is shown for traditional glow discharge diode sources in Fig. 7. In this figure the amplitude of composition modulation determined by x-ray diffraction analysis is plotted as a function of individual layer thickness for InSb–GaSb multilayers sputtered at argon pressures of 1, 3, and 15 mtorr. Layer thicknesses in each sample were equal to half the modulation wavelength. The increase in argon pressure acts to thermalize the sputtered atoms by gas scattering and to reduce substrate bombardment by the same mechanism. The important point is that for fixed layer thickness the composition modulation increases as the argon pressure increases and the degree of ion bombardment decreases. Ion bombardment during synthesis provides knock on mixing between layers. Therefore, compositionally abrupt interfaces cannot be synthesized by using traditional sources as a result of such sputter plasma-deposition surface interactions.

Compositionally abrupt interfaces can be formed if magnetron or other ion-beam sputter deposition sources are used. In such cases it is experimentally possible to isolate the deposition surface so that backscattered argon is the only plasma-generated species other than the depositing species incident onto the

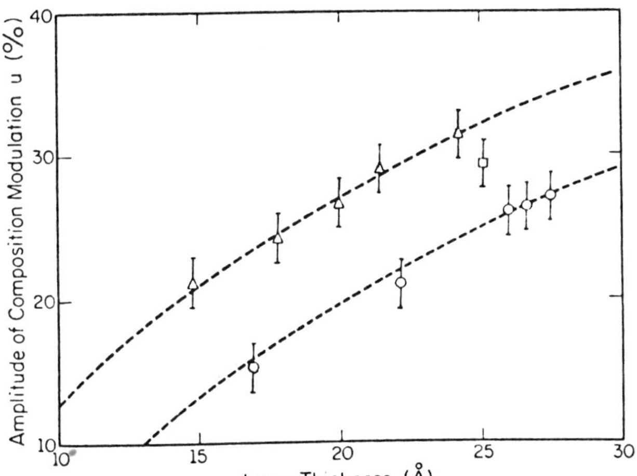

Fig. 7. The amplitude of composition modulation in InSb–GaSb superlattice films grown at 200°C $V_{InSb} = -1100$ V, $V_{GaSb} = -1200$ V) in different Argon sputtering gas pressures (△, 15 mtorr; ○, 1 mtorr; □, 3 mtorr; ——, predicted) using traditional diode sources as a function of layer thickness (Eltoukhy and Greene, 1979).

substrate. Under these conditions the kinetic energy of depositing species is the major contributing external energy source.

The kinetic energy distribution of atoms ejected from the cathode by the incident high-energy-ionized inert gas atom bombardment is dependent on the angle of incidence of the sputter gas ion onto the cathode, the mass ratio of the sputtered atom and the sputter gas, and the sublimation energy of the cathode material and the energy of the incident sputter gas atom. Gold sputtered by using 600-eV argon atoms at normal incidence has a mean energy of approximately 13 eV with a high-energy tail extending to 50–80 eV. This adds substantial energy to the deposition surface and, as shown by Lamont (1978), can cause high bulk substrate deposition temperatures at moderate deposition rates. For example, if gold, sputtered with 1-keV argon, is deposited at a rate of 1000 Å/min, the deposited power flux is 0.021 W/cm^2. This power is directly coupled into the deposition surface, which is no more than a few nanometers thick, resulting in a very large deposition rate of power per unit volume ($\sim 10^5$ W/cm^3 sec) for the sample surface region. It also causes a substantial increase in the bulk substrate temperature for thermally isolated substrates; 180°C with low emissivity and 50°C with high emissivity. The effects of this energy deposition at the deposition surface are not well understood. It has been verified (Lamont, 1978), however, that bulk substrate heating effects in accord with calculation are observed. Also, the sputtered adatoms can be thermalized by gas scattering by operating at high inert gas pressures (30 mtorr) minimizing this effect.

The geometrical arrangement of the sources and substrates is important since this defines the angle of incidence of the depositing species onto the substrate. In standard deposition experiments the sample is fixed relative to the source, deposition occurring by line-of-sight motion of the sputtered species from the target to the substrate. If the substrate is "ideally" smooth and clustering can be ignored, the deposited film will exactly replicate in thickness the deposition profile of the source at all positions. If the substrate is rough on a near atomic level (5–50-Å lateral period), the solid angle subtended by a unit area of substrate will vary with position on the substrate and hence the local deposition rate will spatially vary. This is an unstable growth condition and results in substantial roughness as well as nonuniform layer formation. Such substrate roughness will also strongly influence the in-plane or lateral crystalline coherence of the deposited films. If nucleation occurs on each facet of the rough substrate surface and growth is with the close-packed plane in the plane of deposition, the lateral coherence of the deposited films will most likely reflect the scale of the surface roughness. This places very high requirements on substrate quality relative to surface roughness since with multilayer films the scale of the layering is typically of the order of 10 to 1000 Å.

In the case of sputter deposition these effects are moderated by the motion of the substrates past the sputter sources. Sputter sources are distributed in nature, thus providing deposition at many angles of incidence at the substrate. Additionally, the motion of the substrates past the source results in deposition at nearly all angles of incidence. These two effects minimize the role of substrate roughness in producing films of nonuniform thickness but will have negligible effect on the lateral coherence degradation by substrate roughness.

The preceding discussion has shown that sputtering techniques, though not as advanced as high-vacuum techniques, can be effectively applied to the synthesis of multilayer structures. Substantial development is needed before materials of quality equal to that attainable with UHV techniques will be possible by means of sputtering relative to purity and crystal lattice perfection, though synthesis of Nb/Ta, single-crystal, multilayer films has been reported by Hertel *et al.* (1982).

V. MULTILAYER X-RAY OPTICS

One obvious use of multilayer structures is as dispersion elements for x rays, particularly soft x rays. This area has received increasing attention (Spiller, 1981; Spiller *et al.,* 1980; Underwood and Barber, 1981; Barbee, 1981) and represents a modern attempt to achieve the goals of the earliest work (Koeppe, 1929; Deubner, 1930; Dumond and Youtz, 1935, 1940) in the multilayer field. In this application the superlattice structure formed by the layering is used as the interference structure producing the wavelength selectivity characteristically termed Bragg diffraction. Experience (Spiller, 1981; Spiller *et al.,* 1980; Barbee, 1981; Underwood and Barbee, 1981) has demonstrated that modulation periods less than 15 Å cannot currently be produced with sufficient quality for application as dispersion elements. This limits diffraction angles for photons of energy $E > 1500$ eV to values less than approximately 15°, i.e., near grazing-to-grazing incidence. These simple multilayer structures, which can be termed layered synthetic microstructures (LSMs), can be used as simple dispersion elements and can also be used to form more complex interference structures. In the following results (Barbee and Underwood, 1983) for a more complex structure, a solid Fabry–Perot etalon for x rays is discussed. This represents one step in the development of multilayer x-ray optics.

A more complex LSM (Barbee and Underwood, 1983) in which two identical LSM Bragg diffractors are separated by a low absorption layer much thicker than a single period of the LSMs is exactly analogous to optical Fabry–Perot etalons. In Fig. 8 a schematic diagram of an x-ray etalon is shown and compared to a typical solid etalon for the visible spectral region. The direct correspondence between the two LSM mirrors LSM_1 and LSM_2 and the mirrors M_1 and M_2 on the surface of the optical flat spacer for the long-wavelength etalon is indicated. A low-absorption carbon film was used as the resonance cavity in the x-ray etalon and is directly analogous to the optical flat of the conventional solid etalon.

The spectral resolution of a Fabry–Perot etalon is conventionally expressed

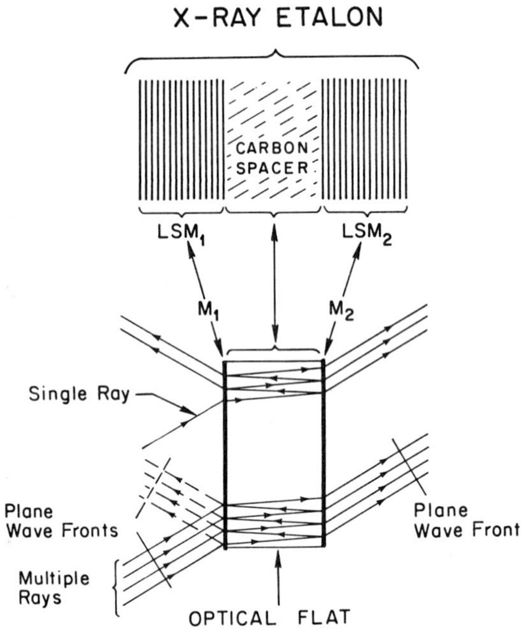

Fig. 8. Schematic diagram of a solid Fabry–Perot x-ray etalon and a solid etalon for the visible spectral region. The x-ray Bragg mirrors LSM_1 and LSM_2 are directly analogous to the optical mirrors M_1 and M_2 that can be either single-metal films or dielectric multilayers. The carbon spacer acts as a resonance cavity in the x-ray etalon and is directly analogous to the optical flat.

(Born and Wolf, 1959) in terms of the fringe sharpness or finesse $\mathcal{F}$, defined as the separation between two adjacent fringes (measured in phase difference Δ) divided by their half-intensity width (in Δ), and is given by

$$\mathcal{F} = \pi\sqrt{R}/(1 - R), \tag{3}$$

where R is the reflectivity of a single LSM mirror. Equation (3) gives the correct value for $\mathcal{F}$ if θ (the angle of incidence) and λ are held constant and t_{Sp} (the carbon spacer thickness) varied. However, when scanning in angle θ, holding both t_{Sp} and λ constant, Eq. (3) gives only a local approximation for $\mathcal{F}$. Therefore, this definition for $\mathcal{F}$ is useful here as an approximation only and provides estimates of the upper limits of the finesse expected in our $\theta/2\theta$ experiments. Detailed numerical computations are necessary for accurate estimates of the finesse to be made.

Experimental diffraction spectra taken by using Cu K_α radiation from three samples in the region of the first-order Bragg peak of the component LSM mirrors are shown in Fig. 9. Curve (a) shows the observed reflectivity (multiplied by 2.4) from a single Bragg mirror LSM (82-005, $N_w = 15$, $t_w = 8.5$ Å, $t_c = 19.1$ Å) identical to the mirrors in the etalons studied [curves (b) and (c)].

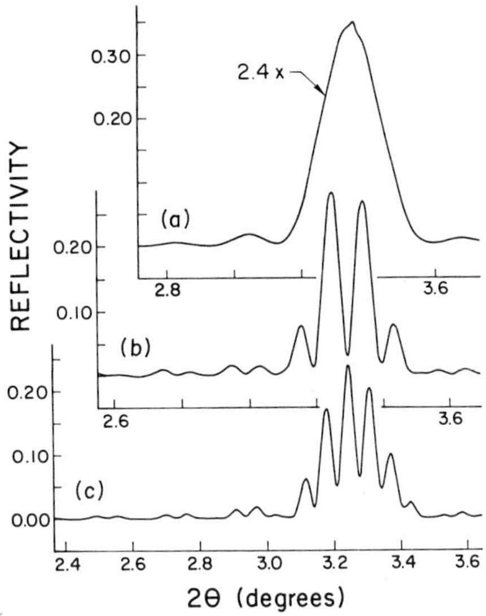

Fig. 9. The experimentally observed reflectivity of Cu K_α ($\lambda = 1.5418$ Å) in the region of the first-order Bragg peak of the LSM mirrors ($d = 27.6$ Å) plotted as a function of 2θ.

These data are plotted in the expanded form as the reflectivity of the single LSM Bragg mirror forms an envelope for the diffraction spectra of the etalons whose absolute reflectivities are larger since two LSM mirrors are active in the interference–reflection process. The value of the multiplier (2.4) was arbitrarily chosen so that Fig. 9 (and Fig. 10) would clearly demonstrate that the calculated diffraction reflection response of a single LSM mirror forms an envelope for the diffraction reflection behavior of the two Fabry–Perot etalon structures examined.

Curve (b) of Fig. 9 shows the reflectivity of an etalon (82-003) structure consisting of two Bragg mirrors as in (a) separated by a carbon spacer cavity thickness $t_{Sp} = 496.6$ Å. Curve (c) of Fig. 9 shows the reflectivity of an etalon (82-001) structure consisting of two Bragg mirrors as in (a) separated by a carbon spacer of thickness $t_{Sp} = 981$ Å. Experimentally observed finesses were approximately 2.3 for both etalons studied.

It is clear in this figure that the observed reflectivity of a single-component LSM forms an envelope for the reflectivity of the etalons when appropriately scaled. It is also clear that the periods of LSM mirrors in these three samples are not identical, since the angular positions of the first-order Bragg peaks vary by approximately 0.088° in 2θ. The periods of the LSM mirrors in the three samples vary by approximately 1.0 Å about an average value of 27.6 Å. This variation is the result of synthesis process variation since these samples were deposited in three separate synthesis runs.

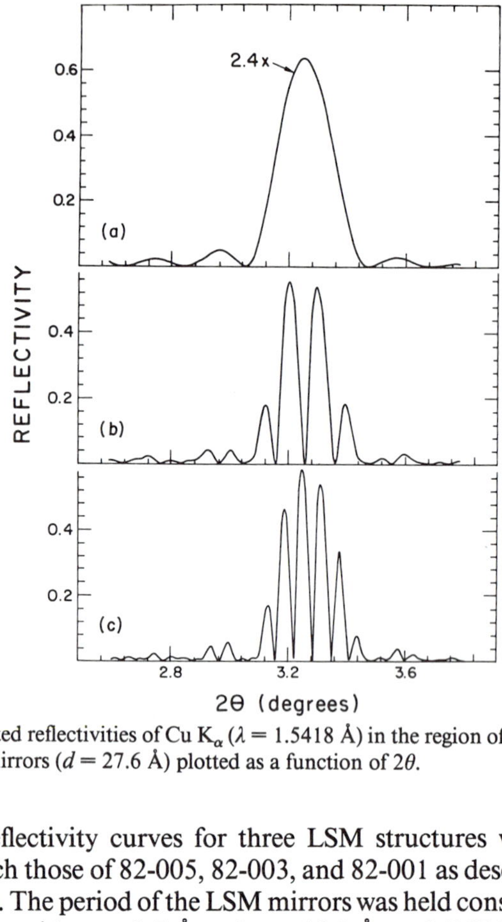

Fig. 10. Calculated reflectivities of Cu K_α ($\lambda = 1.5418$ Å) in the region of the first-order Bragg peaks of the LSM mirrors ($d = 27.6$ Å) plotted as a function of 2θ.

Calculated reflectivity curves for three LSM structures whose structural parameters match those of 82-005, 82-003, and 82-001 as described earlier are shown in Fig. 10. The period of the LSM mirrors was held constant at 27.6 Å in these calculations with $t_w = 8.5$ Å and $t_c = 19.1$ Å, each LSM mirror containing 15 tungsten layers. Curve (a) is again the reflectivity of a single LSM Bragg mirror (multiplied by 2.4). Curve (b) is for an etalon structure having a carbon spacer 496.6 Å thick. Curve (c) is for an etalon structure having a carbon spacer 981 Å thick. Resonance cavity thicknesses were chosen to give best agreement between the experimental and calculated angular dependencies of the reflectivities. Calculated finesses for both etalons were approximately 2.4.

Comparison of the experimental results presented in Fig. 9 and the calculated reflectivities presented in Fig. 10 demonstrates that multiple-beam interference of x rays in a solid Fabry–Perot etalon has been experimentally observed. The primary areas of disagreement between experiment and modeling calculations are that (1) the absolute reflectivities observed are approximately 50% of the predicted values and (2) the contrast of the fringes of equal inclination observed are not as large as predicted. At the present time both these areas of disagreement can be explained in terms of interfacial roughness (~3.2 Å) (Spiller, 1981; Spiller *et al.*, 1980; Barbee, 1981; Underwood and Barbee, 1981) and the divergence of the incident light.

In Fig. 11 an overlay of diffraction curves taken from two areas (1 × 5 mm) separated by 10 cm on sample 82-001 (t_{Sp} = 981 Å) is plotted. The differences observed are shown as shaded areas. If the differences are assumed to be the result of nonuniformity in the carbon spacer, modeling calculations indicate these cannot be larger than 0.4 Å. Therefore, the spacer is t_{Sp} = 981 ± 0.2 Å or uniform to 0.04%. This is a unique result in that the interfacial roughness (~3.2 Å) appears to be substantially larger than the apparent nonuniformity in average thickness of the spacer layer, indicating that the number of atoms per unit area is conserved.

The results obtained for solid x-ray Fabry–Perot etalons provide an indication of the quality of layering possible with sputtering techniques and generally attainable with vapor deposition technology. Interfacial roughness between the spacer and LSM_1 and LSM_2, as well as the thickness uniformity of the spacer, and important structural parameters, the most important being the spacer thickness. This arises from the fact that the reflection process in these Bragg mirrors occurs within the LSM structure and is not strongly affected by the local interfacial roughness at the carbon spacer LSM mirror interfaces. Irrespective of this, interfacial roughness of ~3 Å and spacer thickness uniformity at the

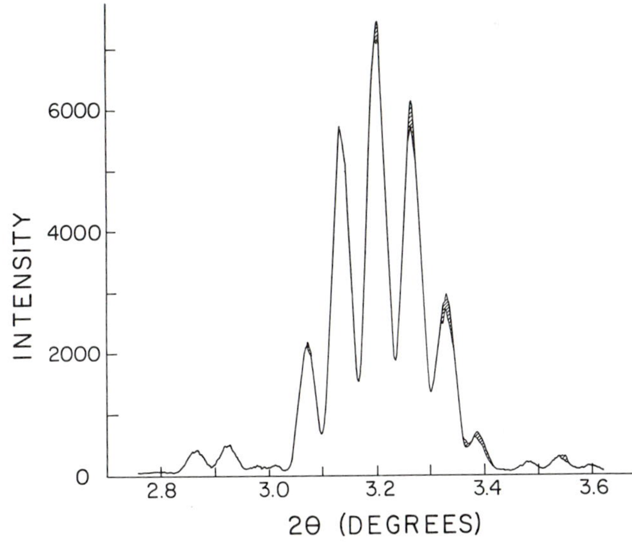

Fig. 11. An overlay of two experimental $\theta/2\theta$ scans taken, using Cu K_α radiation, from regions separated by 1.0 cm on sample 82-001, the Fabry–Perot etalon shown in Fig. 9c. Shaded areas show the difference between the two scans.

atomic plane level are demonstrated. Such structural quality gives an indication of the promise such technology holds for synthesis of x-ray optic devices as well as unique materials having controlled metastable structures.

VI. SUMMARY

In this chapter a short review of the important features of physical deposition processes relative to the synthesis of multilayer structures has been given. This was presented in terms of intrinsic and extrinsic deposition process characteristics. The intrinsic characteristics were primarily concerned with film nucleation and growth. Extrinsic processes were more related to the technological characteristics of specific synthesis approaches. An example taken from x-ray optics work was then discussed and used to demonstrate the quality of multilayer structures that can be formed. The fundamental conclusion of this work is that techniques are now available for the synthesis of multilayer structures of high quality in purity, crystalline perfection and in an x-ray optic sense from elements and compounds that may not be used in other synthesis systems or approaches (for example, molecular-beam epitaxy).

ACKNOWLEDGMENTS

I wish to acknowledge the efforts of D. L. Keith and other colleagues at Stanford University in supporting my efforts in multilayer synthesis research.

Preparation of this article was supported by the Department of Energy and benefited from early research support by the National Science Foundation through the Center for Materials Research at Stanford University.

REFERENCES

Baker, J. C., and Cahn, J. W. (1971). *In* "Solidification," p. 28. American Society of Metals, Metals Park, Ohio.

Barbee, T. W., Jr. (1981). *In* "Low Energy X-Ray Diagnostics—1981" (D. T. Attwood and B. L. Henke, eds.), p. 131. AIP Conf. Proc. No. 75, AIP, New York.

Barbee, T. W., Jr., and Keith, D. L. (1980). *In* "Synthesis and Properties of Metastable Phases" (E. S. Machlin and T. J. Rowland, eds.), p. 93. AIME, New York.

Barbee, T. W., Jr., and Underwood, J. H. (1983). *Opt. Commun.* **48,** 161.

Bevk, J., Harbison, J. P., and Bell, J. L. (1978). *J. Appl. Phys.* **49,** 6031.

Born, M., and Wolf, E. (1959). "Principles of Optics," 1st ed. Chap. VII. Pergamon, New York.

Deubner, W. (1930). *Ann. der Phys.* **5,** 261.

DuMond, J. W. M., and Youtz, J. P. (1935). *Phys. Rev.* **48,** 703.

DuMond, J. W. M., and Youtz, J. P. (1940). *J. Appl. Phys.* **11,** 357.

Durbin, S. M. (1983). Dissertation, University of Illinois, Urbana–Champaign, Illinois.

Durbin, S. M., Cunningham, J. E., and Flynn, C. P. (1982). *J. Phys. F* **12,** L75.

Eltoukhy, A. H., and Greene, J. E. (1979). *J. Appl. Phys.* **50,** 505.

Flevaris, N. K., Baral, D., Hilliard, J. E., and Ketterson, J. B. (1981). *Appl. Phys. Lett.* **38,** 992.

Goodman, F. O. (1962). *J. Phys. Chem.* **23,** 1269, 1491.

Hertel, G., McWhan, D. B., and Rowell, J. M. (1982). *In* "Superconductivity in d- and f-band Metals—1982" (W. Buckel and W. Weber, eds.), p. 299. Kernsforschungsanlage, Karlsruhe, Federal Republic of Germany.
Hilliard, J. E. (1979). In "Modulated Structures—1979" (J. M. Cowley, J. B. Cohen, M. B. Solomon, and B. J. Wuensch, eds.), p. 407. AIP Conf. Proc. No. 53, AIP, New York.
Huggins, H. A., and Gurvitch, M. (1983). *J. Vac. Sci. Tech.* **A1,** 77.
Kaufman, L. (1967). *In* "Phase Stability in Metals and Alloys" (P. S. Rudman, J. Stringer, and R. I. Jaffee, eds.), p. 125. McGraw-Hill, New York.
Koeppe, H. (1929). Dissertation, Giessen, Federal Republic of Germany.
Lamont, L. T., Jr. (1978). *J. Vac. Sci. Tech.* **15,** 171.
Lennard-Jones, J. E. (1937). *Proc. R. Soc. London Ser. A* **163,** 127.
Lennard-Jones, J. E., and Strachan, C. (1935). *Proc. R. Soc. London Ser. A* **150,** 442.
McCarroll, B., and Ehrlich, G. (1963). *J. Chem. Phys.* **38,** 523.
Maissel, L. I., and Glang, R. (eds.) (1979). *"Handbook of Thin Film Technology."* McGraw-Hill, New York.
Matthews, J. W. (1975). *In* "Epitaxial Growth—Part B" (J. W. Matthews, ed.), p. 560. Academic Press, New York.
Shaw, D. W. (1975). *In* "Epitaxial Growth—Part A" (J. W. Matthews, ed.), p. 89. Academic Press, New York.
Spiller, E. (1981). *In* "Low Energy X-Ray Diagnostics—1981" (D. Attwood and B. L. Henke, eds.), p. 131. AIP Conf. Proc. No. 75, AIP, New York.
Spiller, E., Segmuller, A., Rife, J., and Haelbich, R. P. (1980). *Appl. Phys. Lett.* **37,** 1048.
Stringfellow, G. B. (1982). *Rep. Prog. Phys.* **45,** 469.
Tsai, D. H., and McDonald, R. A. (1977). *Bull. Am. Phys. Soc.* **21,** 1301.
Underwood, J. H., and Barbee, T. W., Jr. (1981). *Nature* **194,** 492.
van der Merwe, J. H., and Ball, C. A. B. (1975). *In* "Epitaxial Growth—Part A" (J. W. Matthews, ed.), p. 493. Academic Press, New York.
Venables, J. A., and Price, G. L. (1975). In "Epitaxial Growth, Part A" (J. W. Matthews, ed.), p. 381. Academic Press, New York.
Voorhoeve, R. J. H. (1976). *In* "Treatise on Solid State Chemistry," Vol. 6A, "Surfaces I" (N. B. Hannay, ed.), p. 241. Plenum, New York.
Vossen, J. L., and Kern, W. (eds.) (1978). "Thin Film Processes." Academic Press, New York.
Wright, J. G. (1975). *In* "Epitaxial Growth—Part A" (J. W. Matthews, ed.), p. 73. Academic Press, New York.
Zheng, J. O., Ketterson, J. B., Falco, C. M., and Schuller, I. K. (1981). *Physica B & C* **108,** 945.
Zwanzig, R. W. (1960). *J. Chem. Phys.* **32,** 1173.

9 ELECTRONIC AND MAGNETIC PROPERTIES OF METALLIC SUPERLATTICES

Charles M. Falco

Departments of Physics and of Optical Sciences and Arizona Research Laboratories
The University of Arizona
Tucson, Arizona

Ivan K. Schuller

Materials Science and Technology Division
Argonne National Laboratory
Argonne, Illinois

I. INTRODUCTION

Early work on artificially layered metals was initiated by DuMond and Youtz (1940) to produce layered structures for x-ray diffraction gratings. This goal, as well as producing polarizers and monochromators for neutrons [see Spiller (1981) and references cited therein, also Barbee in the same volume], has motivated a number of workers (Saxena and Schoenborn, 1977a,b; Hamelin, 1976; Lynn *et al.*, 1976). It was very early discovered that interdiffusion greatly affected the structures produced, and many workers have concentrated on these materials for the study of the physics of diffusion (Dinklage, 1967; Dinklage and Frerichs, 1963; Cook and Hilliard, 1969; Berry and Pritchet, 1976; Paulson and Hilliard, 1977; Roll and Reill, 1982). However, as described in several other

ISBN 0-12-170470-X

chapters in this volume, advances in thin-film-deposition technology have motivated interest in artificially prepared materials exhibiting different physical properties than those that occur naturally. This chapter will review work on the electronic transport properties and magnetic properties of artificially layered metals.

II. SAMPLE PREPARATION

By far the greatest amount of published electrical transport properties data is on the Nb–Cu system. The various groups (Schuller and Falco, 1980; Geerk *et al.*, 1982, Lowe *et al.*, 1981) reporting some results on this material all used the same sputtering technique for producing samples. In this technique single-crystal substrates (Schuller and Falco, 1980; Falco and Schuller, 1982a; Geerk *et al.*, 1982; Lowe *et al.*, 1981) or amorphous Kapton tape (Lowe *et al.*, 1982) are placed in a platform that can be rotated at a precisely controlled speed either above (sputter-up configuration) or below (sputter-down configuration) the Nb and Cu sputtering sources. This is shown schematically in Fig. 1. By varying the rotation speed, as well as the sputtering rates, samples of individual layer thickness in the range ~ 5–5000 Å have been prepared. Typically samples with total film thickness of ~ 1 μm are used for transport properties measurements. One disadvantage of this sputtering technique is that argon gas pressures of $\sim 1 \times 10^{-3}$ to 2×10^{-2} torr are necessary during the deposition so that no *in situ* surface characterization probes can be used to monitor the growth during deposition.

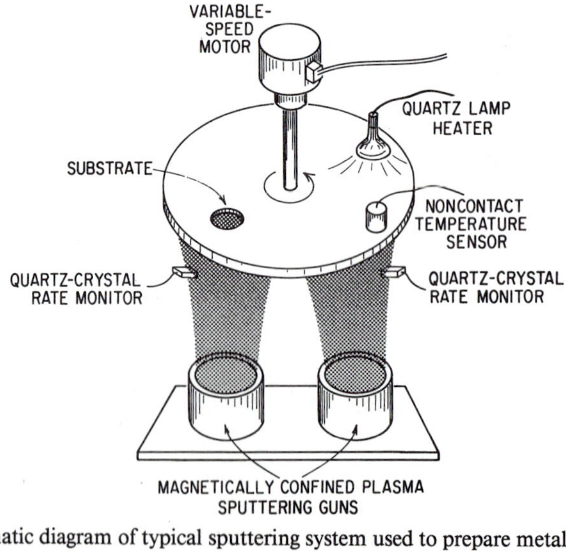

Fig. 1. Schematic diagram of typical sputtering system used to prepare metallic superlattices.

In one case (Nb–Ta) metallic superlattices have also reportly been grown by conventional evaporation in a ultrahigh vacuum (UHV) system at 4×10^{-9} torr (Durbin *et al.*, 1981, 1982). In that case, the substrates were kept fixed on a heated platform (700–900°C) while two electron-beam ovens were alternately shuttered to produce the layered structure. This technique has the advantage that the UHV environment allows the use of surface probes during growth, although growth rates are lower (0.1–10 Å/sec) compared to sputtering (10–200 Å/sec).

III. X-RAY CHARACTERIZATION

A knowledge of the structure of the multilayer films is essential in order to correlate with measured electronic or magnetic properties. We will address here only the x-ray characterization of those materials whose electronic properties have been measured in some detail. This is of importance since some electronic properties have been found to be intimately related to the structural characteristics. Detailed studies (Lowe *et al.*, 1981; Schuller, 1980) of the x-ray diffraction data from Nb–Cu samples grown by several groups (Schuller and Falco, 1980; Geerk, 1982; Lowe, *et al.*, 1981) show that for layer thicknesses greater than 10 Å the samples have long-range structural coherence perpendicular to the layers (z coherent). A lower limit on this perpendicular structural coherence length of ~200 Å can be placed by the x-ray measurements. In the plane of the sample the crystallite size is limited to ~150–200 Å.

An alternate, x-ray method for the determination of the layer thickness relies on small angle scattering around $\theta = 0°$. This method is based on the direct scattering by the difference in x-ray refractive index of the two constituents. The advantage of such a method is that the layer thickness can be determined even for constituents formed of amorphous–amorphous and crystalline–amorphous materials and does not require z coherence. However, this method gives no information regarding the crystal structure and interatomic spacing within each layer. This type of x-ray scattering has been the basis for the production of neutron reflectors (Saxena and Schoenborn, 1977a,b; Hamelin, 1976; Lynn *et al.*, 1976).

For z-coherent samples with layer thicknesses in the range 10–100 Å, the x-ray diffraction patterns show a series of superlattice lines due to the constructive and destructive interference of the x rays by the coherent registry of sequential layers of the sample. This is shown in Fig. 2. The spacing of these superlattice lines can be directly used to determine the layer thickness since (Schuller, 1980)

$$\Lambda = \lambda_x / 2(\sin \theta_i - \sin \theta_{i+1}), \tag{1}$$

where Λ is the superlattice wavelength, λ_x is the x-ray wavelength, and i and

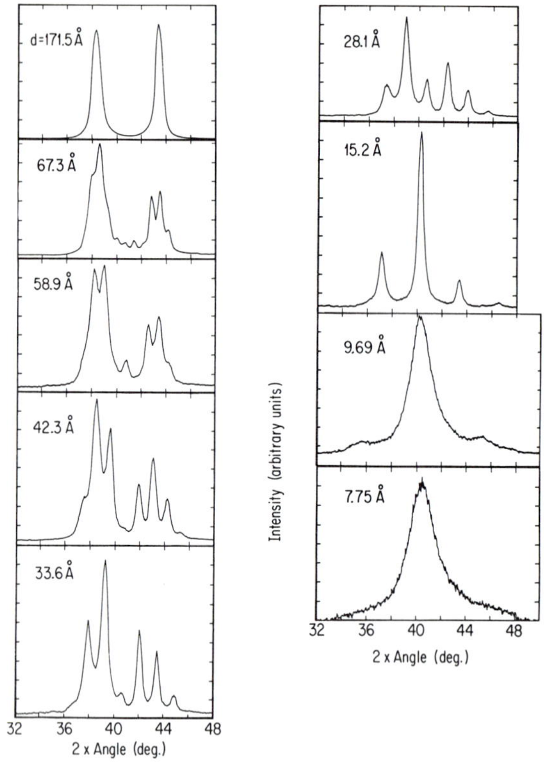

Fig. 2. Bragg x-ray diffraction data for a series of Nb–Cu samples of different layer thicknesses *d*.

$i + 1$ refer the adjacent diffraction peaks in the $\theta - 2\theta$ scan. For thicker layers, the spacing of the superlattice lines decreases, until at about $d \cong 100$ Å they merge with the Cu (111) and Nb (110) x-ray peaks. Because of this, for layer thicknesses beyond this point x rays can no longer be used to determine layer thickness. Two other independent methods are (a) using the size of the particle beam in conjunction with the sputtering rate and the rotation speed to determine a layer thickness and (b) measuring the total thickness and dividing by the total number of revolutions to determine a layer thickness. These four independent methods agree to within 5% in the common thickness range (Schuller and Falco, 1982) and thus any of the methods can be used.

One group has studied Nb–Cu for very thin layers (Werner *et al.*, 1982). Below 10 Å the material becomes progressively more disordered, causing the x-ray linewidths to greatly increase; however, in this regime the layer thickness as determined from the sputtering geometry and rate remains a good parameter for characterizing the elastic electron mean free path.

Extensive x-ray measurements (Durbin *et al.*, 1982) have also been reported on Nb–Ta multilayers. These samples were grown on several substrates in a variety of crystallographic orientations. Structural coherence distances were derived from x-ray rocking curves and in the next section are compared with electron mean free paths determined from resistivity measurements. X-ray measurements were also used to establish that the Nb and Ta interdiffuse over several lattice planes about each interface in the Nb (28-Å)–Ta (22-Å) samples. Durbin *et al.* (1982) concluded from these measurements that "it does not appear possible to choose a growth temperature and (convenient) evaporation rate such that the interfaces remain sharp while the growth perfection (as determined by x-ray rocking curves) remains satisfactory." If an alternate growth technique (such as sputtering) is not found for this material, the diffuseness of the interfaces will limit the growth of this material to layer thickness of $\gtrsim 10-15$ Å. This same diffuseness may also prohibit the observation of superlattice effects in the electronic properties of Nb–Ta.

Work on producing Nb–Ta multilayers by sputtering has also found interdiffusion at the interfaces (Hertel *et al.*, 1982). In this case, samples grown at temperatures of > 810°C were found to interdiffuse over at least 8.5 Å. Samples grown at temperatures of < 810°C were found to be polycrystalline. Further work is necessary to find whether a suitable growth temperature along with appropriate sputtering parameters can be found that gives sharp interfaces and crystalline samples.

IV. ELECTRICAL TRANSPORT

A. Resistivity

Figure 3 shows a plot of the low-temperature (20-K) resistivity ρ versus layer thickness for a series of 25 Nb–Cu samples grown in random order over a period of one year (Werner *et al.*, 1982; Falco and Schuller, 1982b). Below approximately 10-Å layer thickness the resistivity saturates close to the Ioffe–Regel (1960) limit of ~ 150 μohm cm. Above this thickness the resistivity is inversely proportional to the layer thickness as would be predicted for a mean free path that is limited by boundary scattering at the layers. As can be seen from Fig. 4, for very thick layers the resistivity approaches the value expected for bulk Cu and Nb films in parallel. The arrows in Fig. 4 indicate the measured resistivities for ~ 1-μm-thick pure Cu and Nb film grown under the same sputtering conditions as were the layered samples. Also shown is the expected result for Nb and Cu films connected in parallel; corrected by the appropriate geometrical factor of 2 since the individual layer thickness is half the total sample thickness. It should be emphasized that the value of such resistivity measurements should not be underestimated since the electronic mean free path will determine whether superlattice effects on the band structure are likely to be observed.

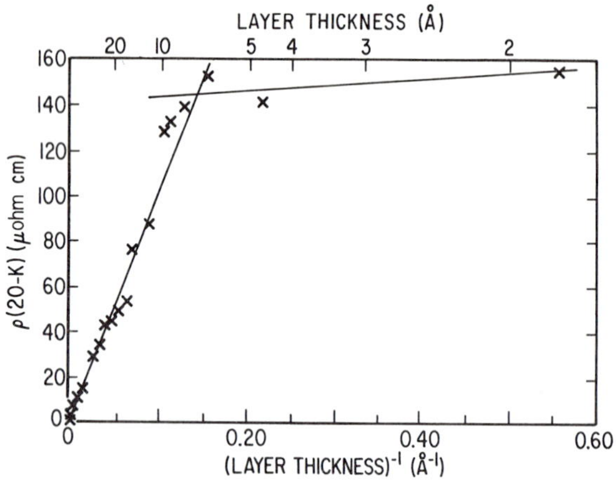

Fig. 3. Resistivity at 20 K versus inverse layer thickness for a series of Nb–Cu samples (Werner *et al.*, 1982).

Since resistivity is expected to scale approximately as the inverse of layer thickness, a useful way to present these data is in a plot of resistivity times layer thickness versus layer thickness. Figure 5 is a plot of $2\rho d$ versus $2d$ for published Nb–Cu data from two groups (Lowe *et al.*, 1981; Werner *et al.*, 1982). Log–log scales are used for convenience in presentation; with the solid curve being a straight line on a linear plot. This straight line behavior is predicted for a mean free path proportional to layer thickness (Mayadas and Shatzkes, 1970). Resis-

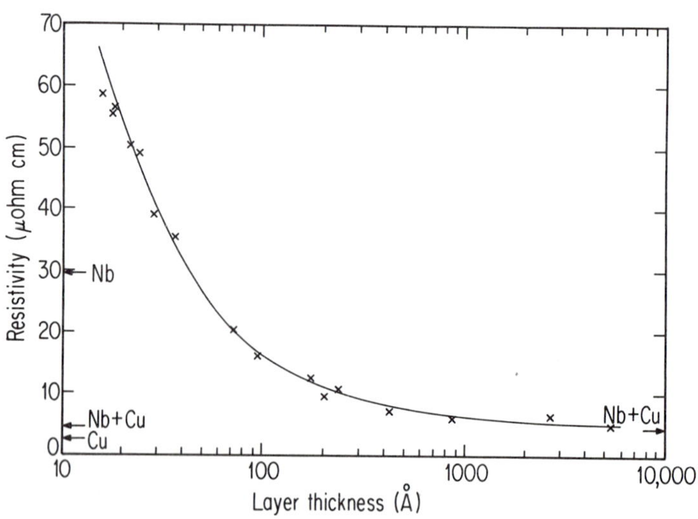

Fig. 4. Room-temperature resistivity for a series of Nb–Cu samples versus layer thickness at 295 K.

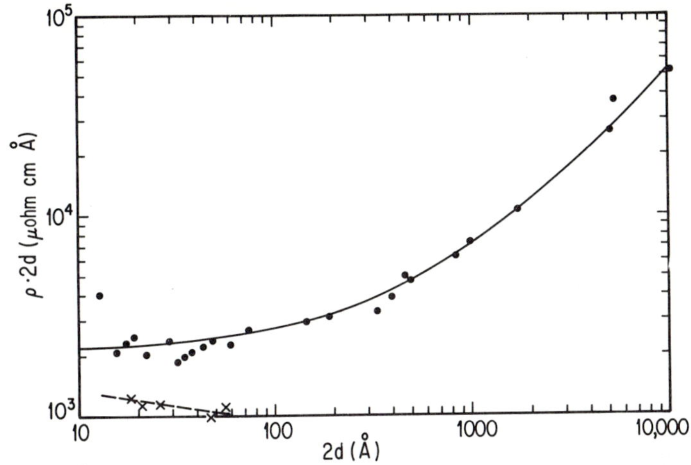

Fig. 5. Resistivity times the bilayer thickness versus bilayer thickness for Nb–Cu samples. [●, from Werner *et al.* 1982; ×, from Lowe *et al.* (1981).]

tivity data on Nb–Ge (Ruggiero, 1981) have not been plotted on this figure since the "spacer" material (Cu in one case and Ge in the other) is considerably different. The data of Lowe *et al.* (1982) in Fig. 6, which shows ρ increasing as d is decreased, perhaps imply increasing amounts of disorder for these particular films, causing ρ to increase faster than d^{-1}. Published values of the superconducting transition temperatures for these Nb–Cu films (Lowe *et al.*, 1981) are much lower than those samples that follow the solid line (Banerjee *et al.*, 1982). This is also consistent with larger amounts of disorder or the presence of impurities.

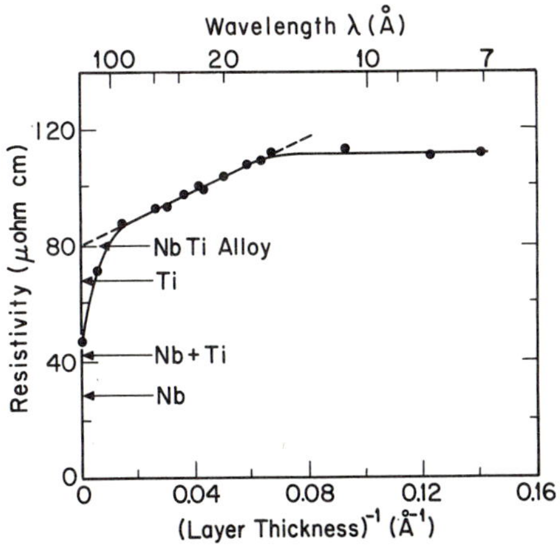

Fig. 6. Resistivity versus inverse layer thickness for a series of Nb–Ti samples.

The behavior of resistivity shown in Figs. 4 and 5 is not unique to Nb–Cu, although there can be differences in details for different materials as discussed below. Figure 6 shows the measured room-temperature resistivity of a series of Nb–Ti samples (Zheng *et al.*, 1981b). In this case the resistivity begins to saturate for ~15-Å layer thicknesses at 115 μohm cm (slightly lower than the case of Nb–Cu although close to values expected from the Ioffe–Regel criterion). For slightly thicker layers the resistivity scales as the inverse of the layer thickness just as it did for Nb–Cu. However, the value of 80 μohm cm obtained by extrapolating to infinite layer thickness is identical to that of a uniform Nb–Ti alloy. This is easily understood since the equilibrium binary phase diagram for Nb–Ti, unlike Nb–Cu, shows that the two materials form continuous sets of solid solutions. [These and other binary phase diagrams can be found in Hansen and Anderk (1958), Elliott (1965), Shunk (1969), and Moffatt (1981).] Thus in this regime of layer thickness there is probably considerable interdiffusion contributing both to the elastic mean free path and causing the "background" resistivity to be that of the uniform alloy. Finally, for layers thicker than ~100 Å in which any contribution from interdiffusion at the interfaces becomes negligible, the resistivity approaches that expected of thick Nb and Ti slabs in parallel.

Mean free path results obtained from resistivity measurements have also been reported for several Nb–Ta superlattices of unequal layer thickness (Durbin *et al.*, 1982) (Nb, 28 Å; Ta, 22 Å). However, as shown in Fig. 7, the reported mean free paths do not scale with the structural coherence distance (derived from x-ray rocking curves). The arrow in Fig. 7 denotes the meanfree path of a

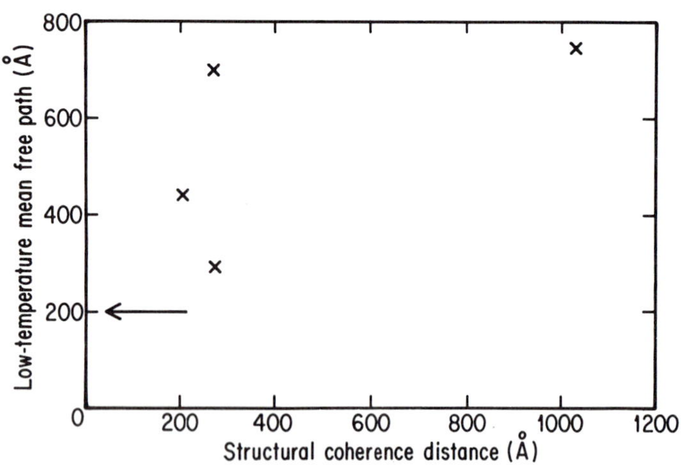

Fig. 7. Low-temperature electronic mean free path versus structural coherence distance determined from x-ray measurements for four Nb–Ta superlattices. [Data taken from Durbin *et al.* (1982).]

"disordered $Nb_{0.47}-Ta_{0.53}$ alloy" (Calverly and Rose-Innes, 1960). Within the uncertainties of the assumptions in determining the mean free path from the resistivity, this value is essentially the same as reported (Durbin *et al.*, 1982) for Nb–Ta superlattices. Without a detailed study of ρ versus d for this material, it is hard to reconcile these observations, as well as to predict whether Nb–Ta films produced in this way can be of sufficient perfection to expect superlattice effects on the electronic properties.

B. Temperature Dependence of the Resistivity

Figure 8 shows the temperature-dependent part of the resistivity versus layer thickness for a series of Nb–Cu samples (Werner *et al.*, 1982). Even in the region below 10 Å, in which the x-ray measurements (Schuller, 1980) (Fig. 2) show this material to be becoming progressively more disordered, the systematic way in which the resistivity changes with layer thickness (determined by using the preparation parameters) shows that it is meaningful to use the substrate platform rotation speed (see Fig. 1) as an indication of the elastic mean free path. While the layer thickness serves as a useful parameter for characterizing the mean free path, in this regime of very short mean free paths the results are independent of the exact structure of the film. All that is important is the fact that it is possible to control artificially the resistivity (i.e., elastic mean free path) of the films by the deposition process.

At any arbitrary fixed temperature in the range $20\ K < T < 300\ K$, the temperature coefficient of resistivity (TCR) for these Nb–Cu samples shows a rather sudden change from positive (metallic) to negative (nonmetallic) as a function of mean free path (Werner *et al.*, 1982). This is shown in Fig. 9, in which $d\rho/dT$ versus layer thickness is plotted at three temperatures. This

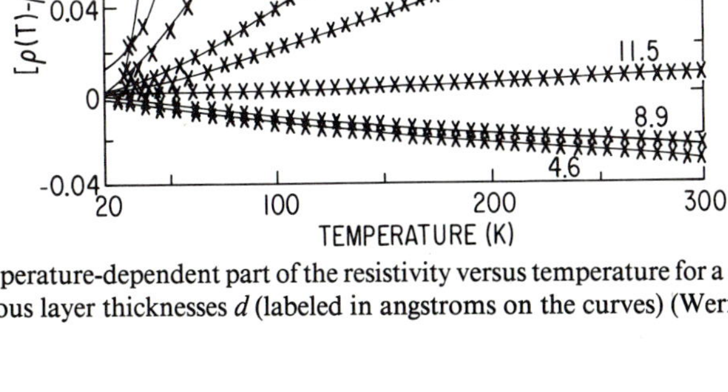

Fig. 8. Temperature-dependent part of the resistivity versus temperature for a series of Nb–Cu samples of various layer thicknesses d (labeled in angstroms on the curves) (Werner *et al.*, 1982).

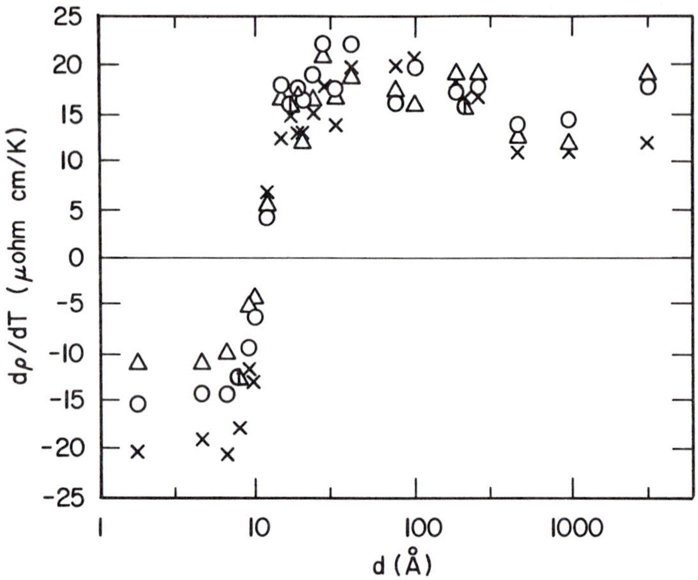

Fig. 9. Plot of $d\rho/dT$ at three temperatures versus layer thickness for a series of Nb–Cu samples (Werner *et al.*, 1982). (△, 250 K; ○, 150 K; ×, 50 K.)

change from positive to negative TCR occurs for a mean free path that is decreased by only 2 or 3 atomic spacings (~5–6 Å).

The point at which the TCR passes through zero has a very useful application. Because $d\rho/dT$ is very weakly dependent on temperature, for a sample of appropriate layer thickness ($d \cong 10$ Å) the TCR can be made to be less than 8 ppm/K over the entire temperature range 300–3 K (at which point the sample becomes superconducting). Thus one can fabricate materials for thin-film resistors that are extremely stable against temperature fluctuations or drifts.

The temperature dependence of the conductivity of a material in the nonmetallic regime (negative TCR) has been the subject of extensive theoretical investigation. [For a review of the experimental and theoretical situation, see Dynes (1982); also, Abrahams *et al.* (1979), Altshuller and Aronov (1979), and Kaveh and Mott (1981, and references cited therein).] It has been shown that due to electron–electron interaction effects (Altshuller and Aronov, 1979) or localization (Abrahams *et al.*, 1979), the temperature-dependent part of electrical conductivity can be written

$$\frac{\delta\sigma}{\sigma} = 1 - \left(\frac{3}{2x}\right) \ln(1+x) \left(\frac{1}{N_0 \pi^3} \frac{k_B}{(hD)^3}\right)^{1/2} T^{1/2}, \qquad (2)$$

where D is the diffusion coefficient for electrons, N_0 is the density of states at the Fermi surface, and $x = (2k_F/\mu)^2$, where k_F is the Fermi momentum and μ is the inverse screening length. Assuming $\delta\sigma \ll \sigma$, as is observed experimentally in the Nb–Cu samples, and using the Einstein relation $\sigma_0 = N_0 e^2 D_0$ leads to

$$\delta\sigma \cong 1 - \left(\frac{3}{2x}\right) \ln(1+x) \left(\frac{k_B e^4}{h^3 \pi^6}\right)^{1/2} \frac{D_0}{D^{3/2}} T^{1/2}. \tag{3}$$

Figure 10 shows a plot of $\delta\sigma$ versus $T^{1/2}$ for a Nb–Cu sample of layer thickness 9.5 Å (Werner *et al.*, 1982). This is just slightly into the negative TCR regime, as seen from Fig. 10. Above ~35 K the temperature dependence is well described by Eq. (3) with a temperature-independent diffusion coefficient D. Similar behavior was reported for all five other Nb–Cu samples that exhibit a negative TCR. However, it should be noted that in the same temperature region the dependence $\exp(-B/T^{1/4})$ predicted by Mott's variable-range hopping model (Mott, 1969) cannot be distinguished from the $T^{1/2}$ dependence predicted by localization and electron–electron correlation theories. Fits to a number of other dependences have been tried, but no other theoretical expression has been found to fit the experimental results.

The prefactor to the $T^{1/2}$ term in Eq. (3) can be extracted from a plot similar to Fig. 10. The diffusion coefficient D_0 can also be obtained from the Einstein relation and the extrapolated zero temperature conductivity σ_0. Comparing these two quantities leads to $1 - (3/2x)\ln(1+x) \cong 10^{-2}$. Solving for x leads to $k_F/\mu \cong \frac{1}{2}$, implying that both localization and electron–electron correlation effects are important in the Nb–Cu system in this regime when the electron

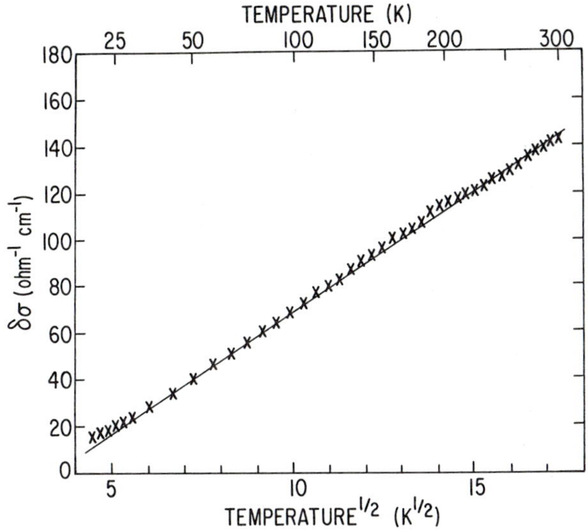

Fig. 10. Plot of $\delta\sigma$ versus $T^{1/2}$ for a Nb–Cu sample of layer thickness 9.5 Å (Werner *et al.*, 1982). Here $\delta\sigma = \sigma(T) - 7762.3$ ohm^{-1} cm^{-1} is the change in conductivity from its value at zero temperature.

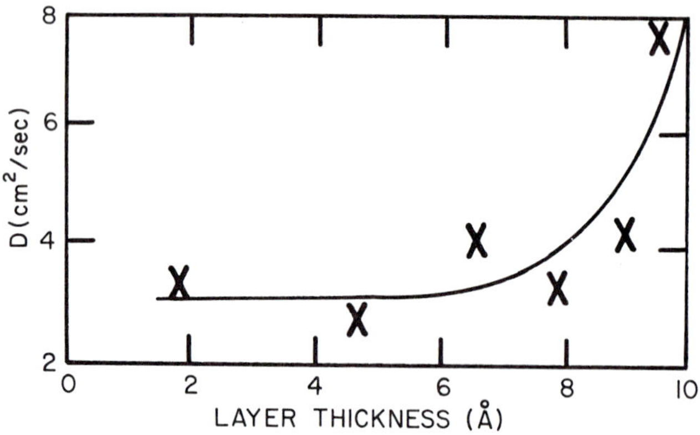

Fig. 11. High-temperature diffusion coefficient D [extracted from Eq. (2)] versus layer thickness for a series of Nb–Cu samples (Werner *et al.*, 1982).

mean free path has been artificially reduced to very small values by the material fabrication process.

It is interesting to note that while the low-temperature diffusion coefficient D_0 obtained from the Einstein relation for these samples is essentially constant (because the conductivities are approximately the same), the high-temperature diffusion constant D is not. The high-temperature diffusion constants extracted from the slope of Eq. (3) increase markedly as the nonmetallic to metallic transition is approached. This is shown in Fig. 11.

C. Correlation of Electrical and Mechanical Properties

Although mechanical properties of artificially layered metals are discussed in detail in other contexts within this volume, at least one such property has been shown to be correlated with an electronic transport property. Since this is the first example observed of such a relationship, we will discuss it here.

Samples of Nb–Cu with individual layers of equal thickness in the range 5 Å to 1 μm were prepared on single-crystal sapphire substrates. Both electrical resistivity and elastic constant measurements were made on the same set of samples. The electrical resistivity was measured by using a four-probe technique over the temperature range 300–10 K. The elastic constants were determined from a Brillouin light-scattering technique at room temperature (Kueny *et al.*, 1982). The details of the experiment are described by Kueny *et al.*, (1982).

A spectrum from one sample is shown in Fig. 12. From such measurements at several values of ω and k, it is possible to obtain the surface sound (Rayleigh) velocity v_R. The measured velocities as a function of twice the layer thickness

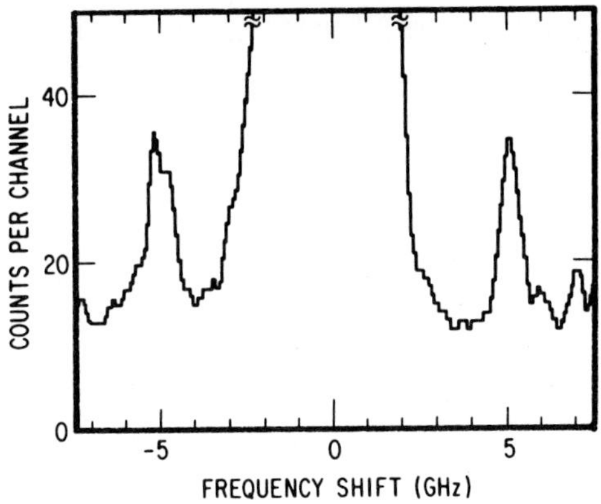

Fig. 12. Brillouin spectrum of a Nb–Cu sample with bilayer thickness 20 Å (Kueny *et al.*, 1982).

are shown in Fig. 13. Note the ~20% drop in sound velocities for layer thickness centered around ~10 Å. Referring to the temperature derivative of resistivity $d\rho/dT$ shown in Fig. 9, we can see that the minimum in v_s corresponds closely to the point where $d\rho/dT$ undergoes a transition from positive (metallike) to negative (nonmetallic). Thus the structural and electronic properties of these materials are correlated.

For an interpretation of the results of Fig. 13 in terms of elastic constants, the reader is referred to Farnell (1969). In the region in which the wavelength of the phonons probed in these experiments (3000 Å $< \lambda_{\text{phonon}} <$ 4300 Å) is much

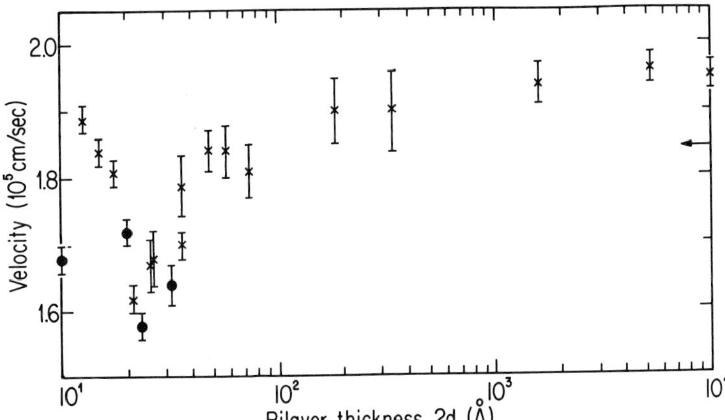

Fig. 13. Surface wave velocity of Nb–Cu superlattices as a function of bilayer thickness. The arrow indicates the calculated velocity for films with bilayer thickness ≲ 100 Å (Kueny *et al.*, 1982).

greater than the bilayer thickness $2d$, the films can be considered to be homogeneous and the velocity of the surface wave is given by (Farnell, 1969)

$$v_R = \beta(C_{44}/\rho_m)^{1/2}, \quad (4)$$

where ρ_m is the mass density of the film and β a rather insensitive function of C_{11}, C_{33}, and C_{13}. Under certain assumptions for the C_{ij}'s, it was found (Kueny *et al.*, 1982) that $0.86 \leq \beta \leq 0.98$. Thus by using Eq. (4) it was concluded that the dominant part of the decrease and subsequent increase in v_R centered about $d = 10$ Å corresponds to a ~35% decrease in C_{44}.

The expected value for v_R, assuming that the Nb and Cu layers kept their bulk properties, can be calculated by following Rytov (1956). Assuming that the β's for Nb and Cu are not very different and making use of the fact that the densities of Nb (8.60 gm/cm) and Cu (8.93 gm/cm) are nearly equal, then

$$v_R^2(\text{Nb-Cu}) \cong 2v_R^2(\text{Cu})v_R^2(\text{Nb})/[v_R^2(\text{Cu}) + v_R^2(\text{Nb})]. \quad (5)$$

For [111] Cu and [110] Nb the calculated velocity $v_R(\text{Nb-Cu}) = 1.84 \times 10^5$ cm/sec, which is shown by the arrow on the right side of Fig. 13. If the artificial layering of this material did not affect the bulk properties of the Cu and Nb layers, this value should be observed for all film for which $2d \ll \lambda_{\text{phonon}}$ since in this case the phonons average over many layers.

Deviations are expected for $2d \gtrsim 1000$ Å since in this case the phonons no longer average over many layers and consequently the topmost layer will dominate in determining the velocity of the surface wave. As can be seen from Fig. 13, in the region $2d = 100-1000$ Å there is reasonable agreement between experiment and the calculated $v_R(\text{Nb-Cu})$. For $2d > 1000$ Å the observed value matches the calculated $v_R(\text{Nb})$, which is also expected since for these particular samples the topmost layer was Nb. However, a ~20% dip in v_R occurs between $2d = 100$ and 12.5 Å, which as discussed above is interpreted according to Eq. (4) as a ~35% decrease in C_{44}.

The question remains as to how to explain the observed decrease in C_{44} due to the layering. This dip in the elastic constants might arise from either structural changes that alter the phonon spectrum or changes in the electronic structure (the subject of this chapter). It has been found that the softening of the elastic constant is correlated with an expansion of the average lattice spacing (Khan *et al.*, 1983). It has been speculated that this expansion might give rise to the changes observed in the elastic constant. These ideas are at this writing, being tested by using molecular dynamics studies (Schuller and Rahman, 1983).

Another possible explanation is the electronic contribution to the elastic moduli. One possible mechanism is analogous to that proposed (Keyes, 1967) to account for effects due to free carriers in semiconductors. The argument is that when a phonon modulates the band structure of a material, thereby lifting certain degeneracies, it allows a redistribution of electrons or holes. This redistribution reduces the free energy and thus effectively lowers the elastic constant. Since degeneracies and a high density of states are most likely to occur either at the boundary or at the center of the Brillouin zone, we would expect a large contribution only if the Fermi energy lies close to one of these regions of the Brillouin zone. A very simple argument shows that the Fermi energy will be folded with a periodicity

$$k_F = n2\pi/d, \tag{6}$$

where k_F is the Fermi wave vector, n an integer, and d is the layer thickness. With the value $k_F = 1.36$ Å^{-1} for Cu, the three lowest solutions are for $2d =$ 9.2, 18.3, and 27.6 Å. This is to be compared to Fig. 14, in which there is only one experimentally observed dip at $2d = 21$ Å, which makes this explanation improbable. In addition a complete check of this simple argument is not possible without a full band structure calculation for such a layered material. Such calculations are extremely time-consuming and expensive due to the large number of atoms in a unit cell.

A third explanation for the decrease in Cu, which has already been alluded to, can be proposed by noting from Figs. 9 and 13 that at the thickness at which the elastic constant has its minimum, the temperature coefficient of resistivity

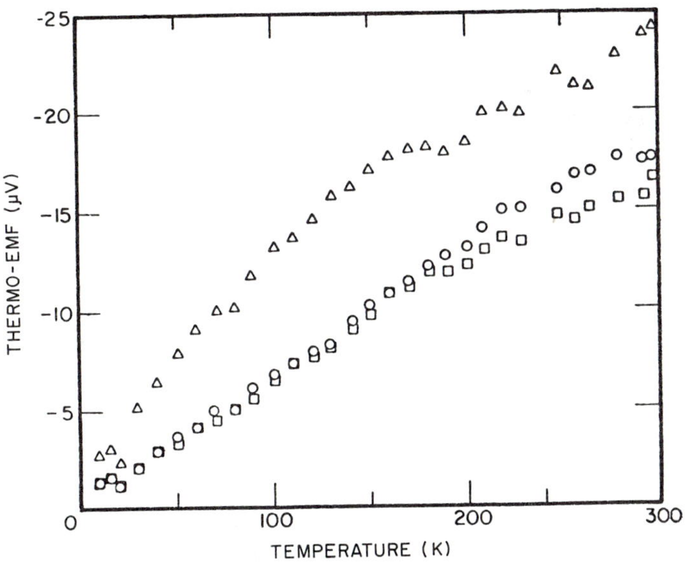

Fig. 14. Thermoelectric power versus temperature for a Cu–Ni sample of modulation wavelength 17 Å. The three sets of data correspond to the sample as prepared with modulation amplitude $A = 0.32$ (□), the sample after being partially annealed to reduce the modulation amplitude to $A = 0.24$ (○), and the sample after being fully annealed to produce a uniform alloy (△) (Schuller *et al.*, 1979).

changes from positive to negative (Werner *et al.,* 1982). This change in sign has been attributed to three-dimensional localization (Abrahams *et al.,* 1979; Imry, 1981) and electron–electron correlation effects (Altshuller and Aronov, 1979; Imry, 1981) as a function of layer thickness. Here, the opening up of a Coulomb correlation gap at E_F would be reflected in an effect on the elastic constants of the material. This explanation is qualitatively similar to that for binary alloys, in which changes in the lattice parameters and crystal structure are attributed to an energy-gap opening up at the Fermi surface when it touches the Brillouin zone. [See Kittel (1957). We are grateful to F. Blatt for pointing this out to us.] A fourth explanation has been advanced in the literature (Chaikin, 1978). In this, the changes in the elastic constants are due to singularities of the dielectric function due to Fermi surface nesting or dimensional effects. Although this explanation is obvious and might seem attractive at first sight, further evidence is needed to substantiate the existence of low-dimensionality behavior in these materials. In addition, it is not clear how these ideas would explain the observed metal–insulator transition.

D. Other Transport Properties (Thermopower, Magnetoresistance, and Hall Effect)

In addition to resistivity, other electronic transport properties such as thermopower and Hall effect should provide important information on the electronic density of states near the Fermi energy in these materials. However, to date there have been only a few reports of such measurements. The problem perhaps lies in the fact that while in principle the thermopower F (within the free-electron model) is simply related to the Fermi energy through

$$F = -2k_B^2 T/|e|E_F, \qquad (7)$$

in practice the situation is much more complicated [see, for example, Barnard (1972)]. In the next-higher degree of approximation when real metals properties are included, Eq. (7) is modified by an additional factor that contains details of the shape of the Fermi surface. In particular, the thermopower is related to the energy derivative of the mean free path. While a great deal of progress has been made in understanding thermopower measurements on pure simple metals, the situation is not so clear for many of the more complicated materials (such as alloys) (Barnard, 1972).

The first measurements of thermopower of a series of compositionally modulated metallic films used Cu–Ni (Schuller *et al.,* 1979). In this material there is a large degree of intermixing due to the mutual solubilities of Cu and Ni. Consequently, a composition modulation exists, although the constituent layers are not perfectly defined and, indeed, for some purposes the interfaces can be thought of as an alloy. For thinner-layer samples there is no portion of the material that contains the pure constituent elements, and the entire sample is a composition modulated alloy (CMA). When thermopower measurements on this particular material are viewed in this way, it is perhaps not surprising that they are not easy to interpret.

As shown in Fig. 14 it was found for Cu–Ni samples with modulation wavelength 17 Å and two different modulation amplitudes (plus a fully annealed sample) measured over the temperature range 20–300 K that the thermopower was larger than that of a pure Cu film made under the same deposition conditions. Thinking of the fully annealed sample as having zero modulation amplitude, an effect was also noted for the thermopower to increase with decreasing amplitude. This is consistent with the idea that most of the temperature dependence of the thermopower comes from impurities.

The effect of varying the composition wavelength on the thermopower of composition-modulated Cu–Ni films was reported by Baral and Hilliard (1979). Here freestanding films approximately 2 μm thick were evaporated and the thermopower relative to pure Cu measured by holding one end of the sample at 0°C and varying the other end between 10 and 200°C in a dc fashion. In agreement with the earlier studies (Schuller *et al.*, 1979), they also found the thermopowers of the composition-modulated and fully annealed samples to be greater than that of pure Cu. In addition, several films with different initial wavelengths were fully annealed and found to give approximately the same thermopowers. By assuming a linear relationship between thermopower and composition amplitude, all samples were normalized to the same amplitude of 0.25, where 1.00 corresponds to full modulation. The result is shown in Fig. 15. A broad maximum located at approximately 20 Å coincides with the point at which the biaxial modulus measured by this same group was found to exhibit a maximum. The conclusion of these latter thermopower measurements was that a critical wavelength exists in composition-modulated Cu–Ni foils. However, we should point out that the existence of a critical wavelength as determined by elastic constant measurements is the subject of some controversy. This result was confirmed independently by Young and torsional moduli (Testardi *et al.*, 1981) measurements, in contrast to the findings of Berry and Pritchet (1976) and those of Itozaki (1981).

Resistivity and magnetoresistance of two composition-modulated Cu–Ni films as well as a fully annealed film have also been reported (Schuller *et al.*, 1979). The magnetoresistance of the sample with the highest degree of composition modulation as well as that of the fully annealed sample (lowest degree of composition modulation) was linear and negative at low fields (< 50 kG), finally becoming positive at high fields. This behavior is typical of a ferromagnet. In contrast, the sample with intermediate composition modulation amplitude was quadratic at low fields (< 10 kG), becoming linear at higher fields up to the maximum measured of 70 kG. This is typical of a two-band metal below saturation. The Hall coefficient of all three samples was typical of pure Ni. The complicated behavior of the transport properties has been briefly addressed by theoretical band structure calculations (Pickett, 1982).

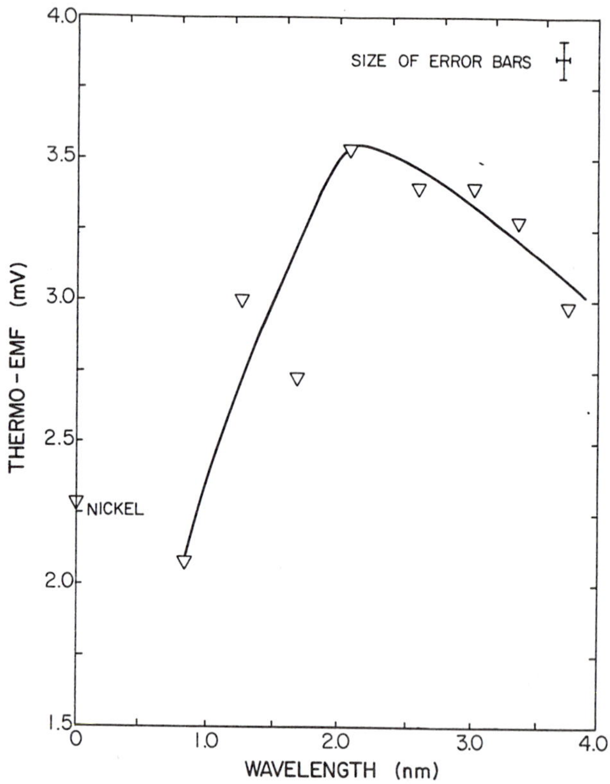

Fig. 15. Thermoelectric power versus modulation wavelength for a series of Cu–Ni samples normalized to the same composition amplitude of $A = 0.25$. [From Baral and Hilliard (1979).]

V. MAGNETIC PROPERTIES

A. Cu–Ni

The first magnetic measurements on compositionally modulated materials were those of Hirsch *et al.* on Cu–Ni in 1964 (Hirsch *et al.*, 1964; Hirsch, 1965). They produced films of individual layer thickness ~10–1000 Å by evaporation and measured the coercive force. They were interested in the behavior of individual thin Ni films but showed that interactions between the Ni layers could not be neglected. About the same time, several groups (Clow, 1962; Friedaender *et al.*, 1965) showed interest in layered samples formed by alternating an alloy of 80% Ni–20% Fe with SiO. It was found (Clow, 1962) that these materials had much lower domain wall coercivities than the pure Ni–Fe alloy.

When discussing the magnetic properties of these composition-modulated materials, it is useful to keep in mind two questions that many of the studies are implicitly addressing: (1) Can layering produce altered electronic properties such that the magnetization of one of the components is enhanced over the value of the pure material? (2) What coupling, if any, exists between the magnetic layers: direct exchange, dipole, or RKKY? Also, it is important to keep in mind whether the possibility of clustering one or both of the component materials has been conclusively ruled out, and if not, how this can affect the conclusions.

Recently, Thaler *et al.* (1978) examined the behavior of compositionally modulated materials of Cu–Ni. One sample ~4500 Å thick containing approximately 41-at. % Ni and with modulation wavelength of about 30 Å was prepared. X-ray measurements showed one central peak (note that the lattice constants of Cu and Ni are almost identical) flanked on both sides by a sidelobe due to the composition modulation. An analysis of this x-ray measurement indicated an essentially sinusoidal composition modulation of amplitude 0.41 ± 0.06-at. % fraction.

A ferromagnetic resonance (FMR) technique was used to study this sample since this should be sensitive to the local magnetization density $M(z)$, where z is the distance into the sample. It was argued that since $M(z)$ should be an oscillatory function, it would probably be essentially zero in the Cu-rich regions, thereby reducing the average magnetization of the sample. Thus a local probe was necessary to look for any possible effects of the composition modulation on the magnetization of Ni. The original interpretation (Thaler *et al.*, 1978) of these FMR measurements was that at temperatures below approximately 180 K the magnetization of the Ni atoms within each layer was *greater* than that of bulk Ni at 0 K. This observation stimulated further experimental and theoretical investigations of the Cu–Ni system, including static magnetization measurements, neutron scattering, band structure calculations as well as reinterpretations of the original FMR results.

White and Herring (1980) showed that an *apparent* enhancement of the magnetization could arise in a modulated system given a complex distribution of the magnetization within a layer. However, this mechanism is apparently not responsible for the behavior of Cu–Ni, and it is now generally agreed that the magnetization of the Ni is *reduced* rather than enhanced.

Jarlborg and Freeman (1980) performed a self-consistent band structure calculation of the magnetization of a layered Ni–Cu sample. The geometry chosen for the calculation consisted of three atomic layers each of Cu and Ni growing with the [111] direction perpendicular to the substrate. Due to the large unit cell of this system, only one geometry was calculated. However, the results are of great interest. Jarlborg and Freeman found that the Ni atoms in the center of their layers suffered a small reduction in moment below the value of pure Ni ($0.54\mu_B$) while the interface Ni atoms experienced a much larger reduction to $0.37\mu_B$.

Gyorgy *et al.* (1980, 1982) repeated the FMR measurements on a number of

Cu–Ni samples of various layer thicknesses. They also measured the average magnetization by using a vibrating sample magnetometer as well as a force method. Figure 16 shows the average magnetization as a function of inverse layer thickness. Since both the magnetization and the anisotropy field depend only on the Ni thickness and not on the Cu thickness, they concluded that the intrinsic Ni properties are not determined by modulation effects. Also, the absence of superparamagnetic effects led Gyorgy *et al.* to conclude that if the Ni were in clusters in their samples, the minimum size of such clusters was approximately 10 × 200 × 200 Å.

There are two possible reasons for the behavior shown in Fig. 16 of the average magnetization as the thickness is reduced. Either the magnetization of the Ni decreases with layer thickness, or there are dead layers of nonmagnetic

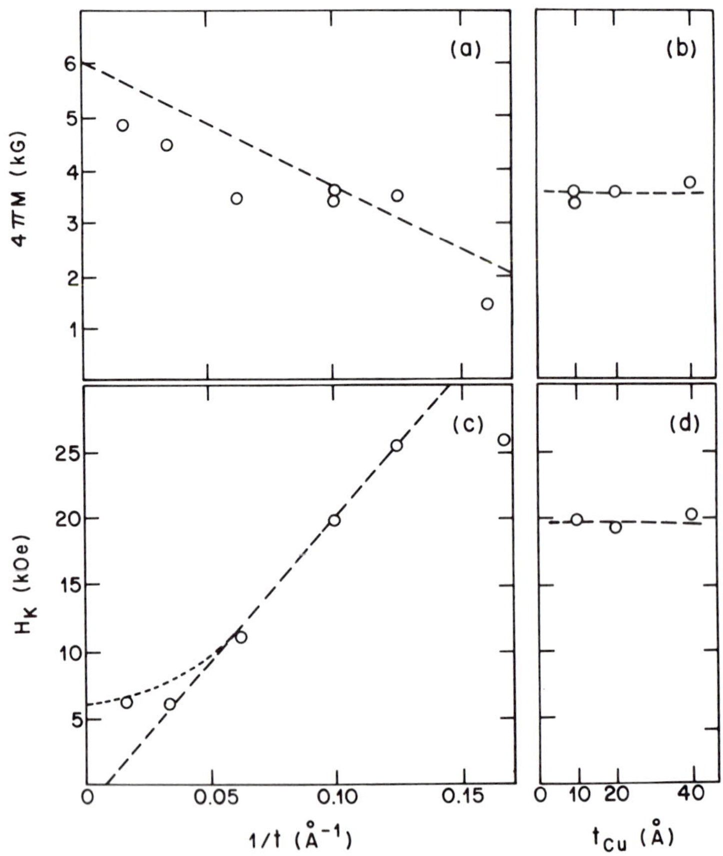

Fig. 16. Static magnetization for a series of Cu–Ni compositionally modulated films. (a) $t_{Ni} = t_{Cu} = t$; (b) $t_{Ni} = 10$ Å; (c) $t_{Ni} = t_{Cu} = t$; (d) $t_{Ni} = 10$ Å. [From Gyorgy *et al.* (1980).]

Ni at the Ni–Cu interfaces. The data of Gyorgy *et al.* did not allow them to unambiguously choose between the two options, but Gyorgy *et al.* speculated that the amplitude of the composition modulation in their films suggested that the reduced magnetization was a better explanation.

Direct current magnetization measurements of Cu–Ni samples as a function of temperature, magnetic field, composition wavelength, and composition amplitude have also been reported (Zheng *et al.*, 1981, 1982). The results found for these samples were that the maximum magnetization was 0.36 μ_B/Ni atom, in agreement with neutron scattering results (Felcher *et al.*, 1980) and with band structure calculations (Jahlborg and Freeman, 1980). Evidence for Ni clustering within the layers was also found.

Polarized neutron scattering (Felcher *et al.*, 1980) has been used to determine the average Ni moment as well as the moment fluctuations about the average for a sample of composition modulation wavelength 22 Å and average composition $Ni_{0.4}Cu_{0.6}$. The maximum low-temperature magnetization found was approximately 0.3 μ_B/Ni atom, in agreement with dc measurements (Gyorgy *et al.*, 1980, 1982; Zheng *et al.*, 1981a, 1982) and with band structure calculations (Jahlborg and Freeman, 1980). From the size of the moment fluctuations, it was concluded that the Ni in this sample tended to form clusters. The size of the clusters varied from about 50 Ni atoms in the Ni-rich regions of the sample to about 20 Ni atoms in the Cu-rich regions.

Work by the Northwestern group on Cu–Ni has also used a torque magnetometer and a superconducting quantum interference device (SQUID) susceptometer in addition to the FMR technique (Flevaris *et al.*, 1982b). They found agreement among all three techniques at room temperature. For one sample consisting of alternating three atomic planes of Ni-rich layers with six atomic planes of Cu-rich layers, they found a magnetization that increased linearly as the temperature was reduced. This was interpreted as evidence for two-dimensional ferromagnetism. However, given the considerable structural data on similarly prepared films, which show large amounts of interdiffusion and clustering (Gyorgy *et al.*, 1980, 1982; Zheng *et al.*, 1981a, 1982; Felcher *et al.*, 1980), further work is clearly needed before such an interpretation can be considered conclusive for this system.

A basic problem in interpreting static magnetization measurements on these materials is that it is not known how the composition modulation might affect some of the average properties. In a model calculation of the effect of composition modulation on the Curie temperature T_c of a modulated sample, Jones (1982) found that Arrot plots led to T_c's that were essentially (within 4%) those corresponding to the maximum T_c of the sample (T_c^{max}), rather than to some average T_c. He found that only the upper 40% of the magnetic composition, i.e., that portion of each layer with the largest magnetic concentration, contributed to T_c^{max}. This is a very useful result for interpreting static magnetization measurements on these materials.

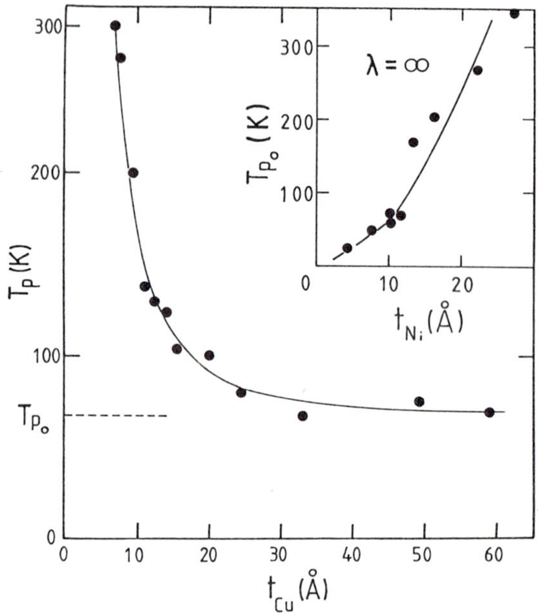

Fig. 17. Temperature of the peak in the ac susceptibility T_p versus thickness of the Cu layer for a series of Ni–Cu samples. [From Zhou *et al.* (1981). Copyright 1981 North-Holland Publishing Co. Amsterdam.]

In measurements of the low-field ac susceptibility on a number of Cu–Ni samples of differing wavelengths, Zhou *et al.* (1981) found a peak in the susceptibility as a function of temperature. The dependence of the magnetic transition temperature (as they defined it by the peak in the susceptibility) on the wavelength of the composition modulation is shown in Fig. 17. In this experiment, Zhou *et al.* kept the average thickness of a Ni layer constant at 10 Å and varied the Cu thickness to alter the modulation wavelength. Their conclusion was that they could not account for the dependence of the peak in the temperature dependence of the ac susceptibility on modulation wavelength if the magnetization density of the Ni planes were uniform. For this reason they feel that the indirect RKKY exchange between Ni layers must be responsible. They also conclude that there is not an extensive (>4-Å) magnetically dead Ni region in these samples. However, the issue of possible clustering in these samples as well as possible shorts connecting adjacent Ni layers and how this would affect the results was not directly addressed.

B. Magnetic Properties of Other Materials

While to date Cu–Ni has been the most extensively studied magnetic modulated material, there have been measurements reported on several other systems. Preliminary attempts have been made to alter the number of itinerant d electrons per atoms in Mn, and thereby to alter its magnetic properties, by alternately layering Mn with Ni and Co (Stearns, 1982). The measurements to date indicate that the materials exhibit a wide variety of complicated magnetic behavior. Part of this is probably due to the various crystallographic orientations obtained for different layer thicknesses. This points out the need for better control over the crystal growth of these materials for a number of studies, as well as very thorough sample characterization.

A torque magnetometer was used by the Northwestern University group to determine the magnetization density and first-order uniaxial anisotropy constant for a series of PD–Ni films (Flevaris *et al.*, 1982a). In general the number of Pd and Ni atoms in each layer was not identical for these studies. It was concluded that for samples with very few atoms per layer (<7) the magnetization density per Ni atom became greater than that of pure Ni. This effect was found to be as large as 25% for the thinnest layer samples studied and was attributed to different exchange-enhanced susceptibilities of Pd in the different local environments of the sample. However, when dealing with samples that contain an *average* of 3 or 4 atoms per layer, sample characterization is critical. It should be kept in mind that the standard $\theta-2\theta$ Bragg diffraction with the scattering vector perpendicular to the films is sensitive *only* to structure perpendicular to the film. Structural differences within a layer (such as clustering) cannot be detected in this way.

Data on the magnetic properties of CMAs incorporating hydrogen have also been reported (Flevaris *et al.*, 1981). Unfortunately, the hydrogen was incorporated at room temperature and quantitative measurements of the amounts remaining after charging the samples were not given, nor were details given of the possible damage the electrolytically incorporated hydrogen caused. Considerable amounts of ac susceptibility and FMR data have been taken on these materials; however, the theories needed to interpret these data contain a number of adjustable parameters. Given that sample characterization is incomplete and given the complex nature of the materials, no clear conclusions can be drawn from these particular measurements. A general comment would be that much more work is needed on "simple" modulated metals before more complex materials can be fruitfully studied.

VI. SUMMARY

The study of electronic and magnetic properties of metallic superlattices is not yet as developed as the study of physical properties of semiconductor superlattices. Much progress has been made in just the past few years, but metallic samples do not yet exhibit the same degree of long-range structural coherence perpendicular to the layers. However, the several metallic systems that have been successfully prepared show a number of interesting physical properties. Transport properties have been shown to be related to elastic properties, and the usefulness of these materials for studying phenomena such as two-dimensional magnetism is now apparent. Judging from the rapid progress made in the past few years in studying electronic and magnetic properties of these materials, the next few years should lead to a number of discoveries.

We have made every attempt to make this chapter a comprehensive review (up to the submission date of Fall 1982) but, as in every rapidly developing field, it is inevitable that some work has escaped our attention. We offer our apologies to our colleagues in this field whose work may not have been covered.

ACKNOWLEDGMENTS

We would like to thank our collaborators who contributed to various aspects of the work described here: I. Banerjee, C. Chun, G. P. Felcher, M. Grimsditch, R. T. Kampwirth, J. B. Ketterson, M. R. Khan, A. Kueny, K. Meyer, A. Rahman, S. K. Sinha, J. Vicent, T. R. Werner, Q. S. Yang, G. G. Zheng, and J. Q. Zheng.

This work was supported by the U.S. Department of Energy and the Office of Naval Research under grant No. N00014-83-F-0031.

REFERENCES

Abrahams, E., Anderson, P. W., Licciardello, D. C., and Ramakrishnan, T. V. (1979). *Phys. Rev. Lett.* **42,** 673.

Altshuller, B. L., and Aronov, A. G. (1979). *Zh. Eksp. Teor. Fiz.* **77,** 2028 [*Sov. Phys.-JETP (Engl. transl.)* **50,** 968 (1979).]

Banerjee, I., Yang, Q. S., Falco, C. M., and Schuller, I. K. (1982). *Solid State Commun.* **41,** 805.

Baral, D., and Hilliard, J. E. (1979). *Appl. Phys. Lett.* **41,** 156.

Barnard, R. D. (1972). "Thermoelectricity in Metals and Alloys," Taylor and Francis, London.

Berry, B. S., and Pritchet, W. C. (1976). *Thin Solid Films* **33,** 19.

Calverly, A., and Rose-Innes, A. C. (1960). *Proc. R. Soc. London* **255A,** 267.

Chaikin, P. M. (1978). *Ann. N.Y. Acad. Sci.* **128,** 127.

Clow, H. (1962). *Nature* **194,** 1035.

Cook, H. E., and Hilliard, J. E. (1969). *J. Appl. Phys,* **40,** 2191.

Dinklage, J. B. (1967). *J. Appl. Phys.* **38,** 3781.

Dinklage, J., and Frerichs, R. (1963). *J. Appl. Phys.* **34,** 2633.

DuMond, J., and Youtz, J. P. (1940). *J. Appl. Phys.* **11,** 357.

Durbin, S. M., Cunningham, J. E., Mochel, M. F., and Flynn, C. P. (1981). *J. Phys. F* **11,** L223.

Durbin, S. M., Cunningham, J. E., and Flynn, C. P. (1982). *J. Phys. F* **12,** L75.

Dynes, R. C. (1982). *Surf. Sci.* **113,** 510.

Elliott, R. P. (1965). "Constitution of Binary Alloys," 1st supplement, McGraw-Hill, New York.

Falco, C. M., and Schuller, I. K. (1982a). *In* "Novel Materials and Techniques in Condensed Matter" (G. W. Crabtree and P. Vashishta, eds.), p. 21, North-Holland Publ., New York.

Falco, C. M., and Schuller, I. K (1982b). *In* "Proceedings of IV Conference on Superconductivity in d- and f-band Metals" (W. Weber and W. Buckel, eds.), p. 283. Kernsforschungsanlage, Karlsruhe, Federal Republic of Germany.

Farnell, G. W. (1969). *In* "Physical Acoustics" (W. P. Mason and R. N. Thurston, eds.), Vol. 6, p. 109, Academic Press, New York.

Felcher, G. P., Cable, J. W., Zheng, J. Q., Ketterson, J. B., and Hilliard, J. E. (1980). *J. Mag. Mag. Mater.* **21,** L198.

Flevaris, N. K., Baral, D., and Hilliard, J. E. (1981). *In* "Metal Hydrides" (G. Bambakidis, ed.), p. 301, Plenum, New York.

Flevaris, N. K., Ketterson, J. B., and Hilliard, J. E. (1982a). *J. Appl. Phys.* **53,** 2439.

Flevaris, N. K., Ketterson, J. B., and Hilliard, J. E. (1982b). *J. Appl. Phys.* **53,** 8046.

Friedlaender, F. J., Kneller, E., and Silva, I. F. (1965). *IEEE Trans. Magn.* **MAG-1,** 251.

Geerk, J., Gurvitch, M., McWhan, D. B., and Rowell, J. M. (1982). *Physica B + C* **109-110,** 1772.

Gyorgy, E. M., Dillon, J. F., Jr., McWhan, D. B., Rupp, L. W., Jr., Testardi, L. R., and Flanders, P. J. (1980). *Phys. Rev. Lett.* **45,** 57.

Gyorgy, E. M., McWhan, D. B., Dillon, J. F., Jr., Walker, L. R., and Waszczak, J. V. (1982). *Phys. Rev. B* **25,** 6539.

Hamelin, B. (1976). *Nucl. Instrum. Meth.* **135,** 299.

Hansen, M. and Ankerko, K. (1958). "Constitution of Binary Alloys," 2nd ed., McGraw-Hill, New York.

Hertel, G., McWhan, D. B., and Rowell, J. M. (1982). *In* "Proceedings of IV conference on Superconductivity in d- and f- band Metals" (W. Weber, and W. Buckel, eds.) p. 299. Kernsforschungsanlage, Karlsruhe, Federal Republic of Germany.

Hirsch, A. A. (1965). *IEEE Trans. Magn.* **MAG-1,** 254.

Hirsch, A. A., Friedman, N., and Eliezer, Z. (1964). *Physica* **30,** 2314.

Imry, Y. (1981). *J. Appl. Phys.* **52,** 1817.

Ioffe, A. F., and Regel, A. R. (1960). *Prog. Semicond.* **4,** 237.

Itozaki, H. (1981). Ph.D. thesis, Northwestern Univ., Evanston, Illinois.

Jahlborg, T., and Freeman, A. J. (1980). *Phys. Rev. Lett.* **45,** 653.

Jones, W. R. (1982). *J. Appl. Phys.* **53,** 2442.

Kaveh, M., and Mott, N. F. (1981). *J. Phys. C* **14,** L183, L177.

Keyes, R. W. (1967). *Solid State Phys.* **20,** 37.

Khan, M., Chun, C. S. L., Felcher, G., Grimsditch, M., Kueny, A., Falco, C. M., and Schuller, I. K. (1983). *Phys. Rev. B* **27,** 7186.

Kittel, C. (1957). "Introduction to Solid State Physics," 2nd ed., p. 322, Wiley, New York.

Kueny, A., Grimsditch, M., Miyano, K., Banerjee, I., Falco, C. M., and Schuller, I. K. (1982). *Phys. Rev. Lett.* **48,** 166.

Lowe, W. P., Barbee, T. W., Jr., Geballe, T. H., and McWhan, D. B. (1981). *Phys. Rev. B* **24,** 6193.

Lynn, J. W., Knems, J. K., Pasell, L., Saxena, A. M., and Schoenborn, B. P. (1976). *J. Appl. Crystallogr.* **9,** 454.

Mayadas, A. F., and Shatzkes, M. (1970). *Phys. Rev. B* **1,** 1382.

Moffatt, G. (1981). "Handbook of Binary Phase Diagrams," General Electric Co., Schenectady, New York.

Mott, N. F. (1969). *Philos. Mag.* **19,** 835.

Paulson, W. M., and Hilliard, J. E. (1977). *J. Appl. Phys.* **48,** 2117.

Pickett, W. E. (1982). *J. Phys. F* **12,** 2195.

Roll, K., and Reill, W. (1982). *Thin Solid Films* **89,** 221.

Ruggiero, S. (1981). Ph.D. thesis, Stanford Univ., Stanford, California.

Rytov, S. M. (1956). *Akust. Zh.* **2,** 71 [*Sov. Phys.-Acoust. (Engl. transl.)* **2,** 67 (1956).]

Saxena, A. M., and Schoenborn, B. P. (1977a). *Acta. Crystallogr. A* **33,** 805.

Saxena, A. M., and Schoenborn, B. P. (1977b). *Acta. Crystallogr. A* **33,** 813.

Schuller, I. K. (1980). *Phys. Rev. Lett.* **44,** 1597.

Schuller, I. K., and Falco, C. M. (1980). *In* "Inhomogeneous Superconductors–1979" (D. U. Gubser, T. L. Francavilla, J. R. Leibowitz, and S. A. Wolf, eds.), p. 197, American Institute of Physics, New York.

Schuller, I. K., and Falco, C. M. (1982). *Surf. Sci.* **113,** 443.

Schuller, I. K. and Rahman, A. (1983). *Phys. Rev. Lett.* **50,** 1377.

Schuller, I. K., Falco, C. M., Thaler, B., Ketterson, J. B., and Hilliard, J. (1979). *In* "Modulated Structures" (J. M. Cowley, J. B. Cohen, M. B. Salamon, and B. J. Wuensch, eds.), p. 417, American Institute of Physics, New York.

Shunk, F. A. (1969). "Constitution of Binary Alloys," 2nd supplement, McGraw-Hill, New York.

Spiller, E. (1981). *In* "Low Energy X-Ray Diagnostics—1981" (D. T. Attwood and B. L. Henke, eds.), p. 124, American Institute of Physics, New York.

Stearns M. B. (1982). *J. Appl. Phys,* **53,** 2436.

Testardi, L. R., Willens, R. H., Krause, J. T., McWhan, D. B., and Nakaha, S. (1981). *J. Appl. Phys.* **52,** 510.

Thaler, B. J., Ketterson, J. B., and Hilliard, J. E. (1978). *Phys. Rev. Lett.* **41,** 336.

Werner, T. R., Banerjee, I., Yang, Q. S., Falco, C. M., and Schuller, I. K. (1982). *Phys. Rev. B* **26,** 2224.

White, R. M., and Herring, C. (1980). *Phys. Rev. B* **22,** 1465.

Zheng, J. Q., Falco, C. M., Ketterson, J. B., and Schuller, I. K. (1981a). *Appl. Phys. Lett.* **38,** 424.

Zheng, J. Q., Ketterson, J. B., Falco, C. M., and Schuller, I. K. (1981b). *Physica B + C* **108,** 945.

Zheng, J. Q., Ketterson, J. B., Falco, C. M., and Schuller, I. K. (1982). *J. Appl. Phys.* **53,** 3150.

Zhou, W. S., Wong, H. K., Owers-Bradley, J. R., and Halperin, W. P. (1981). *Physica* **108B,** 953.

10 SYNTHETICALLY LAYERED SUPERCONDUCTORS

S. T. Ruggiero*

Section of Applied Physics
Yale University
New Haven, Connecticut

M. R. Beasley

Department of Applied Physics
Stanford University
Stanford, California

I. INTRODUCTION

Among metallic multilayer systems, the study of the effects of layering on the properties of superconductors is perhaps the most advanced at the present time. The reason why is very simple. Layering of a superconductor, on almost any

* Present address: Department of Physics, University of Notre Dame, Notre Dame, Indiana.

ISBN 0-12-170470-X

TABLE I

System	Deposition	Typical periodicities (Å)	Dimensionality	
			Layer	System
Al/Ge	E-beam	60–500[a]	2D	2D
Nb/Al	Sputter	20–300[a]	2D	—
Nb/Cu	Sputter	10–10,000[a]	2D[b]	Quasi-2D[b]
Nb/Ge	Sputter	20–200	2D[b]	Quasi-2D[b]
Nb/Ta	E-beam and sputter	20–250[a]	—	—
Nb/Ti	Sputter	6–6000[a]	2D[b]	Quasi-2D[b]
Nb/Zr	Sputter	—	—	—
Nb_3Ge/Nb_3Ir	Sputter	1500, 3000[a]	3D	3D
Nb_3Sn/Y	E-beam	500–20,000	3D	3D
In/Ag	Thermal evaporation	200[a]	2D	—
Sn/Ag	Thermal evaporation	200[a]	2D	—
Pb/Bi	Thermal evaporation	700–8000	3D	—
Pb/Fe	Thermal evaporation	~500[c]	2D	—

[a] Systems in which component layers are always of equal thickness.
[b] For $D_{Nb} \lesssim \xi_{Nb}(0) < 300$ Å.
[c] Nonperiodic structures.

length scale, can significantly affect its properties. Indeed, the characteristic length scale at which layering begins to dramatically modify the intrinsic physical properties of a superconductor is quite large. This length is the superconducting coherence length $\xi_z(0)$ along the layered direction. For extrinsic properties such as flux pinning the length scale is the magnetic field penetration depth λ, which is generally larger than ξ for most materials of interest.

Superconducting coherence lengths typically range from 50 to 1000 Å, and hence layering on an atomic scale and/or single-crystal superlattice formation is not required to achieve striking effects. Concomitantly, considerably less sophisticated syntheses approaches, for example, sequential sputtering in only modest vacuums, have proved quite effective. Of course, fine-scale, epitaxially layered superconductors are of considerable interest, and here the ultrahigh-vacuum (UHV), molecular-beam-epitaxy (MBE) approach using electron-beam thermal evaporation is highly desirable even if perhaps not absolutely necessary. Indeed, epitaxial, single-crystal, superconducting superlattices were first produced in such a system. Nevertheless, the large coherence lengths typical of superconductors make them much more "forgiving." Superconductivity is unusual among cooperative phenomena in having such a large, and therefore readily accessible, characteristic length scale.

It is important to note that there are many naturally occurring layered

Synthetically Layered Superconductors Studied to Date

Superconducting properties studied	Reference(s)
H_{c2}, T_c	Haywood and Ast, 1978
T_c, tunneling	Geerk *et al.*, 1982, McWhan *et al.*, 1983
H_{c2}, T_c, tunneling	Schuller and Falco, 1979; Schuller, 1980; Lowe *et al.*, 1981; Geerk *et al.*, 1982; Banerjee *et al.*, 1982, 1983; Banerjee, 1982; Yang *et al.*, 1983; Chun *et al.*, 1984
H_{c2}, T_c, paraconductivity	Ruggiero *et al.*, 1980, 1982
T_c, tunneling	Durbin *et al.*, 1981, 1982; Hertel *et al.*, 1982
H_{c2}, T_c	Zheng *et al.*, 1981; Qian *et al.*, 1982
T_c, specific heat	Lowe and Geballe, 1984; Claeson *et al.*, 1984; Broussard *et al.*, 1984
T_c	Schmidt *et al.*, 1980
J_c	Howard *et al.*, 1977
T_c	Granqvist and Claeson, 1979
T_c	Granqvist and Claeson, 1979
J_c	Raffy *et al.*, 1972, 1974; Raffy and Guyon, 1981
T_c, tunneling	Claeson, 1980

superconductors as well. These include the quasi-two-dimensional layered dichalcogenides, their intercalates, and intercalated graphite, for example. Another important system is lamellar eutectic alloys. However, in keeping with the theme of this volume, we restrict our attention to synthetically produced systems. Table I lists most of the synthetically layered superconductors studied to date. Also shown is the method of preparation, typical periodicities, dimensionality, and some of the physical and structural properties that have been investigated. The materials are grouped according to the nature of the primary superconducting component. Only the Nb-based systems have received systematic study as part of the current interest in synthetically layered materials. The other materials were, for the most part, studied for specialized purposes. They are of considerable historical interest, however, and serve to illustrate many of the interesting aspects of synthetically layered superconductors generally.

As is clear from Table I, layering can effect many of the physical properties of a superconductor. Many other possibilities are also imaginable. At the coarsest scale of layering, $\lambda \lesssim S \gg \xi_z(0)$, where S is the periodicity of the system, the individual layers essentially retain their bulk properties both from the point of view of the superconducting order parameter and the microscopic properties of the superconductor. The net effect is simply to introduce spatially coherent planar defects into the material. Such a situation is of interest, for example, in the study of flux pinning and critical currents in superconductors, in which a registration of the vortex lattice and the layer superlattice can occur. Such coherent flux pinning has been demonstrated in the Pb/Bi (Raffy *et al.*, 1972, 1974; Raffy and Guyon, 1981), Nb_3Sn/Y (Howard *et al.*, 1977), and Pb–Bi/Cr (Yetter *et al.*, 1982) systems.

For very fine scale layering, $S \ll \xi_z(0)$, the system will, in general, behave as a bulk anisotropic superconductor, a phenomenon that is most easily observed in the upper critical field phase boundary $H_{c2}(T)$. For $S \lesssim \xi_z(0)$, however, the situation is more interesting and has already received considerable attention. Here the spatial structure of the superconducting pair potential $\Delta(r)$ and the related pair wave function (or order parameter) $\psi(r)$ become directly affected. For example, the critical fields, critical currents, and vortex dynamics should all be dramatically different and not just simply anisotropic. If one of the layering constituents is a nonsuperconductor (e.g., an insulator or a normal metal), it is possible to create quasi-two-dimensional S/I or S/N layered superconductors in which the superconducting layers become weakly coupled. Under appropriate conditions such materials form a new class of type II superconductors in which the vortices are Josephson vortices rather than the usual Abrikosov variety. Such effects are manifest in the Nb/Cu (Banerjee, *et al.*, 1983; Chun *et al.*, 1984) Nb/Ge (Ruggiero *et al.*, 1980, 1982), and Nb/Ti (Qian *et al.*, 1982) systems and have perhaps been most dramatically illustrated by the dimensional-crossover behavior observed in the parallel upper-critical fields of the Nb/Ge composites (Ruggiero *et al.*, 1980, 1982). Also, since under these conditions $\Delta(r)$ becomes periodic, changes in the single-particle excitations of the superconductor are expected as well. For example, the superconducting density of states $N(E)$ and related properties will change if the material has a sufficiently long electron mean free path (van Gelder, 1969; Gallagher, 1982; Arnold *et al.*, 1982).

Layering on the scale $S \approx \xi_z(0)$ does not as a rule fundamentally change the superconductivity at a microscopic level, however. That is, the microscopic origins and mechanisms of superconductivity are not necessarily changed. All else being equal, the intrinsic T_c and superconducting material parameters of the individual layers retain their "bulk" values. The behavior of the layered system as a whole is different, of course, but can be understood on the basis of the proximity effect, usually in the context of an equivalent bilayer system with appropriate boundary conditions.

When $S < \xi_z(0)$, the layering becomes fine on the scale of the Cooper pairs themselves, and superconductivity can in principle be modified at the microscopic level. The simple Cooper limit of the proximity effect may or may not be appropriate. Of course, any changes in the microscopic normal-state properties of the material that play a role in superconductivity will also affect its superconducting properties. The intrinsic transition temperature of the layers might be raised (or lowered), for example. Possibilities here include changes in the phonons, electronic band structure and Fermi surface parameters, damage or disorder, interface phenomena, cooperative electronic effects between layers, and stress-induced changes due to epitaxial coherency strains. The stabilization of entirely new phases is an even more dramatic example, not to mention possible exitonic interactions or other novel effects. Indeed, it is perhaps in these latter areas that many of the most unexpected results may turn out to be found.

Along these lines nonequilibrium phases have been obtained in the Nb/Zr (Lowe and Geballe, 1984; Claeson *et al.*, 1984) and possibly the Nb/Ti (Qian *et al.*, 1982) systems. Single-crystal superlattices have been grown in the Nb/Ta (Durbin *et al.*, 1981, 1982; Hertel *et al.*, 1982) system and partially coherent interfaces are observed in Nb/Al (Geerk *et al.*, 1982; McWhan *et al.*, 1983) multilayers. Both of these systems have even been studied via electron tunneling (Hertel *et al.*, 1982; Geerk *et al.*, 1982).

These initial investigations, along with the critical field and critical current studies already mentioned, demonstrate the interesting potential of superconducting multilayers and, perhaps more important at this stage, are establishing the degree of perfection in layering that is possible. In this chapter we review this early work with the dual objectives of demonstrating what has been done and what may be possible in the future. The detailed fabrication of layered superconductors will not be discussed per se since it is covered in the more general articles on metallic layering contained in this volume. We will, however, mention the approaches used in the preparation of the specific systems discussed. Two earlier reviews of artificially layered superconductors are also of interest (Beasley, 1979; Schuller and Falco, 1982).

Our approach is as follows. In Section II we discuss in more specific terms the classification of the layered superconducting systems of interest, making more precise the regimes and distinctions just mentioned, and further explaining the categories shown in Table I. In Section III we review the phenomenological Ginzburg–Landau-like theories of layered superconductors. Such theories are most effective for understanding the effects of layering on the pair potential and the related pair wave function. In Section IV we discuss the effects of layering on the single-particle excitations of the superconductor. In Section V we discuss briefly the various factors that may influence the transition temperatures of layered superconductors. Finally in Section VI we review the experimental situation including a discussion of the material systems that have received the most serious study.

II. CLASSIFICATION OF SYNTHETICALLY LAYERED SUPERCONDUCTORS

There are three basic types of superconducting multilayers, depending on the nature of the materials being layered. They are S/S′ systems, in which both types of layers are intrinsically superconductors; S/N systems, in which one of the materials is a normal metal, or at least sufficiently conducting so as to have a finite conductivity at low temperatures (e.g., a degenerately doped semiconductor); and S/I systems, in which one of the materials is basically insulating and through which the superconducting layers are coupled via tunneling. Such S/I systems have an obvious similarity to the naturally occurring layered compounds and their intercalates. Layered superconductors should also be distinguished with regard to the regularity of their layering, the degree to which their interfaces are structurally coherent (i.e., whether epitaxy has been achieved or not), and the degree to which their interfaces are compositionally sharp. When intermixing or damage is present at the interfaces, the real structure may be more complicated. For example, under these conditions an S/S′ and/or an S/N system may become S/S″/S′ or S/S′/N, respectively.

Layered superconductors can also be usefully distinguished according to whether the superconducting layers are thick or thin compared with their intrinsic (i.e., intralayer) temperature-dependent Ginzburg–Landau coherence length $\xi(T)$. If the superconducting layer thicknesses are such that $d_s > \xi(T)$ [in which case $\xi_z(T) \to \xi(T)$], then the layers themselves are three-dimensional and exhibit the properties of the corresponding bulk material. If $d_s < \xi(T)$, then the individual layers can be thought of as two-dimensional from the point of view of spatial variations of the superconducting pair wave function, even if they are not sufficiently thin to be considered two-dimensional from the point of view of the wave function of the individual electrons. If $d_s < \xi(T)$ and the material is of the S/N or S/I variety, it can be thought of as a quasi-two-dimensional superconductor in which the layers are weakly coupled, as already mentioned. In this case the interlayer coherence length $\xi_z(0)$ is strongly dependent on the coupling, and the relative magnitude of this coupling will determine the dimensional character of the system as a whole.

In Table I the systems studied to date are classified along these lines to the degree that sufficient information is available.

III. EFFECTS OF LAYERING ON THE PAIR WAVE FUNCTION—PHENOMENOLOGICAL THEORIES OF LAYERED SUPERCONDUCTORS

As noted in the introduction, layering of a superconductor can have a dramatic effect on its physical properties. Moreover, these changes can be quite independent of any changes in the basic superconducting properties of the individual layers themselves and reflect solely the introduction of periodicity into the system. To date the consequences of such periodicity have been investigated mostly with regard to their effect on the pair wave function ψ and those physical properties of a superconductor that depend on ψ (e.g., critical fields, critical currents, vortex phenomena, etc.). Since these properties do not as a rule reflect directly on the microscopic aspects of superconductivity, they are most simply understood on the basis of the Ginzburg–Landau (GL) theory of superconductivity and its generalizations. [For general discussions of the GL theory, see Tinkham (1975) and Werthamer (1969).] Here the microscopic properties of the superconductor enter only through the parameters of the phenomenological theories. [For a careful compilation of these relationships, see Orlando *et al.* (1979).] At the same time, of course, determination of the phenomenological parameters associated with such theories from experiment, particularly as the layering parameters of the material are systematically varied, provides considerable insight into the effects of layering on the microscopic material properties of the system.

Because of their relative simplicity and because they have received considerable attention in the literature, in this section we review the phenomenological theories of layered superconductors at some length. We begin with the GL theory and then discuss its applications and generalizations to layered superconductors.

A. The Ginzburg–Landau Theory

Using standard notation, the Ginzburg–Landau (GL) free energy of a superconductor in the presence of a magnetic field can be written as

$$F = \int (f_s - f_n)\, dv$$
$$= \int \left[\alpha|\psi|^2 + \frac{\beta}{2}|\psi|^4 + \frac{1}{2m^*}\left|\left(\frac{\hbar}{i}\nabla - \frac{e^*}{c}\mathbf{A}\right)\psi\right|^2 + \frac{1}{8\pi}B^2 \right] dv, \tag{1}$$

where $\alpha = \alpha'(T - T_c)$ and $m^* = 2m$, $e^* = 2e$, and upon variation with respect to ψ and $\mathbf{A}$, one obtains the familiar GL equations

$$\alpha\psi + \beta|\psi|^2\psi + \frac{1}{2m^*}\left(\frac{\hbar}{i}\nabla - \frac{e^*}{c}\mathbf{A}\right)^2\psi = 0 \tag{2}$$

and

$$\mathbf{J}_s = \frac{\hbar e^*}{2m^* i}(\psi^* \nabla\psi - \psi\nabla\psi^*) - \frac{e^{*2}}{m^* c}|\psi|^2\mathbf{A}. \tag{3}$$

The GL theory can also be generalized to an anisotropic form, in which case the mass m is replaced by a GL effective mass tensor

$$\frac{\mathbf{1}}{\mathbf{m}} \equiv \begin{bmatrix} 1/m & 0 & 0 \\ 0 & 1/m & 0 \\ 0 & 0 & 1/M \end{bmatrix}, \tag{4}$$

where the form given is appropriate for a layered system with the coordinate system shown in Fig. 1 and where following convention we use m not m^* for the mass tensor.

Associated with this anisotropic mass tensor are the anisotropic GL coherence length and magnetic field penetration depth

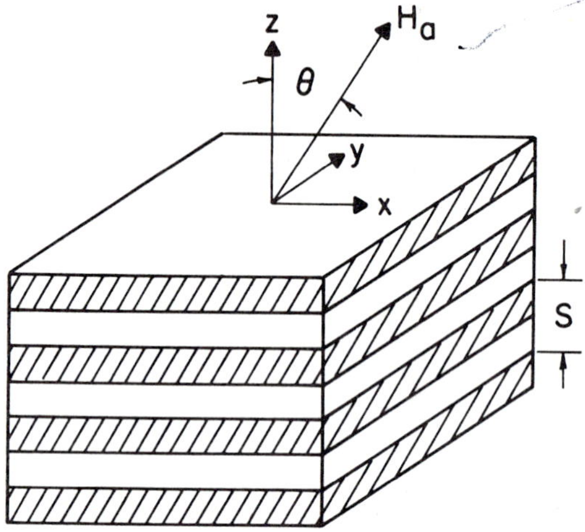

Fig. 1. Geometry of a layered superconductor in the presence of an applied magnetic field H_a. The layer periodicity S (variously denoted λ, Λ, etc.) is also indicated.

$$\xi^2(T) \equiv \frac{\hbar^2}{2|\alpha|}\frac{\mathbf{1}}{m^*} = \begin{bmatrix} \xi^2(T) & 0 & 0 \\ 0 & \xi^2(T) & 0 \\ 0 & 0 & \xi_z^2(T) \end{bmatrix} \tag{5}$$

and

$$\frac{\mathbf{1}}{\lambda^2(T)} = \frac{4\pi e^{*2}|\psi|^2}{c^2}\frac{\mathbf{1}}{m^*} = \begin{bmatrix} 1/\lambda^2(T) & 0 & 0 \\ 0 & 1/\lambda^2(T) & 0 \\ 0 & 0 & 1/\lambda_z^2(T) \end{bmatrix}. \tag{6}$$

Clearly, from Eqs. (5) and (6)

$$\frac{\xi_z(T)}{\xi(T)} = \frac{\lambda(T)}{\lambda_z(T)} = \sqrt{\frac{m}{M}}. \tag{7}$$

It is important to stress that since superconductors undergo second-order phase transitions, $\psi \to 0$ and the superconducting coherence length diverges as $T \to T_c$. Hence $|\psi|^2$ and $|\nabla\psi|^2$ are vanishingly small in this limit, and the Taylor series expansion underlying the GL theory is always valid sufficiently near T_c. For layered superconductors these results have the important implication that as $T \to T_c$ and $\xi_z(T) \to \infty$, the superconductor inherently averages over many layers and (neglecting possible critical phenomena) the anisotropic GL theory should apply. Moreover it is in this limit that the interlayer coherence length $\xi_z(T)$ can always be defined.

It follows from the preceding argument that near T_c all layered superconductors should exhibit simple, anisotropic superconductivity describable by an effective mass tensor [Eq. (4)]. Away from T_c, where $\xi_z(T)$ decreases, it is possible that $\xi_z(T)$ becomes comparable to or even smaller than the layer period S. Under these circumstances the long wavelength approximation of the anisotropic GL theory breaks down and more novel behavior is possible. Provided that this breakdown does not arise too far from T_c, the GL theory still applies in spirit but must be modified to take the layered nature of the material explicitly into account. Even far from T_c where such GL-like theories are not quantitatively justified, generally they are found to predict the correct qualitative behavior.

B. Generalized GL Theories of Layered Superconductors

In generalizing the GL theory to layered superconductors there have been two basic approaches, depending on the type of material being considered. In the first, appropriate for S/S′ and S/N systems, the theory is assumed to apply in each layer separately (with different GL parameters) and the order parameters in each layer are connected via appropriate boundary conditions at the interfaces. In the S/S′ case the two superconductors may have different T_c's and/or differ in other relevant properties such as electron mean free path, which would lead to differences in the GL coherence lengths. This basic approach has been used by Ami and Maki (1975) and Dobrosavljevic (1973).

In the second approach, appropriate for quasi-two-dimensional layered superconductors, the superconductor is modeled as a stacked array of 2D superconductors coupled via the Josephson effect. Kats (1969, 1970) and then Lawrence and Doniach (1970) first investigated such a theory as a model for the intercalated layered compounds. Important extensions of this model have been developed by Bulaevskii (1973), Bulaevskii and Guseinov (1974), and Klemm *et al.* (1974, 1975). Additional work has been reported by Deutscher and Entin-Wohlman (1977). In the model, the GL free energy becomes

$$F = \sum_l \int \Big[\alpha|\psi_l|^2 + \frac{\beta}{2}|\psi_l|^4 + \frac{1}{2m^*}\Big|\Big(\frac{\hbar}{i}\nabla - \frac{e^*}{c}\mathbf{A}\Big)_{xy}\psi_l\Big|^2 + \eta|\psi_l - \psi_{l-1}|^2 \Big]\, dx\, dy, \tag{8}$$

where ψ_l is the pair wave function in layer l, the Laplacian acts in the xy (layer) plane only, and the discrete-difference term represents the Josephson coupling. The connection to the Josephson effect can be seen more explicitly by writing $\psi_l = |\psi_l| \exp(i\phi_l)$, in which case

$$\begin{aligned} \eta|\psi_l - \psi_{l-1}|^2 &= \eta|\psi_l|^2\,[1 - \cos(\phi_l - \phi_{l-1})] \\ &= E_J[1 - \cos(\phi_l - \phi_{l-1})] \end{aligned} \tag{9}$$

which is of the Josephson form, where we have taken $|\psi_l| = |\psi_{l-1}|$ for simplicity.

Variation of this free energy leads to the so-called Lawrence–Doniach (LD) equations originally cast in the form

$$\alpha\psi_l + \beta|\psi_l|^2\psi_l + \frac{1}{2m^*}\left(\frac{\hbar}{i}\nabla - \frac{e^*}{c}A_{xy}\right)^2\psi_l - \eta\left(\psi_{l+1} - 2\psi_l + \psi_{l-1}\right) = 0 \quad (10a)$$

and

$$J_{l+1,l} = \frac{2e\eta S}{i\hbar}(\psi_l^*\psi_{l+1} - \psi_{l+1}^*\psi_l). \quad (10b)$$

As expected, and as shown explicitly by Lawrence and Doniach, taking the long-wavelength approximation of these equations, i.e., assuming that ψ_l varies slowly on the scale of the layer repeat distances S, one obtains the anisotropic GL theory, where

$$M^* = 2M = \hbar^2/2S^2\eta. \quad (11)$$

C. Critical Fields of Layered Superconductors

Let us now turn to the predictions of the various phenomenological theories for the upper critical field H_{c2} of layered superconductors. It is this property of layered superconducting systems that has received the most attention to date. The reasons why are that H_{c2} is quite sensitive to layering, that it is relatively easy to calculate and to measure, and that it provides considerable physical insight into the nature of the superconducting vortices in these materials. In short, it provides an overall qualitative picture of the nature of the type II superconductivity in these materials.

1. Anisotropic GL Theory

As discussed in Subsection III.A, all layered superconductors are expected to exhibit anisotropic GL behavior sufficiently close to T_c. Hence it represents an important limiting behavior. Also, as we shall see, it provides a means of determining the coherence lengths and GL effective mass ratio of the material.

Physically, $H_{c2}(T)$ corresponds to the field at which superconductivity is destroyed via a second-order phase transition. Theoretically it is given by determining the temperature at which, in an applied field H, the linearized GL equation has a nontrivial solution. Mathematically this corresponds to finding the lowest eigenvalue of the linearized GL equation, where one identifies α with the eigenvalue. Thus for an isotropic superconductor one has the equation

$$(\hbar^2/2m^*)[-(d^2/dx^2) + (e^*Hx/c)^2]\,\psi = \epsilon(H)\psi, \quad (12)$$

where $H_{c2}(T)$ is given by the implicit relation

$$\epsilon(H) = \alpha(T) = [\hbar^2/2m^*\xi^2(0)][(T - T_c)/T_c]. \quad (13)$$

Equation (12) is equivalent to that of a free electron in a magnetic field H. Hence it reduces to the harmonic oscillator equation associated with Landau levels in the free-electron model. Correspondingly,

$$\epsilon(H) = \hbar e^* H/2m^* c, \tag{14}$$

and it follows that

$$H_{c2} = [\Phi_0/2\pi\xi^2(0)][(T_c - T)/T_c], \tag{15}$$

where $\Phi_0 = hc/e^*$ is the superconducting flux quantum. At H_{c2} the density of vortices $n_\phi \simeq \xi^{-2}(T)$, which leads to the simple, attractive interpretation of H_{c2} as that field at which the superconductor essentially becomes filled with the normal cores of the vortices.

Obviously, for an anisotropic superconductor the situation is essentially the same except that one gets the equation for an anisotropic harmonic oscillator. As originally shown by Tilley (1965), in this case

$$H_{c2}(T, \theta) = H_{c2\perp}(T)/[(m/M)\sin^2\theta + \cos^2\theta]^{1/2}, \tag{16}$$

where

$$H_{c2\perp}(T) = [\Phi_0/2\pi\xi^2(0)][(T_c - T)/T_c] \tag{17}$$

and θ is the angle of the applied field with respect to the z axis (see Fig. 1).

It follows from Eq. (16) that

$$H_{c2\parallel}(T) = [\Phi_0/2\pi\xi(0)\xi_z(0)][(T_c - T)/T_c] \tag{18}$$

from which it is evident that measurement of $dH_{c2\parallel}/dT|_{T_c}$ and $dH_{c2\perp}/dT|_{T_c}$ in the limit as $T \to T_c$ allows determination of $\xi(0)$, $\xi_z(0)$, and the mass ratio m/M. Moreover, it is also evident that, in general, a linear dependence of H_{c2} on T is expected as $T \to T_c$.

By using the anisotropic GL theory it is also possible to determine the structure of the vortices in an anisotropic type II superconductor. When the applied field is along one of the principal axes, the situation is essentially like that in an isotropic system but with an ellipsoidally shaped normal core and circulating currents. The situation is illustrated in Fig. 2. When the field is in an arbitrary direction, the vortex structure is more subtle, as discussed by Kogan (1981) and Kogan and Clem (1981). In this case, for example, the circulating currents of the vortex do not lie in a plane perpendicular to the vortex axis. For a layered system such behavior has the reasonable physical interpretation that the circulating currents prefer to flow within the plane of the layers even as the applied field is tipped away from the z axis. It also has the implication that the magnetic induction B is not along the applied field H_a as is not uncommon in anisotropic magnetic materials generally. The superconducting case is unusual, however, in that the local magnetic field has a vortex structure.

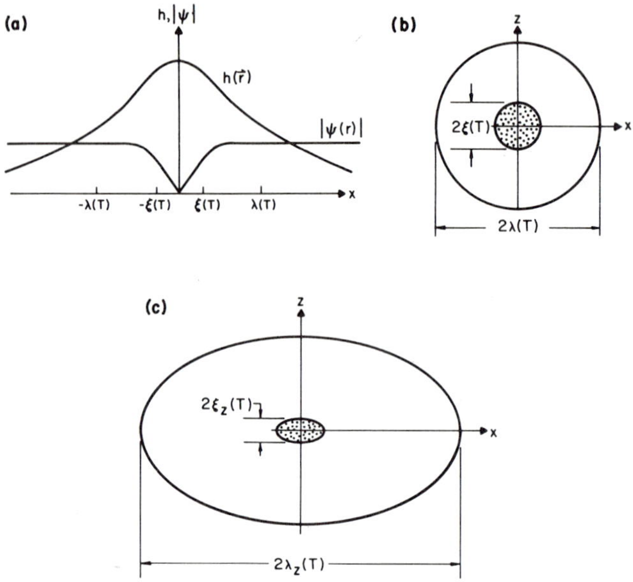

Fig. 2. (a) Profiles of the magnetic field and superconducting order parameter for an isolated vortex. Also shown are the structure of vortices in the (b) isotropic ($M = m$) and (c) anisotropic ($M = 4m$) cases, including the extent of the (darkened) vortex core.

2. Lawrence–Doniach theory

For quasi-two-dimensional layered superconductors the situation is more interesting since as the layers are progressively decoupled one expects some kind of dimensional crossover behavior. The critical fields for this case were first calculated by Klemm *et al.* (1974, 1975) and by Bulaevskii (1973) and Bulaevskii and Guseinov (1974) using the LD theory. This theory is expected to be quantitatively correct in S/I systems in which the coupling coefficient η is field independent but perhaps only qualitatively correct for S/N systems in which η may be field dependent.

Generalizing the phase factors of the Josephson coupling term in Eq. (10a) to include the magnetic field, the linearized LD equation becomes

$$\left\{\frac{\hbar^2}{2m^*}\left[-\frac{d^2}{dx^2}+\left(\frac{2eHx}{\hbar c}\cos\theta\right)^2\right] + \frac{\hbar^2}{M^*S^2}\left[1-\cos\left(\frac{2eHxS}{\hbar c}\sin\theta\right)\right]\right\}\psi = -\alpha\psi, \qquad (19)$$

where θ is the angle of the field H with respect to the z axis.

For $\theta = 0$ (i.e., for H perpendicular to the layers), the second term in Eq. (19) drops out, and we reduce once again to the harmonic oscillator equation.

Therefore $H_{c2\perp}$ is given by Eq. (17), and hence for H perpendicular to the layers, a quasi-two-dimensional superconductor behaves conventionally.

For parallel magnetic fields Eq. (19) reduces to the Mathieu equation

$$\left[-\frac{\hbar^2}{2m^*}\frac{d^2}{dx^2}+\frac{\hbar^2}{M^*S^2}\left(1-\cos\frac{2eHxS}{\hbar c}\right)\right]\psi=-\alpha\psi, \tag{20}$$

which is the Schrödinger equation for electron motion in a potential with both harmonic and periodic components. For small magnetic fields the harmonic oscillator equation is obtained once again, and hence the effective-mass result is immediately recovered as expected. At higher relative field strengths, however, we explore solutions representing systematically weaker layer coupling.

Here the eigenvalue problem corresponds to a particle in a periodic potential, the wavelength of which is proportional to H^{-1}. A perturbation expansion for large H reveals that

$$H_{c2\parallel}(T)\cong\frac{\xi(0)\xi_z(0)/S^2}{[1-S^2/2\xi_z^2(T)]^{1/2}}H_{c2\perp}(T). \tag{21}$$

We are led, therefore, to the remarkable conclusion that as the temperature is reduced and $H_{c2\parallel}(T)$ increases, there is a temperature T^* defined by the relation

$$\xi_z(T^*)=(m/M)^{1/2}\,\xi(T)=S/\sqrt{2} \tag{22}$$

at which $H_{c2\parallel}$ diverges. This behavior is illustrated in Fig. 3, in which we summarize the predictions for $H_{c2}(T)$ within the anisotropic GL and LD theories.

This remarkable result can be interpreted as follows. For $T^* < T < T_c$, the interlayer coherence length $\xi_z(T)$ extends over many layers and the bulklike anisotropic GL theory results pertain. Correspondingly $H_{c2}(\theta)$ is given by the effective mass model. For $T < T^*$, where $\xi_z(T) < S/\sqrt{2}$, the normal cores of the vortices can effectively fit between the superconducting layers where they have no pair-breaking effect. More broadly speaking, we surmise that the vortices have gone from being Abrikosov-like to Josephson-like (Likharev, 1975, 1979). The situation is illustrated in Fig. 4, in which we show schematically the structure of the vortices above and below T^*. Since even in the Josephson limit the penetration depth of the vortices still extends over many layers, and hence the vortices continue to interact as in an Abrikosov type II superconductor, it is perhaps physically appealing to think of T^* as a crossover between Abrikosov and Josephson type II superconductivity.

As we discuss in greater detail later, the divergence in $H_{c2\parallel}$ just described is, of course, unphysical. It will be mitigated by any intralayer pair-breaking effect in a magnetic field. Discussions of such effects can be found in Klemm *et al.* (1975) for Pauli paramagnetic limiting (spin pair breaking) and Deutscher and Entin-Wohlman (1978) for finite layer thickness (orbital pair breaking). More

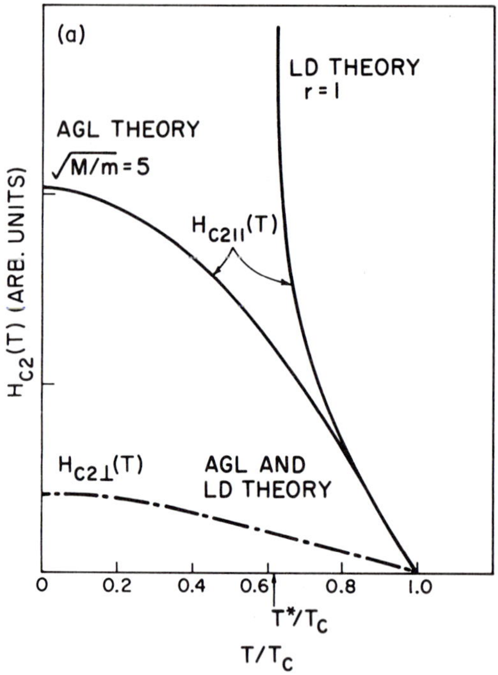

Fig. 3. (a) Comparison of the upper critical fields in the anisotropic (Ginzburg–Landau) and Josephson-coupled (Lawrence–Doniach) theories. The LD theory predicts an upturn in $H_{c2\parallel}(T)$ at a temperature T^* defined by $\xi_z(T^*) = S/\sqrt{2}$. The parameter $r = 4/\pi\,[2\xi_z(0)/S]^2$ is a measure of the layer coupling. (b) Parallel upper critical fields for various interlayer coupling strengths in the KBL (Klemm–Beasley–Luther) theory. Coupling is weaker as $r \to 0$ where r is defined as in (a). [From (Klemm *et al.* (1975).]

realistic critical field curves are shown in Fig. 3b for the Pauli limiting case. Note, however, that the crossover from Abrikosov- to Josephson-like vortices in these systems remains valid since it is solely an interlayer effect.

3. Ginzburg–Landau theory for S/N and S/S′ systems

Dobrosavljević (1973) has considered the critical fields of S/N and S/S′ multilayered superconductors within the GL theory. In the S/N case he assumes that $T_{cS'} = T_{cS}$ but that the mean free path is different in the two materials. By working in the dirty limit and using the de Gennes–Werthamer boundary conditions, when the individual layers are two-dimensional, his results can be written in the suggestive form

$$\left.\frac{dH_{c2\perp}}{dT}\right|_{T=T_{cNS}} = \frac{\Phi_0}{2\pi\xi^2(0)}\frac{1}{T_{cNS}} = k\bar{\gamma}\rho_{x,y}, \tag{23}$$

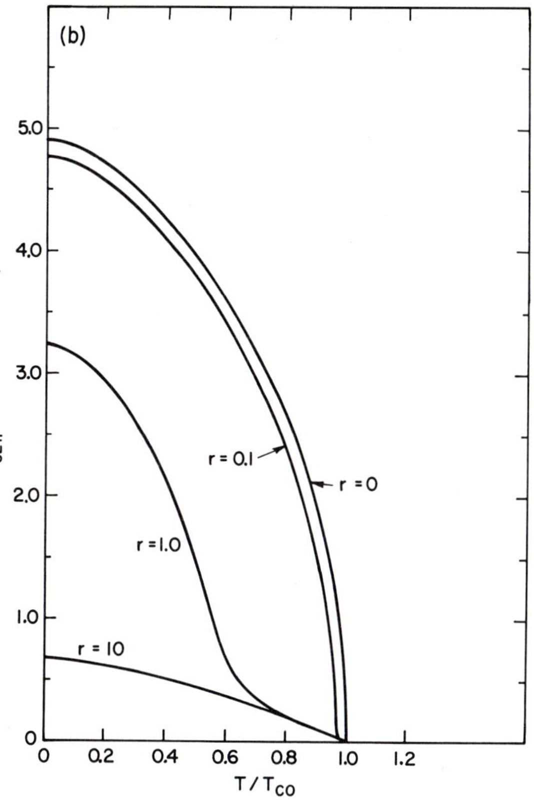

Fig. 3. (*Continued*)

$$\left.\frac{dH_{c2\parallel}}{dT}\right|_{T_{cNS}} = \frac{\Phi_0}{2\pi\xi(0)\xi_z(0)}\frac{1}{T_{cNS}} = k\bar{\gamma}(\rho_z\rho_{x,y})^{1/2}, \tag{24}$$

where

$$\bar{\gamma} = (D_S\gamma_S + D_N\gamma_N)/(D_S + D_N), \tag{25}$$

$$\rho_{x,y} = 1/\sigma_{x,y} = (D_S + D_N)/(D_S\sigma_S + D_N\sigma_N), \tag{26}$$

$$\rho_z = (D_S\rho_S + D_N\rho_N)/(D_S + D_N) \tag{27}$$

are the average electronic heat capacity coefficient, in-plane and z-axis resistivities, respectively, and k is a numerical constant identical to that appropriate to the isotropic case [see Orlando *et al.* (1979)]. From these results it follows that under these assumptions, the mass ratio

$$M/m = \rho_z/\rho_{x,y}. \tag{28}$$

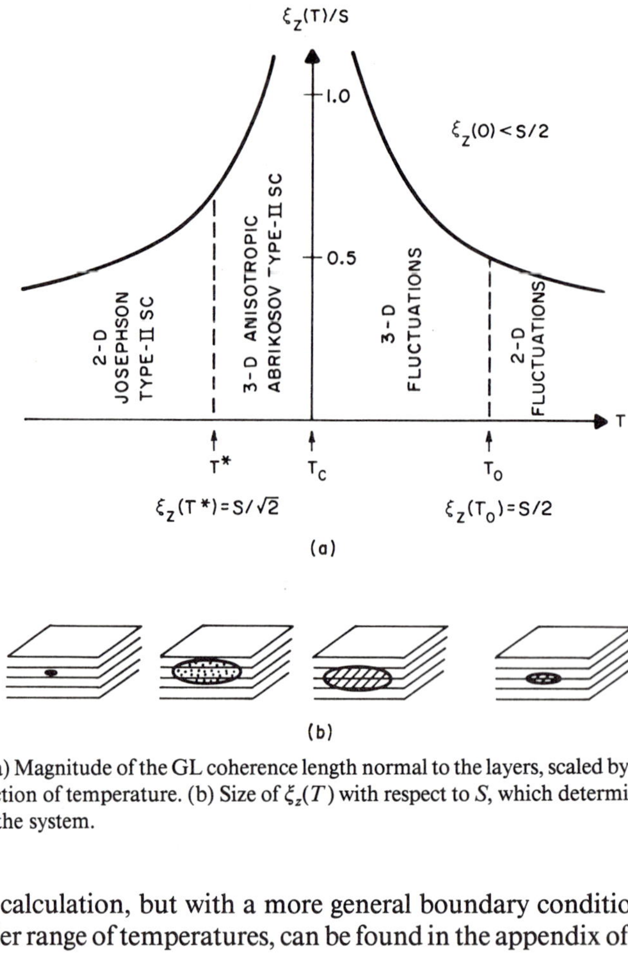

Fig. 4. (a) Magnitude of the GL coherence length normal to the layers, scaled by layer periodicity, as a function of temperature. (b) Size of $\xi_z(T)$ with respect to S, which determines the dimensionality of the system.

A similar calculation, but with a more general boundary condition and valid over a wider range of temperatures, can be found in the appendix of Ruggiero *et al.* (1982).

Dobrosavljević and Kulic (1978) following the earlier work by Ami and Maki (1975) have considered the case of a superconductor in which the electron mean free path (and hence the diffusion constant) is sinusoidally modulated. Calculation of the critical field reduces to solving the GL equations in the case in which the effective mass is periodically modulated. These authors find a kink in $H_{c2\parallel}(T)$ at a temperature T_1 given by the relation

$$\bar{\xi}(T_1) = \sqrt{2}S/2\pi, \tag{29}$$

where S is the period of the variations and $\bar{\xi}$ the GL coherence length of a uniform alloy with the average mean free path.

The authors point out that for $T_1 < T < T_c$ the vortices are centered in regions of minimum scattering (large ξ), whereas for $T < T_1$ they are centered in regions of maximum scattering (small ξ). Upward curvature reminiscent of the Josephson-coupled model is found at $T \simeq T_2$ defined by the relation

$$0.32\bar{\xi}(T_2) = S/2\pi. \tag{30}$$

Characteristic anomalies are also found in the perpendicular critical field and the surface critical field H_{c3}.

D. Microscopic Generalizations

With the exception of the work by Dobrosavljević and Kulić (1978), the calculations described in the preceding discussion have all been carried out in the Ginzburg–Landau limit, that is, close to T_c and in terms of phenomenological parameters. In order to predict the behavior at lower temperatures and to establish the relationship between the phenomenological parameters and microscopic material parameters, it is necessary to use one of the formulations of the microscopic Bardeen–Cooper–Schrieffer (BCS) theory.

Several microscopic treatments of multilayered superconductors have been reported. Klemm *et al.* (1975) have generalized the calculation of H_{c2} in the Josephson-coupled model including intralayer pair breaking due to Pauli paramagnetic limiting. Bulaevskii has reviewed the Soviet literature in this regard in the context of layered compounds (Bulaevskii, 1975). In the dirty limit the result can be expressed concisely in terms of the general theory of pair breaking in superconductors. Specifically, H_{c2} is given by the universal pair-breaking equation

$$\ln(T/T_c) = \psi(\tfrac{1}{2}) - \psi(\tfrac{1}{2} + \rho/2\pi k_B T), \tag{31}$$

where $\psi(x)$ is the digamma function and

$$\rho = \epsilon(H) + [3\tau_{so}(\mu_B H)^2/2\hbar] \tag{32}$$

is the pair-breaking parameter in the limit as $\tau_{so} \to 0$. (The subscript so stands for spin orbit.) Here $\epsilon(H)$ is the pair-breaking parameter associated with the orbital motion of the pairs and is given by the eigenvalue of the GL equation (19). The second term in Eq. (32) is due to Pauli paramagnetic (spin pair breaking) and serves to remove the unphysical divergence found in the LD model. Curves of $H_{c2\|}$ including this intralayer pair breaker are shown in Fig. 3b. Thus the novel critical field behavior of layered superconductors is seen to arise exclusively from the effects of layering on the orbital response of the pairs to an applied field. The orbital pair breaker for simple anisotropic superconductors is given by Eq. (14). Expressions for $\epsilon(H)$ in the S/N and S/S′ cases can be found in the paper by Dobrosavljević and Kulić (1978).

In yet even more general cases, Galaiko and Bezuglyi (1973) have considered the microscopic theory of multilayered superconductors in which the layers differ in their BCS interaction parameter and their normal-state electron densities of state. Nabutovskii and Shapiro (1981) and Nabutovskii (1982) have carried out a similar calculation but including only variations in the BCS interaction parameter. Other work in this area has been reported by Xing and Gong (1981).

E. Fluctuation Effects

As is well known, systems with reduced dimensionality generally show large effects due to fluctuations. Quasi-two-dimensional layered superconductors are no exception, and it is in this context that fluctuations are interesting in multilayered superconductors.

1. Fluctuations above T_c

Fluctuations above a second-order transition in quasi-two-dimensional systems are well understood in principle. They remain an excellent diagnostic to confirm reduced dimensional behavior in a particular system, however. They also can provide quantitative tests of specific theories. The general picture as it applies to multilayered superconductors is shown in Fig. 4. In real space the fluctuations are roughly ellipsoids with diameter $\xi(T)$ in the x, y plane and $\xi_z(T)$ in the z direction. Thus, near T_c where the coherence lengths diverge, the superconductor behaves like a three-dimensional anisotropic system. On the other hand, providing $\xi_z(0) < S/2$ as in the figure, for T well above T_c, $\xi_z(T) < S$, and the fluctuations are confined to a single layer in which case they become two-dimensional. Hence dimensional crossover behavior is expected in the fluctuations at a crossover temperature T_0 defined by the relation

$$\xi_z(T_0) = S/2. \tag{33}$$

Comparing the situation above and below T_c in Fig. 4, we see that the dimensionality of a quasi-two-dimensional superconductor is governed by the characteristic size and shape of the "foreign" phase—superconducting fluctuations in the normal state above T_c and normal vortex cores in the superconducting state below T_c. The asymmetry in the figure results from the precise definition of the coherence length and has no particular physical significance.

The first calculation of the fluctuation behavior of a quasi-two-dimensional superconductor was carried out by Lawrence and Doniach in their original paper, which neglected the anomalous Maki contributions. Using the Josephson-coupled model they obtained for the fluctuation conductivity

$$\sigma_{\parallel}(\tau) = \frac{e^2}{16\hbar\tau S}\frac{1}{[1 + 4\xi_z^2(T)/S^2]^{1/2}}, \tag{34}$$

where $\tau = (T - T_c)/T_c$ is the reduced temperature difference from T_c. The

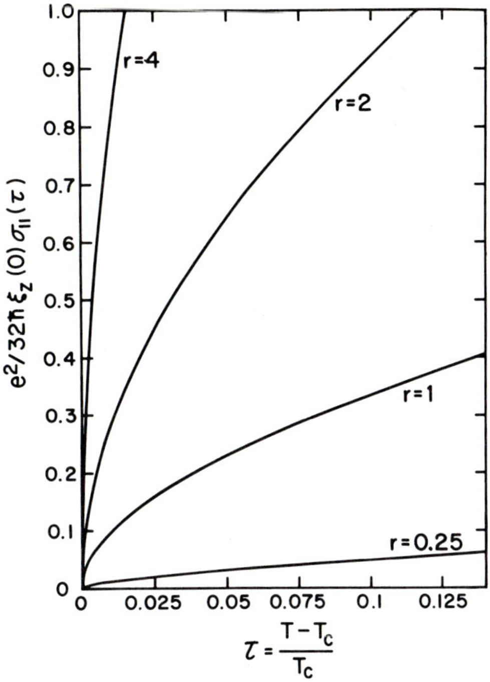

Fig. 5. Temperature dependence of the zero-field fluctuation conductivity of a layered superconductor above T_c as calculated from the LD theory (Lawrence and Doniach, 1970). A more 2D (3D) behavior is evident for smaller (larger) values of *r*.

result is plotted in Fig. 5. Note that Eq. (34) has the simple limits

$$\sigma_{\parallel}(\tau) = \sigma_{2D}(\tau) = \frac{e^2}{16\hbar S}\tau^{-1}, \qquad T_0 \ll T, \tag{35}$$

$$\sigma_{\parallel}(\tau) = \sigma_{3D}(\tau) = \left(\frac{M}{m}\right)^{1/2} \frac{e^2}{32\hbar\xi(0)}\tau^{-1/2}, \qquad T_c \precsim T < T_0, \tag{36}$$

thereby explicitly demonstrating the crossover behavior derived earlier on physical grounds. A treatment of the fluctuation conductivity using the microscopic theory has been presented by Klemm (1974).

Unfortunately, the simple results presented here become much more complicated when the anomalous Maki–Thompson (Maki, 1968a,b; Thompson, 1970) corrections are present. These effects can be directly included in the purely 2D and 3D limits. [For a general review, see Skocpol and Tinkham (1975).] They have also been calculated for particular quasi-two-dimensional superconductors by Aslamasov and Varlamov (1980).

2. Fluctuation effects below T_c

The role of possible fluctuation effects below T_c in quasi-two-dimensional superconductors is much more interesting. The interest arises from the realization that two-dimensional superconductors [i.e., thin films with a thickness $D < \xi(T)$] should exhibit a Kosterlitz–Thouless-type phase transition at a temperature $T_{2D} < T_{c0}$, where now T_{c0} has the meaning of the BCS transition temperature neglecting fluctuations. In this transition vortex–antivortex pairs are produced by thermal fluctuations that are bound for $T < T_{2D}$ but that progressively unbind, leading to finite resistance in the superconductor via flux flow beginning at $T = T_{2D}$.

As shown by Beasley *et al.* (1979) for homogeneous films the Kosterlitz–Thouless transition temperature for a 2D superconductor can be simply estimated

$$T_{2D} = T_{c0} \frac{1}{1 + 0.173\, R_\square / R_c}, \qquad R_\square \ll R_c, \tag{37}$$

where $R_c = \hbar / e^2 = 4.12$ kohm.

Moreover, as first discussed by Huberman and Doniach (1979), and later by Fisher (1979, 1980), in a perpendicular applied field, the usual vortex lattice associated with the mixed state may melt via two successive Kosterlitz–Thouless-type transitions at temperatures T_M and T_H, this time involving dislocation–antidislocation pairs in the 2D vortex lattice. This melting is a special case of the general theory of melting in 2D solids proposed by Halperin and Nelson (1978), Nelson and Halperin (1979), and Young (1979) on the basis of the ideas of Kosterlitz and Thouless (1973). The applicability of this theory has been drawn into question because of suspected first-order transitions at a lower temperature. Nevertheless, T_M remains an upper limit for the melting temperature, and hence some kind of melting of the vortex lattice must take place in a 2D superconductor. As shown by Huberman and Doniach

$$\frac{T_M}{T_{2D}} = \frac{1}{4\pi\sqrt{3}}. \tag{38}$$

Hence $T_M < T_{2D}$ and both should be accessible in films of moderate to large sheet resistances.

The net result of all these theoretical developments is to render the phase diagram of a 2D superconducting film to be quite different from that previously supposed. The revised phase diagram is shown in Fig. 6. The dotted line indicates the phase boundary $H_{C2\perp}(T)$ [Eq. (15)] neglecting the fluctuations.

The importance of these novel fluctuation effects in multilayered, quasi-two-dimensional superconductors remains an open question. Coupling between the layers will quench the two-dimensional fluctuations producing the

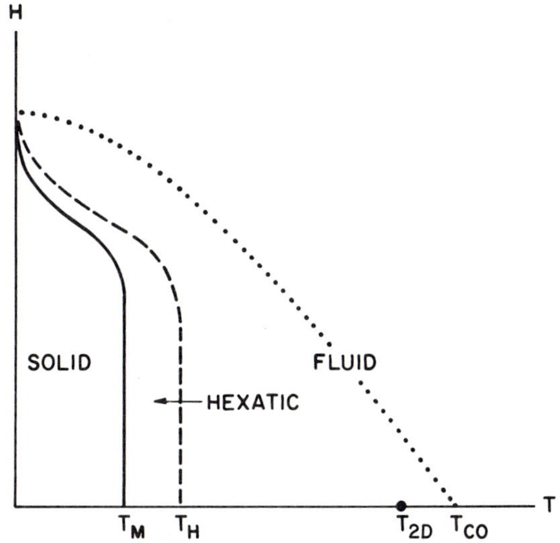

Fig. 6. Phase diagram for the vortex lattice of a thin-film superconductor. Shown are the solid, hexatic, and fluid phases. The zero-field resistive critical temperature is indicated as T_{2D}. The (usual) bulk upper critical field phase boundary, $H_{c2}(T)$ is indicated by the dotted line. [From Fisher (1979).]

changes. The details of the crossover as the layer coupling is varied have not yet been studied quantitatively.

IV. EFFECT OF LAYERING ON THE SINGLE-PARTICLE EXCITATIONS

In Section III we discussed some of the novel consequences of layering on the superconducting pair wave function ψ. Layering can also affect the single-particle excitations (i.e., the "normal" electrons) in a superconductor. These effects have received much less attention so far. The physical origin of such effects is the well-known phenomenon of Andreev reflection (Andreev, 1964) at discontinuities of the pair potential $\Delta(r)$. In Andreev reflection an electronlike single-particle excitation incident on a discontinuity in Δ is reflected as a hole-like excitation. The equation of motion for these quasiparticles is analogous to Schrödinger's equation for a particle in a potential $\Delta(r)$. For example, the pair potential appropriate for an S/N multilayer and the related equivalent SN bilayer are shown schematically in Fig. 7. (Here we have assumed $T_{CN} = 0$, so that $\Delta_N = 0$, and we have taken $d_{N,S} \simeq \xi_{N,S}$.)

Not surprisingly, the energy-level spectrum and the associated density of states of the quasiparticles under these circumstances is closely analogous to those of a particle in a box and the related Kronig–Penney model. The de Gennes–Saint-James bound state (de Gennes and Saint-James, 1963) in SN

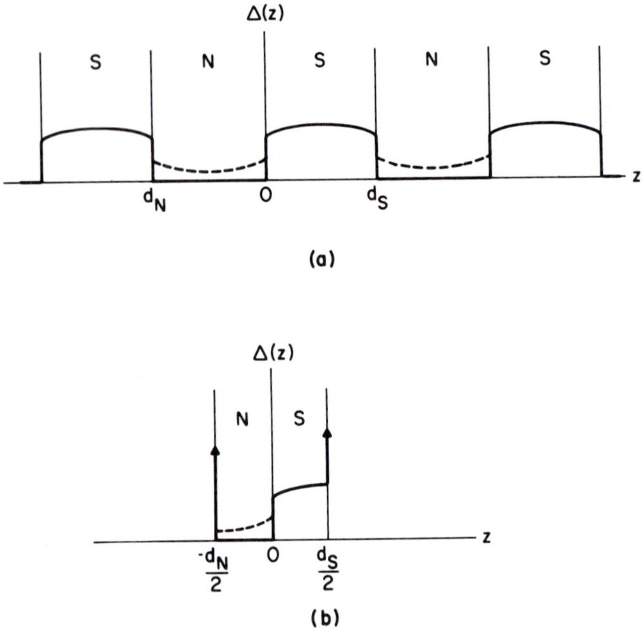

Fig. 7. (a) The variation of the superconducting pair potential of a composite of superconducting films layered with a normal metal with a zero pair potential. The abrupt change of $\Delta(z)$ at the S/N boundary (detailed below) gives rise to quasiparticle reflections or "Andreev scattering." (b) The equivalent bilayer problem.

and SNS layers is a consequence of Andreev reflections, as are the Tomasch (1965, 1969), Shih *et al.* (1981), and Rowell–McMillan (1966) oscillations seen in tunneling experiments on SN bilayers. The energy scale of the structure in the one-dimensional density of states perpendicular to the layers due to Andreev reflection is $\Delta E = \hbar^2k^2/2m - \hbar^2k_F^2/2m = v_F\,\Delta k$, where $\Delta k = k_\perp - k_F$ is fixed by the size of the box. Here $k_\perp$ is the wave vector normal to the interface. Hence for the lowest bound states, $\Delta E = \hbar v_{FN}/2d_N \simeq \Delta_S(\xi_N/d_N)$. For $\Delta E > \Delta_S$ and for $d \ll \xi_N$, the states become a continuum. Clearly such structure is most prominent when $d_N \simeq \xi_N$. In the case of a periodic potential, energy bands occur in the density of states as first pointed out by van Gelder (1969). A more modern (and theoretically complete) treatment of this phenomenon has been developed by Gallagher (1980, 1982), who noted that the periodic case is equivalent to that of a single bilayer (with S and N half as thick as in Fig. 7b), if one assumes perfectly reflecting boundary conditions at the vacuum interfaces (i.e., at $-d_N/2$ and $d_S/2$ in the figure). The evolution of bands from the bound states in N and in the continuum above Δ_S in S is illustrated for a particular case in Fig. 8, taken from the work of Gallagher. For quasi-particles moving parallel to the layers, Andreev reflections are not present and the density of states is unchanged. We also note that for intermediate angles these banding effects will only be seen by tunneling particles with a narrow angular distribution. The effects would therefore tend to give rise to very anisotropic properties.

It is important to stress that such banding effects will only arise in very clean materials, i.e., where the electron mean free path $l_{N,S} \gg \xi_{N,S}$. Energy bands arise from interference of the quasi-particle wave functions, and, hence, any scattering in the bulk or at the interfaces will lead to dephasing of the interfering waves. (Note that some such lifetime broadening has been included in Fig. 8.) Scattering will also tend to make the density of states more isotropic, as in any anisotropic superconductor. Consequently, the observation of such effects has presently been limited to bilayers of such soft-metal systems as Cd, Pb, Sn, and Zn with relatively thick ($\sim 1\ \mu$m) and clean ($l \gtrsim d$) layers. Hence, bound states and band formation in the density of states are to be expected only in very well-made layered structures.

Unfortunately, the physical implications of band formation in the density of states of a layered superconductor have not yet been thoroughly examined. Thus it is not known in detail what further changes in the physical properties of a superconductor might arise. We note, however, that in response to the tunneling measurements on Nb/Ta multilayers by the Bell group (Hertel *et al.*, 1982), Arnold *et al.* (1982) have investigated the tunneling density of states in very thin superconducting multilayers in the Cooper limit including the effects of strong coupling so as to be able to explain the observed phonon structure $[\alpha^2 F(\omega)]$ in the density of states of this system at high energies. Although their theory is tailored to the Nb/Ta multilayers, the results are of general interest.

V. EFFECTS OF LAYERING ON T_c

The mechanisms by which layering can influence the transition temperature of a superconductor fall into several categories.

(1) *The Proximity Effect.* In this well-known effect the nonlocal nature of the pairing phenomenon results in a kind of averaging of the superconductivity on adjacent sides of an interface. [A comprehensive review of this subject is given by Silvert (1975).] It is always present to some degree because of the relatively long coherence lengths typical of superconductivity. In general, a simple juxtaposition of two superconductors, or a superconductor and a normal metal, will lead to an overall T_c intermediate between those of the two materials. The transition temperature can not be raised above that of the material with the higher T_c by this effect alone. The effect is well understood in principle, although calculation in specific cases is not simple. The relevant integral equation is nonlocal and the behavior is sensitive to the specific nature of electron transmission through the interface.

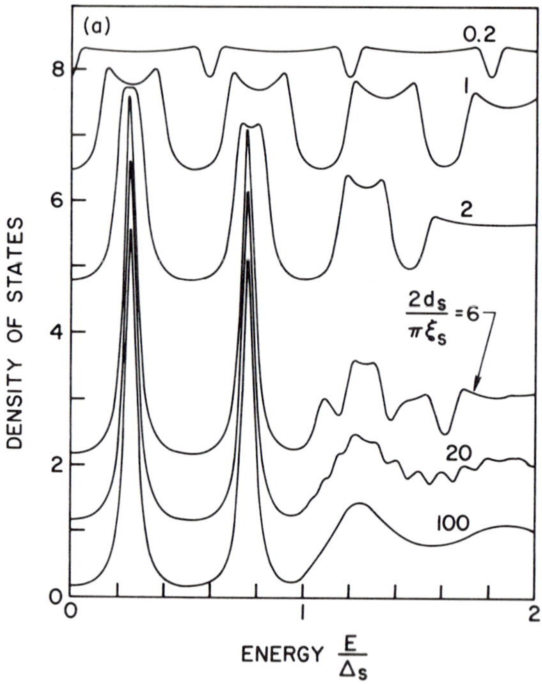

Fig. 8. Local density of states for $\mathbf{K}_{\|}=0$ at the surfaces of (a) a normal metal backed by superconductors of various thicknesses and (b) a superconductor backed by thick and thin normal metals (vertical axes displaced for clarity). Resonant structure is due to multiple quasiparticle reflections. Note strong deviation from BCS behavior in (b). [From Gallagher (1980).]

(2) *Changes in the Electron–Electron Interaction.* If layering changes the basic electronic and/or phonon properties of the layers (for example, through large lattice constant changes due to epitaxial growth), the transition temperature will certainly be affected. Similarly, a high density of interfaces (with localized electronic or phonon states) might introduce additional channels for the pairing interaction. Interlayer electron interactions can apparently influence T_c either up or down (Kulik, 1976). On the other hand, except under special circumstances or unless near-perfect layering on the atomic level can be achieved, it seems unlikely that the introduction of additional periodicity per se into metallic systems will have a sufficiently pronounced effect on the basic electronic structure or phonon modes of the material to affect radically T_c.

(3) *New Materials or Phases.* Formation of fundamentally new materials, as opposed to the juxtaposition of known ones, will certainly be possible by layering and will lead to new superconducting materials. For example, coherent interfaces may stabilize new or otherwise only metastable phases in

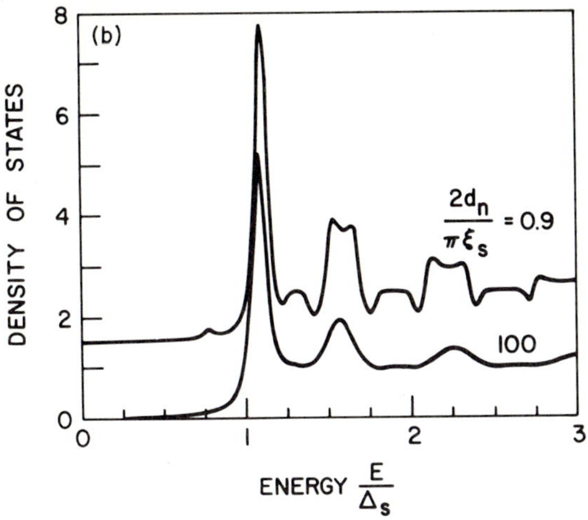

Fig. 8. (*Continued*)

the individual layers. Similarly, the greater importance of surface (i.e., interface) free energy in phase formation may alter the equilibrium phases of the system. The possibility of a deep connection between high-T_c superconductivity and phase instability is an important theme in present-day superconductivity research. The connection of this theme to possibilties in layered superconductors has been discussed by Geballe and Rowell (1982).

(4) *Changes in the Cooper Pair State.* When the individual layers of a multilayer superconductor are very thin ($d \ll \xi$), fundamental changes in the nature of the Cooper pairs themselves are possible. One specific example that has been proposed is a greater propensity for *p*-wave pairing in quasi-two-dimensional layered superconductors. According to Efetov and Larkin (1974) such pairing might occur if there was an attractive interaction between electrons on adjacent layers; *p*-wave superconductivity could arise even if the interaction on a given layer was repulsive. The layered nature of the material enters explicitly by keeping the pairing electrons physically apart, thereby ameliorating the inhibiting effect of the Pauli exclusion principle.

(5) *New Mechanisms.* Entirely new mechanisms of superconductivity are possible, even if improbable. For layered materials the excitonic mechanism is the obvious candidate when one of the layers has a large electronic polarizability. The detailed requirements for a favorable interaction of this sort have been discussed in detail by Allender *et al.* (1973). At a minimum these calculations provide considerable qualitative guidance on how the excitonic mechanism might be implemented in layered superconductors.

TABLE II

System	Deposition substrate[a]	Deposition conditions	Layer structure
Al/Ge	—	$T_D = 77$ K	Granular Al
Nb/Al	Sapphire	$T_D > 30°$C	bcc Nb(110)/fcc Al(111)
Nb/Cu	Sapphire	$T_D > 30°$C	bcc Nb(110)/fcc Cu(111)
Nb/Ge	Sapphire	$T_D > 30°$C	bcc Nb(110)/a-Ge
Nb/Ta	[100] MgO	$T_D = 780°$C	bcc Nb(100)/bcc Ta(100)
Nb/Ta	$[01\bar{2}0]$ sapphire	$T_D = 780°$C	bcc Nb(110)/bcc Ta(110)
Nb/Ta	[0001] sapphire	$T_D = 780°$C	bcc Nb(111)/bcc Ta(111)
Nb/Ta	$[10\bar{1}0]$ sapphire	$T_D = 780°$C	bcc Nb(211)/bcc Ta(211)
Nb/Ta	Sapphire	$T_D > 810°$C	bcc Nb(100)/bcc Ta(100)
Nb/Ta	Sapphire	$T_D \leq 810°$C	bcc Nb(110)/bcc Ta(110)
Nb/Ti	Sapphire	$T_D > 30°$C	bcc Nb(110)/bcc Ti(110)[b]
Nb/Zr	Sapphire	$T_D > 30°$C; $\Lambda < 30$ Å	bcc Nb(110)/bcc Zr(110)
Nb/Zr	Sapphire	$T_D > 30°$C; $\Lambda > 30$ Å	bcc Nb(110)/hcp Zr(002)
Nb_3Ge/Nb_3Ir	Sapphire	$T_D = 1090°$C	A15/A15
Nb_3Sn/Y	Hastelloy	$T_D = 690°$C	A15 Nb_3Sn
In/Ag	Quartz	$T_D \cong 0.5$ K	—
Sn/Ag	Quartz	$T_D \cong 0.5$ K	—
Pb/Bi	Quartz	—	—
Pb/Fe	Quartz	$T_D \cong 0.5$ K	—

[a] All single-crystal except for Hastelloy.
[b] Fractional mixing with hcp(0001) Ti indicated.
[c] For $\Lambda \lesssim 30$ Å systems exhibit electron mean free paths comparable to one lattice constant indicative of strong electron scattering at or within the interfacial region.
[d] Nondiscrete layers. Systems have a sinusoidally modulated Bi concentration.

VI. EXPERIMENTS TO DATE

As can be seen in Table I, a wide variety of layered superconducting systems have been made to date and studied from various points of view. Most have been produced by using advanced sputtering techniques, although electron-beam deposition has been employed in some cases. In this section we review this early work. Unfortunately, few generalizations appear possible at this stage. As a result we simply present a brief summary of the results obtained with

Deposition and Structural Information

Interfacial structure	Type	Reference(s)
—	S/I	Haywood and Ast, 1978
Strong scattering,[c] limited coherence	S/S′/S″	Geerk *et al.*, 1982; McWhan *et al.*, 1983
Strong scattering,[c] incoherence, strong layering	S/S′/N	Schuller and Falco, 1979; Schuller, 1980; Lowe *et al.*, 1981; Geerk *et al.*, 1982; Banerjee *et al.*, 1982, 1983; Banerjee, 1982; Yang *et al.*, 1983; Chun *et al.*, 1984
Disordered Nb at interface	S/S′/I	Ruggiero *et al.*, 1980; 1982
Interdiffusion, coherence, single-crystal growth	S/S′/S″	Durbin *et al.*, 1981, 1982
Interdiffusion, coherence, single-crystal growth	S/S′/S″	Durbin *et al.*, 1981, 1982
Interdiffusion, coherence, single-crystal growth	S/S′/S″	Durbin *et al.*, 1981, 1982
Interdiffusion, coherence, single-crystal growth	S/S′/S″	Durbin *et al.*, 1981, 1982
Interdiffusion, coherence, single-crystal growth	S/S′/S″	Hertel *et al.*, 1982
Limited coherence	S/S′/S″	Hertel *et al.*, 1982
Strong scattering,[c] incoherence, strong layering	S/S′/S″	Zheng *et al.*, 1981; Qian *et al.*, 1982
Interdiffusion, limited coherence	S/S′/S″	Lowe and Geballe, 1984; Claeson *et al.*, 1984; Broussard *et al.*, 1984
Interdiffusion, limited coherence	S/S′/S″	Lowe and Geballe, 1984; Claeson *et al.*, 1984; Broussard *et al.*, 1984
Lattice match without epitaxy	S/S′	Schmidt *et al.*, 1980
Limited interdiffusion	S/N	Howard *et al.*, 1977
—	S/N	Granqvist and Claeson, 1979
—	S/N	Granqvist and Claeson, 1979
Intermixing[d]	S/N	Raffy *et al.*, 1972, 1974; Raffy and Guyon, 1981
—	S/N	Claeson, 1980

various material systems. In each case we describe the specific technique used to produce the layering and the structural information available, particularly as regards the degree of perfection achieved in the layering and at the layer interfaces. In most cases, transport data and systematic T_c measurements are available. In some cases more extensive superconducting measurements (e.g., H_{c2}, fluctuation effects, or tunneling) have also been carried out. Deposition and structural information on these systems has been presented in Table II.

A. Nb/Ge

High-quality layered Nb/Ge composites have been produced and studied by Ruggiero *et al.* (1980, 1982). The materials were prepared by sequential sputter deposition in a standard diffusion-pumped vacuum system (base pressure $< 1 \times 10^{-6}$). The samples were placed on a rotating turntable and alternately passed under physically isolated magnetron sputtering heads containing Nb and Ge targets. The substrates were thermally floating and rose to an estimated temperature of ~80°C during deposition. Individual layer thicknesses D_{Nb} and D_{Ge} ranged from 5 to 100 Å and were held at a fixed ratio for any given run. Total sample thicknesses were typically ~5000 Å. Since the Ge was nonconducting at low temperatures, these multilayers are of the S/I type.

Low-angle x-ray diffractometer scans show well-defined superlattice lines demonstrating a high degree of regularity in the layering of these materials at all

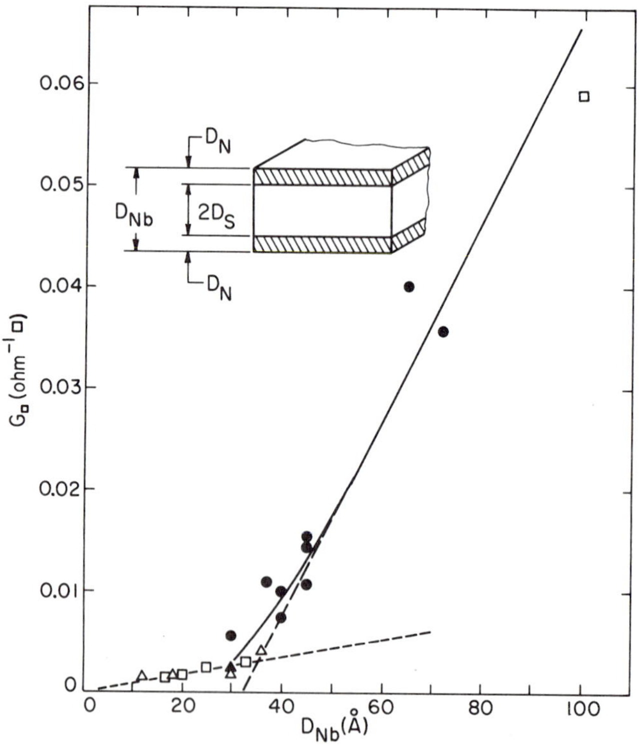

Fig. 9. Low-temperature conductivity per square versus niobium layer thickness in the Nb–Ge system. [From Ruggiero *et al.* (1982).] The data for thicker Nb layers are distinctly offset from the origin by ~30 Å. ▲, △, ●: Nb–Ge composites; □: Ge–Nb–Ge trilayer. The data have been interpreted on the basis of a "two-component" model where it is assumed that each Nb layer is composed of a central region of well-ordered, bulk-like material of 9.8 μohm cm on either side of which is a total of 26 Å of disordered, interfacial material of 114 μohm cm (see inset).

layer thicknesses. The absence of Ge lines indicates that the Ge layers are amorphous. In addition, Read camera photographs show that, independent of D_{Ge}, there is a sharp change in the structure of the Nb layers at $D_{Nb} \simeq 30$ Å. For $D_{Nb} > 30$ Å the Read photos indicate that the individual layers are ordered with a well-defined (110) texture. For $D_{Nb} < 30$ Å, however, the Nb layers appear distinctly disordered.

Similar conclusions can be drawn from the resistivity data for these samples shown in Fig. 9. The behavior of the data is revealing. As noted by Ruggiero *et al.*, G versus D_{Nb} approaches a linear behavior at large D_{Nb}, as expected for a film of uniform resistivity but with a displaced origin. For $D_{Nb} \lessapprox 30$ Å, linear behavior is also seen, but the slope implies a much lower conductivity. Although a depression in conductivity is expected due to boundary scattering in just this region, Ruggiero *et al.* have shown that simple boundary scattering cannot account for the observed behavior. In fact, to first order the data suggest a simple offset in Nb layer thickness of ~30 Å, independent of the Ge layer thickness.

On the basis of these x-ray and transport data Ruggiero *et al.* have adopted a two-component model of the Nb layers as pictured in Fig. 9. They propose that each Nb layer consists of 30 Å of disordered Nb ($\rho_{15\,K} \simeq 114\ \mu$ohm cm) distributed between the two Nb/Ge interfaces (or perhaps localized at one or the other interface) in between which there is $D_{Nb} - 30$ Å of relatively pristine crystalline Nb ($\rho_{15\,K} \simeq 10\ \mu$ohm cm). The origin of this disordered Nb at the interfaces is not known with certainty but may be due to interfacial mixing of the components, lattice mismatch or acute incorporation of residual gas as samples pass from beneath one sputtering head to the other.

The critical temperatures T_c of the Nb/Ge multilayers are observed to depend on both the Nb and Ge layer thicknesses. These data are presented in Fig. 10. The figure shows T_c versus $1/D_{Nb}$, where T_c has been extrapolated to both zero and very thick (i.e., 100-Å) Ge layer thicknesses. In both limits there is a systematic decrease in T_c with decreasing D_{Nb}. In the limit of large Ge thickness the rapid depression in T_c with $1/D_{Nb}$ is in close accord with the T_c's of single, similarly prepared Nb films. Interestingly, the depression of T_c versus $1/D_{Nb}$ in the limit of zero Ge thickness is considerably more gradual. In fact, for fixed D_{Nb}, T_c was always observed to increase as D_{Ge} was reduced. This effect is not understood. Under the assumptions of the two-component model of the individual Nb films, the reduction of T_c with $1/D_{Nb}$ can be understood, however, as resulting from a simple proximity effect between the pristine and damaged interfacial Nb. As shown by the dashed line in the figure, the data in this limit ($D_{Ge} \rightarrow 100$ Å) can be fit by using one adjustable parameter ($\eta = 0.28$) on the basis of the de Gennes–Werthamer theory and measured film properties. As discussed by Ruggiero *et al.* (1982), the model is also consistent with systematic changes observed in the perpendicular upper critical fields of these systems using the same value of the adjustable parameter.

The primary motivation of Ruggiero *et al.* was to study dimensional cross-

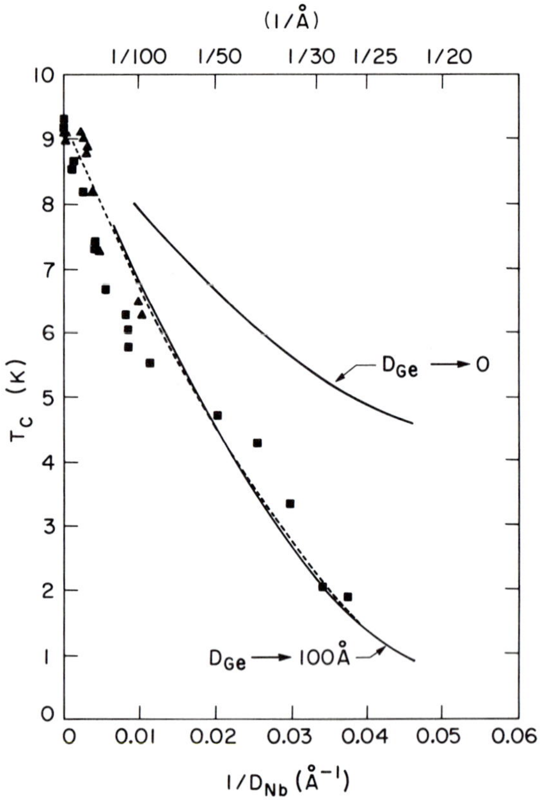

Fig. 10. Solid lines depict T_c versus $1/D_{Nb}$ for layered Nb–Ge composites where the data for a series of samples at fixed D_{Nb} have been extrapolated to zero and 100 Å Ge layer thickness. [From Ruggiero *et al.* (1982).] The solid points (▲,■) are the results for single, sputter-deposited Nb films. Also shown are the proximity effect results for the two-component model (dashed line) as discussed in the text.

over effects in quasi-two-dimensional superconductors. These Nb/Ge multilayers proved highly successful in this regard because the degree of interlayer coupling could be adequately controlled by varying the Ge thickness. As shown by Ruggiero *et al.* these S/I multilayers exhibit 2D, 2D → 3D, and anisotropic 3D behavior in both their fluctuation superconductivity above T_c and in their parallel critical field behavior below T_c.

The parallel and perpendicular upper critical fields for a series of these multilayers of varying Ge thickness is shown in Fig. 11. We first note the high degree of anisotropy for all the materials, which increases with increasing D_{Ge}, as expected. We see that with a relatively large separation between the Nb layers ($D_{Ge} = 50$ Å) the entire $H_{c2\parallel}(T)$ phase boundary exhibits a 2D-like behavior [$H_{c2\parallel}(T) \sim (1 - T/T_c)^{1/2}$] near T_c. This explicitly confirms the premises that

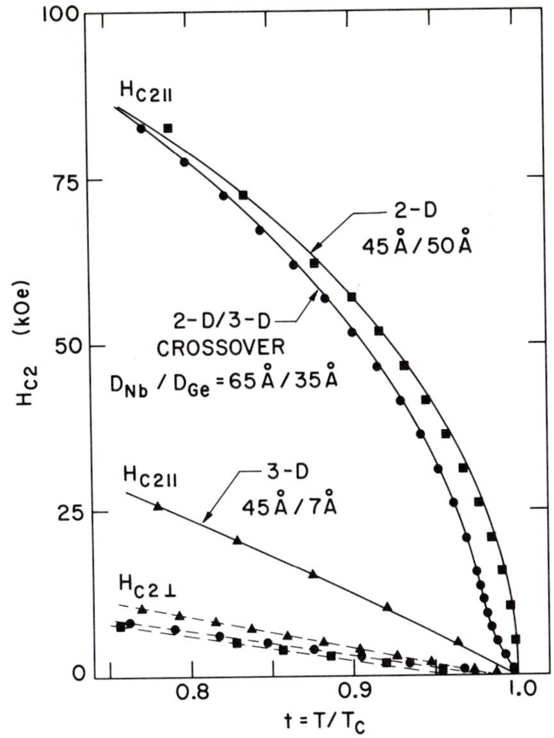

Fig. 11. Upper critical fields near T_c for Nb–Ge composites [from Ruggiero *et al.* (1980)] including a sample exhibiting an upturn in $H_{c2\parallel}(T)$. Systematically increasing the Ge-layer thickness between the 2D Nb layers effects a progression from anisotropic 3D behavior [$H_{c2\parallel}(T) \sim (1 - t)$] to "crossover" behavior and finally to decoupled, 2D behavior [$H_{c2\parallel}(T) \sim (1 - t)^{1/2}$]. Solid lines are from the Josephson-coupling (Klemm–Beasley–Luther) theory.

$\xi(T) > D_{Nb}$ for all temperatures and that the Nb layers are 2D. In this case, the characteristic temperature $T^* \simeq T_c$, demonstrating that the coupling between layers is vanishingly small. In contrast, when the Ge layer thickness is very thin ($D_{Ge} = 7$ Å) the Nb layers couple together strongly and T^* is not defined. More explicitly, $\xi_z(T) > S$ at all temperatures, and 3D behavior is observed as expected. This is reflected by the GL-like behavior [$H_{c2\parallel}(T) \sim (1 - T/T_c)$] near T_c exhibited by this sample. Most important is the data of the sample that exhibited intermediate coupling ($D_{Ge} = 35$ Å). Here T^* lies at an observable temperature (0.98T_c). For temperatures $T > T^*$ a linear 3D behavior is exhibited which crosses over, precisely at T^*, to a 2D dependence [$H_{c2\parallel}(T) \sim (1 - T/T_c)^{1/2}$]. Therefore in this sample all regimes of dimensional behavior can be explored by simply varying temperature. Note that the solid lines in Fig. 11 are theoretical fits to the Josephson-coupled model described in Section III.

It is seen that the overall degree of Josephson coupling in these materials varies rapidly with D_{Ge}. A more quantitative measure of this transition can be seen in Fig. 12a in which the effective-mass ratio

$$M/m \equiv \lim_{T \to T_c} [H_{c2\parallel}(T)/H_{c2\perp}(T)]^2 \tag{39}$$

is plotted versus D_{Ge}. It is clear that in the vicinity of $D_{Ge} = 30$ Å interlayer coupling strength varies rapidly with layer separation. Figure 12b shows the interlayer coupling parameter $r = (4/\pi)(m/M)(2\xi(0)/S)^2$ derived by Ruggiero *et al.* from this data. The exponential increase in r with D_{Ge} confirms that the interlayer coupling energy η decreases as $\exp(-D_{Ge}/D_0)$, where $D_0 \simeq 8$ Å.

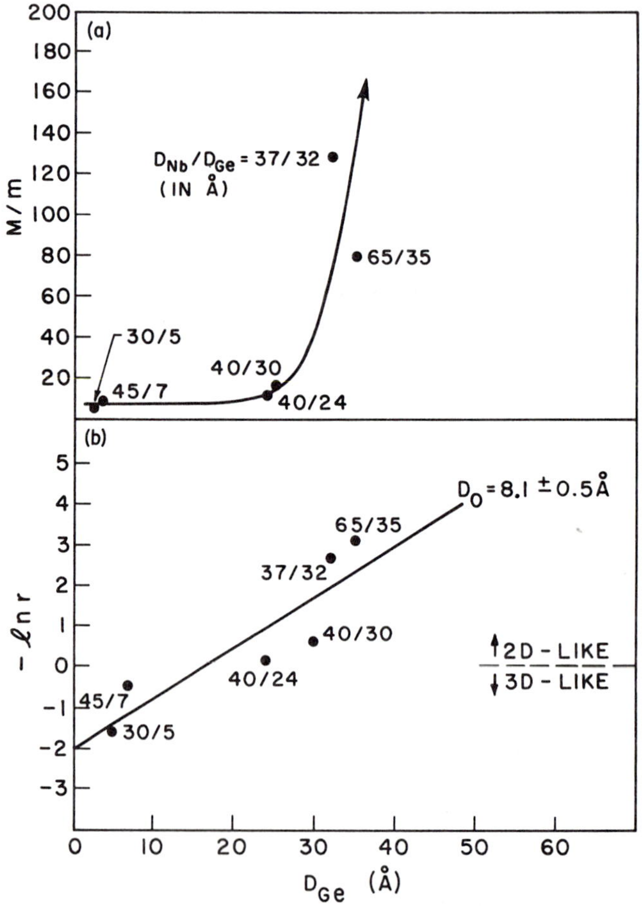

Fig. 12. (a) Effective-mass ratio M/m and (b) logarithm of the coupling parameter $r = (4/\pi)(m/M)[2\xi(0)/S]^2$ as a function of D_{Ge} for the Nb–Ge composites. The Josephson coupling strength rapidly decreases in the vicinity of $D_{Ge} \sim 30$ Å. The linear dependence of ln r with D_{Ge} implies an 8-Å characteristic tunneling length in the amorphous Ge layers. [From Ruggiero *et al.* (1982).]

Measurements of tunnel junctions formed with amorphous Si yield a similar characteristic length. It should be noted, however, that work by Meservey *et al.* (1982) has shown that tunneling through amorphous Si (and presumably amorphous elemental semiconductors in general) is not describable by simple tunneling barrier models. These authors also note a deterioration of the tunneling density of states for thick a-Si tunnel barriers. Halbritter (1982) has discussed a possible model for tunneling through such materials in which the tunneling is via localized states. The situation is in a state of flux, and both the nature of the interlayer coupling in superconductor/amorphous-semiconductor multilayers and the role of possible interactions between the semiconducting layers and the metal must be regarded as open issues.

Whereas the parallel critical field data of the composites (if not the details of the coupling) appear well understood, some interesting anomalous behavior was observed in the perpendicular upper critical fields. According to the GL theory the temperature dependence of the perpendicular upper critical field should be linear near T_c. Nevertheless, as can be seen in Fig. 11, these Nb/Ge multilayers show a small, but definite, positive curvature in the temperature dependence of $H_{c2\perp}$ near T_c. The universality of this behavior in the naturally occurring superconducting layered compounds was first pointed out by Woollam *et al.* (1974). Such positive curvature is also commonly observed in multilayer composites, not just the Nb/Ge system as we shall see next.

It is important to note, however, that in the Nb/Ge system, this curvature was observed exclusively in the perpendicular upper critical fields. In fact, a systematic increase in positive curvature was observed (at roughly fixed D_{Nb}), with increasing D_{Ge}. This systematic increase in positive curvature with D_{Ge} suggests that this anomaly may, in part, have its origin in some phenomenon intrinsic to the individual layers. A possible (and definitely speculative) interpretation of this phenomena as a manifestation of Kosterlitz–Thouless melting of the vortex lattice has been discussed by Ruggiero *et al.* (1980). Positive curvature in 2D superconductors due to localization and coulomb effects has been found theoretically by Maekawa *et al.* (1983).

B. Al/Ge

The Al/Ge multilayers of Haywood and Ast (1978) are conceptually similar to the Nb/Ge multilayers described in the preceding subsection. In fact, the work of Haywood and Ast preceded that of Ruggiero *et al.* but did not achieve significant Josephson coupling. These workers prepared their samples on a rotating liquid-nitrogen-cooled substrate holder using dual electron-beam evaporation and alternately exposing the substrate to the Al and Ge sources via a rotating shutter. In all cases the layer thicknesses D_{Ge} and D_{Al} were maintained equal. No x-ray studies of these multilayers were reported. Transmission electron microscopy studies, however, showed that the grain size of the Al films is uniform within a given sample and increases linearly with layer thickness according to the empirical relation $D_{grain} = D_{Al} + 35$ Å.

Transport measurements provide some additional information regarding the structure of these multilayers. In particular, as shown in Fig. 13, the resistivity of these samples at 4.2 K shows a sharp break at ~80–90 Å as a function of layer thickness. The authors claim that above this critical thickness the in-layer resistivity can be explained in terms of boundary scattering off the surfaces of the individual layers. Below 80–90 Å the Al layers apparently become discontinuous. Thus the integrity of the layers themselves appears to be good only for relatively thick layers. The degree of regularity of the layering is not known for these very thin layer thicknesses.

The critical temperatures observed in these multilayers are shown in Fig. 14. Unlike Nb films, but as seen in other Al systems, T_c is not a monotonic function of D_{Al}. In fact, T_c peaks at a value of ~2.2 K for D_{Al} ~ 80 Å, just in the vicinity of the thickness at which the break in ρ occurs. The overall behavior of T_c is reported to be in good accord with theoretical modeling in which the electron–phonon interaction is enhanced on the surface of the grains (Garland *et al.*, 1968; Pettit and Silcox, 1976).

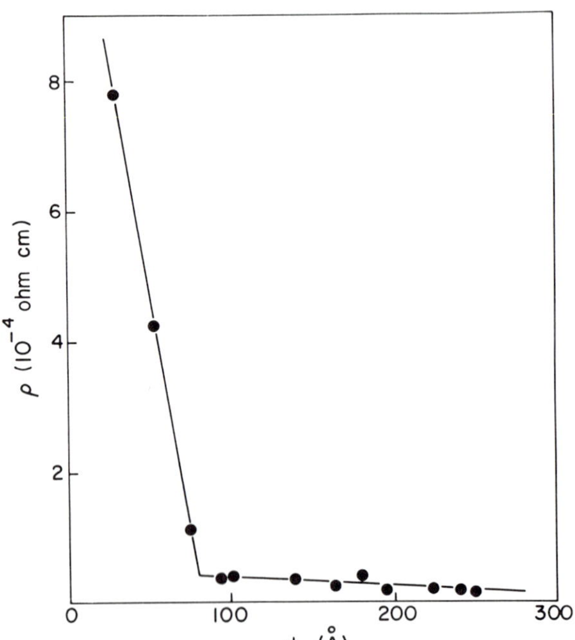

Fig. 13. Resistivity $\rho_{4.2\text{-K}}$ as a function of Al thickness for the Al–Ge composites. Sharp break in behavior in the vicinity of $d_{Al} = 80$ Å is attributed to the approach of the percolation threshold for conduction. [From Haywood and Ast (1978).]

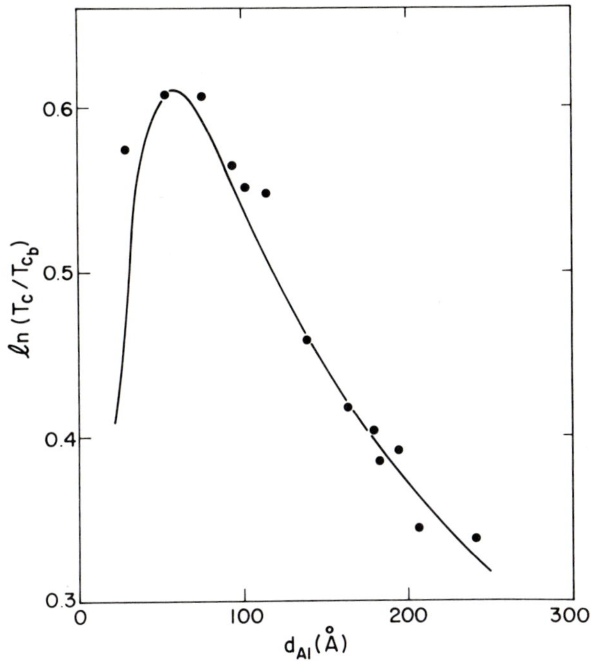

Fig. 14. Plot of T_c (scaled to the bulk transition temperature) as a function of aluminum thickness in the Al–Ge system. The solid line is based on a theory incorporating grain-structure effects on the electron–phonon coupling strength λ. [From Haywood and Ast (1978).]

As in the case of Nb/Ge composites, investigations of the Al/Ge system focused on their reduced dimensional nature. In this case, however, it was observed that $H_{c2\parallel}(T) \propto (1 - T/T_c)^{1/2}$ near T_c for all samples indicating that the individual layers, although intrinsically 2D [$D_{Al} < \xi(T)$], were not Josephson coupled over the thickness range studied. Representative data are shown in Fig. 15. As also noted by the authors and seen clearly in the figure, there is a large, systematic positive curvature in the perpendicular upper critical fields observed in this system. Moreover, the positive curvature was found to increase with decreasing Al layer thickness and as the number of Al/Ge layer pairs was increased. It was also found that, as in the case of the Nb/Ge systems, the parallel upper critical fields exhibited no such anomalous curvature. The increase in positive curvature with decreasing layer thickness, which is especially prominent in the range $D_{Al} < 80$ Å in which the films break up, immediately suggests an explanation based on granularity. It was also found, however, that although the degree of positive curvature changed with the number of layer pairs, there was no coincident effect upon the grain size of the individual layers (or T_c). Hence, although the degree of positive curvature does correlate with the individual layer structure for these Al/Ge systems, other evidence suggests the

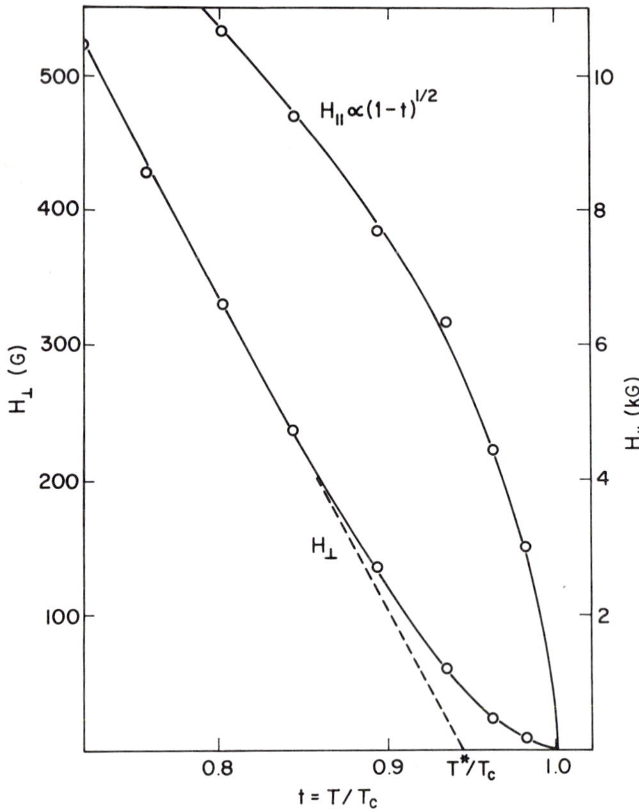

Fig. 15. Perpendicular and parallel upper critical field behavior for an Al–Ge layered composite with each constituent layer approximately 140 Å thick. The decoupled (2D) nature manifest in $H_\parallel$ is common to all samples studied ($d_{Al} = d_{Ge} \geq 30$ Å). [From Haywood and Ast (1978).]

possibility of a more intrinsic origin. In short, a fully satisfactory understanding of this ubiquitous phenomenon is not yet at hand.

C. Nb/Cu

The Nb/Cu S/N system has also received considerable experimental attention to date. The structural and superconducting properties of Nb/Cu multilayers have been studied by Schuller and Falco (1979), Schuller (1980), Banerjee (1982), Banerjee *et al.* (1982, 1983), and Chun *et al.* (1984). Further important clarifying studies of the structure of these materials, particularly as regards coherence at the interfaces, were carried out by Lowe *et al.* (1981). Since Nb and Cu are immiscible, the potential for very sharp compositional interfaces is high. The preparation technique employed by both groups has been to rotate the

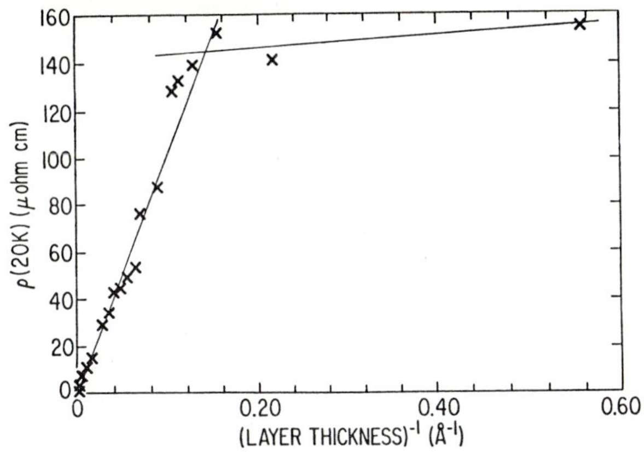

Fig. 16. Plot of low-temperature resistivity of Nb–Cu composites as a function of layer thickness ($D = D_{Nb} = D_{Cu}$). [From Banaerjee (1982).]

substrates on a table alternately past two magnetron sputtering heads. Substrate temperatures are typically $<80°C$. X-ray diffraction studies have shown that for layer thicknesses ($D_{Nb} = D_{Cu}$) greater than 10 Å the samples are very regularly layered in the growth direction, showing both superlattice lines at low angles and superlattice satellite peaks around the main Nb and Cu diffraction lines at high angles. These multilayers do not exhibit true interfacial coherency associated with epitaxial growth, however. For layer thicknesses less than 10 Å the material becomes progressively disordered.

The parallel resistivity of these Nb/Cu multilayers at low temperatures (20 K) as measured by the Argonne group is shown in Fig. 16 as a function of inverse layer thickness ($D_{Nb} = D_{Cu}$). For layer thickness greater than 10Å the resistivity is proportional to the inverse thickness, as expected for a layer thickness limited mean free path. (Presumably in this regime the conductivity of the multilayers is dominated by the Cu layers.) Below 10 Å, where x-ray data indicate that the layers become disordered, the resistivity saturates at ~150 μohm cm.

The T_c's of the Nb/Cu composites of Banerjee and co-workers (1982) are shown in Fig. 17. In order to interpret their data and attempt to extract the intrinsic properties of the Nb layers, Banerjee *et al.* (1982) used the de Gennes–Werthamer (Hauser *et al.*, 1964; Werthamer, 1963) theory of the proximity effect to analyze their T_c data. A straightforward application of this theory assuming a thickness-independent value of T_c for the Nb films (see solid line in Fig. 17) was not successful, however, and it was found necessary to assume that the T_c of the Nb layers decreases with decreasing layer thickness. These inferred values of T_c, which presumably represent the intrinsic value for the Nb layers themselves in the absence of the Cu layers, are shown in Fig. 18 along with data from the literature on single Nb films.

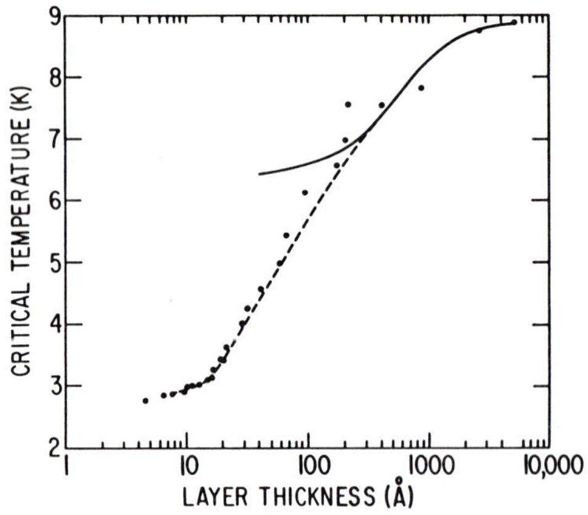

Fig. 17. Plot of T_c of Nb–Cu composites versus layer thickness ($D = D_{Nb} = D_{Cu}$). The solid line is a calculation from the de Gennes–Werthamer theory (dGW) with no adjustable parameters. The dotted line is the dGW fit with the T_c of the Nb layers as a continuously adjustable parameter. [From Banerjee *et al.* (1982).]

Interestingly, the T_c's for an isolated Nb layer (i.e., as $D_{Cu} \rightarrow 0$) inferred from the Nb/Cu results are in close correspondence with the experimental limit as $D_{Ge} \rightarrow 0$ in the Nb/Ge system. Neither group has definitively identified the mechanism responsible for the residual depression of T_c in this limit. In the case of the two-component model of the Nb layers used by Ruggiero *et al.* (1982),

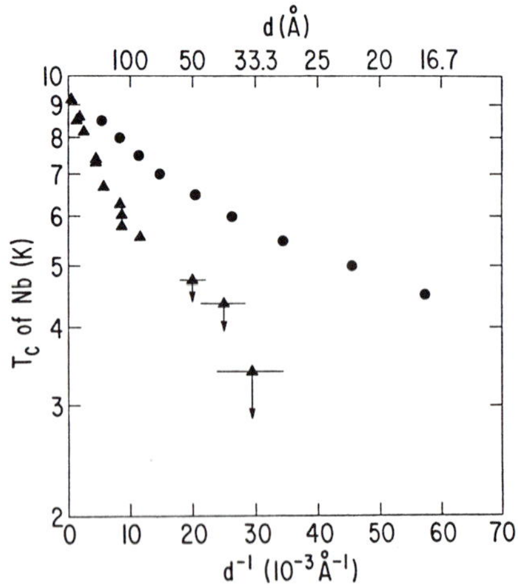

Fig. 18. Plot of the T_c of the Nb films comprising Nb–Cu composites. Circles are values inferred from a de Gennes–Werthamer fit of the T_c's of Nb–Cu composites where T_c(Nb) was an adjustable parameter (see Fig. 17). Triangles are the results for single Nb films. [From Banerjee *et al.* (1982).]

this limiting residual T_c value at $D_{Ge} \rightarrow 0$ could correspond to a somewhat thinner and/or higher-T_c disordered interfacial layer (than that with $D_{Ge} \rightarrow 100$ Å). Banerjee *et al.* (1982) have speculated that a reduced density of states associated with increased disorder as the Nb layers become thinner may account for this behavior.

As discussed in Section II, interlayer Josephson coupling is also possible in S/N systems. Here, the coupling is expected to take place via the proximity effect in which a finite superconducting pair wave function is induced in the normal layers. In this case, much longer characteristic Josephson coupling lengths, on the order of $\xi_N(0) \gtrsim 100$ Å, are expected. Banerjee (1982) reported results on the upper critical fields of Nb/Cu multilayers, for which $\xi_{Cu}(0) \geq 300$

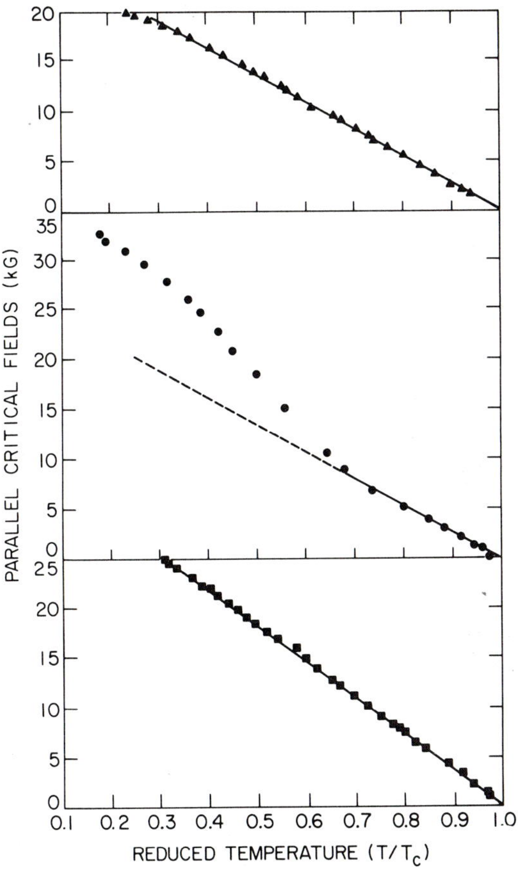

Fig. 19. Plot of the parallel upper critical fields of Nb–Cu composites as a function of reduced temperature for various layer periodicities. ▲, $D = 42.3$ Å; ●, $D = 171.5$ Å; ■, $D = 867.5$ Å. [From Banerjee *et al.* (1983).]

Å, although in his samples D_{Nb} and D_{Cu} were not varied independently, which makes separation of intralayer and interlayer effects awkward.

Figure 19 shows the parallel critical fields measured by Banerjee *et al.* (1983) for samples of varying layer thickness ($D = D_{Nb} = D_{Cu}$). For $D = 171.5$ Å, there is an upturn in the critical field qualitatively like that expected from dimensional crossover. These authors also report that for smaller layer thickness where the layers are well coupled, $H_{c\parallel}/H_{c\perp} \simeq 1.69$, suggesting that surface superconductivity is present in these thin multilayers. In layer thicknesses where the layers are decoupled, they find $H_{c\parallel}/H_{c\perp} \to 1$, presumably due to quenching of surface superconductivity on the individual Nb layers by the Cu. Thus for these samples both dimensional crossover behavior and surface superconductivity (on internal interfaces) apparently play a role in the critical field behavior.

Indeed, reported work of Chun *et al.* (1984) has specifically addressed the issue of surface effects and shown dimensional crossover in $H_{c2\parallel}(T)$ and an effective-mass behavior in the angular dependence of $H_{c2}(T)$.

Also of interest are tunneling results in Nb/Cu (Yang *et al.*, 1983) where the range 8 Å $< D_{Nb} <$ 5000 Å was explored. Among the results reported were transverse- and longitudinal-acoustic Nb phonon peaks unshifted in energy for $D_{Nb} \geqslant 30$ Å and a distinct broadening of the peaks for $D_{Ta} \sim 10$ Å. In light of these results the authors suggest the participation of large changes in the phonon dispersion with layer thickness.

D. Nb/Ti

In contrast to the Nb/Cu system in which the individual components are immiscible, in the Nb/Ti system considerable miscibility is present on both the Nb- (bcc) and Ti- (hcp) rich sides of the phase diagram. This system has been studied by the Northwestern group (Qian *et al.*, 1982) with some preliminary work reported earlier in collaboration with the Argonne group (Zheng *et al.*, 1981). Multilayers were deposited by using the standard sputtering approach on 90° sapphire substrates at ambient temperature. The layer thicknesses were maintained equal and varied from 6 to 6000 Å. The samples were strongly layered (except for the thinnest layers) as revealed by conventional diffractometer x-ray studies. Additional x-ray analysis using transmission Laue photographs and off-normal diffractometer scans demonstrated a good (110) texture but no evidence for epitaxy. An interesting aspect of the data was the indication that the Ti layers might be in a metastable bcc phase. Evidence for hcp and fcc phases was also present. The data are difficult to interpret, however, and both groups conclude only that the samples are a multiphase system containing bcc, hcp, and fcc phases.

Room-temperature resistivity measurements for these Nb/Ti multilayers are shown in Fig. 20. Three distinct regimes of behavior are indicated as a

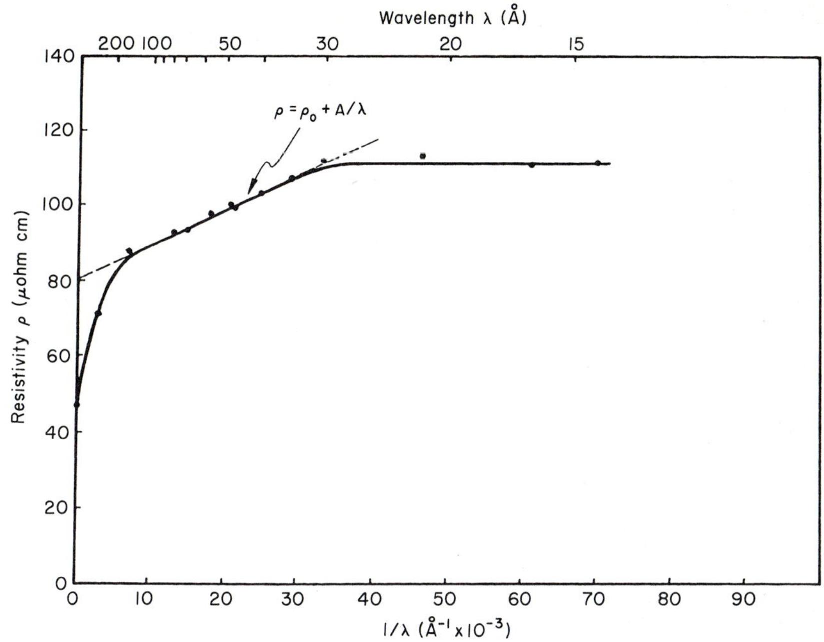

Fig. 20. The dependence of the room-temperature resistivity on inverse wavelength in the Nb–Ti system where $D_{Nb} = D_{Ti}$. Data show three distinct regimes of behavior as discussed in the text. [From Zheng *et al.* (1981). Copyright 1981 North-Holland Publishing Co., Amsterdam.]

function of the layer periodicity Λ. For long wavelengths, the resistivity is reported to extrapolate to the value for parallel bulk slabs of Nb and Ti as prepared by the authors. Their extrapolated value, however, seems somewhat large compared with the bulk resistivities of pure Nb ($\rho_{300\ K} = 12.5$ μohm cm) and Ti($\rho_{300\ K} = 42.0$ μohm cm). However, for 30 Å $\leq \Lambda \leq$ 200 Å, a linear dependence of ρ with $1/\Lambda$ is observed of the form $\rho = \rho_0 + A/\Lambda$. Interestingly, the value of $\rho_0 = 80$ μohm cm represents that expected for a 50% Nb–Ti alloy. Finally, for $\Lambda < 30$ Å the saturation effect (also observed in Nb/Cu) occurs in which there is a sharp break in the dependence of ρ with $1/\Lambda$. Here ρ saturates at a value of ~115 μohm cm.

The T_c data for these Nb/Ti multilayers are shown in Fig. 21. As Λ is decreased, there is the expected decrease in T_c due to the proximity effect with the Ti layers. For Λ in the range 20–200 Å, T_c is constant. According to Qian *et al.* (1982), the value of T_c in this regime is lower than expected from the Cooper limit of the proximity effect theory. Given the uncertainties of the T_c of thin Nb layers and the presence of unconventional phases in the material, no definite interpretation is apparently possible. For $\Lambda < 20$ Å, T_c again rises and saturates at the value expected for a bulk 50/50 Nb–Ti alloy. These T_c data along with the resistivity data discussed earlier clearly imply that for $\Lambda < 20$ Å there is consid-

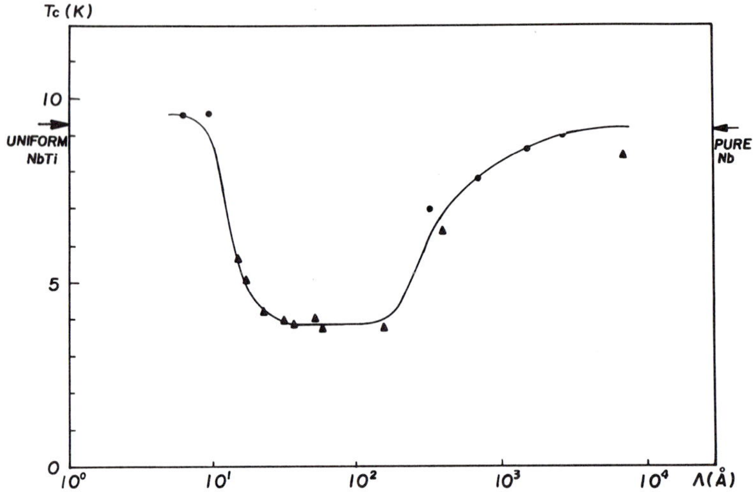

Fig. 21. The dependence of T_c on sample periodicity in the Nb–Ti system where $D_{Nb} = D_{Ti}$. Arrows indicate the T_c's for both pure Nb and a uniform 50/50 NbTi film as prepared by the authors. The depression of T_c in intermediate thickness regimes is attributed to the proximity effect. Samples prepared and run under identical conditions: ▲, group A; ●, group B. [From Qian *et al.* (1982).]

erable intermixing (i.e., alloying) of the layers in the Nb/Ti system. Presumably this intermixing is also present at the interfaces of the larger Λ multilayers as well. Depending on what is assumed about the crystallographic form of the Ti and about the importance of the intermixing of the Nb/Ti interface, these multilayers are possibly S/N, S/S′, or S/S″/S′, where S″ refers to the 50/50 Nb–Ti alloy.

The layer thickness dependence of the parallel critical field slope for these Nb/Ti multilayers is shown in Fig. 22. There are two peaks present on top of a generally rising background as Λ decreases. Presumably this background rise is due to a general decrease in the mean free path as Λ decreases, since $dH_{c2}/dT \propto \xi(0)^{-2} \propto l^{-1}$. The lower peak clearly corresponds to the point at which T_c starts to rise and the intermixing of the layers is playing an important role, although since ρ does not change in this regime and T_c rises, it is not obvious why the critical field decreases unless there is a dramatic decrease in the electronic density of states. The peak near $\Lambda \simeq 300$ Å is reminiscent of the peak observed in Nb/Cu, but no additional, more direct evidence for dimensional crossover behavior has been reported. Clearly a more definitive interpretation of these data must await more precise chemical identification of the observed phases and a careful consideration of the influence of a high-T_c solid solution alloy phase at the interfaces. Because of the likely interfacial mixing, these multilayers probably can be best thought of as S/S′/N multilayers where S = Nb, S′ = Nb–Ti, and N = Ti and where $T_{cS'} \lesssim T_{cS} > T_{cN}$.

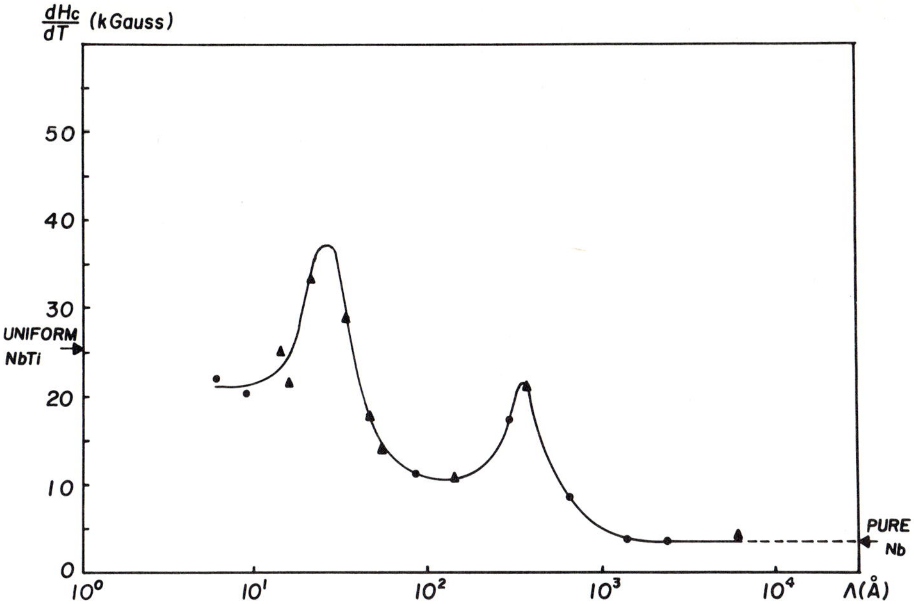

Fig. 22. The dependence of the parallel upper critical field as a function of layer periodicity in the Nb–Ti system. The right-hand peak appears attributable to a quasi-two-dimensional regime similar to that observed in the Nb–Cu system. The peak at lower periodicity may involve alloying effects at the layer interfaces. Samples prepared and run under identical conditions: ▲, group A; ●, group B. [From Qian *et al.* (1982).]

E. Nb/Zr

The Nb/Zr system of Lowe and Geballe (1984) was also prepared with the usual magnetron sputtering technique. Like Nb/Ti the components are miscible, and again some interfacial mixing is anticipated. The principle interest in this system has focused on the structure of the layers and T_c. X-ray data for small Λ indicate an epitaxial stabilization of bcc β-phase Zr on the bcc Nb, instead of the usual hcp α phase. In the intermediate thickness range 10 Å $<$ $\Lambda \lesssim$ 100 Å, small-area coherent lattices of (110) textured bcc Nb and Zr are also observed. Finally, for thick layers, $\Lambda \gtrsim$ 100 Å, bulk equilibrium behavior is observed for both components with α-phase (002) textured Zr.

Measurements of T_c have also been performed on this system and show a systematic increase in T_c from ~6 K at $\Lambda = 200$ Å to 9.75 K extrapolated to $\Lambda = 0$, a value greater than for either component. To explain these data, and in light of the above discussed structural nature of this system, Lowe and Geballe have postulated the existence of 10 Å of a 50% Nb–Zr solid solution at each Nb/Zr interface. In fact, a fit to the T_c versus $1/\Lambda$ data using the Cooper theory and assuming T_c's of 9.2, 9.7, and 0.5 for the Nb, Nb–Zr interfacial mixture and Zr, respectively, has yielded good overall consistency. Furthermore, detailed x-ray and EXAFS studies of this system by Claeson *et al.* (1984), in collaboration with the Stanford group, have definitively established the existence of such an interfacial layer. Thus the Nb/Zr and Nb/Ti systems appear to have many qualitative features in common. Critical field data are not yet available for this system, however.

At this writing, Broussard *et al.* (1984) have also reported measurements of the specific heat of Nb/Zr multilayers.

F. Nb/Al

Multilayers of Nb/Al have also been fabricated by using the standard sputtering technique by Geerk *et al.* (1982). These researchers note that the crystallographic integrity of the system appears to fall between the good layering but incoherent interfaces of the Nb/Cu system and the coherent, single-crystal growth of the Nb/Ta systems to be discussed in the next subsection. In particular, the films show a Nb(110) and Al(111) texture, a 9°-wide rocking curve, and random in-plane microcrystalline orientation. The detailed crystallographic analysis of McWhan *et al.* (1983) on these systems implies that they are most accurately described as strained superlattices. These x-ray studies have also revealed more detailed information about the layer interfaces. In particular, the compositional profile of these systems (deposited at nominally room temperature) has an asymmetric, trapezoidal shape implying an interfacial region ~11 Å in thickness displaced ~4 Å toward the Al layer.

Resistivity measurements for the thinnest samples ($\Lambda \lesssim 20$ Å) yield values on the order of 100 μohm cm as observed in the thinnest layers of other Nb-based multilayers. The mean free paths of these samples are thus noted to be on the order of Λ despite the coherent interfaces with a given microcrystallite and thus appear to be limited by interfacial scattering or by diffusion or strain between the layers.

Measurements of the critical temperatures of these materials indicate, as for all previous systems, an intrinsic depression of the T_c of the Nb films as their thickness is reduced. Critical field data have not been reported for these materials.

These Nb/Al multilayers were the first superconducting multilayer systems in which tunneling studies were carried out (Geerk *et al.*, 1982). The results reported by Geerk *et al.* are shown in Fig. 23 for a set of samples for which the last (i.e., top) layer was Al. As can be seen in the figure, both Nb and Al phonon peaks are evident, changing in relative strength as the layer thickness $\Lambda/2$ was increased. Also evident is unaccounted-for structure below 15 mV. Even more remarkable, evidence for structure near the Nb phonon energies was seen even at temperatures above T_c. These striking results will require additional study before a full interpretation can be made.

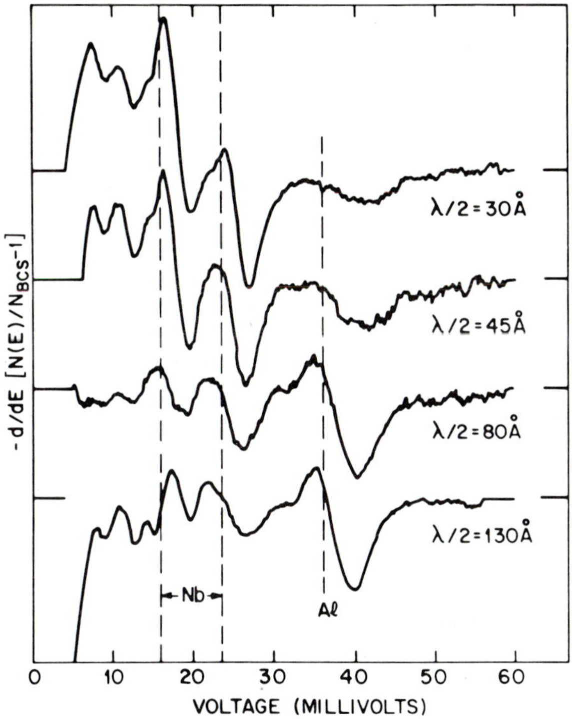

Fig. 23. Electron tunneling results for Nb–Al multilayers, top layer Al, showing the derivative of the normalized density of states. For these materials $D_{Nb} = D_{Al} = \lambda/2$. The positions of the transverse and longitudinal Nb phonon modes and the longitudinal Al mode are indicated. [From Geerk *et al.* (1982).]

An interesting aspect of these tunneling studies is the observation that the normal-state characteristics of the tunneling barrier were found to depend on whether the top layer was Al or Nb. Because of the excellent tunneling properties of Al oxide tunnel barriers compared to those of Nb, Geerk *et al.* then systematically investigated how thick an Al layer would yield a good, Al-like tunnel barrier on thick Nb films. It was found that the characteristic began to change significantly for a nominal Al thickness of only 1.8 Å (Geerk *et al.*, 1982). Indeed, the general utility of (thicker) Al overlayers had been earlier demonstrated by Wolf *et al.* (1980). Now Al overlayers form the basis for technologically significant artificial barriers for Josephson junctions fabricated from Nb (Gurvitch *et al.*, 1981, 1983). Magnesium (Kwo *et al.*, 1983) rare earth metals (Umbach *et al.*, 1982; Kwo *et al.*, 1983), Ta (Ruggiero *et al.*, 1983), and Zr (Celaschi *et al.*, 1982, 1983; Lowe *et al.* 1983) are among other metals found to produce similar results. While these artificial tunnel barriers are not a consequence of periodic multilayering per se, they clearly represent the first technologically significant outgrowth of research in superconducting multilayers.

G. Nb/Ta

The superconducting multilayer system with the highest crystallographic perfection to date is Nb/Ta. Pioneering work and the achievement of the first true epitaxial single-crystal multilayers (28-Å Nb/22-Å Ta) was carried out by Durbin *et al.* (1981, 1982) in a UHV MBE-type chamber using e-gun vapor deposition onto oriented sapphire substrates held at 700–900°C. Hertel *et al.* (1982) have reported superconducting tunneling properties of Nb/Ta multilayers of nearly comparable perfection deposited by means of the sputtering technique like that used with the fabrication of the other Nb-based multilayers. Following Durbin *et al.* elevated (~850°C) substrate temperatures were employed. We should note here that although a comparison of the MBE-type and sputter technique involves a number of trade-offs for a particular system of interest, deposition rates for a given background impurity level can be comparable.

Although Nb and Ta are miscible and there is clear evidence for interfacial diffusion, both groups find that the Nb/Ta system grows as a true single crystal with a coherent Nb/Ta interface. Such high crystal perfection no doubt reflects the closely matched lattice parameters of Nb and Ta (mismatch <0.2%). Also various orientations are possible depending on the substrate: [100] on [100] cleaved MgO, [110] on [11$\bar{2}$0] sapphire, [111] on [0001] sapphire, and [211] on [10$\bar{1}$0] sapphire. As noted by the authors (Durbin *et al.*, 1981, 1982), these results demonstrate that close-packed planes are not a necessary condition for metallic superlattice growth and that substrate symmetry is important.

Careful x-ray analysis of the multilayers of Durbin *et al.* (1981, 1982) indicates that for their samples interdiffusion at the interfaces was typically limited to ≃4.5 Å, that very little ($<10^{-3}$) differential strain was present, and that the lateral domain size over which coherent x-ray scattering was obtained could be as large as ~1000 Å in the most favorable case reported. This coherence length was sensitive to the substrate temperature during growth and a trade-off between improved coherence length and greater interdiffusion as the substrate temperature increases was noted.

In addition to their structural studies, Durbin *et al.* report an in-plane, low-temperature, transport electron mean free path ranging from 12 to 30 times the layer separation and a residual resistance ratio of ~11 for multilayers with layer thicknesses ≃25 Å. The transition temperatures of their samples were 7.2–7.4 K depending on the particular growth direction.

The sequentially sputtered Nb/Ta multilayers of Hertel *et al.* have been made over a wider range of periods and are of almost comparable structural quality to those of Durbin *et al.* It is seen here that for deposition temperatures below ~810°C the films have the usual (110) orientation and rocking curves of ~8°. However, higher temperatures give a (100) orientation and much narrower (~1°) rocking curves; hence a much more ideal—virtually single crystal

—system is obtained. These authors also find that the room-temperature resistivity of their thicker layer samples is 14 μohm cm, representative of bulk Nb and Ta. Moreover, the residual resistance ratio of these systems is found to decrease systematically from ~14 for $\Lambda = 250$ Å to 3–4 at $\Lambda = 18$ Å. The latter is in the range observed for 50% Nb–Ta solid solutions made on the same system. Under the assumption of isotropic scattering the authors conclude that the ratio of the layer thickness to mean free path is $\lesssim 10$ for the thinnest layers and decreases to 3 to 4 for $\Lambda = 250$ Å. The attainment of such a high degree of crystalline perfection and the observation of mean free paths many times in excess of the layer periodicity for Nb/Ta systems is certainly a breakthrough of considerable importance.

Hertel *et al.* have also studied the tunneling properties of their Nb/Ta multilayers by using a thin (<20-Å) Al overlayer to form the tunneling barrier as was done originally on Nb/Al multilayers. They observe a BCS-like density of states near the energy gap and electron–phonon coupling to both the Nb and the Ta. See Fig. 24. For multilayer periods greater than ~40 Å they report that the tunneling appears to probe predominantly only the top layer, either Nb or Ta. For the smallest periods (~18 Å) there is some smearing of the two phonon densities of state reminiscent of that seen in Nb–Ta solid solutions, apparently reflecting the interdiffusion mentioned earlier. Depending on the importance of this intermixing at the interfaces, these Nb/Ta multilayers can be regarded as S/S′ or S/S′/S″ systems. In any event, clearly these Nb/Ta results are of considerable intrinsic interest and have stimulated the model density-of-state calculations of Arnold, Gallagher, and Wolf, which were discussed in Section IV.

H. Pb/Bi

A rather different type of superconducting multilayer system that has been under study for many years by Raffy and co-workers (Raffy *et al.*, 1972, 1974; Raffy and Guyon, 1981) is that formed by codepositing Pb and Bi. They have been studied principally from the point of view of coherent flux pinning by the periodic stucture in the material. The materials were prepared by exposing substrates to a constant flux of Pb and a sinusoidally modulated flux of Bi from feedback-controlled electron-beam sources. The Bi concentration varied as $c(x) = \bar{c} - (\Delta c/2)\cos(2\pi x/a)$, where $2a$ is the sample periodicity and 2 at. % $< \Delta c <$ 20 at. %. Therefore these materials are perhaps more accurately described as compositionally modulated structures as opposed to a stack of discrete layers. Since Pb and Bi interdiffuse readily at room temperature, systematic studies as Δc varies are quite convenient with this system at the expense of dealing with unstable samples. Note also that the periods in these systems can be relatively large (~500–8000 Å) but so is the superconducting coherence length 650 Å $< \xi <$ 250 Å for PbBi alloys with 2 < at. % Bi < 20.

In any case it was found that for $\Delta c <$ 10 at. % multiple peaks in the critical

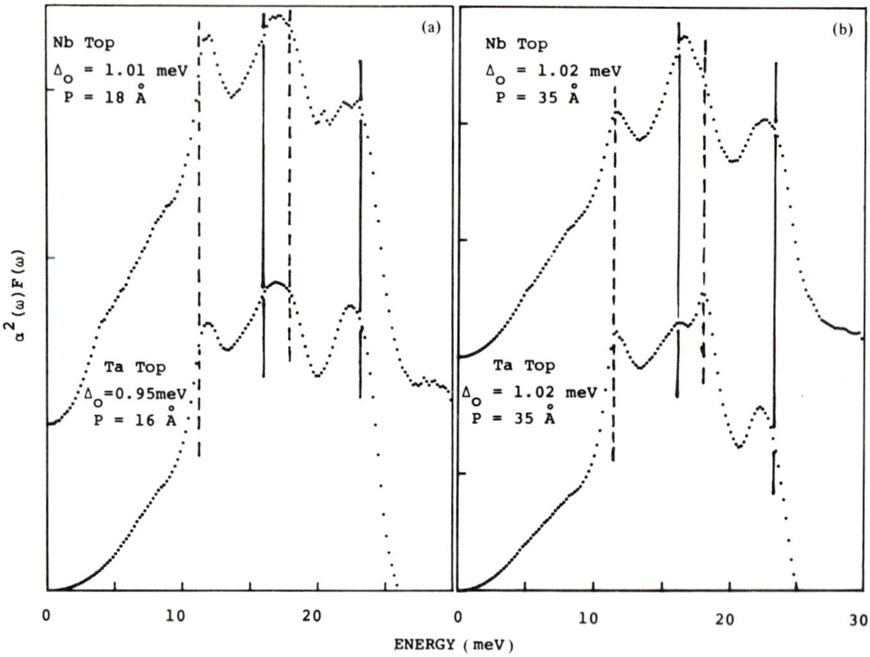

Fig. 24. The electron–phonon coupling parameter $\alpha^2(\omega)F(\omega)$ versus energy, determined from eight Nb–Ta superlattices. The vertical scale is 0.2 dimensionless units/division. The samples are arranged in pairs of similar period, with Nb or Ta top layers. The superlattice periods and gaps are shown; relative thicknesses of Nb and Ta layers have not yet been determined but can be judged from the gap values. The energies of transverse and longitudinal phonon modes in Ta are shown by vertical dashed lines, those in Nb by vertical solid lines. The plots are offset for clarity. [From Hertel *et al.* (1982).]

current $J_c(H)$ could be due to observed registry of the periodicities of the vortex lattice. For a given periodicity $2a$ multiple peaks occurred at $B(2a) = n^2B(2na)$, where $B(2a) = \frac{1}{2}\sqrt{3}\ (\Phi_0/(2a)^2)$, where the second equation gives the field required to produce the smallest interplanar spacing in a triangular vortex lattice with lattice constant a.

I. Other Systems

In addition to the systems already discussed, there are several other systems in which either layering was attempted for particular reasons but for which systematic study was not carried out or only a few layers were involved but the results are pertinent. For the sake of completeness we briefly discuss some of these results here.

To study the stabilization of the A15 phase of Nb_3Ge by epitaxy, Schmidt *et al.* (1980) prepared multilayers of Nb_3Ge/Nb_3Ir. The layers were sputter depos-

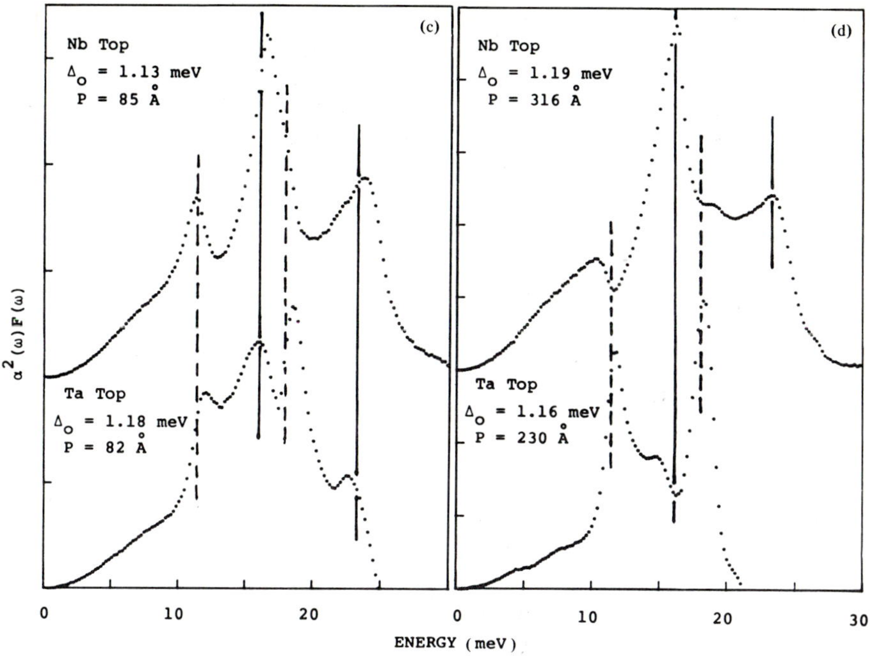

Fig. 24. *(Continued)*

ited from arc-melted targets at deposition temperatures ~1090 K. An initial 2000 Å of Nb_3Ir was first deposited followed by alternating layers 750 Å in thickness. It was found, however, that T_c was roughly independent of total film thickness and was not greater than single-film values. Similar studies of Nb_3Si also did not meet with success in this regard. However, as noted by Geballe and Rowell (1982) work by Kwo and Geballe (1981) on self-epitaxy in Nb_3Al implies that the metastable phase of that system cannot be maintained over thickness greater than ~200 Å. This suggests a renewed attempt at multilayer growth in this thickness regime.

In an attempt to increase flux pinning and hence J_c in Nb_3Sn, Howard *et al.* (1977) produced multilayered Nb_3Sn/Y composites by using electron-beam deposition of both the Nb_3Sn (codeposition of Nb and Sn) and the Y on cylindrical Hastelloy tubes. Yttrium was chosen as the barrier material because Sn and Y are essentially insoluble. The cylindrical geometry was chosen in order to allow inductive critical current measurements in these very strong flux pinning superconductors. It was found that a clear enhancement of critical current by a factor of 3 could be gained by the introduction of the Y layers, with the additional and perhaps more important advantage of a tenfold reduction of the so-called "surface barrier" to flux entry important in considerations of the ac loss of the materials.

The effects of magnetic layers in superconductors were studied by Claeson (1980). The materials were prepared at 0.5 K by thermal evaporation. Both the T_c and tunneling dV/dI curves of the samples were measured. Although the layering was not strictly periodic in these samples, a multiple layering of the constituent components was found to be an efficient experimental vehicle. By studying electron tunneling through ultrathin Fe layers between relatively thick (~250-Å) Pb layers, it was determined that some tunneling could proceed through $\lesssim 2$ atomic layers of Fe but was rapidly depressed for thicker films. In addition, and interestingly, it was found that a layer of Fe has a stronger pair-breaking effect than the same amount of statistically distributed atoms. In some more recent work the effect of thin Cr layers on flux pinning in multilayered Pb–Bi/Cr superconductors was investigated by Yetter *et al.* (1982).

In addition, Granqvist and Claeson (1979) have studied Sn/Ag and In/Ag multilayers in an effort to test the predictions of Kulik (1976) regarding T_c enhancements due to interlayer electron–electron interactions as a function of the number of layer pairs. Note that the effect is quite distinct from the usual proximity effect and from coupling between two, like superconducting layers due to Josephson tunneling. The latter does not influence T_c.

In their work Granqvist and Claeson deposited multiple layers of Sn/Ag and In/Ag onto liquid-helium-cooled substrates to prevent interdiffusion of the layers. The critical temperatures of up to 17 layer pairs were continuously monitored as the multilayers were deposited. They clearly observed increases in T_c with the number of layer pairs that compared well with the theory of Kulik. Of particular interest was the fact that T_c's in excess of the intrinsic single-film values were reported. However, the authors cite evidence for the influence of edge effects that cannot be easily eliminated because of the nature of their deposition process that may have played an important role in their observations.

We also note that the previously discussed Al/Ge composites of Haywood and Ast (1978) revealed no systematic change of T_c with the number of layers, although, perhaps not incidentally, the upper critical field measurements of these materials indicated that from the standpoint of Josephson interactions, the Al layers were completely decoupled.

Finally, we note the intriguing results of Brodsky *et al.* (1982), who have prepared epitaxial trilayers of the form Au/Cr/Au that reportedly showed superconducting critical temperatures up to 3 K.

ACKNOWLEDGMENTS

We would like to thank our many colleagues at Stanford, Yale, and elsewhere from whom we have learned so much about the status and the promise of synthetically structured superconductors. While we shall resist the temptation to name them all, we cannot in clear conscience fail to single out T. W. Barbee, Jr., who, most of all, has helped us understand the materials aspects of this subject, and J. M. Rowell for many helpful comments and for his critical reading of this chapter. Finally, the support during the preparation of this chapter of the National Science Foundation (MRB) and Office of Naval Research (STR) is gratefully acknowledged.

REFERENCES

Allender, D., Bray, J., and Bardeen, J. (1973). *Phys. Rev. B* **7,** 1020.

Ami, S., and Maki, K. (1975). *Prog. Theor. Phys.* **53,** 1.

Andreev, A. F. (1964). *Zh. Eksp. Teor. Fiz.* **46,** 1823. [*Sov. Phys.-JETP* (Engl. Transl.). **19,** 1228 (1964).]

Arnold, G. B., Gallagher, W. J., and Wolf, E. L. (1982). *In* "Superconductivity in d- and f-Band Metals" p. 305, Kernforschungszentrum, Karlsruhe, Germany.

Aslamasov, L. G., and Varlamov, A. A. (1980). *J. Low Temp. Phys.* **38,** 223.

Banerjee, I. (1982). Ph.D. thesis, Northwestern Univ., Evanston, Illinois.

Banerjee, I., Yang, Q. S., Falco, C. M., and Schuller, I. K. (1982). *Solid State Commun.* **41,** 805.

Banerjee, I., Yang, Q. S., Falco, C. M., and Schuller, I. K. (1983). *Phys. Rev. B* **28,** 5037.

Beasley, M. R. (1979). *In* "Inhomogeneous Superconductors," *AIP Conf. Proc.* **58,** 186.

Beasley, M. R., Mooij, J. E., and Orlando, T. P. (1979). *Phys. Rev. Lett.* **42,** 1165.

Brodsky, M. B., Marikar, P., Friddle, R. J., Singer, L., and Sowers, C. H. (1982). *Solid State Commun.* **42,** 675.

Broussard, P. R., Mael, D., and Geballe, T. H. (1984). *Phys. Rev. B.* **30,** 4055.

Bulaevskii, L. N. (1973). *Zh. Eksp. Teor. Fiz.* **64,** 2241. [*Sov. Phys.-JETP* (Engl. Transl.) **37,** 1133 (1973).]

Bulaevskii, L. N. (1975). *Usp. Fiz. Nauk* **116,** 449. [*Sov. Phys.-Usp.* (Engl. Transl.) **18,** 514 (1976)].

Bulaevskii, L. N., and Guseinov, A. A. (1974). *Zh. Eksp. Teor. Fiz., Pis'ma Red.* **19,** 742. [*JETP Lett.* (Engl. Transl.) **19,** 382 (1974.]

Celaschi, S., Geballe, T. H., and Lowe, W. P. (1982). *Proc. of Material Research Society,* p. 241 (unpublished).

Celaschi, S., Hammond, R., Geballe, T. H., Lowe, W. P., and Green, A. (1983). *Bull. Am. Phys. Soc.* **28,** 423.

Chun, C. S. L., Zheng, G., Vincent, J. L., and Schuller, I. K. (1984). *Phys. Rev. B.* **29,** 4915.

Claeson, T. (1980). *Thin Solid Films* **66,** 151.

Claeson, T., Boyce, J. B., Lowe, W. P., and Geballe, T. H. (1984). *Phys. Rev. B* **29,** 4969.

de Gennes, P. G., and Saint-James, D. (1963). *Phys. Lett.* **4,** 151.

Deutscher, G., and Entin-Wohlman, O. (1977). *J. Phys. C* **10,** L433.

Deutscher, G., and Entin-Wohlman, O. (1978). *Phys. Rev. B* **17,** 1249.

Dobrosavljevic, L. (1973). *Phys. Status Solidi B* **55,** 773.

Dobrosavljevic, L., and Kulic, M. (1978). *J. Low Temp. Phys.* **32,** 505.

Durbin, S. M., Cunningham, J. E., Mochel, M. E., and Flynn, C. P. (1981). *J. Phys. F* **11,** L223.

Durbin, S. M., Cunningham, J. E., and Flynn, C. P. (1982). *J. Phys. F* **12,** L75.

Efetov, K. B., and Larkin, A. I. (1974). *Zh. Eksp. Teor. Fiz.* **68,** 155. [*Sov. Phys.-JETP (Engl. Transl.)* **41,** 76 (1975)].

Fisher, D. S. (1979). *AIP Conf. Proc.* **58,** 95.

Fisher, D. S. (1980). *Phys. Rev. B* **22,** 1190.

Galaiko, V. P., and Bezuglyi, E. V. (1973). *Zh. Eksp. Teor. Fiz.* **65,** 2027. [*Sov. Phys.-JETP* (Engl. Transl.) **38,** 1013 (1974).]

Gallagher, W. J. (1980). *Phys. Rev. B* **22,** 1233.

Gallagher, W. J. (1982). *Bull. Am. Phys. Soc.* **27,** 215.

Garland, J. W., Bennemann, K. H., and Mueller, F. M. (1968). *Phys. Rev. Lett.* **21,** 1315.

Geballe, T. H., and Rowell, J. M. (1982). *Thin Solid Films* **91,** 33.

Geerk, J., Gurvitch, M., McWhan, D. B., and Rowell, J. M. (1982). *Physica* **109, 110***B,* 1775.

Granqvist, C. G., and Claeson, T. (1979). *Solid State Commun.* **32,** 531.

Gurvitch, M., Rowell, J. M., Huggins, H. A., Washington, M. A., and Fulton, T. A. (1981). *Proc. Int. Electron Devices Meeting,* p. 115.

Gurvitch, M., Washington, M. A., and Huggins, H. A. (1983). *Appl. Phys. Lett.* **42,** 472.

Halbritter, J. (1982). *Surf. Sci.* **122,** 80.

Halperin, B. I., and Nelson, D. R. (1978). *Phys. Rev. Lett.* **41,** 121, 519(E).

Hauser, J. J., Theuerer, H. C., and Werthamer, N. R. (1964). *Phys. Rev.* **136,** A637.
Haywood, T. W., and Ast, D. G. (1978). *Phys. Rev. B* **18,** 2225.
Hertel, G., McWhan, D. B., and Rowell, J. M. (1982). *In* "Superconductivity in d- and f-Band Metals," p. 299, Kernforschungszentrum, Karlsruhe, Germany.
Howard, R. E., Beasley, M. R., Geballe, T. H., King, C. N., Hammond, R. H., Norton, R. N., Salem, J. R., and Zubeck, R. B. (1977). *IEEE Trans. Magn.* **MAG-13,** 138.
Huberman, B. A., and Doniach, S. (1979). *Phys. Rev. Lett.* **43,** 950.
Kats, E. I. (1969). *Zh. Eksp. Teor. Fiz.* **56,** 1675. [*Sov. Phys.-JETP* (Engl. Transl.) **29,** 897 (1969).]
Kats, E. I. (1970). *Zh. Eksp. Teor. Fiz.* **58,** 1471. [*Sov. Phys.-JETP* (Engl Transl.) **31,** 787 (1970).]
Klemm, R. A. (1974). *J. Low Temp. Phys.* **16,** 381.
Klemm, R. A., Beasley, M. R., and Luther, A. (1974). *J. Low Temp. Phys.* **16,** 607.
Klemm, R. A., Luther, A., and Beasley, M. R. (1975). *Phys. Rev. B* **12,** 877.
Kogan, V. G. (1981). *Phys. Rev. B* **24,** 1572.
Kogan, V. G., and Clem, J. R. (1981). *Phys. Rev. B* **24,** 2497.
Kosterlitz, J. M., and Thouless, D. J. (1973). *J. Phys. C* **6,** 1181.
Kulik, I. O. (1976). *Solid State Commun.* **19,** 535.
Kwo, J., and Geballe, T. H. (1981). *Phys. Rev. B* **23,** 3230.
Kwo, J., Wertheim, G. K., Gurvitch, M., and Buchanan, D. N. E. (1983). *IEEE Trans. Magn.* **MAG-19,** 795.
Lawrence, W. E., and Doniach, S. (1970). *Proc. Int. Conf. on Low Temperature Physics, 12th,* (E. Kanda, ed.), p. 361. Academic Press, Kyoto, Japan.
Likharev, K. K. (1975). *Fiz. Tverd. Tela* **17,** 2787. [*Sov. Phys. Solid State* (Engl. Transl.) **17,** 1856 (1975)].
Likharev, K. K. (1979). *Rev. Mod. Phys.* **51,** 101.
Lowe, W. P., and Geballe, T. H. (1984). *Phys. Rev. B* **29,** 4961.
Lowe, W. P., Barbee, T. W., Jr., Geballe, T. H., and McWhan, D. B. (1981). *Phys. Rev. B* **24,** 6193.
Lowe, W. P., Celaschi, S., and Geballe, T. H. (1983). *Bull. Am. Phys. Soc.* **28,** 508.
McWhan, D. B., Gurvitch, M., Rowell, J. M., and Walker, L. R. (1983). *J. Appl. Phys.* **54,** 3886.
Maekawa, S., Ebisawa, H., and Fukuyama, H. (1983). *J. Phys. Soc. Jpn.* **532,** 1352.
Maki, K. (1968a). *Prog. Theor. Phys.* **39,** 897.
Maki, K. (1968b). *Prog. Theor. Phys.* **40,** 193.
Meservey, R., Tedrow, P. M., and Brooks. J. S. (1982). *J. Appl. Phys.* **53,** 1563.
Nabutovskii, V. M. (1982). *Solid State Commun.* **43,** 405.
Nabutovskii, V. M., and Shapiro, B. Y. (1981). *Solid State Commun.* **40,** 303.
Nelson, D. R., and Halperin, B. I. (1979). *Phys. Rev. B* **19,** 2457.
Orlando, T. P., McNiff, E. J., Jr., Foner, S., and Beasley, M. R. (1979). *Phys. Rev. B* **19,** 4545.
Pettit, R. B., and Silcox, J. (1976). *Phys. Rev. B* **13,** 2865.
Qian, Y. J., Zheng, J. Q., Sarma, B. K., Yang, H. Q., Ketterson, J. B., and Hillard, J. E. (1982). *J. Low Temp. Phys.* **49,** 279.
Raffy, H., and Guyon, E. (1981). *Physica* **108***B,* 947.
Raffy, H., Renard, J. C., and Guyon, E. (1972). *Solid State Commun.* **11,** 1679.
Raffy, H., Guyon, E., and Renard, J. C. (1974). *Solid State Commun.* **14,** 427, 431.
Rowell, J. M. (1973). *Phys. Rev. Lett.* **30,** 167.
Rowell, J. M., and McMillan, W. L. (1966). *Phys. Rev. Lett.* **16,** 453.
Ruggiero, S. T., Barbee, T. W., Jr., and Beasley, M. R. (1980). *Phys. Rev. Lett.* **45,** 1299.
Ruggiero, S. T., Barbee, T. W., Jr., and Beasley, M. R. (1982). *Phys. Rev. B* **26,** 4894.
Ruggiero, S. T., Face, D. W., and Prober, D. E. (1983). *IEEE Trans. Magn.* **MAG-19,** 960.
Schmidt, P. H., Vandenberg, J. M., Hamm, R., Rowell, J. M. (1980). *In* "Superconductivity in d- and f-Band Metals" (H. Suhl and M. B. Maple, eds.), p. 57, Academic Press, New York.
Schuller, I. K. (1980). *Phys. Rev. Lett.* **44,** 1597.
Schuller, I. K., and Falco, C. M. (1979). *In* "Inhomogeneous Superconductors," *AIP Conference Proc.* **58,** 197.

Schuller, I. K., and Falco, C. M. (1982). *In* "*VLSI* Electronics" (N. G. Einspruch, ed.), Vol. 4. pp. 183–205. Academic Press, New York.
Shih, S., Khim, Z. G., Arnold, G. B., and Tomasch, W. J. (1981). *Phys. Rev. B* **24,** 6440.
Silvert, W. (1975). *J. Low Temp. Phys.* **20,** 439.
Skocpol, W. J., and Tinkham, M. (1975). *Rep. Prog. Phys.* **38,** 1049.
Thompson, R. S. (1970). *Phys. Rev. B* **1,** 327.
Tilley, D. R. (1965). *Proc. Phys. Soc. London* **85,** 1177.
Tinkham, M. (1975). "Introduction to Superconductivity," McGraw-Hill, New York.
Tomasch, W. J. (1965). *Phys. Rev. Lett.* **15,** 672.
Tomasch, W. J. (1969). *In* "Tunneling in Solids" (E. Burstein and S. Lundqvist, eds.), p. 315, Plenum Ps, New York.
Umbach, C. P., Goldman, A. M., and Toth, L. E. (1982). *Appl. Phys. Lett.* **40,** 81.
van Gelder, A. P. (1969). *Phys. Rev.* **181,** 787.
Werthamer, N. R. (1963). *Phys. Rev.* **132,** 2440.
Werthamer, N. R. (1969). *In* "Superconductivity" (R. D. Parks, ed.), Vol. 1, p. 321, Dekker, New York.
Wolf, E. L., Zasadinski, J., Osmun, J. W., and Arnold, G. B. (1980). *J. Low Temp. Phys.* **40,** 19.
Woollam, J. A., Somoano, R. B., and O'Connor, P. (1974). *Phys. Rev. Lett.* **32,** 712.
Xing, D., and Gong, C. (1981). *Physica* **108***B,* 987.
Yang, Q. S., Falco, C. M., and Schuller, I. K. (1983). *Phys. Rev. B* **27,** 3867.
Yetter, W. E., Kramer, E. J., and Ast, D. G. (1982). *J. Low Temp. Phys.* **49,** 227.
Young, A. P. (1979). *Phys. Rev. B* **19,** 1855.
Zheng, J. Q., Ketterson, J. B., Falco, C. M., and Schuller, I. K. (1981). *Physica* **108***B,* 945.

11 DIFFUSION

A. L. Greer* and F. Spaepen

Division of Applied Sciences
Harvard University
Cambridge, Massachusetts

I. INTRODUCTION

DuMond and Youtz (1940) were the first to produce an artificial compositionally modulated material and the first to observe diffusion effects in such a material. They made a thin metallic film by evaporating gold and copper on to a glass substrate by using a pendulum clock and a relay to modulate the heating current to the gold source while maintaining a constant current to the copper source. The resulting layered structure had a wavelength of ~ 10 nm, and DuMond and Youtz hoped that the layers would act as artificial Bragg planes for x-ray reflection. Their intention was to use the artificial reflector to calibrate the x-ray wavelength. Although they found that the control of their deposition process was not adequate for this task, they did observe a Bragg reflection attributable to the composition modulation. The intensity of the reflection decreased with time, becoming about half of its initial value after two days at

* Present address: Department of Metallurgy and Materials Science, University of Cambridge, Cambridge, United Kingdom.

ISBN 0-12-170470-X

room temperature. The decay, which was due to interdiffusion in the material, was analyzed by DuMond and Youtz according to the simple diffusion equation

$$\partial c/\partial t = D\, \partial^2 c/\partial x^2, \tag{1}$$

where c is the atom fraction of one of the components, x a spatial coordinate (normal to the film plane in this case), t time, and D a composition-independent diffusion coefficient. They derived that for a sinusoidal composition modulation of wavelength λ, the intensity I of the corresponding Bragg reflection should decrease according to

$$\frac{d}{dt}\left[\ln\left(\frac{I}{I_0}\right)\right] = -\frac{8\pi^2}{\lambda^2} D, \tag{2}$$

where I_0 is the initial intensity. It is seen from Eq. (2) that smaller wavelength modulations will decay more rapidly. DuMond and Youtz pointed out that a general periodic composition modulation can be described by a Fourier series. The higher harmonics in that series will decay rapidly, so that after some interdiffusion the fundamental will be the only significant term. That is, the composition modulation will become sinusoidal, only the first order Bragg reflection will be observed (as was verified experimentally), and Eq. (2) can be used directly. With this simple analysis, it was possible to determine lower diffusivities than by any other technique. As we shall see later (Subsection IV.A), the technique is even more sensitive than the short diffusion distances alone would suggest.

While the procedure of DuMond and Youtz remains the basis for studies of diffusion effects in compositionally modulated materials, they erred in supposing D in Eq. (2) to be the macroscopic interdiffusion coefficient. In the presence of steep composition gradients, such as are found in modulated materials, Eq. (1) is no longer valid. The local free energy of a nonuniform system is dependent not only on the composition at a given point, but also on the spatial derivatives of the composition at that point. The consequence is that D in Eq. (2) is dependent on λ and that only by measuring D as a function of λ can the true macroscopic interdiffusion coefficient be obtained. A simple modification of the diffusion analysis to take account of composition gradient effects is described in Subsection II.A and is applied to a regular solution with only nearest neighbor interactions.

Another factor ignored in the DuMond and Youtz analysis is the presence of strain. Strain in the artificial layers can arise from differential thermal contraction (or expansion), but it is especially important when coherency is maintained in epitaxial materials. Strain and coherence effects are considered in Subsection II.B.

As the analysis of strain effects emphasizes, it is important to take account of the structure of the modulated materials in interpreting interdiffusion measurements. The materials may be amorphous, polycrystalline, or single crystal with coherent or incoherent interfaces. In crystalline materials there will be Bragg reflections from the lattice planes. A composition modulation gives rise to satellites about these peaks. The satellites about the zero-order reflection are the reflections that can be attributed, in the manner of DuMond and Youtz, to the artificial Bragg planes. A detailed analysis of the diffraction effects arising from the composition modulation itself and from the structural modulation it produces is presented in Subsection II.C.

In addition to composition gradient and strain effects, Eqs. (1) and (2) may be invalid because D is composition dependent. DuMond and Youtz suggested that this problem could be overcome by measuring D in nearly homogeneous films. It remains true that the analysis of interdiffusion is easiest when the amplitude of the composition modulation is small. Refinements of the theory to take account of the composition dependence of D are described in Subsection II.D. Consideration is also given to more realistic solution models than that of the nearest neighbor regular solution.

It is important to know the relationship between the interdiffusivity in modulated materials and conventional tracer diffusivities. The latter are measured at higher temperatures and extrapolation is necessary if a comparison is to be made. This problem is considered in Subsection II.E.

In Section III, experimental results on crystalline and amorphous metals and on semiconductors are presented and discussed. Although the majority of the work is on metallic systems, it must be emphasized that the theory of interdiffusion in compositionally modulated materials (Section II) is of general applicability and does not depend on the type of bonding involved. It is to be hoped that more information on nonmetallic systems will become available. On annealing, not all compositionally modulated materials tend to homogenize. In some systems the interdiffusion coefficient is negative; i.e., there is a tendency to phase separation. There are experimental results on such systems, and it should be pointed out that the theory is capable of dealing with these cases also.

In studying interdiffusion in artificial compositionally modulated materials there are two main sources of interest. The first is that extremely low interdiffusivities can be measured: no other technique is as sensitive. Second, an assessment of the stability of artificial modulated materials is of importance. These materials have many fascinating physical properties; if these properties are to be studied at length or applied technologically, the materials must be stable, both compositionally and morphologically, preferably for longer periods than the films of DuMond and Youtz. Section IV is a discussion of the theory of, and experimental results on, interdiffusion in artificial modulated materials, with particular reference to these two areas.

II. THEORY

A. Diffusion in a Steep Concentration Gradient

The first analysis of diffusion in a steep concentration gradient was given by Hillert (1956). He considered a crystalline binary solid solution with compositional variations in one dimension. The free energy of the nonhomogeneous system was calculated by assuming only nearest-neighbor interactions between the atoms and an ideal configurational entropy (i.e., a regular solution). Adapting an analysis of Becker (1938), Hillert showed that the excess local free energy due to a concentration gradient is proportional to the square of that gradient. In the diffusion equation [Eq. (1)], this leads to an extra term in $\partial^4 c/\partial x^4$ on the right-hand side. Hillert considered discrete atomic planes, and obtaining the numerical solution of a set of difference equations, he found equilibrium compositional variations in systems exhibiting spinodal decomposition and ordering. Later (Hillert, 1961), again considering discrete atomic planes, he used the modified diffusion equation to calculate the kinetics of these processes.

A different approach for obtaining the free energy of a nonhomogeneous system was used by Cahn and Hilliard (1958). They considered a continuum in which three-dimensional composition variations might arise. Expressing the local free energy as the sum of Taylor series of local composition derivatives and taking just a few terms, Cahn and Hilliard also found that the local free energy depends on the square of the local composition gradient. The continuum diffusion equation resulting from this was derived by Cahn (1961, 1968), who obtained analytic solutions for small departures from homogeneity.

The work of Cook *et al.* (1969) combined aspects of both the previous treatments. They considered three-dimensional compositional variations on a discrete lattice and did not make any assumptions about the thermodynamics of the binary solution. All these treatments of the free energy of nonhomogeneous systems can be used to derive the equilibrium form and energy of the interface between two domains of different composition. This is one example in a general class of problems in calculating the variation of a parameter at an interface. In particular, the continuum analysis of Cahn and Hilliard (1958) is similar to earlier treatments of the density variation across a single-component liquid–gas interface (van der Waals, 1890), magnetic domain walls (Bloch, 1932), ferroelectric domain walls (Mitsui and Furuichi, 1953), and the interface between normal and superconducting domains in a metal (Ginsburg and Landau, 1950; Bardeen, 1954).

We present here an analysis of diffusion in a steep concentration gradient. In order to obtain simple analytic solutions a continuum model is used, in the manner of Cahn and Hilliard (1958). We restrict our attention to compositional variations in one dimension, because that is relevant to artificial modulated (layered) materials. Consider a solid solution consisting of A atoms and B atoms on a fixed array of sites. Suppose that c is the fraction of atoms that are A. Then the Helmholtz free energy per unit volume of the homogeneous solution f_0 will be a function of c. If the solution is not homogeneous in the x direction, the local free energy per atom will be a function of x. One might expect that the total free energy F of the system would be given by

$$F = \mathcal{A} \int f_0(c)\, dx, \tag{3}$$

where $\mathcal{A}$ is the cross-sectional area of the system normal to the x axis. This is incorrect. Although in a homogeneous system f_0 is the local free energy, in a nonhomogeneous system the local free energy depends not only on the local composition, but also on the composition of the surrounding volume. It is convenient to express the local free energy per volume f in Taylor series form as

$$f\left(c, \frac{\partial c}{\partial x}, \frac{\partial^2 c}{\partial x^2}, \ldots\right) = f_0(c) + \kappa_{11} \frac{\partial c}{\partial x} + \kappa_{12}\left(\frac{\partial c}{\partial x}\right)^2 + \cdots$$
$$+ \kappa_{21} \frac{\partial^2 c}{\partial x^2} + \cdots. \tag{4}$$

We will consider just these terms, assuming that higher order terms are negligible (but see Subsection II.D). If the free energy is to be invariant with respect to reflection, κ_{11} must be zero. Then the free energy of the nonhomogeneous system is given by

$$F = \mathcal{A} \int \left[f_0(c) + \kappa_{12}\left(\frac{\partial c}{\partial x}\right)^2 + \kappa_{21} \frac{\partial^2 c}{\partial x^2}\right] dx.$$

Integrating by parts and taking into account the integration limits, this reduces to

$$F = \mathcal{A} \int \left[f_0(c) + \kappa_{12}\left(\frac{\partial c}{\partial x}\right)^2 - \frac{\partial \kappa_{21}}{\partial c}\left(\frac{\partial c}{\partial x}\right)^2\right] dx$$
$$= \mathcal{A} \int \left[f_0(c) + \kappa \left(\frac{\partial c}{\partial x}\right)^2\right] dx, \tag{5}$$

where $\kappa = \kappa_{12} - (\partial \kappa_{21}/\partial c)$ is called the *gradient energy coefficient*. We will now examine the influence of the gradient energy term on the interdiffusion of A and B atoms. We choose, as is always possible, a frame of reference such that the flux of B atoms is equal and opposite to the flux of A atoms. The interdiffusion flux $\tilde{J}$ is then equal to the flux of A atoms and is given by

$$-\tilde{J} = M \frac{\partial}{\partial x}(\mu_A - \mu_B). \tag{6}$$

The mobility M is always positive, and μ_A and μ_B are the partial free energies per unit volume of A and B, i.e., the chemical potentials per molar volume. The quantity $(\mu_A - \mu_B)$ is the change in free energy on reversibly replacing a unit volume of B atoms with A atoms. From Eq. (5),

$$\mu_A - \mu_B = f_0' - 2\kappa \frac{\partial^2 c}{\partial x^2}, \tag{7}$$

where $f_0' = \partial f_0/\partial c$. The interdiffusion flux is then found to be

$$-\tilde{J} = Mf_0'' \frac{\partial c}{\partial x} - 2M\kappa \frac{\partial^3 c}{\partial x^3}, \tag{8}$$

where $f_0'' = \partial^2 f_0/\partial c^2$. Assuming that M, f_0'', and κ are independent of composition, and taking the divergence, the following linearized diffusion equation is

$$\frac{\partial c}{\partial t} = \tilde{D} \frac{\partial^2 c}{\partial x^2} - \frac{2\tilde{D}}{f_0''} \kappa \frac{\partial^4 c}{\partial x^4}, \tag{9}$$

where $\tilde{D}$ is the interdiffusion coefficient and is Mf_0''. The assumption that M, f_0'', and κ are independent of composition is reasonable when the deviations from homogeneity are small; the effects that arise when this is not the case are considered in Subsection II.D. A particular solution to Eq. (9) of interest here is

$$c - c_0 = \exp\left[-\tilde{D}\beta^2\left(1 + \frac{2\kappa\beta^2}{f_0''}\right)t\right]\cos \beta x, \tag{10}$$

where c_0 is the composition of the system if it were homogeneous, t time, and $\beta = 2\pi/\lambda$ the wave number of the composition wave. This cosine wave in composition may decay or grow in amplitude A, depending on the sign of the *amplification factor R*:

$$\frac{d}{dt}(\ln A) = R = -\tilde{D}\left(1 + \frac{2\kappa\beta^2}{f_0}\right)\beta^2 = -\tilde{D}_\lambda \beta^2, \tag{11}$$

where $\tilde{D}_\lambda$ is an effective interdiffusion coefficient, dependent on the wavelength λ, of the composition modulation under consideration. If there were no gradient energy effects, the interdiffusion coefficient would be independent of λ. This simple case, valid in the absence of steep concentration gradients, is described by Eqs. (7)–(11) if $\kappa = 0$ is substituted. It should be noted that in the analysis of DuMond and Youtz, Eq. (2) is correct if $\tilde{D}_\lambda$ is substituted for D.

Equation (10) describes the evolution of a simple harmonic composition wave. In an artificial modulated material the periodic composition variation can be described by a sum of such waves. The advantage of the linearized diffusion equation [Eq. (9)] is that this sum is also a solution and that the components of the sum can be considered independent. Equation (9) is thus very useful in analyzing the annealing behavior of a compositionally modulated material, but it has the disadvantage that it is valid only for composition modulations of small amplitude. The restriction to small amplitude also applies to the analytic treatment of Cook *et al.* (1969). Hillert's numerical treatment

(1956), however, is valid for any amplitude (see Subsection II.D). In examining the various types of annealing behavior we will use Eqs. (9) and (10) and will for simplicity consider a single harmonic modulation. We need to know κ and f_0'', and we can obtain these only from a model for the binary solution.

The model we have chosen is that of the *regular solution,* i.e., a solution with ideal configurational entropy and an energy varying parabolically with composition. The variation of the energy is derived here by considering nearest-neighbor interactions only. The drawbacks of this model [also known as the "zeroth-order approximation" (Guggenheim, 1952)] have been assessed by Christian (1975). Nonetheless, the model does give good qualitative descriptions of the behavior of binary solutions. We use it here because of its simplicity and because of the difficulty in experimental verification of more complex models.

In the binary solution we assume that each atom has z nearest neighbors. The bond energy (a negative quantity) between two A atoms is V_{AA}, between two B atoms V_{BB}, and between an A and a B atom V_{AB}. Then, given a random distribution of atoms on the sites, the internal energy of a system of N sites is $\frac{1}{2}Nz[c^2V_{AA} + (1-c)^2V_{BB} + 2c(1-c)V_{AB}]$, where c is the fraction of A atoms. The internal energy of the pure components is $\frac{1}{2}Nz[cV_{AA} + (1-c)V_{BB}]$, so that the change in internal energy ($\approx$ enthalpy, for a condensed system) on mixing is given by

$$\Delta U = Nzc(1-c)\Delta V, \tag{12}$$

where $\Delta V = V_{AB} - \frac{1}{2}(V_{AA} + V_{BB})$. The configurational entropy of mixing in a system of N atoms is

$$\Delta S = -k_B N[c \ln c + (1-c)\ln(1-c)], \tag{13}$$

where k_B is Boltzmann's constant. Depending on the sign of the interaction energy ΔV, ΔU can be positive or negative, but ΔS is always positive. The behavior of a binary solution can be understood by considering the free energy of mixing, $\Delta F = \Delta U - T\,\Delta S$. When ΔV is negative, A–B bonds are preferred and the system has a tendency to order. Provided that an ordered compound does not form, however, the regular solution model can be used, and the variation of ΔF with composition is as shown schematically in Fig. 1a. At all temperatures f_0'' $(=(1/N_V)\,\partial^2\,\Delta F/\partial c^2$, where N_V is the number of atoms per unit volume) is positive and the equilibrium state is homogeneous. The variation shown in Fig. 1a also applies if ΔV is positive when the temperature is high enough. In that case, the entropy [Eq. (13)] dominates and the equilibrium state is homogeneous, despite the fact that A–B bonds are not favored. At low temperature, however, the bond energy dominates and the free energy of mixing is as shown in Fig. 1b. It is now favorable for the system to separate into two phases. The sign of f_0'' is a function of c. When f_0'' is negative, the homogeneous system is unstable to arbitrarily small composition fluctuations; i.e., *spinodal decomposition* can occur.

Having dealt with the free energy of homogeneous solutions, we must now

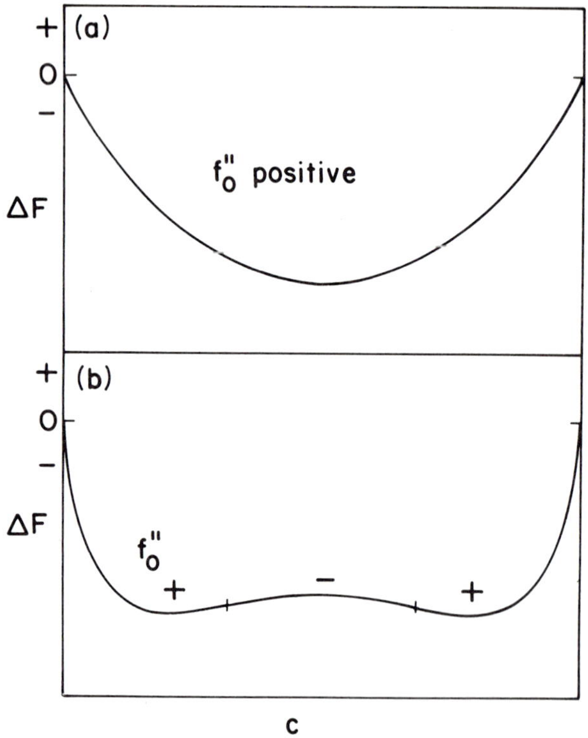

Fig. 1. The free energy of mixing of a solution of A and B atoms as a function of the fraction c of A atoms: (a) applies at any temperature if ΔV is negative or at high temperature if ΔV is positive; (b) applies at low temperature if ΔV is positive.

calculate the effect of a composition gradient. We consider a system of parallel atomic planes normal to the x axis. Each plane is homogeneous, but the compositions of the planes c_x are a function of x. Each plane contains N_A atoms per unit area. The excess free energy of the system due to inhomogeneity arises entirely from the bond energy, and not the entropy of mixing (Cahn and Hilliard, 1958). We wish to calculate the energy of bonding between an atomic plane of composition c and a neighboring plane of composition $c + \Delta c$. Assuming a random distribution of atoms on each plane and taking y bonds per atom out of the planes (on each side), the energy between the planes per unit area is

$$E_c^{c+\Delta c} = N_A yc(c + \Delta c)V_{AA} + N_A y(1 - c)(1 - c - \Delta c)V_{BB} + N_A yc(1 - c - \Delta c)V_{AB} + N_A y(1 - c)(c + \Delta c)V_{AB}. \quad (14)$$

The energy of the bonding if each plane were of composition c would be

$$E_c^c = N_A yc^2 V_{AA} + N_A y(1 - c)^2 V_{BB} + 2N_A yc(1 - c)V_{AB}. \quad (15)$$

There is thus an excess energy associated with the change of composition Δc. We assign half of the bonding energy between the planes to each plane. The excess energy per unit area of the first plane (of composition c), due to its neighbor having a composition $c + \Delta c$, is

$$\Delta E_1 = \tfrac{1}{2} N_A y \, \Delta c [c V_{AA} + (c - 1) V_{BB} + (1 - 2c) V_{AB}]. \tag{16}$$

Note that this excess is linear in Δc. Consider a plane of composition c lying in a uniform composition gradient so that one neighboring plane is of composition $c + \Delta c$ and the other of composition $c - \Delta c$. The excess energy of the central plane due to the composition gradient is zero. In Eq. (5), then, $\kappa_{12} = 0$, a result first derived by Cahn and Hilliard (1958).

Suppose that the variation of composition with distance has some curvature, as represented for example by a plane of composition c, between neighbors both of composition $c + \Delta c$. Now the excess energy of the central plane is

$$\Delta E_2 = N_A y \, \Delta c [c V_{AA} + (c - 1) V_{BB} + (1 - 2c) V_{AB}]. \tag{17}$$

The excess energy per unit volume is obtained by dividing by the distance between the planes d and is $\kappa_{21} \, \partial^2 c/\partial x^2$ [Eq. (5)]. In this case, $\partial^2 c/\partial x^2 = 2 \, \Delta c/d^2$, so that

$$\kappa_{21} = (N_A y d/2)[c V_{AA} + (c - 1) V_{BB} + (1 - 2c) V_{AB}]. \tag{18}$$

The gradient energy coefficient $\kappa = \kappa_{12} - \partial \kappa_{21}/\partial c$ is then given by

$$\kappa = N_A y \, \Delta V = N_V y d^2 \, \Delta V, \tag{19}$$

where N_V is the number of atoms per unit volume. For (111) planes in a face-centered cubic material, for example, $y = z/4$ and $d = \sqrt{\tfrac{2}{3}} a_0$, where a_0 is the atomic diameter. Therefore,

$$\kappa = N_V z a_0^2 \, \Delta V/6. \tag{20}$$

Cahn and Hilliard (1958) did not assume just nearest neighbor interactions in their regular solution model. They expressed κ in terms of a rms interaction distance λ, which is sensitive to the nature of the long-range interactions. For nearest-neighbor interactions only, $\lambda = a_0/\sqrt{3}$, and their expression for κ is identical to Eq. (20). From Eq. (20) it can be seen that κ is positive for phase-separating systems and negative for ordering systems. A consequence of the positive κ in phase-separating systems is that the interface between two phases in equilibrium is diffuse. The equilibrium width of the interface is found by minimizing the integral in Eq. (5). The gradient energy contribution is reduced by having a broad interface, but the f_0 contribution (see Fig. 1b) is reduced by having a narrow interface.

Knowing κ and f_0'' for a regular solution, we are now able to calculate the annealing behavior of a harmonic composition modulation. Qualitatively, the behavior depends on the signs of κ and f_0''. Figures 2–4 show schematically the

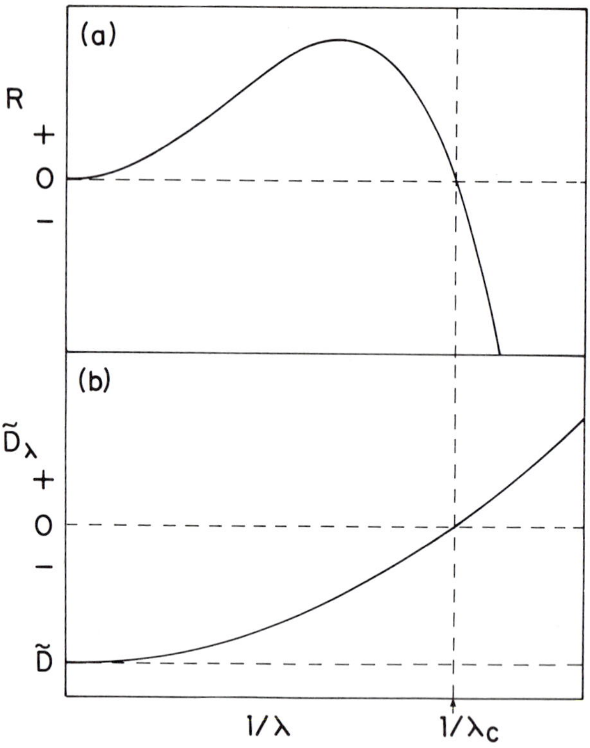

Fig. 2. The amplification factor R and the effective interdiffusivity $\tilde{D}_\lambda$ as a function of $1/\lambda$ for a system in which κ is positive and f_0'' is negative. (Applies to spinodal decomposition.)

amplification factor R and the effective interdiffusion coefficient $\tilde{D}_\lambda$ as a function of wavelength [Eq. (11)] for three important cases: κ positive and f_0'' negative; κ positive and f_0'' positive; and κ negative and f_0'' positive. The fourth combination, κ and f_0'' both negative, is very unlikely to arise and is impossible in a regular solution with nearest-neighbor interactions.

Diffusion in a steep concentration gradient was first studied in connection with spinodal decomposition, and we will begin with that case. For phase separation to be favorable, ΔV must be positive (κ positive) and the temperature must be low enough. If f_0'' is negative (Fig. 1b), the system, in the absence of gradient energy effects, is unstable to small composition modulations of any wavelength. With gradient energy effects, R and $\tilde{D}_\lambda$ are as shown in Fig. 2. When λ is large, gradient energy effects are negligible and $\tilde{D}_\lambda$ is negative as expected in phase separation. As λ is decreased, however, $\tilde{D}_\lambda$ becomes positive. There is, thus, a critical wavelength λ_c below which R and $\tilde{D}_\lambda$ change sign and the system, although it could lower its free energy by macroscopic phase separation, is stable to fluctuations of small amplitude. The existence of this critical wavelength was first noted by Hillert (1956), and it shows simply that when λ is small

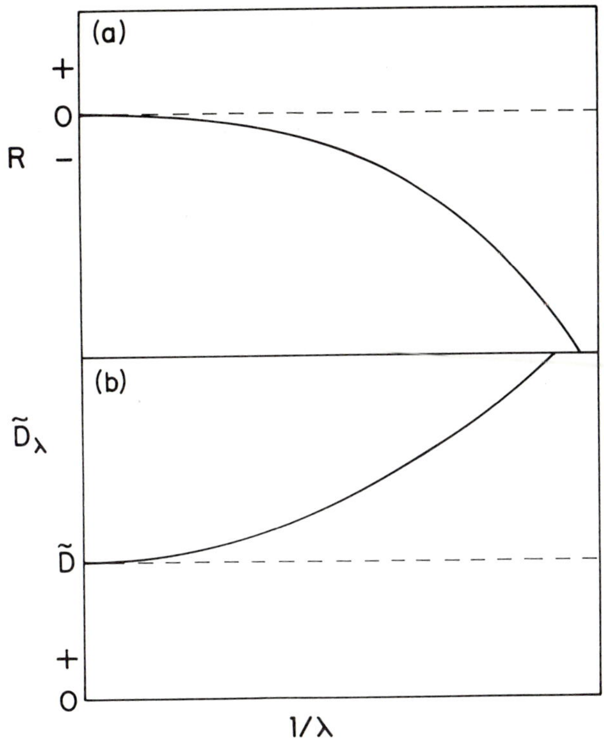

Fig. 3. The amplification factor R and the effective interdiffusivity $\tilde{D}_\lambda$ as a function of $1/\lambda$ for a system in which κ is positive and f_0'' is positive. (Applies to a phase-separating system outside the spinodal.)

the interfacial energy in the system would be too great to permit spinodal decomposition. Hillert showed that with $\lambda < \lambda_c$, modulations of large amplitude would still grow. There exists, however, a lower critical value of λ below which the system is stable to any modulation. With the present analysis we can examine only the stability of the system to infinitesimal composition modulations.

On annealing an initially homogeneous sample, a number of composition modulations with $\lambda > \lambda_c$ will develop, the most favored being the one for which $\lambda = \sqrt{2}\lambda_c$. We are, however, interested in artificial modulations in which a particular wavelength is imposed on the system. If that wavelength is greater than λ_c, the modulation will grow in amplitude. In this experiment it would be desirable to have a small initial amplitude. If $\lambda < \lambda_c$, the modulation will decay. The rate of decay will not be given accurately by Eq. (11) because the continuum approach used here breaks down at very short wavelengths, as discussed later.

We consider next the same system (ΔV and κ positive), but with a composi-

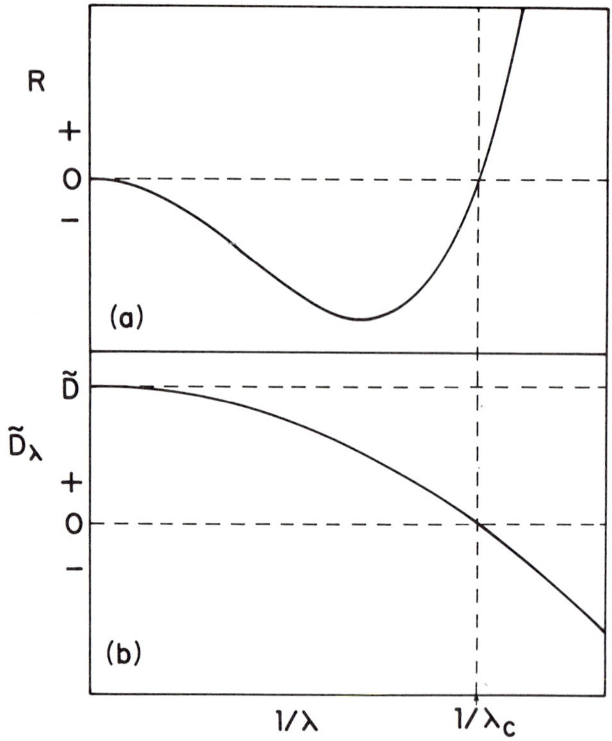

Fig. 4. The amplification factor R and the effective interdiffusivity $\tilde{D}_\lambda$ as a function of $1/\lambda$ for a system in which κ is negative and f_0'' is positive. (Applies to ordering.)

tion outside the spinodal, i.e., with $f_0'' > 0$. Such a solution might or might not be able to lower its free energy by phase separation, but in any case it is stable to small composition fluctuations. As shown in Fig. 3, any composition modulation will decay. This is true also of a phase-separating system at high temperature when the free energy as a function of composition is as shown in Fig. 1a.

In a system with a tendency to order, κ is negative and f_0'' is positive (Fig. 1a). In this case, as can be seen from Fig. 4, large-wavelength modulations will decay. This is expected, since the equilibrium state of the system is homogeneous. At small wavelengths, however, a modulation will grow. This apparent phase separation on such a fine scale is in fact a one-dimensional ordering. Again, it must be stressed that the kinetics of this ordering are not well described by the continuum approximation.

We have now seen the categories of behavior predicted by the continuum model. The model has the advantage of simplicity, but it breaks down at low wavelengths. This problem can be solved only by taking a fully microscopic approach, such as in the work of Cook *et al.* (1969). They derived the analog of

Eq. (9) for atomic interchange between discrete sites and considered diffusion in a general direction in simple cubic, bcc and fcc structures. In the special cases of diffusion along the ⟨100⟩ or ⟨110⟩ directions in bcc structures, or along ⟨100⟩ or ⟨111⟩ in fcc structures, their treatment is simplified, and the amplification factor R for a harmonic wave is given by an expression of the form of Eq. (11) in which the wave number squared β^2 is replaced by the function B^2, given by the dispersion relation

$$B^2 = (2/d^2)[1 - \cos(2\pi d/\lambda)], \qquad (21)$$

where d is the spacing between the atomic planes perpendicular to the diffusion direction. Note that for $\lambda \gg d$, the cosine can be expanded and $B^2 = 4\pi^2/\lambda^2$, the value of β^2 in the continuum model.

The two models are compared in Fig. 5. In the discrete model, the behavior is symmetric about $\lambda = 2d$. There are, in the range of d/λ shown in the figure, two values of λ for a given d that give the same value of B^2 and hence the same

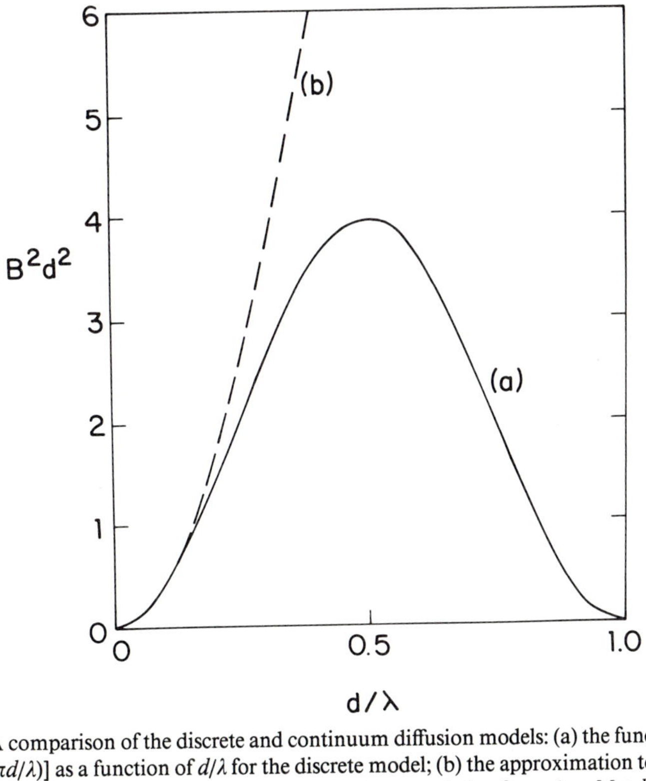

Fig. 5. A comparison of the discrete and continuum diffusion models: (a) the function $B^2d^2 = 2[1 - \cos(2\pi d/\lambda)]$ as a function of d/λ for the discrete model; (b) the approximation to B^2d^2 in the continuum model, i.e., $\beta^2 d^2 = (2\pi d/\lambda)^2$. [Reprinted with permission from *Acta Metall.* **17**, Cook *et al.* Copyright 1969, Pergamon Press, Ltd.]

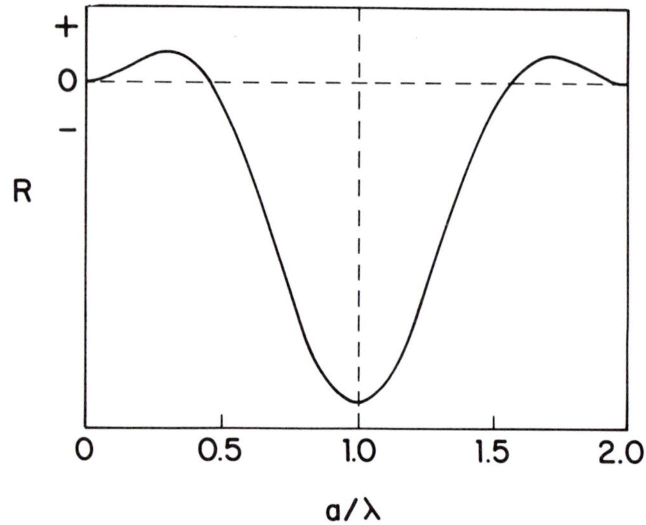

Fig. 6. The amplification factor R versus a/λ for a phase-separating system inside the spinodal. [Reprinted with permission from *Acta Metall.* **17,** Cook *et al.* Copyright 1969, Pergamon Press, Ltd.]

amplification factor. These wavelengths correspond to composition waves in a discrete lattice that are physically equivalent. It can be seen that the continuum approach is a good approximation for $\lambda > \sim 6d$. The general characteristics of the discrete analysis can be illustrated by considering diffusion normal to the {200} planes in bcc and fcc structures. Figures 6–9, redrawn from Cook *et al.*

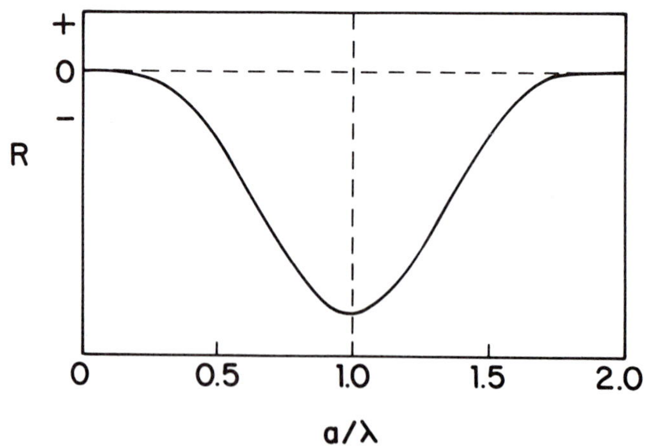

Fig. 7. The amplification factor R versus a/λ for a phase-separating system outside the spinodal. [Reprinted with permission from *Acta Metall.* **17,** Cook *et al.* Copyright 1969, Pergamon Press, Ltd.]

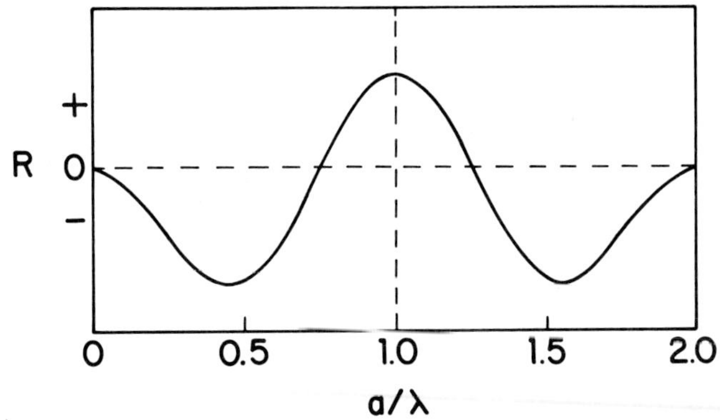

Fig. 8. The amplification factor R versus a/λ for an ordering system below the critical temperature. [Reprinted with permission from *Acta Metall.* **17**, Cook et al. Copyright 1969 Pergamon Press, Ltd.]

(1969), show the amplification factor calculated from Eqs. (11) and (21) with $d = a/2$, where a is the lattice parameter.

Figure 6 is for a phase-separating system inside the spinodal. Note that in the range $a < \lambda < \infty$ the behavior is similar to that in Fig. 2a. As the critical wavelength λ_c tends to infinity, the ratio of the most-favored wavelength (for which R is a maximum) to λ_c approaches $\sqrt{2}$, as in the continuum model. Figure 7 applies to a phase-separating system outside the spinodal. Note that the amplification factor is negative for any composition fluctuation. For an ordering system below its critical temperature, the variation of R is shown in Fig. 8. As

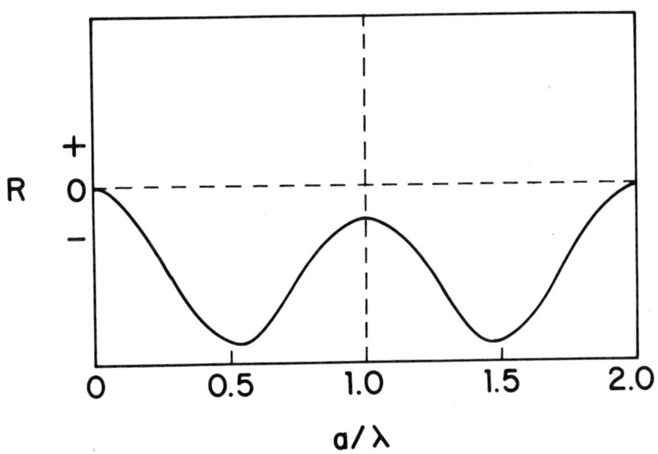

Fig. 9. The amplification factor R versus a/λ for an ordering system above the critical temperature. [Reprinted with permission from *Acta Metall.* **17**, Cook *et al.* Copyright 1969 Pergamon Press, Ltd.]

seen from the continuum model, composition waves of small wavelength will grow. The favored wavelength is $\lambda = a$, corresponding to an alternating composition of successive {200} planes. In the continuum model, R becomes infinite as λ goes to zero. In the discrete model this is avoided, and as a consequence the maximum value of R (at $\lambda = a$) may be negative. This is shown in Fig. 9, which applies to an ordering system above its critical temperature; the system is stable to small composition fluctuations of any wavelength.

Most measurements of interdiffusion coefficicnts from artificial compositionally modulated materials are made in systems that homogenize on annealing. We have seen that for moderate wavelengths ($\lambda \geq 2 \times$ atomic spacing) this can occur in ordering systems or in phase-separating systems outside the spinodal. The two cases can be distinguished from the λ dependence of $\tilde{D}_\lambda$. For single-crystalline materials, an analysis based on that of Cook *et al.* (1969) is most appropriate; for amorphous and polycrystalline materials the continuum analysis is useful.

Yamauchi and Hilliard (1972), starting from a microscopic model for diffusion, derived an equation for diffusion on a lattice. They showed that, even in the absence of gradient energy effects (for an ideal solution with $\Delta V = 0$, $\kappa = 0$), a continuum diffusion equation does not provide an accurate description. Their treatment leads to a correction term in the amplification factor for a harmonic composition wave, which has the same λ dependence as the gradient energy and so acts as a correction to κ in the general case. The correction term is anisotropic, and for cubic lattices Yamauchi and Hilliard showed that it has a magnitude $\frac{1}{6}$ to $\frac{1}{2}$ that of the gradient energy term, depending on the type of lattice and the direction of diffusion. Although the correction is significant, it does not extend the range of validity of the continuum approach. The implausible divergence of R and $\tilde{D}_\lambda$ at low wavelength, seen in Figs. 2 and 4, is still present.

None of the treatments described so far (Hillert, 1956; Cahn and Hilliard, 1958; Cook *et al.*, 1969) has taken account of differential stresses in the layers of a compositionally modulated material. Stress effects are analyzed in the following section. They are most important in single-crystalline materials composed of atoms of different sizes. The simple treatments just described, however, are likely to be applicable to systems that are amorphous or where the atomic size difference is small.

B. Strain Effects

The first treatment of strain effects on the free energy of a compositionally modulated solid was by Cahn (1961). He considered a three-dimensional isotropic solid with no long-range elastic strain fields. In a crystal of this type, with no structural imperfections, coherency strains will arise if the molar volume is a function of composition. Provided that the amplitude of the composition modulation is small, it is reasonable to suppose that no dislocations will arise as a consequence of the modulation. Also for small amplitude, the stress-free molar volume $V(c)$ is assumed to vary linearly with composition c. For a modulation of the form

$$c - c_0 = A \cos \beta x, \tag{22}$$

$V(c)$ is given by

$$V(c) = V(c_0)(1 + 3A\eta \cos \beta x), \tag{23}$$

where η is the linear expansion per unit composition change. In a perfect crystal the distortion perpendicular to the x axis due to the composition modulation is zero. (This condition would be relaxed only if the sample width were as small as the modulation wavelength.) On the other hand, the distortion parallel to the x axis is unrestricted and there is no stress component in that direction. The elastic strain in the crystal is the difference between the total strain and the stress-free strain, and its components are

$$\begin{aligned} \epsilon_{xx} &= A\eta[2\nu/(1-\nu)] \cos \beta x, \\ \epsilon_{yy} &= \epsilon_{zz} = -A\eta \cos \beta x, \\ \epsilon_{xy} &= \epsilon_{yz} = \epsilon_{zx} = 0, \end{aligned} \tag{24}$$

where ν is Poisson's ratio. The stress components are

$$\begin{aligned} \sigma_{xx} &= 0, \\ \sigma_{yy} &= \sigma_{zz} = -[AE\eta/(1-\nu)] \cos \beta x, \\ \sigma_{xy} &= \sigma_{yz} = \sigma_{zx} = 0, \end{aligned} \tag{25}$$

where E is the Young's modulus for the average composition c_0. From these components, Cahn (1961) derived that the average elastic energy per unit volume is independent of the wavelength of the composition modulation and has the value

$$A^2\eta^2 E/2(1-\nu). \tag{26}$$

The elastic energy due to a general composition modulation is then found by analyzing that modulation into its Fourier components. The total elastic energy of the system is

$$\frac{\eta^2 E}{1-\nu} \mathcal{A} \int (c - c_0)^2 \, dx, \tag{27}$$

where, as before, $\mathcal{A}$ is the cross-sectional area of the system perpendicular to the x axis. Note that the elastic energy is independent of the details of the modulation, depending only on the integrated square deviation from the average composition. The total free energy of the system F is given by

$$F = \mathcal{A} \int \left[f_0(c) + \kappa \left(\frac{\partial c}{\partial x} \right)^2 + \frac{\eta^2 E}{1 - \nu} (c - c_0)^2 \right] dx. \tag{28}$$

It is necessary to define the free energy of the homogeneous, stress-free material $f_0(c)$ with some care. It is the free energy of the number of atoms that would occupy unit volume if the composition were c_0. In the case in which the molar volume is independent of composition ($\eta = 0$), Eq. (28) reduces to the earlier Eq. (5).

Cahn (1961) applied this analysis to spinodal decomposition in phase-separating binary systems ($\kappa > 0$). Ignoring strain effects, the amplification factor R for a harmonic composition modulation is given by Eq. (11). With f_0'' positive, R is negative; i.e., the solution is stable to fluctuations of all wavelengths. With f_0'' negative, the solution is unstable to fluctuations with $\lambda = 2\pi/\beta > (8\kappa\pi^2/f_0'')^{1/2}$. In an infinite system the absolute limit of stability is $f_0'' = 0$. The correction for noninfinite samples is negligible. Thus, the gradient energy term leaves the limit of stability unaffected. When strain effects are present, however, the expression for the amplification factor becomes

$$R = -\tilde{D}\{1 + (2\kappa\beta^2/f_0'') + [2\eta^2 E/f_0''(1 - \nu]\}\beta^2, \tag{29}$$

and the limit of stability is altered to

$$f_0'' + 2\eta^2 E/(1 - \nu) = 0. \tag{30}$$

Thus, strain effects tend to stabilize the homogeneous solution, which must be undercooled below the spinodal temperature ($f_0'' = 0$) to obtain decomposition. Cahn (1961) computed that the required undercooling could be as great as 200 K in a case such as Au–Ni, where the atomic sizes are quite different.

In a subsequent paper, Cahn (1962) removed the restriction of his analysis to isotropic materials. In the analysis of spinodal decomposition in cubic crystals, the only factor with anisotropy is the elastic strain energy. Cahn showed that amplification factor for a harmonic wave is given by

$$R = -\tilde{D}[1 + (2\kappa\beta^2/f_0'') + (2\eta^2/f_0'')Y_{\langle \mathrm{UVW} \rangle}]\beta^2, \tag{31}$$

in which $Y_{\langle \mathrm{UVW} \rangle}$ is a modulus, dependent on the direction [UVW] of the wave vector of the composition modulation. In an isotropic material, $Y_{\langle \mathrm{UVW} \rangle}$ has the value $E/(1 - \nu)$ and as such appears in Eqs. (28)–(30). For the high-symmetry directions $\langle 100 \rangle$, $\langle 110 \rangle$, and $\langle 111 \rangle$ in a cubic crystal, $Y_{\langle \mathrm{UVW} \rangle}$ is given by

$$Y_{\langle \mathrm{UVW} \rangle} = \frac{1}{2}(C_{11} + 2C_{12}) \times \left[3 - \frac{C_{11} + 2C_{12}}{C_{11} + 2(2C_{44} - C_{11} + C_{12})(l^2m^2 + m^2n^2 + n^2l^2)} \right], \tag{32}$$

where C_{11}, etc., are the stiffness coefficients and l, m, and n are the direction cosines of the directions $\langle UVW \rangle$ with respect to the cubic axes. With stiffness coefficients $(2C_{44} - C_{11} + C_{12}) > 0$, the elastic energy is minimum for composition modulations with the wave vector parallel to $\langle 100 \rangle$ directions. In the case with $(2C_{44} - C_{11} + C_{12}) < 0$, the energy is minimum for wave vectors parallel to $\langle 111 \rangle$. The former case is more common for metals, and indeed a microstructure arising from the superposition of $\{100\}$ composition waves is often observed in spinodal decomposition. This is a confirmation of the role of elastic energies. The anisotropy of these energies suggests that in artificially modulated thin films, the diffusion results should be dependent on the crystal orientation of the films.

The continuum approach just used is no longer valid when the elastic strains vary over distances similar to interatomic spacings in the crystal. In these cases, notably in order–disorder reactions, a discrete analysis must be used. Cook and deFontaine (1969, 1971) used the microscopic theory of elasticity, as formulated by Born and Huang (1954), to derive the elastic free energy of binary solid solutions. In contrast to the results for the continuum, it was found that the excess free energy of a system due to a harmonic composition modulation depends not only on the direction, but also on the magnitude of the wave vector of the modulation. This wavelength dependence can be described by using an effective modulus to replace $Y_{\langle UVW \rangle}$ in Eq. (31). For high-symmetry directions in cubic crystals, the effective modulus for a particular direction $M_{\langle UVW \rangle}$ is dependent only on d/λ, where λ is the modulation wavelength and d the spacing between the atomic planes perpendicular to $\langle UVW \rangle$. In the long-wavelength limit, the moduli $M_{\langle UVW \rangle}$ tend to the values $Y_{\langle UVW \rangle}$ given by the continuum analysis, but for short wavelengths inconsistencies arise if the continuum analysis is used. For example, as pointed out by Cook and deFontaine (1971), the ordered β-brass structure can be obtained by imposing on a homogeneous bcc structure a harmonic composition wave, either along $\langle 100 \rangle$ with $\lambda = a$, or along $\langle 111 \rangle$ with $\lambda = a/\sqrt{3}$. In this case the spacings of the relevant atomic planes are $d_{\langle 100 \rangle} = a/2$ and $d_{\langle 111 \rangle} = a/2\sqrt{3}$, where a is the lattice parameter. These two composition waves, with the same amplitude, correspond to the same structure and must therefore have the same elastic energy. As the wavelength of the composition modulations is decreased from values at which the continuum approach is valid ($M_{\langle UVW \rangle} = Y_{\langle UVW \rangle}$), the moduli $M_{\langle 100 \rangle}$ and $M_{\langle 111 \rangle}$ must approach each other, becoming equal at $d/\lambda = \frac{1}{2}$. This behavior is illustrated in Fig. 10.

The dependence of $M_{\langle UVW \rangle}$ on λ can be described by using a Taylor expansion about $1/\lambda = 0$. The first term in the expansion is simply $Y_{\langle UVW \rangle}$, and since the first derivative is zero, the second term is proportional to the second derivative. There is, then, in effect a gradient energy coefficient κ_E for the elastic energy. This coefficient is analogous to the "chemical" gradient energy coefficient, κ_c ($\kappa_c = \kappa$ in Subsection A). Cook and deFontaine (1971) calculated κ_E as

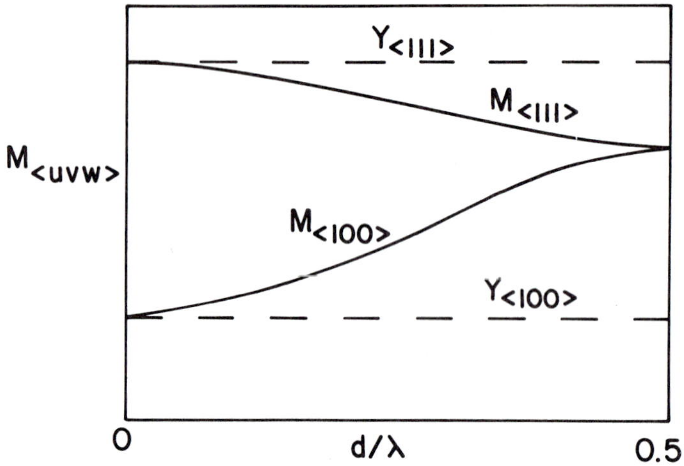

Fig. 10. A schematic diagram of the variation of elastic moduli with d/λ for a bcc lattice. The effective moduli in a discrete lattice are $M_{\langle 100\rangle}$ and $M_{\langle 111\rangle}$. The corresponding moduli in a continuum are $Y_{\langle 100\rangle}$ and $Y_{\langle 111\rangle}$. [Reprinted with permission from *Acta Metall.* **19**, Cook and deFontaine. Copyright 1971, Pergamon Press, Ltd.]

a function of the distortion parameter η $[=(1/a)(da/dc)]$ by assuming only nearest neighbor interactions. They draw the general conclusion for metals that κ_E will be negligible, unless $\eta > 0.1$, i.e., unless the atomic size difference exceeds $\sim 10\%$.

Although κ_E is small, the contribution of elastic energy to the bulk free energy of a solid solution $f_0(c)$ is significant, at least from $\eta \gtrsim 0.04$. This has two important consequences. The first is that in calculating the chemical gradient energy coefficient from the total energy of the system, following Cahn and Hilliard, the elastic contribution must first be subtracted from the total energy. Otherwise, an overestimate of κ_c will be obtained (Cook and deFontaine, 1971). Second, the elastic energy contributions to $f_0(c)$, and hence to $f_0''(c)$, favor phase separation. This is in apparent conflict with the earlier statement, derived from Eq. (30), that strain effects tend to stabilize the homogeneous solution. Although the strain effects embodied in the second term in Eq. (30) do inhibit phase separation, they are outweighed by the elastic energy contributions to f_0''. Ordering reactions are also promoted by elastic energy effects. Using a nearest-neighbor interaction model, Cook and deFontaine (1969, 1971) showed that, although only a small decrease in elastic energy will occur on going from a random solute distribution to an ordered Cu_3Au structure, the elastic energy is completely removed on forming an ordered β-brass structure.

The contribution of elastic energy to the bulk free energy of the system can be ignored in comparing the behavior of a compositionally modulated coherent structure with that of a structure in which dislocations, in partially coherent or incoherent interfaces, can relieve the strain. Since the effective moduli $M_{\langle UVW \rangle}$ are positive, homogenization will always be relatively favored in the coherent structure.

In all of the treatments described in this subsection, the dependence of the elastic moduli on composition has been ignored. Including this effect destroys the linearity of the analyses. This is discussed in Subsection D.

C. X-Ray Measurements

The presence of a one-dimensional variation, such as a compositional or displacement modulation that is a function of one position coordinate only, causes diffraction effects only in that direction in reciprocal space (Guinier, 1963a). In order to investigate these effects, it is therefore sufficient to use scattering vectors **k** that are parallel to this direction. Measurements can therefore conveniently be performed in $\theta - 2\theta$ scans on a conventional diffractometer.

A harmonic binary compositional modulation, such as that of Eq. (10),

$$c(x) = c_0 + A \sin(\beta x) \tag{33}$$

corresponds to a modulation in the average atomic scattering factor:

$$\mathcal{F}(x) = \mathcal{F}_0 + \Delta\mathcal{F} \sin(\beta x), \tag{34}$$

where $\mathcal{F}_0 = c_0\mathcal{F}_1 + (1 - c_0)\mathcal{F}_2$ and $\Delta\mathcal{F} = A(\mathcal{F}_1 - \mathcal{F}_2)$, with $\mathcal{F}_1$ and $\mathcal{F}_2$ the atomic scattering factors of the two species. Since, in a crystalline material, the lattice parameter depends on the composition, the spacing of the planes perpendicular to the modulation direction is also a function of position:

$$d(x) = d_0[1 + \epsilon \sin(\beta x)], \tag{35}$$

where d_0 is the average spacing. A strain parameter ϵ is defined by

$$\epsilon = AW\eta, \tag{36}$$

where η is the linear expansion defined in Subsection B [Eq. (23)],

$$\eta = 1/a \; da/dc \tag{37}$$

(in which a is the lattice parameter), and W is a numerical factor, equal to unity for incoherent modulations, and a function of the elastic constants of the material for coherent modulations [e.g., $W = (1 + \nu)/(1 - \nu)$ for isotropic materials; see Eq. (24)]. The modulation in $d(x)$ implies that each lattice plane is displaced from its position in the unmodulated material. The new position is then

$$x_n = \sum_{i=0}^{n} d_i = nd_0 + \sum_{i=0}^{n} \epsilon d_0 \sin(\beta i d_0), \tag{38}$$

where d_i is the spacing between the ith and $(i-1)$th planes. This sum can be approximated by an integral,

$$x_n = nd_0 + \epsilon \int_{x_0}^{nd_0} \sin(\beta x)\, dx = nd_0 - \frac{\epsilon}{\beta}\cos(\beta nd_0), \tag{39}$$

where the reference position x_0 has been chosen to keep the average displacement of the planes equal to zero.

The intensity of the diffraction effects due to the combined modulations of scattering factor and plane spacing was first calculated by Preston (1938). In the following discussion Guinier's (1963b) approach will be used since it is simple. A detailed examination of these effects, including analog modeling, has been given by deFontaine (1966) and can easily be extended to noncrystalline materials.

The amplitude of the scattered radiation from an infinitely thick modulated structure is the Fourier transform of the function

$$\mathcal{F}(x) \sum_{n=-\infty}^{+\infty} \delta(x - x_n). \tag{40}$$

Note that in this problem the transform is one-dimensional, since only scattering vectors $\mathbf{k}$ in the x direction need be considered.

The transform of the position function is calculated first:

$$\begin{aligned} g_1(k) &= \int \sum_n \delta(x - x_n)\exp(ikx)\, dx \\ &= \int \sum_n \delta\left[x - nd_0 + \frac{\epsilon}{\beta}\cos(\beta nd_0)\right]\exp(ikx)\, dx \\ &= \sum_n \exp\left\{ik\left[nd_0 - \frac{\epsilon}{\beta}\cos(\beta nd_0)\right]\right\} \\ &= \int \sum_n \delta(x - nd_0)\exp\left[-\frac{ik\epsilon}{\beta}\cos(\beta x)\right]\exp(ikx)\, dx. \end{aligned} \tag{41}$$

Equation (41) shows that $g_1(k)$ is therefore the transform of the product of two functions, the transforms of which can be calculated separately as

$$g_2(k) = \int \sum_n \delta(x - nd_0)\exp(ikx)\, dx = \frac{2\pi}{d_0}\sum_n \delta\left(k - \frac{2\pi n}{d_0}\right) \tag{42}$$

and

$$g_3(k) = \int \exp\left[-\frac{ik\epsilon}{\beta}\cos(\beta x)\right]\exp(ikx)\, dx. \tag{43}$$

For $k\epsilon/\beta \ll 1$, the first exponential can be expanded as

$$g_3(k) \approx \int \left[1 - \frac{ik\epsilon}{\beta}\cos(\beta x)\right] \exp(ikx)\, dx$$

$$= \delta(k) - \frac{ik\epsilon}{2\beta}[\delta(k+\beta) + \delta(k-\beta)]. \tag{44}$$

The convolution of these two transforms thus gives the desired result:

$$g_1(k) = g_2(k) * g_3(k)$$

$$= \frac{2\pi n}{d_0} \sum_n \delta\left(k - \frac{2\pi n}{d_0}\right)$$

$$- \frac{\pi i k\epsilon}{\beta d_0} \sum_n \left[\delta\left(k - \frac{2\pi n}{d_0} + \beta\right) + \delta\left(k - \frac{2\pi n}{d_0} - \beta\right)\right]. \tag{45}$$

This function represents the reciprocal lattice points at $(2\pi n/d_0)$, surrounded by a symmetrical pair of satellites at $\pm\beta$, with an amplitude $(k\epsilon/2\beta)$ times that of the central peak.

The transform of $\mathcal{F}(x)$ is now calculated as

$$g_4(k) = \int [\mathcal{F}_0 + \Delta\mathcal{F}\sin(\beta x)] \exp(ikx)\, dx$$

$$= \mathcal{F}_0\delta(k) - \frac{i\,\Delta\mathcal{F}}{2}[\delta(k+\beta) - \delta(k-\beta)]. \tag{46}$$

This represents a central peak surrounded by an *antisymmetrical* pair of satellites at $\pm\beta$. The total transform of Eq. (40) is then the convolution of $g_1(k)$ and $g_4(k)$. Since these are both sharply peaked functions, this operation can easily be performed by centering $g_4(k)$ at each peak of $g_1(k)$, appropriately weighted, and summing. Since ϵ and $(\Delta\mathcal{F}/\mathcal{F}_0)$ are both small, their products can be neglected. The required convolution then becomes

$$g_5(k) = g_1(k) * g_4(k)$$

$$= \sum_n \frac{2\pi\mathcal{F}}{d_0}\delta\left(k - \frac{2\pi n}{d_0}\right)$$

$$- \frac{\pi i\mathcal{F}_0}{d_0} \sum_n \left[\left(\frac{\epsilon k}{\beta} - \frac{\Delta\mathcal{F}}{\mathcal{F}_0}\right)\delta\left(k - \frac{2\pi n}{d_0} - \beta\right)\right.$$

$$\left. + \left(\frac{\epsilon k}{\beta} + \frac{\Delta\mathcal{F}}{\mathcal{F}_0}\right)\delta\left(k - \frac{2\pi n}{d_0} + \beta\right)\right]. \tag{47}$$

This represents the reciprocal lattice points surrounded by an *asymmetrical* pair of satellites at $\pm\beta$. It should also be kept in mind that $\mathcal{F}_0$ and $\Delta\mathcal{F}$ are also functions of k. The modulation wavelength can therefore be determined from the satellite positions, whereas ϵ and $\Delta\mathcal{F}$ can be determined from measurements of the intensities of both satellites (Fleming *et al.*, 1980; Tsakalakos, 1977).

It should be noted that the satellites about the (000) reflection, or forward scattering ($k = 0$), *are* symmetrical and are proportional to $\Delta\mathcal{F}^2$ and hence also to the square of the composition amplitude A^2. The modulation in the interplanar spacings has no effect on the forward scattering satellites. In fact, these satellites are independent of the particular atomic structure of the system. For example, in a noncrystalline material, the sum $\Sigma_n\, \delta(x - x_n)$ in Eq. (40) no longer refers to a sum over lattice planes, but to a sum over the x coordinates of individual atom positions. The Fourier transform of this function $g_2(k)$ is still well defined (its square is the "structure function" of the material) and still has a δ function at $k = 0$. Convolution of $g_2(k)$ with $g_4(k)$ [Eq. (46)] therefore still gives the same satellites around the forward scattering. {The convolution with $g_3(k)$ [Eq. (44)] is, of course, superfluous for this reflection.}

A general, nonharmonic, composition modulation can be Fourier analyzed as

$$c(x) = \sum_{m=-\infty}^{+\infty} A_m \sin(m\beta x + \phi_m), \tag{48}$$

where β is still equal to $2\pi/\lambda$ (λ is the repeat distance). This linear superposition gives a series of satellites at $\pm m\beta$ around each sharp reflection; the intensities of the satellites around the forward scattering I_m are again proportional to A_m^2.

If the diffusion process is linear [i.e., obeys Eq. (9), with $\tilde{D}_{\lambda/m}$ independent of concentration], the Fourier components evolve independently with time. In the continuum approximation, leading to Eqs. (11) and (29), the slope of $\ln I_m(t)$ is

$$\frac{d}{dt}(\ln I_m) = -2\tilde{D}_{\lambda/m}(m\beta)^2, \tag{49}$$

from which the interdiffusion coefficients can be determined directly. In the discrete approach of Cook *et al.* (1969) [see Eq. (21)], this becomes

$$\frac{d}{dt}(\ln I_m) = -\frac{4\tilde{D}_{\lambda/m}}{d_0^2}[1 - \cos(m\beta d_0)] \tag{50}$$

for diffusion in the [200] direction in fcc or bcc, with (200) plane spacing d_0.

If the diffusion process is nonlinear, the observed time dependence of the coefficients A_m is more complex and is usually analyzed by direct comparison with numerical simulations (Fleming *et al.*, 1980; Tsakalakos, 1977).

D. Advanced Developments

In the analysis of interdiffusion in Subsections A and B, a *linearized* diffusion equation was used. The Fourier components of an arbitrary composition variation could then be considered independently and an amplification factor R for each could be defined [Eqs. (11) and (31)]. To do this it is necessary to assume that the parameters M, f_0'', κ, and, if strain effects are included, $\eta^2 Y_{\langle \mathrm{UVW} \rangle}$ are independent of composition. This assumption is clearly inadequate for composition modulations of large amplitude, such as might arise in the later stages of spinodal decomposition, or in the early stages of annealing an artificially modulated material. The nonlinear analysis, which becomes necessary in these cases, will be examined in this subsection. We will consider also the effects of higher terms in the expansion [Eq. (4)] for the free energy of a solid solution in a composition gradient. Finally, electronic contributions to the free energy will be considered.

Consider an initially homogeneous system that undergoes spinodal decomposition. The early stages of the reaction are satisfactorily described by the linear analysis, assigning to each composition-dependent parameter the value it would have at the average composition. Beyond the early stages, however, the linear analysis [Eqs. (11) and (31)] predicts that the composition modulations will continue to grow exponentially. In reality, of course, the growth must slow down in the later stages as the maxima and minima of the modulation reach the equilibrium compositions. The primary reason for the failure of the linear analysis is the assumption of constant f_0'', an assumption that becomes particularly bad as the spinodal is approached ($f_0'' \rightarrow 0$). None of the other parameters changes sign as the composition is altered. Indeed, on the basis of simple models, such as the regular solution (Subsection A), it is expected that the chemical gradient energy coefficient κ_c (which is the dominant term in κ) would be independent of composition.

The earliest treatment of nonlinear diffusion in compositionally modulated materials was by Hillert (1956, 1961). He considered composition modulations in one dimension; his approach has already been outlined in Subsection A. Strain effects were ignored, and the quantity $M/c(1 - c)$ and the gradient energy coefficient κ were taken to be constant. On the other hand, the free energy was given by the nearest-neighbor regular solution model, and no restrictive assumptions were made about f_0''. The development of random composition fluctuations was modeled. It was demonstrated that growth slowed down as the equilibrium compositions were approached and that "stationary states" would be reached at infinite time. The stationary states are the minimum energy configurations for a given modulation wavelength. Fluctuations of large wavelength exhibited an appreciable amount of "squaring" during growth, and in the resulting stationary states the composition variations were not harmonic.

Tsakalakos (1977) derived analytical expressions describing the stationary states, which are in excellent agreement with Hillert's numerical solution. He demonstrated that stationary states with a periodicity about the critical wavelength λ_c show a negligible amount of squaring. Both Hillert and Tsakalakos showed that the free energy of the stationary states is maximum at λ_c and decreases at larger wavelengths. It is then favorable for a composition modulation to coarsen, i.e., for the average wavelength to increase. This feature was observed in Hillert's numerical modeling (1961).

The coarsening demonstrated by Hillert is relevant for one-dimensional composition modulations such as are found in artificial materials. His model is, however, unlikely to apply to the coarsening of three-dimensional modulations produced by natural spinodal decomposition. This problem was treated by Cahn (1966). He considered modulations only of a single wavelength in ⟨100⟩ directions and analytically solved the nonlinear diffusion equation by successive approximations. Cahn showed that after the initial stages, harmonics of the original wavelength appear. The odd harmonics cause squaring of the modulation and the even harmonics produce distortions that reduce the spatial extent of one phase while increasing the deviation of its composition from the average. This latter feature is expected for an asymmetric alloy and is consistent with the lever rule. The main drawback of Cahn's approach is the assumption of a single initial wavelength. This may be appropriate for artificial materials but is clearly not valid in spinodal decomposition. It has been pointed out that, somewhat surprisingly, the development of fluctuations with a spectrum of wavelengths is qualitatively well modeled in one dimension (Tsakalakos, 1977; Hilliard, 1970).

A thorough one-dimensional analysis was undertaken by deFontaine (1967) and Hilliard (1970), who used parameters for the Al–Zn system. They described the variation of the free energy with composition by a fourth-degree polynomial. This is the simplest expression that will model nonlinear behavior. (In the linearization of the diffusion equation, the assumption of constant f_0'' is equivalent to describing the free energy by a quadratic function.) The interdiffusion coefficient was described by a quadratic function of composition, fitted to experimental high-temperature data. DeFontaine and Hilliard obtained numerical solutions to the nonlinear diffusion equation by successive iterations. Their results for the asymmetric alloy Al–22.5 at. % Zn are shown in Fig. 11. This shows the development of a regular structure followed by coarsening.

Using a similar numerical technique, deFontaine (1967) also studied the evolution of a one-dimensional homogenizing modulated system governed by a nonlinear diffusion equation. Figure 12 shows the evolution of the first three Fourier components of an Ag–Au-layered structure, for which the gradient energy term κ is assumed zero and the interdiffusivity $\tilde{D}$ a quadratic function of composition. The first Fourier component decays almost exponentially, except for a slight deviation at the very early stages of annealing; the effect of the

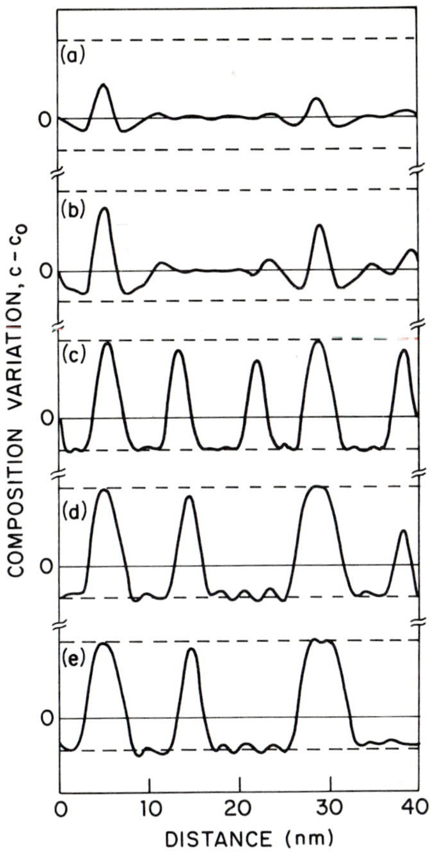

Fig. 11. Composition profiles obtained from a computer solution for an Al–22.5% Zn alloy aged at 100°C for (a) 10, (b) 13.3, (c) 60, (d) 400, and (e) 416 min. The broken horizontal lines indicate equilibrium compositions. [From J. E. Hilliard, *Phase Transformations,* American Society for Metals (1970), and de Fontaine (1967).]

nonlinearity on this component is evidently small. The second and third Fourier components, however, are strongly affected by the nonlinearity. One of the most striking features is the emergence of the second component, which is not present at all in the original rectangular profile. This reflects the developing asymmetry in the composition profile, due to more diffusion occurring in the Au-rich than in the Ag-rich region. Tsakalakos (1977) has given an approximate analytic solution to the nonlinear diffusion equation, which is in good agreement with deFontaine's numerical results.

The higher-order terms in the expansion of the free energy of a nonhomogeneous solution [Eq. (4)] lead to deviations in the wavelength-dependence of the interdiffusivity $\tilde{D}_\lambda$, as given by Eq. (11), (29), or (31). Tsakalakos (1977,

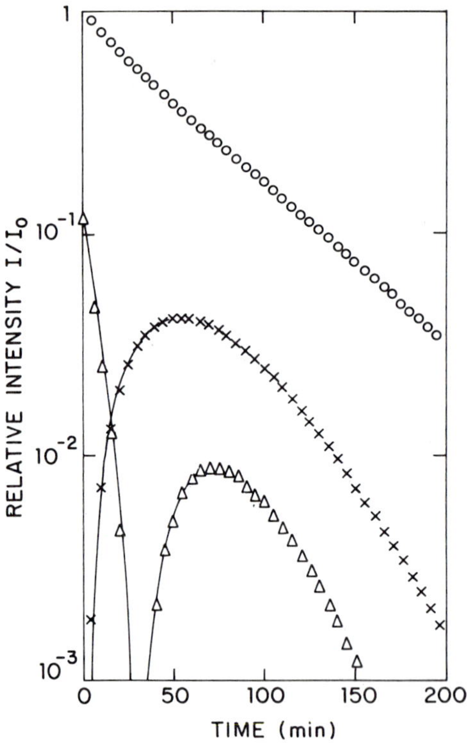

Fig. 12. Intensity (amplitude squared) versus annealing time for the first (○), second (×), and third (△) Fourier components of an Ag–Au layered structure governed by a nonlinear diffusion equation. [After de Fontaine (1967).]

1981a) has investigated these effects by considering the Fourier energy of a concentration wave:

$$\Psi(h) = k_B T/c_0(1 - c_0) + 2\eta^2 Y_{\langle UVW \rangle} + 2V(h), \tag{51}$$

where h is the position coordinate in reciprocal space and $V(h)$ the Fourier transform of the interatomic potential $V(\bar{r})$, which hence also depends on long-range interactions.

Alternatively, the Fourier energy can also be expressed as

$$\Psi(h) = f_0'' + 2\eta^2 Y_{\langle UVW \rangle} + 2 \sum_{\mu=1}^{\infty} K_\mu B^{2\mu}, \tag{52}$$

where B is a function of h according to the dispersion relation of Eq. (21), which reduces to $\beta = 2\pi/\lambda$ in the continuum limit ($\lambda \to \infty$; β, B, $h \to 0$). The interdiffusion coefficient corresponding to this free energy is

$$\tilde{D}_\lambda = \tilde{D}\left[1 + \frac{2\eta^2}{f_0''} Y_{\langle UVW \rangle} + \frac{2}{f_0''} \sum_{\mu=1}^{\infty} K_\mu B^{2\mu}\right]. \tag{53}$$

The long-wavelength limit of this expression is Eq. (31), with $K_1 = \kappa$, the gradient energy coefficient derived by Cahn and Hilliard (1958) (see Subsection A). The other coefficients K_μ therefore result from the higher-order terms in Eq. (4) and become important at large B or short wavelengths. A comparison of Eqs. (51) and (52) allows determination of the K_μ in terms of the interatomic potentials, including interactions beyond the nearest neighbors. For the first coefficient, the general result of Cahn and Hilliard (1958) is recovered as

$$K_1 = \frac{1}{6} \sum_i z_i r_i^2 V_i, \tag{54}$$

where z_i, r_i, and V_i are, respectively, the coordination number, radius, and potential of the ith coordination shell. Notice that the long-range interactions are heavily weighted in Eq. (54). Tsakalakos (1981a) has suggested that this may explain the observation of a negative K_1 in a system with a positive heat of mixing (such as Cu–Ni), since taking into account only nearest-neighbor interactions [as in Eq. (20)] results in K_1 being proportional to the heat of mixing. Using interatomic potentials V_i derived from short-range-order measurements (Moss and Clapp, 1968), he demonstrated that the higher-order gradient terms K_μ can lead to nonlinearities in $\tilde{D}_\lambda$ versus B^2 [Eq. (53)], as is observed in certain systems (see Subsection III.A).

The interaction of the Fermi surface with the Brillouin zone boundaries in compositionally modulated metals has been proposed as an explanation for abrupt changes in $\tilde{D}_\lambda(B)$ at specific modulation wavelengths. Purdes (1976) has pointed out that a (111) modulation creates two additional Brillouin zone boundaries at $\pm\beta$ from the (111) zone face of the average crystal. When this new zone boundary touches the Fermi sphere, it develops a flat portion with an area proportional to the strength of the new zone face. This in turn produces a discontinuity in the energy $\Psi(h)$ and hence in the interdiffusivity $\tilde{D}_\lambda$. Tsakalakos (1977) used this approach to account for position of a sharp peak in $\tilde{D}_\lambda(B)$ for Cu–Ni modulated films.

E. Relation between Interdiffusivity and Tracer Diffusivities

Darken (1948) showed that the interdiffusivity $\tilde{D}$ in a binary system is related to the tracer diffusivities D_1^* and D_2^* by

$$\tilde{D} = [cD_1^* + (1-c)D_2^*]\left[1 + \frac{\partial(\ln \gamma_2)}{\partial(\ln c)}\right], \tag{55}$$

where c and γ_2 are, respectively, the concentration and activity coefficient of component 2. Since $\tilde{D}$ in compositionally modulated films is usually measured at much lower temperatures than in bulk diffusion couples, care must be taken to extrapolate the data correctly. Hilliard has discussed this problem in detail (Hilliard, 1957; Huston *et al.*, 1966; Hilliard, 1970). The temperature dependence of the thermodynamic factor in Eq. (55) must be taken into account. This can be done by recognizing that

$$1+\frac{\partial(\ln \gamma_2)}{\partial(\ln c)}=\frac{c(1-c)f''(T)}{N_V k_B T}. \tag{56}$$

If thermodynamic data are available, $f''(T)$ can then be determined directly. Otherwise, approximate methods must be used. Hilliard (1957) points out that the temperature dependence of the thermodynamic factor in homogenizing systems is weak and contributes only a few kilocalories per mole to the total activation energy of $\tilde{D}$. In spinodally decomposing systems, however, the thermodynamic factor is very temperature dependent, since it vanishes at the spinodal temperature T_s and has a different sign on either side. An estimation of $f''(T)$ can thus be made by expanding it about T_s, since $f''(T_s)=0$:

$$f''(T)=(T-T_s)\frac{\partial f''}{\partial T}=-(T-T_s)s'', \tag{57}$$

where $s''=\partial^2 s/\partial c^2$ (s is the entropy per unit volume) has negligible temperature dependence. Equation (55) can then be rewritten as

$$\frac{\tilde{D}T}{T-T_s}=-\frac{c(1-c)s''}{N_V k}[cD_1^*+(1-c)D_2^*] \tag{58}$$

If the activation energies of the tracer diffusivities are not too different, an Arrhenius plot of $\tilde{D}T/(T-T_s)$ will be linear, and such a plot should be used to compare data over a large temperature range, especially inside and outside the spinodal. Figure 13 is a schematic comparison of the temperature dependence of $\tilde{D}$ and Eq. (58).

III. EXPERIMENTAL RESULTS

A. Crystalline Metals

The first measurement of interdiffusion in an artificially modulated material was made in a crystalline metal sample. This experiment, by DuMond and Youtz (1940), has been described briefly in Section I. The sample consisted of copper with a modulated gold content. It was produced by vapor deposition and presumably had a fine-grained polycrystalline structure. The total thickness of the deposited film was ~ 1000 nm and the modulation wavelength was ~ 10 nm. X-ray diffraction was performed with a special spectrograph, but only a qualitative assessment of intensities was possible. They found that the sample tended to homogenize, the satellite intensity having a half-life of about two days at room temperature. Using Eq. (2) they calculated the effective diffusivity to be $D \approx 5 \times 10^{-24}\ \mathrm{m^2\ sec^{-1}}$. This is a remarkably high diffusivity. Extrapolating from tracer data for the bulk diffusion of gold in copper, determined in the range 400–1050°C (Askill, 1970), the diffusivity at 25°C is found to be 1.7 ×

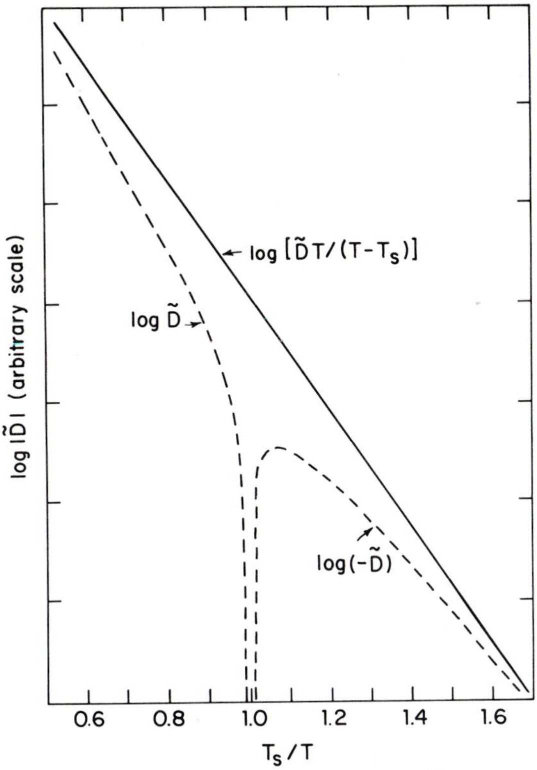

Fig. 13. Schematic plot illustrating that an Arrhenius plot of $\tilde{D}T/(T-T_s)$ should be used to extrapolate high-temperature data into the spinodal (solid line). The diffusivity $\tilde{D}$ itself displays a singularity at T_s. [From J. E. Hilliard, *Phase Transformations,* American Society for Metals (1970).]

10^{-37} m^2 sec^{-1}. In a polycrystalline thin film "short-circuit" diffusion along grain boundaries is expected. On the basis of the measured self-diffusivities of gold in grain boundaries (Gupta and Asai, 1974), the grain size in DuMond and Youtz's film would have to be ~10 nm to explain their high value for D. This grain size seems rather small, and it is likely that another mechanism of fast diffusion can operate in such films. Such an alternative mechanism, proposed by Cook (1966) and Cook and Hilliard (1969), is discussed later in this section.

The next published work using the technique of DuMond and Youtz was that of Dinklage and Frerichs (1963) at Northwestern University. They vapor deposited films consisting of 140 layers of lead, 0.25 nm thick, separated by magnesium layers 2.66 nm thick, with an 0.5-μm magnesium layer on top for protection. Only first-order diffraction peaks were observed. Dinklage and Frerichs attributed the absence of higher-order peaks to the rapid decay of the higher harmonics of the composition modulation. Indeed, rapid interdiffusion was observed on annealing at temperatures in the range 0–61.5 °C. The intensities of the diffraction peaks due to the modulation decayed exponentially with time at any temperature, as expected from Eq. (2) if D is constant. The values of the effective diffusivity D exhibited an Arrhenius dependence on temperature, $D = D_0 \exp(-Q/RT)$, with $D_0 = 5.2 \times 10^{-10}$ m^2 sec^{-1} and $Q = 103$ kJ mole^{-1}. The diffusivity at 0 °C was 1.65×10^{-26} m^2 sec^{-1}, which was probably the lowest value of any diffusivity measured up to that time. Dinklage and Frerichs developed their films for selective diffraction in a soft x-ray spectrometer. Such an application, they concluded, would be possible if it were not for the rapid interdiffusion in the films. The half-life of the intensity of their diffraction peaks was only about two days at room temperature (25 °C).

The work of DuMond and Youtz and of Dinklage and Frerichs demonstrated the feasibility of sensitive measurements of interdiffusivities in compositionally modulated films that tend to homogenize. In each case, however, the study of interdiffusion was secondary to the original purpose of the experiment. The measured diffusivities are very hard to interpret because the microstructures of the thin films were not characterized. In addition, there is a possibility that intermetallic compounds could arise; indeed, Dinklage and Frerichs (1963) detected the phase Mg_2Pb in their films. In neither experiment was the interdiffusivity measured as a function of the modulation wavelength λ. Thus, it was not possible to study the gradient energy and strain effects discussed in Subsections II.A and II.B. The first studies in compositionally modulated films produced specifically for the study of diffusion, gradient energy and strain effects, were made under the direction of Hilliard at Northwestern University, and the remainder of the work cited in this section is from that group. They have studied almost exclusively face centered cubic modulated films. They were produced by dual source vapor deposition on heated mica substrates. This usually results in a strong texture with the $\langle 111 \rangle$ directions perpendicular to the substrate. The control and characterization of the microstructure of their films were an improvement over the early experiments and were very important in analyzing the diffusion results.

The first alloy system to be studied by the Hilliard group was gold–silver. Important criteria in this choice were complete solid solubility above room temperature, eliminating the concern about compound formation; availability of good thermodynamic and tracer diffusivity data; and widely different electron densities in gold and silver, ensuring a large x-ray scattering from the modulation. Gold–silver is a system in which the interaction parameter (ΔV in Subsection II.A) is negative, and there is a predicted ordering reaction at -113 °C (Norman and Warren, 1951). It is, therefore, expected to homogenize on annealing and to exhibit a negative gradient energy coefficient κ. Furthermore, since the atomic sizes in gold and silver are very similar ($\Delta a/a = 0.15\%$), strain effects (Subsection II.B) are expected to be negligible, with a consequent simplification of the interpretation of the diffusion results. Cook (1966) produced Au–Ag films of total thickness 200–500 nm, of selected average compositions in the range 18–50 at. % Au, and containing composition modulations with wavelengths 0.7–3.5 nm.

The intensities of the first-order 000^+, 111^+, 111^-, 222^+, and 222^- satellite peaks were measured (Cook and Hilliard, 1969) ($^+$ and $^-$ denote the peaks at higher angle and at lower angle than the Bragg reflection). An example of the decay of the satellite intensities, $\ln[I(t)/I(0)]$, on annealing is shown in Fig. 14. According to Eq. (49), the slope of the 000^+ line is proportional to the interdiffusivity $\tilde{D}_\lambda$ since the intensity of that satellite is a function of the composition modulation only. The intensity of the satellites of the higher Bragg peaks, however, also depends on the modulation of the lattice spacing and on the intensity of the central Bragg peak itself, which in turn depends on the texture. Both the lattice spacing modulation and the texture changes may give rise to a nonlinearity of the lines in Fig. 14. These effects are expected to become negligible as annealing proceeds, as can be observed from the constant slopes of the lines after an intensity decrease of about 40%. Cook and Hilliard (1969) derived $\tilde{D}_\lambda$ for each film by applying Eq. (50), which incorporates the discrete analysis of

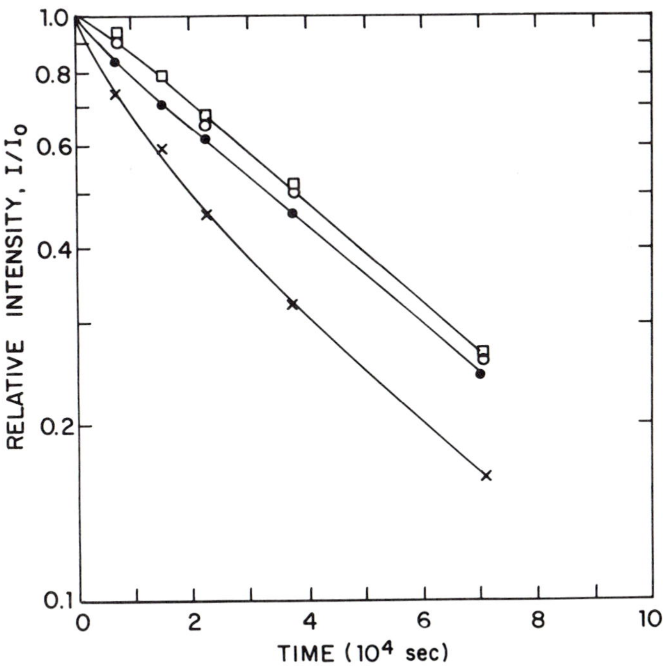

Fig. 14. Plots of $\log I/I_0$ versus time for the 000^+ (●), 111^+ (○), 222^- (×), and 222^+ (□) satellite intensities in a silver–gold modulated film. The film was of average composition 32 at. % Au, with a modulation wavelength of 3.35 nm, and was annealed at 228.3°C. [From Cook and Hilliard (1969).]

Cook *et al.* (1969), to the later stages of the anneals. The plane spacing d was $a/\sqrt{3}$ (a is the lattice parameter) in these experiments since diffusion was parallel to $\langle 111 \rangle$. The values of $\tilde{D}_\lambda$ were strongly dependent on the average film composition. For a given composition, $\tilde{D}_\lambda$ was observed to vary with λ according to Eq. (11). From this relationship κ/f_0'' could be evaluated. Since for the Ag–Au system f_0'' could be calculated from thermodynamic data, the gradient energy coefficient κ could be determined. It was found to be fairly independent of temperature and composition, as expected from the simple regular solution model discussed in Subsection II.A. The average value of κ was $-(2.6 \pm 0.5) \times 10^{-11}$ J m^{-1}. This value is negative, as expected for an ordering system, and is in excellent agreement with the value -2.5×10^{-11} J m^{-1}, computed by Cook and Hilliard on the basis of nearest-neighbor interactions [Eq. (20) in Subsection II.A]. By extrapolating the $\tilde{D}_\lambda$ values to $\lambda = \infty$ by using Eq. (11), the macroscopic interdiffusivity $\tilde{D}$ was determined as a function of temperature and average composition. The values of $\tilde{D}$ were found to be in good agreement with the extrapolation of high-temperature bulk interdiffusion measurements (Johnson, 1942), as illustrated in Fig. 15.

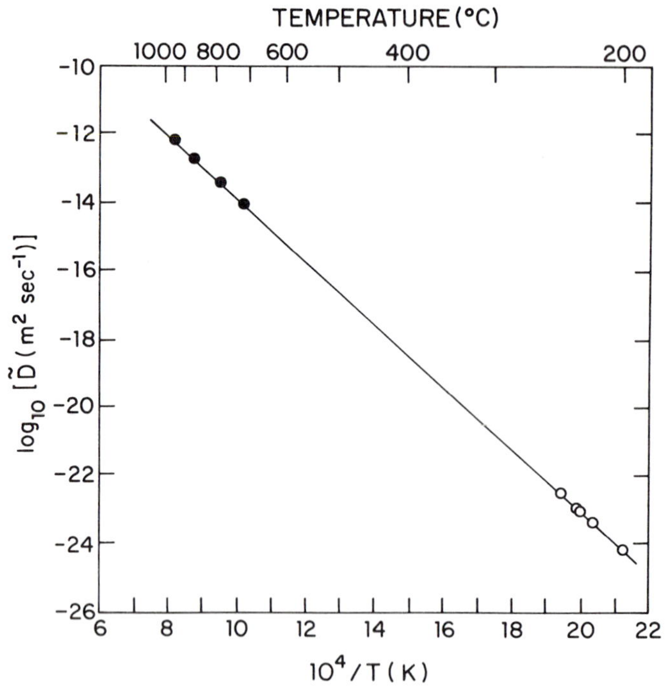

Fig. 15. Arrhenius plot of the interdiffusivity $\tilde{D}$ in equiatomic silver–gold alloys. The high-temperature data are from conventional bulk measurements (Johnson, 1942). The low-temperature data were determined from $\tilde{D}_\lambda$ measured in modulated films, extrapolated to $\lambda = \infty$. [From Cook and Hilliard (1969).]

In conventional measurements in macroscopic diffusion couples, it is not possible to determine bulk diffusivities below about two-thirds of the absolute melting temperature because the transport at low temperatures is dominated by "short-circuit" diffusion along grain boundaries or dislocations. This appears not to be the case in the modulated films. The reason is that the diffusion distances in the modulated films ($\lambda = 0.7-3.5$ nm) are small compared to the distances between short-circuit paths. As described earlier Cook was able to analyze his diffusion results with a model for bulk diffusion in single crystals. This was possible because of the favorable microstructure of his films, in particular the strong $\langle 111 \rangle$ texture and the fairly large grain size. By measuring the intensity of the (111) Bragg peak, Cook found that the degree of $\langle 111 \rangle$ texture increased during the diffusion anneals. This increase could be achieved either by growth of favorably oriented grains or by recrystallization, but in either case the moving interfaces provide high-diffusivity paths through the film thickness. The increase in texture was particularly rapid during the early annealing treatments and correlated with the initial rapid decay of the 000^+ satellite in the films (Fig. 14). Cook therefore attributed the initial high diffusivity to short-circuit diffusion along *moving* grain boundaries. A comparison of the initial rates of decay of the various satellites in Fig. 14 suggested that in the volume swept out by a moving boundary the composition modulation is completely diffused out (Cook, 1966). This may explain the high diffusivities measured by DuMond and Youtz (1940) and by Dinklage and Frerichs (1963).

By analyzing the asymmetry of the 222^- and 222^+ peaks due to the lattice parameter modulation (see Subsection II.C), Cook determined da/dc at various compositions. His results are in fair agreement with precision lattice parameter measurements on bulk alloys. In addition, since d^2a/dc^2 is positive, the average lattice parameter in an Ag–Au modulated film should decrease as it homogenizes; Cook confirmed this experimentally.

The work of Cook (1966) and of Cook and Hilliard (1969) was the first to show directly that the gradient energy effect proposed by Hillert (1956) and by Cahn and Hilliard (1958) is significant for short diffusion distances and that, in the absence of strain effects, it can be described rather well by the treatments in Subsection II.A.

The effects of strain on interdiffusion in modulated films were first studied by Philofsky (1968) and by Philofsky and Hilliard (1969). The copper–palladium system was chosen because of the significant difference in atomic size ($\Delta a/a \approx 7\%$) and because, despite this size difference, the system shows no intermediate compounds when the palladium content exceeds 60 at. %. Films of average composition 70–90 at. % Pd were produced, containing composition modulations with wavelengths in the range 1.0–4.0 nm. Like gold–silver, copper–palladium is an ordering system that is expected to homogenize on annealing with a negative gradient energy coefficient κ

Philofsky analyzed his results by using Cahn's continuum treatment of the strain effects [Eq. (31)], modified by the discrete treatment of Cook *et al.* (1969)

at short wavelengths. The resulting expression for the interdiffusion coefficient is

$$\tilde{D}_{\lambda\langle UVW\rangle} = \tilde{D}\left(1 + \frac{2\kappa B^2}{f_0''} + \frac{2\eta^2}{f_0''} Y_{\langle UVW\rangle}\right), \tag{59}$$

where B is the dispersion relation of Eq. (21). Notice that incorporation of the strain effects makes the interdiffusivity dependent on the diffusion direction $\langle UVW\rangle$. Philofsky (1968) attempted to demonstrate this anisotropy by measuring the interdiffusivity in films with $\langle 100\rangle$ texture and comparing the results with those in the usual $\langle 111\rangle$ textured films. Unfortunately, the attempts to produce $\langle 100\rangle$ texture on NaCl substrates were unsuccessful.

The films with $\langle 111\rangle$ texture were annealed in the range 355–440°C. As found by Cook (1966), the degree of texture increased, especially in the early anneals. Philofsky found that the average lattice parameter in the films increased on annealing, as expected, since in the copper–palladium system d^2a/dc^2 is negative. The intensities of the 000^+, 111^-, 111^+, 222^-, and 222^+ satellites were measured, and a representative set of decay curves is given in Fig. 16. Initially, the curves are not linear, which may be due to microstructural changes

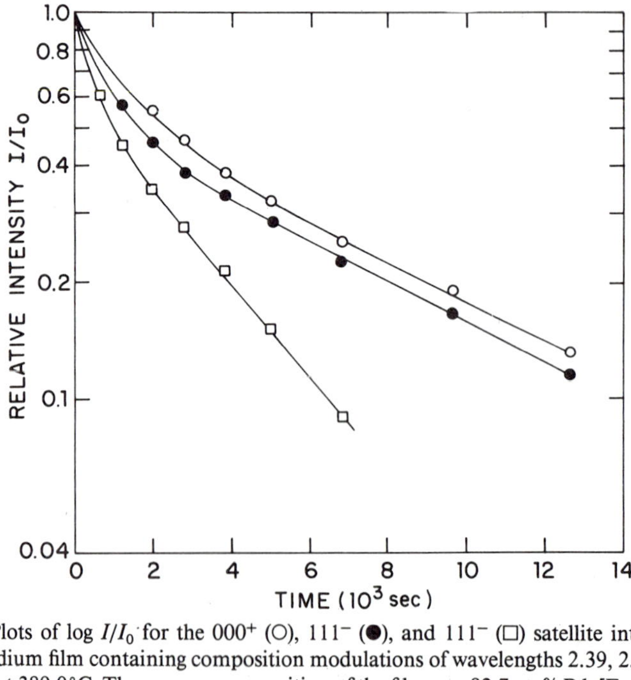

Fig. 16. Plots of log I/I_0 for the 000^+ (○), 111^- (●), and 111^- (□) satellite intensities in a copper–palladium film containing composition modulations of wavelengths 2.39, 2.39, and 1.21 nm annealed at 389.0°C. The average composition of the film was 82.7 at. % Pd. [From Philofsky and Hilliard (1969).]

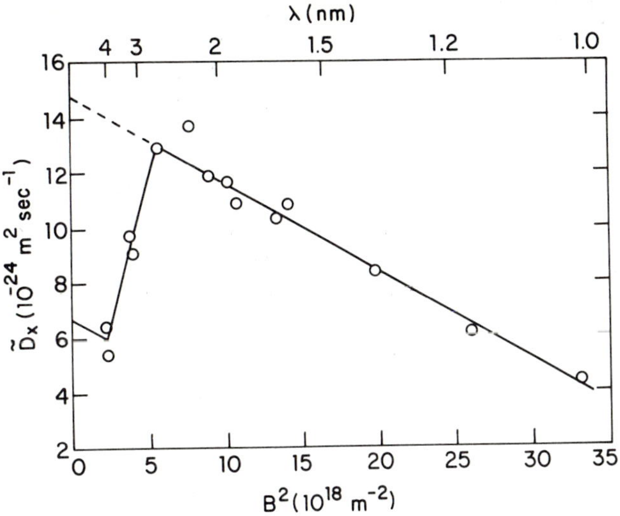

Fig. 17. The variation of the effective interdiffusivity $\tilde{D}_\lambda$ in copper–palladium modulated films (82 at. % Pd) with B^2 and λ. The measurements were made at 389°C. The composition modulations were coherent for $\lambda < 2.8$ nm and incoherent for $\lambda > 3.8$ nm. [From Philofsky and Hilliard (1969).]

as the texture increases or the variation of interdiffusivity $\tilde{D}$, the distortion parameter η, or the modulus $Y_{\langle UVW \rangle}$ with composition. All of these effects become negligible for small composition fluctuations, and the plots in Fig. 16 become linear after the satellite intensities have decayed to ~30% of their initial values. Furthermore, for a particular film, the gradients of the linear portions of the curves are the same for all the satellites. The interdiffusivity $\tilde{D}_\lambda$ was determined from the terminal gradients of these decay curves, according to Eq. (50). The variation of $\tilde{D}_\lambda$ with B^2 at a particular temperature and average composition is shown in Fig. 17. According to Eq. (59), the plot should be linear with a gradient of $2\kappa\tilde{D}/f_0''$. On Fig. 17 the plot is linear for $\lambda < 2.8$ nm, but at larger wavelengths $\tilde{D}_\lambda$ is approximately halved. Such a drop in the interdiffusivity suggests a loss of coherency in the films since, if the layers are incoherent, the positive strain energy contribution to $\tilde{D}_\lambda$ is lost.

In the continuum treatment of strain effects in coherent modulated films, it is found that the elastic energy due to the modulation is independent of its wavelength (Subsection II.B). On the other hand, if the difference in lattice parameter between the layers in a film is accommodated by misfit dislocations, the excess strain energy due to the dislocations will be proportional to the number of interfaces and therefore inversely proportional to the wavelength. It is expected, then, that as the wavelength is increased, it will at some point become energetically favorable for the layers in a modulated film to be partially coherent or incoherent. Using low-angle grain boundary data for copper, Philofsky (1968) estimated from a simple energy balance that coherency should be lost for wavelengths greater than 2.5 nm. The degree of coherency in the copper–palladium films was determined by measuring the ratio R of the intensity of the 111^- satellite to that of the 111^+ satellite in as-deposited films. The experimental values of the ratio are plotted in Fig. 18, together with calculated values for incoherent and coherent modulations. The curve for incoherent modulations was obtained from the treatment of deFontaine (1966) with no adjustable parameters. The curve for coherent modulations was obtained from a treatment similar to that of Guinier (1963b), but, in the absence of elastic coefficient data, the quantity $Y_{\langle 111 \rangle}/c_{44}$ had to be treated as an adjustable parameter to fit the observed values of R. It is clear from Fig. 18 that the modulations were coherent for $\lambda < 2.8$ nm and incoherent for $\lambda > 3.8$ nm. It is interesting to note that there is a gradual loss of coherency. The partial coherency at intermediate λ may be an equilibrium state since the corresponding values of $\tilde{D}_\lambda$ appear to remain constant in the terminal stages of annealing. The behavior illustrated in Fig. 18 supports the interpretation of Fig. 17.

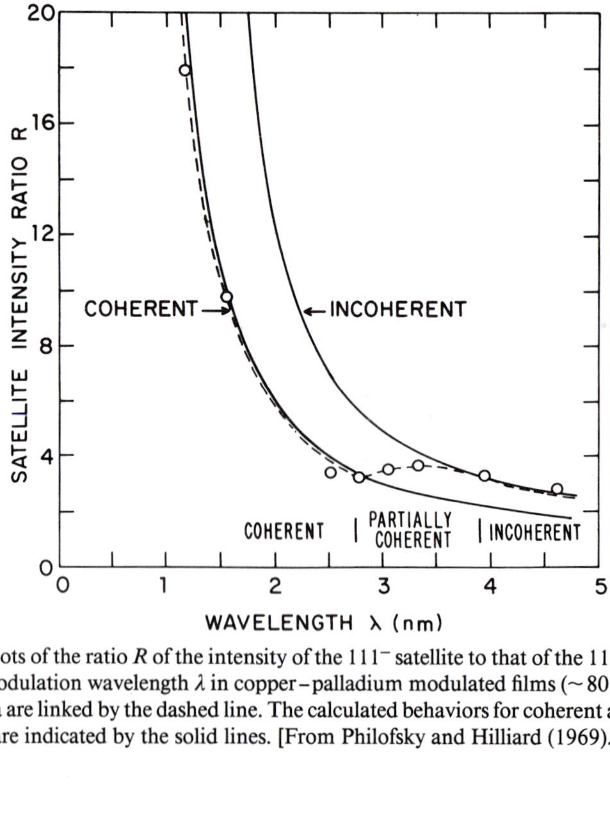

Fig. 18. Plots of the ratio R of the intensity of the 111^- satellite to that of the 111^+ satellite as a function of modulation wavelength λ in copper–palladium modulated films (~80 at. % Pd). The measured data are linked by the dashed line. The calculated behaviors for coherent and incoherent modulations are indicated by the solid lines. [From Philofsky and Hilliard (1969).]

Although the plot in Fig. 17 could be used to derive the gradient energy coefficient κ, Philofsky and Hilliard (1969) determined κ from a comparison of $\tilde{D}_\lambda$ for two modulations of different wavelengths in the same specimen. In this way, they eliminated errors due to uncertainties in composition and in annealing treatment. Considering a short wavelength p and a long wavelength q, where both modulations are coherent, Eq. (59) gives

$$\kappa/(f_0'' + 2\eta^2 Y_{\langle 111\rangle}) = (\tilde{D}_p - \tilde{D}_q)/2(\tilde{D}_q B_p^2 - \tilde{D}_p B_q^2). \tag{60}$$

From determinations in four specimens an average value of $\kappa/(f_0'' + 2\eta^2 Y_{\langle 111\rangle})$ was obtained. If the modulation of long-wavelength q is incoherent and the modulation of short-wavelength p is coherent, Eqs. (11) and (59) can be combined to give

$$2\eta^2 Y_{\langle 111\rangle}/f_0'' = (1/\tilde{D}_q)[(\tilde{D}_p - \tilde{D}_q) + (2\kappa/f_0'')(\tilde{D}_p B_q^2 - \tilde{D}_q B_p^2)]. \tag{61}$$

An average value of $2\eta^2 Y_{\langle 111\rangle}/f_0''$ was calculated from the solutions to Eq. (61) for three specimens. The solutions were obtained by successive approximations and used a value of f_0'' calculated from existing thermodynamic data. Combining the results of Eqs. (60) and (61) and the value of f_0'' gave $\kappa = -10.3 \times 10^{-11}$ J m^{-1}. Note that κ is negative as expected for an ordering system. Philofsky and Hilliard (1969) estimated κ by using the nearest-neighbor solution model of Subsection II.A and obtained a value of -6.3×10^{-11} J m^{-1}, in fair agreement with experiment. Later, Cook and deFontaine (1971) pointed out that if the elastic contribution to the total heat of mixing is subtracted and just the chemical contribution is kept, a value of -8.2×10^{-11} $J\,m^{-1}$ is obtained, in better agreement with experiment. Even in the copper–palladium system, with its relatively large strain effects, the elastic gradient energy coefficient (Subsection II.B) $\kappa_E\langle 111\rangle$ is negligible. Using the elastic constants given by Philofsky and Hilliard (1969), Cook and deFontaine (1971) derived $\kappa_E = -3.6 \times 10^{-13}$ J m^{-1}.

Using Eq. (11) for incoherent modulations and Eq. (59) for coherent modulations, Philofsky (1968) determined the values of the bulk interdiffusivity $\tilde{D}$ as a function of temperature and average composition. A strong composition dependence was found; for example, at 389°C, $\tilde{D}$ decreased from 2.3×10^{-23} to 6×10^{-24} m^2 sec^{-1} between 70 and 90 at. % Pd. This is not surprising since conventional measurements of $\tilde{D}$ (Thomas and Birchenall, 1952) show a decrease by a factor of 20 between pure Cu and pure Pd. For a particular wavelength and composition, $\tilde{D}_\lambda$ was found to have an Arrhenius temperature dependence. In Fig. 19, the results for $\tilde{D}$ are compared to high-temperature measurements. Although the extrapolation of the modulated film data lies somewhat above the high-temperature data, the agreement is still reasonable.

The work of Philofsky confirmed the existence of gradient energy effects at short modulation wavelengths and demonstrated the existence of elastic energy effects due to coherency. These effects were also demonstrated by Yang (1971),

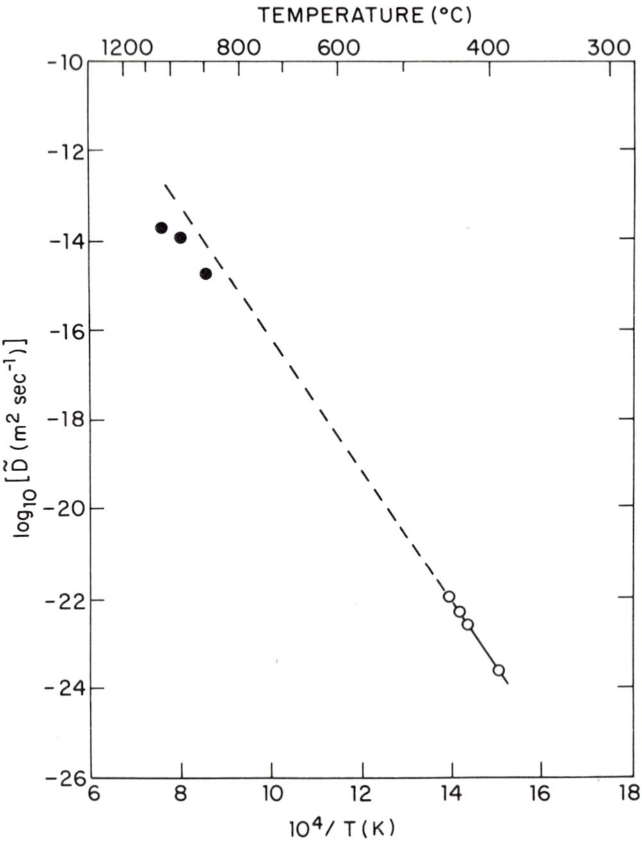

Fig. 19. Arrhenius plot of the interdiffusivity $\tilde{D}$ in Cu–89.8 at. % Pd alloys. The high-temperature data are from conventional bulk measurements (Thomas and Birchenall, 1952). The low-temperature data are from measurements on modulated films. [From Philofsky and Hilliard (1969).]

in the first study of a phase-separating system, gold–nickel. The difference in the atomic sizes of gold and nickel is quite high ($\Delta a/a = 15.7\%$), and it was expected that in this system, unlike Cu–Pd, the elastic gradient energy coefficient would be significant. It was hoped that the interdiffusion data would enable the determination of the spinodal temperature (in a coherent system, the temperature at which $f_0'' + 2\eta^2 Y_{\langle UVW \rangle} = 0$) at a number of compositions. The intensities of the 000^+, 111^-, and 111^+ x-ray satellites were measured. The intensity ratio R of the 111^- and 111^+ satellites was determined as a function of modulation wavelength and compared with values calculated for coherent and incoherent modulations. For all film compositions, results similar to those shown for Cu–Pd in Fig. 18 were obtained. For wavelengths greater than ~3.0 nm in 44.0 at. % Ni films and greater than ~2.2 nm in 30 at. % Ni films,

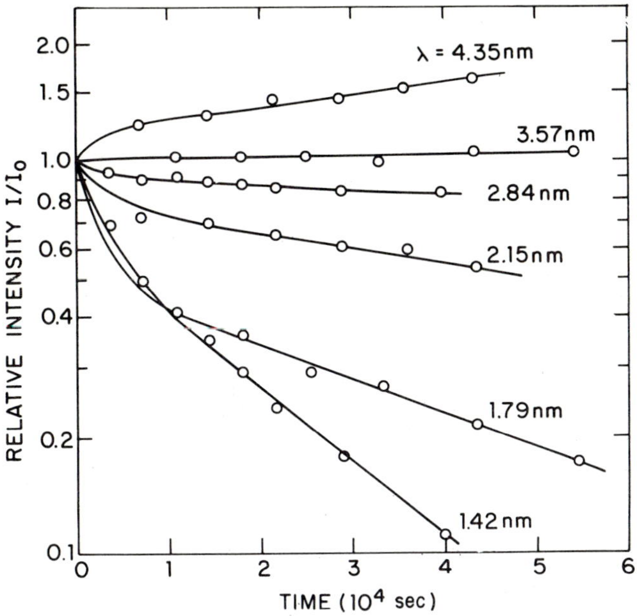

Fig. 20. X-ray satellite (000^+ or $11 1^-$) intensities plotted as $\ln(I/I_0)$ versus annealing time at 275°C for modulations of various wavelengths in specimens of composition Au–44 at. % Ni. [From Yang (1971).]

there was a gradual loss of coherency. In fitting the intensity ratio for coherent modulations, the modulus $Y_{\langle 111 \rangle}$ was taken to be 12% lower than the value calculated from elastic constants measured at room temperature (Golding *et al.*, 1967). These values of the modulus are still reasonably close, but unfortunately variations of this order have a large effect on the calculated spinodal temperatures.

Yang (1971) studied the effect of annealing on the 000^+ and $11 1^-$ satellite intensities. In this system, as can be seen in Fig. 20, it was found that some satellite intensities increase on annealing, as expected for the growth of composition modulations due to spinodal decomposition. The spinodal temperature for the Au–44 at. % Ni alloy is therefore higher than the annealing temperature 275°C. The initial nonlinearity in the intensity plots was ascribed to the same causes as in earlier work, and the amplification factor *R*, defined by Eq. (31), was determined in each case from the gradients of the linear portions of the curves. The variation of *R* with modulation wavelength for the 44 at. % Ni alloy at 250°C is shown in Fig. 21. It can be seen that there is a critical wavelength λ_c and corresponding wavenumber B_c [defined in Eq. (21)] for which *R* is zero. Due to the gradient energy, modulations of smaller wavelength, in this case $\lambda \lesssim 3.0$ nm, will decay, even though the annealing temperature is below the

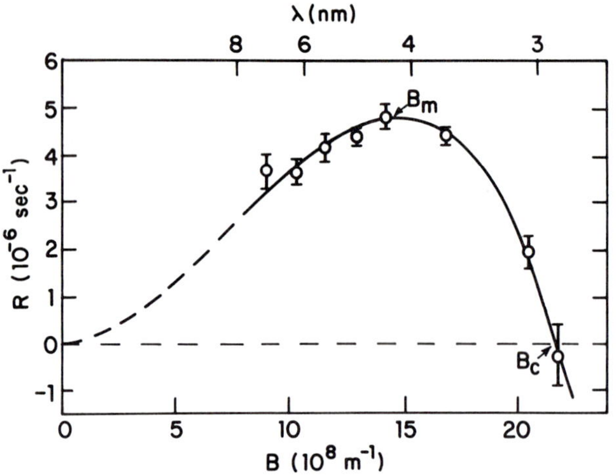

Fig. 21. The amplification factor R as a function of B or λ for composition modulations in Au–44 at. % Ni films annealed at 250°C; $B_c/B_m = 1.44$. [From Yang (1971).]

spinodal temperature. For wavelengths in the range of Fig. 21, the films were predominantly incoherent, so that $R(B)$ obeys Eq. (11). The ratio of B_m for which R is maximum, to B_c is then expected to be $\sqrt{2}$, which is indeed close to the experimental value, 1.44, on Fig. 21. Since the degree of coherency varies

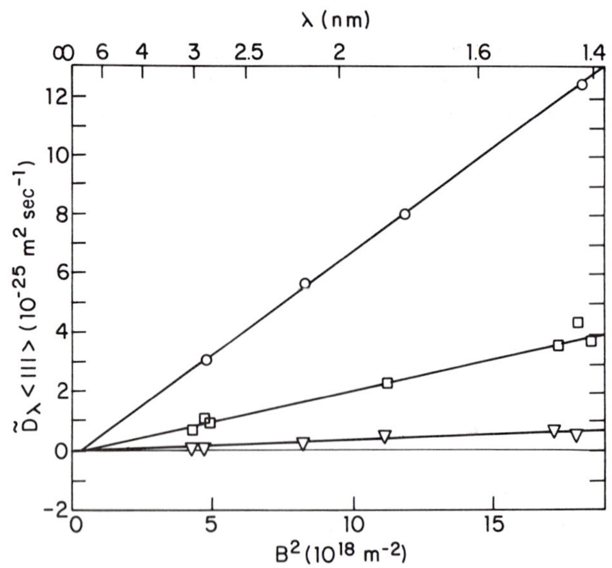

Fig. 22. Plot of effective interdiffusivity $\tilde{D}_\lambda\langle 111\rangle$, versus B^2 for coherent modulations in Au–44 at. % Ni films annealed at 275 (○), 250 (□), and 225°C (▽). [From Yang (1971).]

with λ, this good agreement may be fortuitous, and indeed Yang (1971) found different ratios in analyzing other data.

A more rigorous analysis is possible if only data from coherent films are considered. Strain effects are then present, and Eq. (59) applies. A plot of the effective diffusivity $\tilde{D}_\lambda$ versus B^2 should be linear with a slope $2\kappa\tilde{D}/f_0''$ and an intercept at $\tilde{D}[1 + (2\eta^2/f_0'')Y_{\langle 111\rangle}]$. Figure 22 shows such plots of measurements on coherent films of 44 at. % Ni, and they are indeed linear. This linearity breaks down, as shown in Fig. 23, at longer wavelengths when coherency is lost. With the loss of coherency, the strain energy term, which favors homogenization (positive $\tilde{D}_{\lambda\langle 111\rangle}$), disappears. The critical wavelength for coherent modulations can be obtained from the plots in Fig. 22 by extrapolating to $\tilde{D}_\lambda = 0$. The values so obtained, 5.9–9.7 nm, are significantly higher than the value for incoherent modulations in Fig. 21. Using Eq. (59) and expanding the coherent free energy

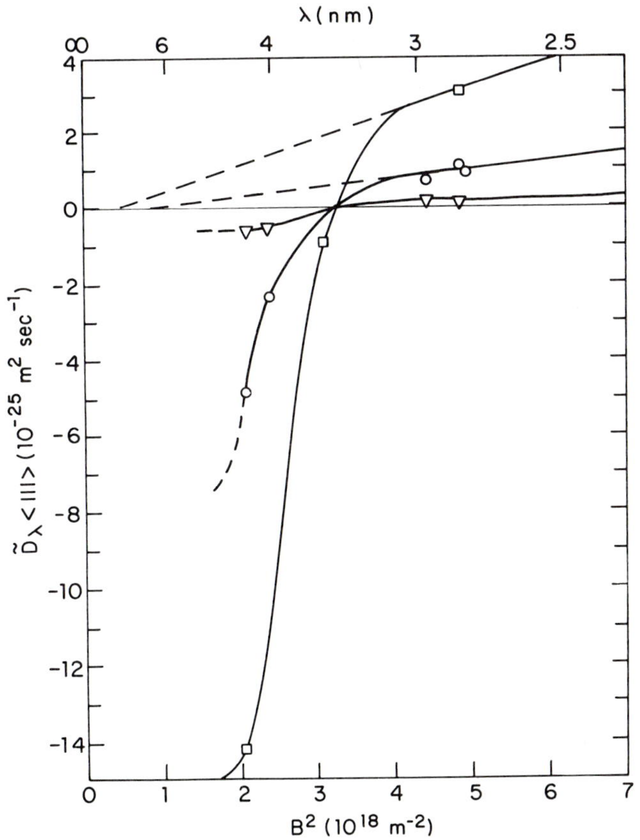

Fig. 23. As in Fig. 22, but for partially coherent modulations annealed at 225 (▽), 250 (○), and 275°C (□). [From Yang (1971).]

similarly to Eq. (57), Yang (1971) argued that B_c^2 should vary with temperature according to

$$(T_s - T)s'' + 2\kappa B_c^2 = 0, \tag{62}$$

where T_s is now the coherent spinodal temperature. The linear dependence on temperature is indeed observed in Fig. 24. Extrapolating the lines to $B_c = 0$ gives $T_s = 284 \pm 3$°C for 44.0 at. % Ni and 306 ± 1°C for 59.0 at. % Ni. These values are about 420 and 500°C lower than the corresponding calculated chemical spinodal temperatures. The large suppression is due to the substantial strain effects in this system.

Using measured values of elastic constants (Golding *et al.*, 1967) and coherent spinodal temperatures and calculated chemical spinodal temperatures, Yang derived values of s''. Since the slopes of the plots in Fig. 24 are $2\kappa/s''$, this allowed him to evaluate κ as 3.73×10^{-10} J m^{-1} for 44 at. % Ni and 1.27×10^{-10} J m^{-1} for 59.0 at. % Ni. The elastic gradient energy coefficient κ_E was

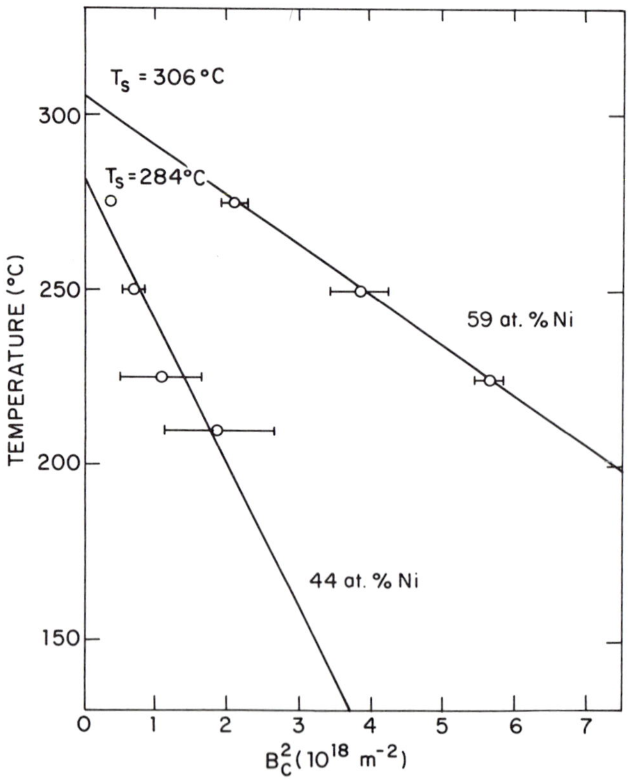

Fig. 24. Plots of temperature versus B_c^2 for Au–44 at. % Ni and Au–59 at. % Ni films. The extrapolation to $B_c = 0$ yields the coherent spinodal temperature T_s. [From Yang (1971).]

calculated from the expression of Cook and deFontaine (1969), who used measured elastic constants, and was found to be 0.27×10^{-10} J m^{-1} and 0.26×10^{-10} J m^{-1} at the same compositions. As expected for a system in which $\eta > 0.1$, the values of κ_E are not negligible. (At 50 at. % Ni, $\eta \approx 0.13$.) Subtracting κ_E from κ, Yang obtained values of the chemical gradient energy coefficient κ_c. At 44 at. % Ni, $\kappa_c = 3.46 \times 10^{-10}$ J m^{-1} and at 59 at. % Ni, $\kappa_c = 1.01 \times 10^{-10}$ J m^{-1}. This strong dependence on composition was not expected. The value of κ_c calculated on the basis of nearest-neighbor interactions [Eq. (20)] is 0.19×10^{-10} J m^{-1}. Thus it appears that the simple model is not a good approximation in this case and that long-range interactions are important.

On annealing a modulated film, it would normally be expected that the Bragg peaks would sharpen because of strain relief and increase in grain size. Yang (1971) found, in contrast, that the base of the peaks broadened. He also found that satellites that grew on annealing tended to move closer to the main peak. He attributed these effects to the growth of long-wavelength composition modulations. For the 44 at. % Ni alloy, he calculated that the optimum temperature for spinodal decomposition is about 250°C and that the dominating modulation wavelength would be ~10 nm. He noted, however, that the decomposition would be so slow that it would be masked by the competing mechanism of cellular precipitation. Spinodal decomposition in bulk Au–Ni alloys is observed (Woodilla and Averbach, 1968), but the modulation develops along $\langle 100 \rangle$ directions, not along $\langle 111 \rangle$. Since Yang produced films only with $\langle 111 \rangle$ texture, it was not possible to study the anisotropy or to compare his results directly with those from bulk alloys.

Paulson (1972) and Paulson and Hilliard (1977) studied interdiffusion in an ordering system with very short modulation wavelengths and tested the discrete analysis of Cook *et al.* (1969). Copper–gold films were made with (111) texture and modulation wavelengths of 0.8 to 2.6 nm. The average composition of the films was 16 at. % Au, which is outside the ordered Cu_3Au region on the phase diagram. The intensities of the 000^+ satellites were measured as a function of annealing time. The satellites decayed in the same manner as shown, for example, in Fig. 16, and values of the amplification factor R were obtained from the linear portions of the curves. The amplification factors at 200, 225, and 260°C are plotted as a function of d/λ in Fig. 25 (d is the interplanar spacing along $\langle 111 \rangle$). It can be seen that R is always negative and is of greatest magnitude for $\lambda \approx 1.7$ nm. The shape of these experimental curves is in qualitative agreement with the predictions of the discrete analysis, as shown in Fig. 9, over the corresponding range $0 \leq a/\lambda \leq 1.0$. Dividing R by B^2 gives the effective interdiffusion coefficient $\tilde{D}_{\lambda\langle 111 \rangle}$, which is plotted as a function of B^2 in Fig. 26. As discussed before, such plots should be linear, with a slope $2\kappa/f_0''$ and an intercept $\tilde{D}[1 + (2\eta^2/f_0'')Y_{\langle 111 \rangle}]$. The experimental data exhibit approximately linear behavior for $\lambda > 1.0$ nm. By extrapolating the linear portions of the curves to $\tilde{D}_{\lambda\langle 111 \rangle} = 0$, the critical wavelength was found to be ~1.0 nm. Paulson and

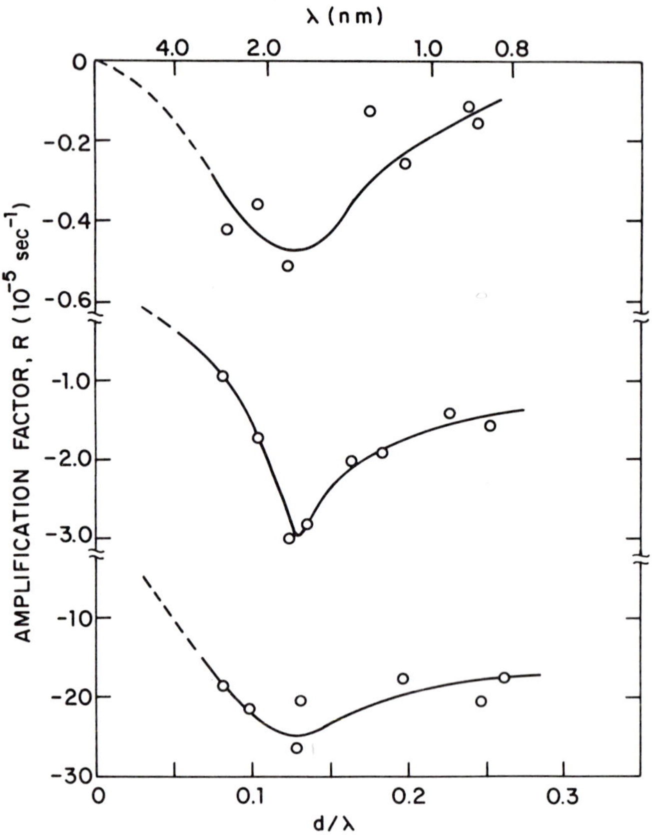

Fig. 25. The amplification factor R as a function of the ratio d/λ (d is the interplanar spacing and λ the modulation wavelength) in Cu – 16 at. % Au films annealed at (from top to bottom) 200, 225, and 260°C. [From Paulson and Hilliard (1977).]

Hilliard (1977) suggested that the nonlinearity at short wavelengths could be due to higher-order terms in the diffusion equation or, perhaps more likely, due to the flux from random solute movements (i.e., Brownian motion). Neither explanation was experimentally verified.

From high-temperature tracer diffusion measurements, Paulson calculated $\tilde{D}/f_0''$, and this value, combined with the slopes of the plots in Fig. 26, gave estimates for the gradient energy coefficient κ. The average value of κ was -4.68×10^{-11} J m^{-1}. From elastic constant and lattice parameter data, it was calculated that the elastic gradient energy coefficient is 1.79×10^{-11} J m^{-1}. Using an estimate of f_0'' from thermodynamic data and the measured value of κ, Paulson and Hilliard (1977) concluded that the chemical gradient energy coefficient κ_c must be in the range -6.5×10^{-11} to -9.0×10^{-11} J m^{-1}. For disor-

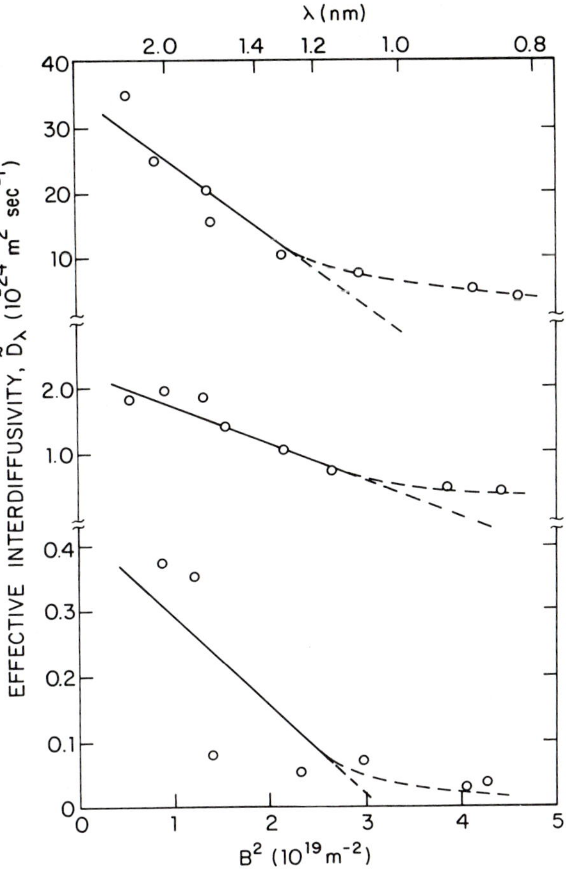

Fig. 26. Variation of the effective interdiffusion coefficient with B^2 in Cu–16 at. % Au films annealed at (from bottom to top) 200, 225, and 260°C. [From Paulson and Hilliard (1977).]

dered Cu_3Au alloys, pairwise interaction energies out to third-nearest neighbors have been derived from x-ray diffuse scattering experiments (Moss and Clapp, 1968). Using these interaction energies and the theory of Yamauchi (1971), Paulson and Hilliard calculated that $\kappa_c = -7.17 \times 10^{-11}$ J m^{-1}. If only nearest-neighbor interactions are considered, Eq. (20) can be used to derive that $\kappa_c = -3.84 \times 10^{-11}$ J m^{-1} (Paulson, 1972). Thus, there is good agreement between measured and theoretical gradient energy coefficients, provided that elastic energy effects are taken into account and that interactions out to third-nearest neighbors are considered.

Extrapolating the plots in Fig. 26 to $B^2 = 0$ gives $\tilde{D}[1 + (2\eta^2/f_0'')Y_{\langle 111\rangle}]$. Paulson evaluated η from lattice parameter measurements, f_0'' from thermodynamic data and $Y_{\langle 111\rangle}$ from elastic constant measurements in Cu and Cu_3Au.

The values of the bulk interdiffusivity $\tilde{D}$ obtained in this way varied from 2.4×10^{-25} m^2 sec^{-1} to 1.95×10^{-23} m^2 sec^{-1} over the temperature range 200 to 260°C. The activation energy, 141 kJ mole^{-1}, was in good agreement with values obtained by Hilliard from conventional interdiffusion measurements at high temperatures, and on an Arrhenius plot a single straight line could be drawn through both sets of data. This confirms that measurements in modulated films with preferred crystal orientation do give *bulk* interdiffusivities. It is interesting to note that the interdiffusivities are very much lower than the one determined by DuMond and Youtz (1940) in polycrystalline Cu–Au films.

The copper–nickel system is of interest, since it is thought to exhibit a miscibility gap at temperatures at which kinetic limitations prevent its direct observation. Clustering measurements, such as those of Vrijen and Radelaar (1978), indicate a critical temperature around 300°C. Since the modulated film technique is the most sensitive one available for measuring diffusion, it seemed ideally suited for investigating unmixing in this system. Tsakalakos (1977, 1981a,b) studied this system by using $\langle 111 \rangle$ textured films containing more than 35 at. % copper. Outside this range, the texture was unsatisfactory, and a 40-nm Cu underlayer was required in all cases. Upon annealing, a decrease in the average lattice parameter was observed, in qualitative agreement with the positive sign of d^2a/dc^2; quantitatively, however, the observed decrease was six times larger than the predicted value. The decay of the satellite intensities as a function of time was similar to that of Fig. 16, including the initial nonlinearity, and the effective interdiffusivity $\tilde{D}_\lambda$ was determined from the terminal slopes of the decay curves, as before. The composition and wavelength dependence of the data were analyzed by an expression of the form of Eq. (53) in Subsection II.D

$$D_\lambda(c, B) = D(c)\left(1 + \sum_{\mu=1}^{4} K'_\mu B^{2\mu}\right) \tag{63}$$

with the generalized interdiffusivity given by

$$D(c) = D_0 + (c - c_0)D_1 + (c - c_0)^2 D_2 \tag{64}$$

and

$$K'_\mu = 2K_\mu/(f''_0 + 2\eta^2 Y_{\langle 111 \rangle}), \tag{65}$$

where the composition dependence of $(f''_0 + 2\eta^2 Y)$ has been calculated from data in the literature. Figure 27 is a plot of the wavelength dependence of $\tilde{D}_\lambda$ in a 50 at. % Cu film at 400°C. The line, with the dashed portion, is the fit to the data of Eq. (63) for $c = c_0$. Three important observations can be made about this figure.

(i) The slope at $B = 0$ is proportional to the gradient energy coefficient κ, which is clearly negative here. This is in contradiction with the positive κ predicted on the basis of nearest-neighbor interactions [Eq. (20)] in a phase-sep-

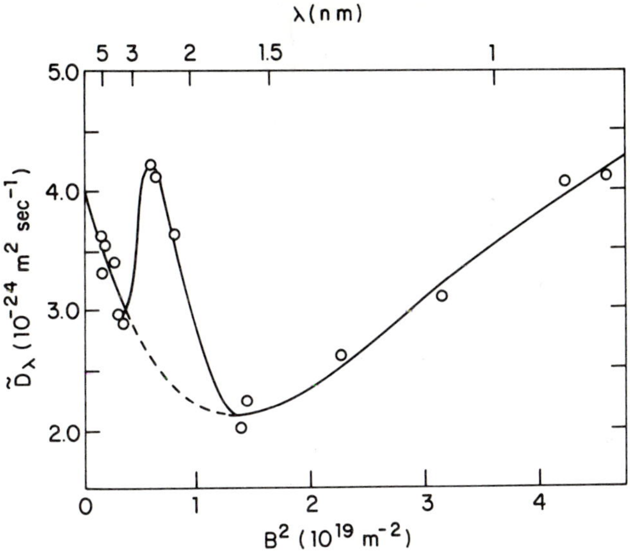

Fig. 27. Variation of the effective interdiffusion coefficient $\tilde{D}_\lambda$ with B^2 in equiatomic Cu–Ni films annealed at 400°C. [From Tsakalakos (1981a).]

arating system such as Cu–Ni. Tsakalakos (1977, 1981a) attributes this to the longer-range interactions in the general expression of Eq. (54). Using six interatomic potentials, derived from experimental short-range order parameters (Vrijen and Radelaar, 1978), he calculates $\kappa = -4.6 \times 10^{-12}$ J m^{-1}, which is indeed negative, but an order of magnitude below the experimental value $\kappa = -4.3 \times 10^{-11}$ J m^{-1}; the discrepancy is attributed to the limited number of potentials and to experimental uncertainty in the determination of the short-range-order parameters.

(ii) The plot is highly nonlinear [as opposed to the prediction of Eq. (31)], with a minimum at $\lambda = 1.6$ nm. Tsakalakos attributes this to the higher-order terms K_μ in the expansion of Eq. (63). His potential-based calculated values of K_μ are in qualitative agreement with his experimental determinations (correct sign) but are again about an order of magnitude low; they still give a minimum in $\tilde{D}_\lambda(B)$ at $\lambda = 1.6$ nm, however.

(iii) The peak at $\lambda = 2.5$ nm is attributed to the interaction of the Fermi surface with the new Brillouin zone face associated with the modulation, as discussed in Subsection II.D. From the position of the Fermi surface, determined by interpolating positron annihilation results on 78.7 and 58.1 at. % Cu alloys (Nanao *et al.*, 1972), Tsakalakos (1977) could account for the peak position quantitatively

As explained in Subsection II.E, in a phase-separating system, such as Cu–Ni, extrapolation of the diffusion data with temperature requires correction for the thermodynamic factor. An Arrhenius plot of $\tilde{D}T/f''$ has a slope corresponding to an activation energy of 270 kJ mole^{-1}, and is in excellent agreement with bulk high-temperature measurements of Monma *et al.* (1964).

The composition dependence of $\tilde{D}(c)$ in Eq. (63) leads to nonlinear diffusion effects, as explained in Subsection II.D. Tsakalakos (1977) therefore also measured the intensity of the second- and third-order satellites of the (111) Bragg peak as a function of annealing time. An example is illustrated on Fig. 28. The qualitative resemblance to the theoretical result of Fig. 12 is obvious. Application of Tsakalakos's approximate analytic solution to the nonlinear diffusion equation (1977) (see Subsection II.D) allows the gradient energy coefficient to be estimated from the features of Fig. 28: $\kappa = -2.3 \times 10^{-11}$ J m^{-1}, which is a factor of 2 less than the one obtained from the wavelength dependence of $\tilde{D}_\lambda(B)$.

Tsakalakos (1981b) studied phase separation in the Cu–Ni system in a $c_0 = 55$ at. % Cu film, with a wavelength of 5 nm, long enough to ensure positive amplification inside the spinodal. After a preanneal at 450°C for 40 min, to decrease the composition amplitude $(c - c_0)$ to 8 at. %, the specimen was annealed at 325°C for several hours, which resulted in an *increase* in

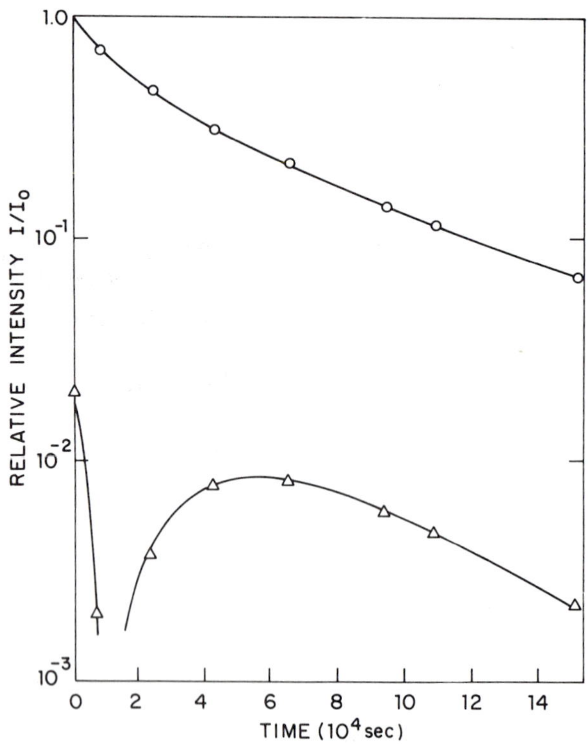

Fig. 28. Variation of the intensities of the first- (○) and second-order (△) (111) satellites in an Ni–52 at. % Cu modulated film, with a wavelength of 4.56 nm. [From Tsakalakos (1977).]

amplitude of about 5 at. %. Further annealing at 340°C resulted in decay of the amplitude. The coherent spinodal temperature therefore lies between these last two annealing temperatures, in agreement with the suggestion of Vrijen and Radelaar (1978) based on their clustering measurements. The slow growth of the amplitude at 325°C was attributed to the coherent spinodal composition, estimated at 64 at. % Cu, being close to the maximum Cu content in the film ($c = 63$ at. % Cu), so that nonlinear effects must be taken into account to describe the approach to a stationary state. The observed *decay* of *large* amplitude ($c - c_0 = 28.3$ at. %) modulations at 325°C was attributed also to parts of the specimen having concentrations outside the coherent spinodal.

The silver–palladium system is a homogenizing one (negative heat of solution) and was investigated by Henein (1979), primarily for modulus studies. Interdiffusion was studied in equiatomic films with ⟨111⟩ texture and modulation wavelengths between 0.94 and 5.08 nm. Interdiffusion coefficients $\tilde{D}_\lambda(B)$ determined from the terminal slopes of the intensity decay curves were analyzed by using Tsakalakos's (1977) expressions [Eqs. (63)–(65)]; Fig. 29 shows the results, along with the fit at $c = c_0$. The gradient energy coefficient was determined to be $\kappa = -5.0 \times 10^{-10}$ J m^{-1}, in qualitative agreement with the negative sign of the heat of mixing and the nearest-neighbor expression [Eq. (20)]; long-range interactions do not have to be invoked here. The curvature in

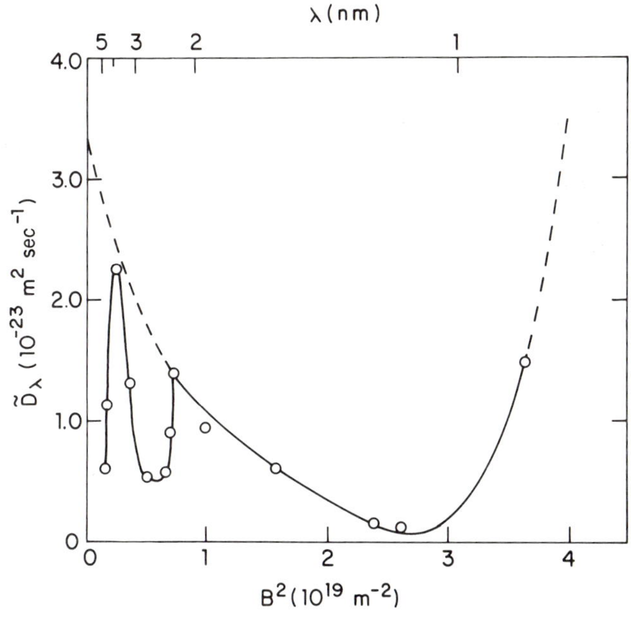

Fig. 29. Variation of the effective interdiffusivity $\tilde{D}_\lambda$ with B^2 in equiatomic Ag–Pd films annealed at 375°C. [From Henein (1979).]

the $\tilde{D}_\lambda(B)$ curve again results from the higher-order terms K_μ in Eq. (63). The sharp decrease in $\tilde{D}_\lambda$ in the range 2.3–2.7 nm was shown not to be a direct result of the enhanced modulus, observed in separate experiments in this wavelength range, but was tentatively attributed to Fermi surface effects similar to those used to explain the Cu–Ni results (see Fig. 27). The sharp drop between 3.9 and 5.1 nm is attributed to a gradual loss of coherency, as seen by Philofsky and Hilliard (1969) in Cu–Pd (see Fig. 17).

Spreng *et al.* (1972) studied the enhancement of the interdiffusivity at liquid-nitrogen temperature in gold–silver modulated films ($\lambda = 1-2$ nm) as a result of electron irradiation. Since *in situ* diffraction patterns are difficult to obtain during irradiation, the change in electrical resistivity $\rho(t)$ was measured to determine the change in the modulation amplitude $A(t)$ based on a free electron calculation by Kolomets and Smirnov (1962)

$$\rho_F - \rho(t) = CA^2(t), \tag{66}$$

where C is a constant and ρ_F the resistivity of a homogeneous sample of the same average composition. The interdiffusivity was found to increase proportionally to the electron current density and to increase with increasing electron energy. For example, for $2\mu A$ cm^{-2} at 2 MeV, $\tilde{D}$ was approximately 10^{-26} m^2 sec^{-1}. The atomistic mechanism of diffusion was thought to be motion of interstitialcies. On heating above $-70\,^\circ$C, additional interdiffusion was observed and attributed to migration of vacancies.

B. Amorphous Metals

Amorphous metals, or metallic glasses, are alloys produced by rapid quenching of the melt or by other techniques, including vapor or sputter deposition. Structurally, they can be considered liquidlike, while mechanically they are solids in that their viscosity η is greater than 10^{15} N sec m^{-2}. The compositionally modulated film technique is very useful for studying diffusion in these materials for two main reasons.

(i) Amorphous metals are metastable with respect to the crystalline phase. Above their glass transition temperature T_g (i.e., the temperature below which $\eta > 10^{12}$ N sec m^{-2}), their atomic mobility is high enough to permit rapid crystallization. Diffusion measurements therefore must be made below 10^{-20} m^2 sec^{-1}. Some other profiling and sectioning techniques can be used to measure D in this range, but none of them has the sensitivity of the modulated film technique (see Subsection IV.A).

(ii) Amorphous metals below T_g are internally unstable in that they continuously transform to amorphous states of lower free energy. This process, called structural relaxation, leads to changes in all physical properties. The most basic of these is a slight densification (<0.5%); most strongly affected are the atomic transport properties such as η and D, which can change by many orders of magnitude (Taub and Spaepen, 1980). It is therefore crucial that the relaxation phenomenon is taken account of in diffusion measurements. The modulated film technique is ideally suited to this, in that the measurements are nondestructive and allow continuous monitoring necessary for time-resolved studies of D with annealing.

Work on compositionally modulated amorphous metals has been conducted at Harvard University for a number of years. The films were produced by dc sputtering from two targets of different compositions. The substrate was mechanically positioned alternately under each target for periods between 1–20 sec, resulting in modulations between 0.2 and 7 nm. Special care had to be taken to adjust the deposition conditions so that incorporation of the argon sputtering gas was minimized. This technique was developed by Rosenblum (1979), and he also made the first diffusion measurements (Rosenblum *et al.*, 1980). His films, of composition $(Pd_{80}Au_7S_{13})_{70}/Fe_{30}$ showed only a first-order satellite (with only 000 satellites present, as discussed in Subsection II.C), indicating a roughly sinusoidal modulation. On annealing, the satellite intensity decayed and $\tilde{D}_\lambda$ was derived in the usual way [Eq. (49)]. It was found that $\tilde{D}_\lambda$ decreased with annealing time, rapidly at first, and then tended to stabilize. This behavior was attributed to structural relaxation. Because of problems with crystallization, new films, of composition $(Pd_{85}Si_{15})_{61}/(Fe_{85}B_{15})_{39}$, were produced, in which each type of layer was of glass-forming composition. Interdiffusion measurements on these films confirmed the earlier results.

Greer *et al.* (1982) and Cammarata and Greer (1984) made measurements on films of composition $(Pd_{85}Si_{15})_{50}/(Fe_{85}B_{15})_{50}$. Films with a total thickness of ~ 5 μm and modulation wavelengths of 0.22 to 7.46 nm were produced. Only a weak first-order satellite was observed for the smallest-wavelength film, while for the $\lambda = 7.46$ nm film, peaks out to fourth order were detected. The metals dominate the x-ray scattering, and the measured $\tilde{D}_\lambda$ is the Pd–Fe interdiffusivity.

Two films with $\lambda = 2.06$ and 3.9 nm were annealed at 250°C. For each film a rapid initial drop in intensity was observed; after about 15 h, however, the decay curve $I(t)$ decreased linearly with time. This initial drop could have a number of causes. The composition dependence of $\tilde{D}$, with a large initial modulation amplitude, can give rise to nonlinear decay curves. As discussed in Subsection II.D, however, this effect is small for the first-order decay curve (see, for example, Figs. 12 and 28). The small angle x-ray scattering from the films decays in about 2 h, probably owing to elimination of voids or reduction of the argon content. The contribution of moving grain boundaries to rapid interdiffusion, discussed by Cook (1966) and Cook and Hilliard (1969) (see Subsection III.A), is also not present in amorphous materials. The continuing decrease in $\tilde{D}_\lambda$ after 2 h is therefore attributed to *structural relaxation* of the amorphous state. Since the annealing temperature was the same for both films, the degree of structural relaxation at a given time is the same. The interdiffusivity $\tilde{D}_\lambda$ was determined from the final gradients of the decay curves and found to be 7.09×10^{-26} m^2 sec^{-1} and 1.37×10^{-24} m^2 sec^{-1} for the 2.06 and 3.94-nm wavelengths, respectively. The wavelength dependence of $\tilde{D}_\lambda$ was analyzed by fitting Eq. (11) to these two data points, giving $\tilde{D} = 1.86 \times 10^{-24}$ m^2 sec^{-1} and $\kappa/f_0'' = -5.18 \times 10^{-20}$ m^2. Since $\tilde{D}$ is positive, f_0'' is positive and hence κ must be negative. A negative gradient energy coefficient is characteristic of an ordering system, which is what one would expect for this system, on the basis of the behavior of Pd–Fe solid solutions. The critical wavelength λ_c is 2.02 nm, which is considerably larger than in crystalline systems.

The kinetics of structural relaxation in this system were studied by annealing a film with $\lambda = 3.22$ nm at 300°C. The decay curves of the first- and second-order satellites are shown in Fig. 30. The behavior of the second-order satellite is similar to that seen in Figs. 12 and 28 and is therefore attributed to the composition dependence of $\tilde{D}$. The dependence of $\tilde{D}$ on time was determined from the instantaneous slope of the first-order decay curve and the value of κ/f_0'' obtained above. It was found that $(1/\tilde{D})$ increased *linearly* with annealing time, from 1×10^{22} sec m^{-2} to 8×10^{22} sec m^{-2} in 120 min.

Since structural relaxation has such a large effect on the diffusivity, a physically meaningful activation energy can only be determined under *isoconfigura-*

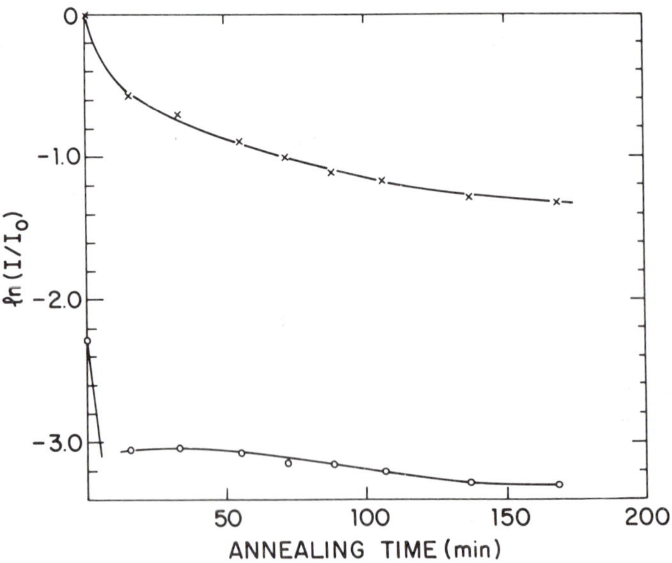

Fig. 30. Decay of the first- (X) and second-order (O) satellite intensities from an amorphous $(Pd_{85}Si_{15})_{50}/(Fe_{85}B_{15})_{50}$ modulated film with a wavelength of 3.22 nm, annealed at 300°C. [From Greer *et al.* (1982).]

tional conditions. This means that a negligible amount of relaxation should occur over the entire course of measurements at different temperatures; this is verified experimentally by cycling the temperature and reproducing the results. The modulated film technique lends itself very well to this verification, and an isoconfigurational experiment on a film with $\lambda = 2.06$ nm gave an Arrhenius temperature dependence in the range 210–270°C, with an activation energy of 195 kJ mole^{-1} and a pre-exponential of 5.6×10^{-5} m^2 sec^{-1}. The latter is similar to the value for substitutional diffusion in fcc metals.

The relationship between viscosity η and diffusivity is of interest in understanding atomic transport mechanisms. In general, it can be stated (Spaepen, 1981) as

$$\eta D = k_B T/L, \tag{67}$$

where L is a characteristic length. For liquid metals, $L = 6\pi r$, where r is the ionic radius, and Eq. (67) then corresponds to the Stokes–Einstein relation. Direct comparison of η and D in the glassy state is made difficult by the uncertainty about the degree of structural relaxation in the two experiments. However, it is of interest to note that the isoconfigurational activation energy for the viscosity of $Pd_{82}Si_{18}$ is 192 kJ mole^{-1} (Taub and Spaepen, 1980), which is very close to the value observed for $\tilde{D}$ in the modulated film. Furthermore, the viscosity increase due to relaxation is also linear with time, and the rate is therefore obviously independent of the degree of relaxation. Consequently, the values of $d\eta/dt$ and $d(1/\tilde{D})/dt$ can be compared and checked against Eq. (67). It is found that $\tilde{D}$ is about 160 times greater than would be predicted from η by the Stokes–Einstein relation. It seems, therefore, that the atomic jumps leading to flow and diffusion are related (similar activation energy and annealing kinetics) but that there are diffusional jumps that do not lead to flow.

In amorphous materials it is possible to produce composition modulations with arbitrarily short wavelengths, whereas in crystalline materials one is limited by the plane spacing normal to the direction of modulation. Figure 31 shows the intensity of the first-order satellite in an Fe–B/Pd–Si film with $\lambda = 0.227$ nm. The initial rise is expected for an ordering system with a modulation below the critical wavelength. The subsequent drop is attributed to ordering within the layers, which removes the one-dimensional order and homogenizes the film (Greer, 1983).

In conclusion, it should be pointed out that the modulated film technique is not only very useful in studying diffusion in amorphous metals, but that the use of amorphous alloys can also provide advantages in the study of modulation effects and applications of modulated films.

(i) There is no interference between crystal periodicity and modulation wavelength, which can therefore be adjusted continuously.

(ii) Coherency strains do not occur, which simplifies the analysis; the

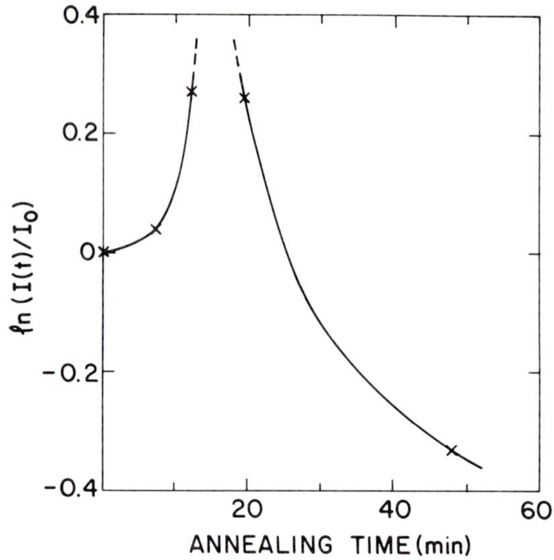

Fig. 31. Annealing behavior of the first-order satellite intensity from an amorphous $(Pd_{85}Si_{15})_{50}/(Fe_{85}B_{15})_{50}$ modulated film with a wavelength of 0.227 nm, annealed at 250°C. [From Greer (1983).]

absence or reduction of internal strains may also affect the dependence of $\tilde{D}_\lambda$ and the elastic moduli on the wavelength.

(iii) Amorphous alloys can be produced homogeneously over a continuous composition range, where crystalline materials are often multiphase.

(iv) Short-circuit diffusion paths, such as dislocations and grain boundaries, especially mobile boundaries (Cook, 1966), do not occur in amorphous metals; this enhances their stability and useful lifetime.

C. Semiconductors

The study of interdiffusion in artificially modulated semiconductors has not been developed as far as in metals. The only detailed investigations so far have been on the (Ga, Al)As and the (In, Ga)Sb systems.

Fleming *et al.* (1980) investigated the interdiffusion in $(GaAs)_n(AlAs)_m$ structures produced by molecular beam epitaxy. They annealed a sample with $(n, m) = (12, 9)$, corresponding to a wavelength of $\lambda = 5.9$ nm, at 860°C, and measured the decay of the first through fourth satellites of the (002) Bragg peak. The results are shown on Fig. 32. The nonlinear behavior of the higher harmonics indicates a concentration-dependent diffusivity (see Subsection II.D and Figs. 12 and 28). Fleming *et al.* fit the data by numerical integration of the diffusion equation and used the expression for the diffusivity

$$D(c, T) = D_0(T) \exp[\alpha(T)c], \tag{68}$$

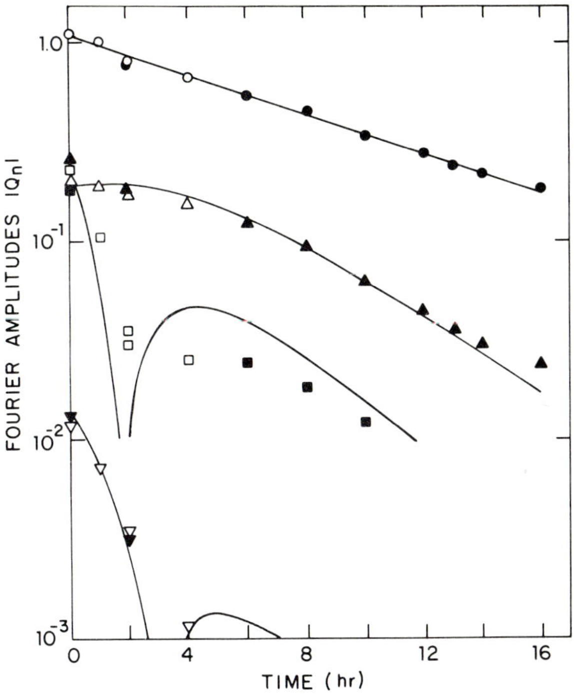

Fig. 32. Fourier amplitudes of the first four satellites of the (002) Bragg reflection in a $(GaAs)_{12}(AlAs)_9$ modulated structure as a function of annealing time at 860°C. ●, ○; ▲, △; ■, □; ▼, ▽; correspond to $m = 1, 2, 3, 4$, respectively. The $m = 4$ curve is times 10^{-1}. The open and closed symbols are for two different samples. The lines are from a theoretical fit with a concentration-dependent diffusivity. [From Fleming *et al.* (1980).]

where c is the Ga concentration,

$$\alpha(T) = (\Delta F_{Al} - \Delta F_{Ga})/k_B T \tag{69}$$

and

$$D_0(T) = D(T) \exp(-\Delta F_{Al}/kT), \tag{70}$$

with ΔF_{Al} and ΔF_{Ga} the vacancy formation energies in AlAs and GaAs, respectively. The gradient energy term was ignored. The results are plotted in Fig. 33; the extremes, i.e., $c = 0$ and $c = 1$, correspond to Ga diffusion in AlAs and Al diffusion in GaAs, respectively. The parameter α was found to be 2.06, which corresponds to a formation energy difference of 0.2 eV. The interdiffusion in a sample corresponding to $(n, m) = (1.1, 1.1)$, annealed at 800°C, was also investigated. Since only the first-order satellite could be observed, the composition dependence of $\tilde{D}$ could not be determined. The average value of $\tilde{D}$ determined from the decay of the first satellite is indicated on Fig. 33. This figure also

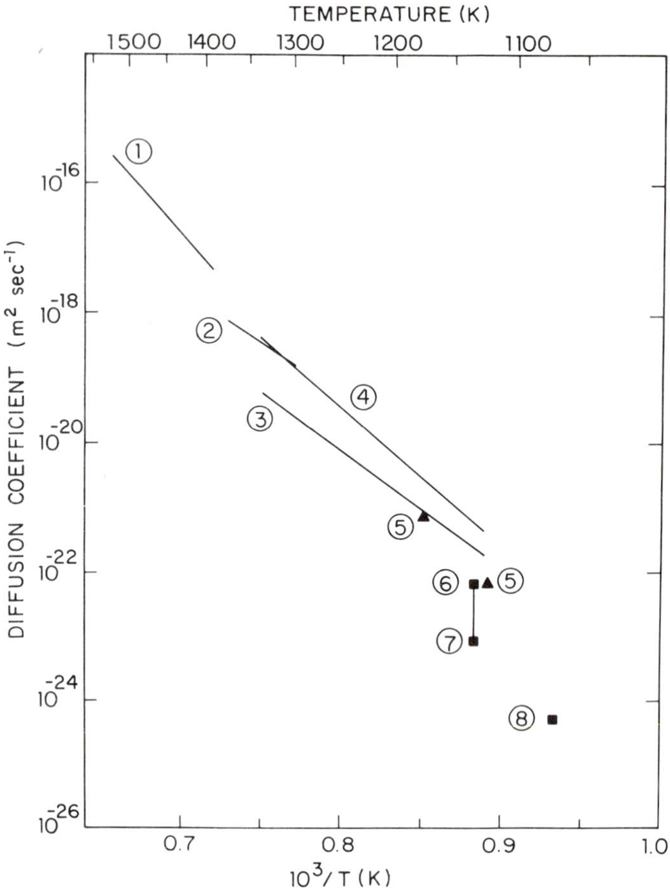

Fig. 33. Comparison of various diffusion coefficients in the (Ga, Al)As system: (1) Ga in GaAs (Goldstein, 1961); (2) Ga in GaAs (Palfrey *et al.*, 1981); (3) Ga in AlAs (Chang and Koma, 1976); (4) Al in GaAs (Chang and Koma, 1976); (5) interdiffusion in (Ga, Al)As (Petroff, 1977); (6) Al in GaAs (Fleming *et al.*, 1980); (7) Ga in AlAs (Fleming *et al.*, 1980); (8) interdiffusion in (Ga, Al)As (Fleming *et al.*, 1980).

shows comparisons with other diffusion data in the (Ga, Al)As system. Petroff (1977) investigated interdiffusion in (Ga, Al)As modulated structures by electron microscopy by using either the decrease in imaging contrast or the disappearance of the satellite reflections, and his result at 850°C agrees roughly with those of Fleming *et al.* at 860°C. Chang and Koma (1976) used Auger electron profiling to determine the composition-dependent interdiffusivity in a single AlAs layer sandwiched between GaAs. The composition extremes are shown on Fig. 33. At high temperatures, their results for GaAs agree with the self-diffusion data of Palfrey *et al.* (1981) and the extrapolation of Goldstein's (1961) data; the activation energy measured by Chang and Koma, 4.3 eV, is intermediate between the high-temperature values 2.6 and 5.6 eV, respectively. At lower temperatures, the interdiffusion coefficients of Chang and Koma are significantly greater than those reported by Fleming *et al.* One possible explanation for this discrepancy may be the occurrence of sputter-enhanced diffusion during Auger profiling. The value of $\tilde{D}$ for the (1.1, 1.1) sample of Fleming *et al.* is lower than one would expect from an extrapolation of the data for the (12, 9) sample. One way to illustrate this is the following comparison: the value of $\tilde{D}$ obtained by extrapolating the data taken at 800°C for the (1.1, 1.1) sample to 860°C, using Chang and Koma's activation energy of 4.3 eV, is 6.3×10^{-24} m^2 sec^{-1}; the value obtained for the (12, 9) sample, with $c = 0.5$, is, according to Eq. (68), 2.4×10^{-23} m^2 sec^{-2}. If this difference is attributed to the difference in modulation wavelength ($\lambda = 0.62$ and 5.9 nm, respectively), it can be analyzed by using Eqs. (11) and (21), where the latter one corresponds to the discrete analysis necessary here due to the very short periodicity of one of the samples. The two adjustable parameters in this analysis can be determined from the two data points: the bulk interdiffusivity $\tilde{D}$ is 2.44×10^{-23} m^2 sec^{-1} and $\kappa/f_0'' = -1.5 \times 10^{-20}$ m^2. The latter value is similar to those obtained for crystalline metallic systems and slightly lower than that for amorphous metallic films. It seems, therefore, that the gradient energy term in semiconductor systems is not negligible and should be included in future analyses. The wavelength of the (12, 9) sample, however, is long enough for its interdiffusivity to correspond to the bulk value. Inclusion of a gradient energy term, therefore, does not explain the discrepancy between the results of Fleming *et al.* and those of Chang and Koma.

Eltoukhy and Greene (1979) have studied interdiffusion in compositionally modulated GaSb/InSb films prepared by sputter deposition, with a (110) texture and modulation wavelength greater than 3.6 nm. The interdiffusion coefficients $\tilde{D}$ were determined from the terminal slopes of the decay curves of the first satellites of the (220) Bragg peak and were analyzed by using Eqs. (21) and (31). The gradient energy term in the latter equation was neglected because the modulation wavelengths were thought to be long enough; however, no wavelength-dependent data were presented to substantiate this. The elastic modulus Y and the free-energy second derivative f_0'' in Eq. (31) were estimated from literature values. Since no measurements of higher-order satellites were made, the composition dependence of $\tilde{D}$ could not be investigated. The results in the temperature range 230–340°C are presented in Fig. 34; an activation energy of 77 kJ mole^{-1} and a pre-exponential equal to 4×10^{-18} m^2 sec^{-1} were observed. Figure 34 also shows various other diffusion data in the GaSb and InSb systems. Although a precise comparison is not possible, the values of $\tilde{D}$ determined by Eltoukhy and Greene (1979) nevertheless appear to fall inside the broad range predicted by extrapolation of the high-temperature data. The temperature dependence of $\tilde{D}$, however, seems questionable: the activation energy is much

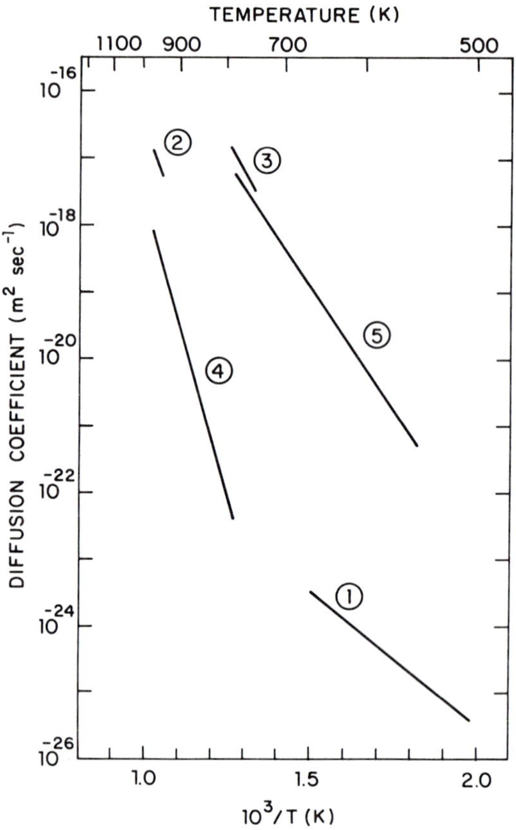

Fig. 34. Comparison of various diffusion coefficients in the (Ga, In)Sb system: (1) interdiffusion in (Ga, In)Sb (Eltoukhy and Greene, 1979); (2) Ga in GaSb (Eisen and Birchenall, 1957); (3) In in InSb (Eisen and Birchenall, 1957); (4) In in GaSb (approximate) (Mathiot and Edelin, 1980); (5) diffusion coefficient controlling creep in InSb (Siebhoff, 1979).

lower than that of any of the other measurements (144–302 kJ mole^{-1}) and, combined with the low pre-exponential, gives a value of the diffusivity at the average melting point of GaSb and InSb of only 10^{-22} m^2 sec^{-1}. The high-temperature data give diffusion coefficients of about 10^{-17} m^2 sec^{-1} at the melting temperature of the respective compounds. Brown and Ashby (1980) have shown that the diffusivity at the melting temperature has a characteristic value that depends on the structure and bonding type. Silicon and germanium, for example, also have diffusivities at their melting point of about 10^{-16} m^2 sec^{-1}. An extrapolation of the interdiffusion data in (Ga, Al)As (Chang and Koma, 1976; Fleming *et al.*, 1980; Petroff, 1977) in Fig. 33 also gives a value around 10^{-16} m^2 sec^{-1} at the melting point of the compounds, indicating that a temperature dependence of D similar to that of the other diffusion data can be expected in the modulated films.

Eltoukhy and Greene (1978, 1979, 1980) and Eltoukhy *et al.* (1977) also characterized the enhancement of interdiffusion in the (Ga, In)Sb system due to low-energy ion bombardment during the rf sputter deposition of the layers. They measured an enhancement of five orders of magnitude over their data of Fig. 34, which sets a lower limit on the width of the interfaces in modulated films prepared by this method.

IV. DISCUSSION

A. Comparison with Other Techniques

The sensitivity of the x-ray diffraction method for measuring the interdiffusivity in compositionally modulated films is illustrated on Fig. 35, in which it is compared to that of other techniques for measuring the diffusivity D. The limiting factor in all the other techniques is the spatial resolution of the profiling technique. By comparing this resolution to the diffusion distance $(Dt)^{1/2}$ over a reasonably long annealing time (one day), a lower limit of D detectable by each technique can be estimated. A review of some of these techniques, applied specifically to amorphous metals, has been given by Cantor and Cahn (1983).

In mechanical profiling, slices of material are removed by microtome or by lapping, and their composition is measured by chemical analysis or radiotracer counting. For example, measurements 1 on Fig. 33 and 2 and 3 on Fig. 34 have been made by these methods. Anodic oxidation, in which successive layers of controlled thickness are oxidized, dissolved off, and analyzed, has a slightly higher sensitivity; measurements 2 on Fig. 33 are a typical example. Removal of material by sputtering, combined with Auger analysis of the exposed surface

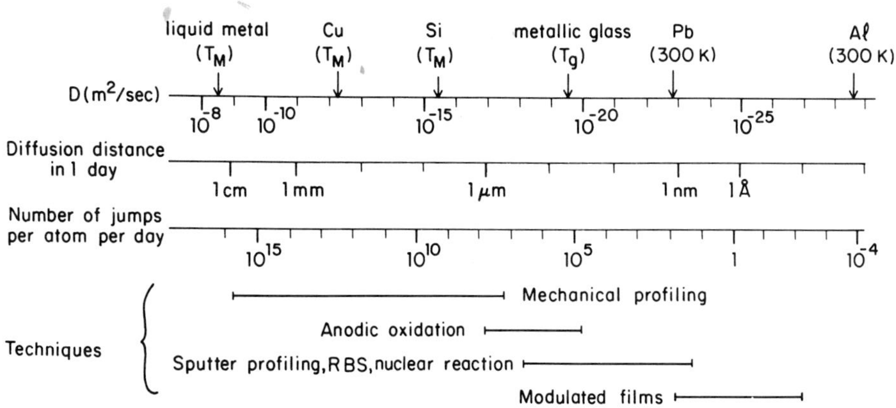

Fig. 35. Comparison of the sensitivity of various techniques for measuring diffusivities.

(Luborsky and Bacon, 1982; Röll and Reill, 1982), mass spectrometry on the sputtered atoms (SIMS) (Cahn *et al.,* 1980; Edelin and Tete, 1981), or radiotracer counting on the removed material (Gupta *et al.,* 1975, 1982) allows measurements of D down to 10^{-23} m^2 sec^{-1}. Similar sensitivity can be obtained by nondestructive profiling techniques, such as Rutherford backscattering (RBS) (Chen *et al.,* 1978) and nuclear reaction profiling (Birac and Lesueur, 1976). Examples of those techniques are measurements 3 and 4 in Fig. 33, where Auger spectroscopy was used, and 4 in Fig. 34, for which SIMS was used. Of special interest to this chapter are sputter profiling experiments on modulated films. Ludeke *et al.* (1974) have demonstrated that this technique, combined with Auger spectroscopy, can resolve composition modulations very well. Röll and Reill (1982) used this method to determine the interdiffusivity in Cu–Ni modulated films with a 90-nm wavelength. They found a diffusivity of 10^{-20} m^2 sec^{-1} at 675 K and an activation energy equal to 106 kJ mole^{-1}. This is in sharp disagreement with Tsakalakos's (1977) results on the same system (see Subsection III.A): he has found an interdiffusivity of 4×10^{-24} m^2 sec^{-1} at 673 K and an activation energy (with correction for the temperature dependence of the chemical driving force term) equal to 270 kJ mole^{-1}. The discrepancy is most likely due to the grain boundary diffusion effects in the films of Röll and Reill, which presumably did not have as good a texture as did those of Tsakalakos.

The x-ray diffraction method for determining the interdiffusivity in compositionally modulated films derives its sensitivity from its ability to sample changes in the composition profile of many interfaces at once. This is why, as illustrated on Fig. 35, profile changes involving atomic jumps of only one atom in a hundred can be detected. If, however, the profile of only a single interface is being determined, as is the case for all the other techniques, each atom has to perform at least a hundred jumps for the broadening of the compositional profile to be detectable. A further advantage of the modulated film technique is its being nondestructive, so that measurements can be made as a function of thermal history. This advantage has been crucial in a number of experiments, for example, in ascertaining isoconfigurationality in experiments on metallic glasses (see Subsection III.B) or in demonstrating phase separation in the Cu–Ni system (see Subsection III.A).

B. Stability of Modulated Structures

Artificial compositionally modulated materials are not equilibrium structures. In particular, they have high interfacial densities—higher than can be attained by any other preparation technique. Given sufficient atomic mobility, changes in the composition profile are expected to occur. The stability of the material, i.e., its resistance to these changes, is obviously of importance in practical applications. In this subsection we describe the present understanding of the stability, based on diffusion studies. Materials, as-deposited, typically have a large-amplitude, approximately square-wave composition profile, and we take this to be the initial state.

First, we consider cases in which the amplification factor R for the fundamental component of the composition modulation is negative, i.e., cases in which the system tends to homogenize. Such cases arise in an ordering system above the critical temperature (Fig. 9) or below the critical temperature, but with $\lambda > \lambda_c$ (Fig. 8), and in a phase-separating system outside the spinodal (Fig. 7) or inside the spinodal, but with $\lambda < \lambda_c$ (Fig. 6). In all except the last example, the decay of the modulation occurs for arbitrarily long wavelengths. For wavelengths large compared to the interatomic spacing, gradient energy effects are small and higher harmonics of the composition modulation decay more rapidly than the fundamental. Thus the first effect of diffusion in such a material is a rounding of the composition profile; in the linear approximation, the profile would become sinusoidal. Although nonlinear effects are present for large amplitudes (Subsection II.D), the decay of the first Fourier component of the composition modulation is well described by the linear analysis, Eqs. (11) and (21). These equations can be used to predict the lifetime of the modulated material. As demonstrated experimentally (Subsection III.A), the interdiffusivity $\tilde{D}$ in Eq. (11) is a bulk property unaffected by short-circuit paths provided by extended lattice defects such as dislocations and grain boundaries. This is true if the modulation wavelength is small compared to the distances between defects. An important exception was pointed out by Cook (1966): moving grain boundaries may have a significant effect because of the large volume they can sweep through. In addition to lattice defects, there may be defects in the composition modulation: thickness variations or even discontinuities in the deposited layers. These are expected to have a negligible effect for the same reason as given for extended lattice defects.

The amplification factor is positive for a phase-separating system inside the spinodal, with $\lambda > \lambda_c$ (Fig. 6). In this case, the first effect of diffusion is that the composition profile tends to the stationary state, i.e., the profile of minimum energy for the given wavelength. The attainment of the stationary state involves significant rounding of the composition profile only when the modulation wavelength is so small as to be similar to the critical wavelength (Subsection II.D). When the stationary state has been reached, the system can still lower its energy by increasing the modulation wavelength. For a system with initially random composition fluctuations in one dimension, numerical modeling has shown coarsening, i.e., an increase in the average wavelength (Hillert, 1961; de Fontaine and Hilliard, 1970). For an artificial system in which the initial wavelength is well defined, such coarsening, which can occur fairly rapidly, is unlikely. Nevertheless, the system is not stable or even metastable. Langer (1971) has shown that the stationary states are unstable to thermal fluctuations. The most common instability mode is one in which there is thinning and thickening of alternate layers of the same composition; this leads ultimately to a doubling of the wavelength. This coarsening, which would occur even with a perfectly periodic composition modulation, is very slow. Expressing Langer's result in terms of the parameters used in Subsection II.A, the average layer thickness l should increase with time t according to

$$l(t) = l_0 + \sqrt{\frac{-\kappa}{f_0''}} \ln\left[1 - \frac{8\tilde{D}f_0''}{\kappa} t \exp\left(-l_0\sqrt{\frac{-f_0''}{\kappa}}\right)\right], \tag{71}$$

where l_0 is the initial thickness. Extended lattice defects are again expected to have a negligible effect on the coarsening kinetics. Defects in the composition modulation, on the other hand, are very important. Thickness variations in the deposited layers may initiate coarsening modes additional to those described by Langer (1971). In lamellar eutectics, coarsening occurs only by the migration of faults in the lamellar structure (Graham and Kraft, 1966). Similarly, in an artificially modulated system the coarsening is expected to be dominated by the migration of faults, such as layer edges, in the modulation sequence. A partial layer, i.e., one not extending across the entire sample area, will shrink laterally and the surrounding layers will thicken. This reaction may be quite rapid since, unlike the other coarsening modes, it has a strong driving force arising from interfacial curvature. In some materials, though, the fault density may be very low.

A positive amplification factor is also obtained in an ordering system below the critical temperature, with $\lambda < \lambda_c$ (Fig. 8). In this case, the modulation wavelength is similar to the interatomic spacing and it is essential to take the crystal structure into account. The crystallographic direction normal to the layers is important. In β-brass, for example, the ordered structure can be obtained by imposing on the bcc lattice a composition wave along $\langle 100 \rangle$ with $\lambda = a$, or along $\langle 111 \rangle$ with $\lambda = a/\sqrt{3}$. On the other hand, there is no composition variation along $\langle 110 \rangle$ in the ordered structure. Consider first the case in which the layers are normal to a direction such as $\langle 100 \rangle$. It is possible that the imposed modulation wavelength is the correct one for the ordered structure; if so, perhaps after some change in the amplitude of the modulation, the structure is in stable equilibrium. If the imposed modulation wavelength is slightly in error, a stack of antiphase domains is the result. Domain coarsening is expected to be dominated by the behavior of faults in the stacking sequence. If the imposed wavelength is substantially in error, it is unlikely that the amplification factor is positive. Consider now the case in which the layers are normal to a direction such as $\langle 110 \rangle$. Whether the amplification factor for the imposed modulation wavelength is positive or not, the modulation will be destroyed by ordering within the layers to produce the normal ordered structure. The kinetics of this ordering are rapid, since only single-atom jumps are required. Ordering can also occur in amorphous systems, and some preliminary data on stability have been presented in Subsection III.B.

It is now possible to summarize the factors controlling the stability of artificial compositionally modulated materials. In all cases, a low bulk interdiffusivity $\tilde{D}$ is desirable for stability. The effective $\tilde{D}$ is unaffected by extended lattice defects, except moving grain boundaries. This latter problem can be removed by using amorphous materials or crystalline materials with a well-developed texture to eliminate surface-anisotropy-driven grain growth. For imposed composition modulations with a negative amplification factor R, the perfection of the composition modulation is unimportant. Phase-separating systems with a positive amplification factor, i.e., systems inside the spinodal with $\lambda > \lambda_c$, are significantly more stable than systems with $R < 0$, but their coarsening kinetics are markedly enhanced by imperfections in the deposited layers.

V. CONCLUSIONS

The theory of interdiffusion in artificial compositionally modulated materials is well developed. Both phase-separating and homogenizing systems have been considered. The gradient energy and strain effects, which become important at small modulation wavelengths, have been analyzed for diffusion both in a continuum and on a discrete lattice. In the simplest form of the theory, the equations are linearized by treating several parameters as composition independent; in this form, the theory is valid for systems close to homogeneity. The interdiffusion kinetics, when nonlinear effects become important, have been treated by numerical modeling.

Experimentally, the interdiffusion is most easily studied by using x-ray diffraction. Analysis of the diffraction pattern can yield the wavelength and amplitude of the composition modulation and, in crystalline materials, the degree of coherency of the layers and the dependence of the lattice parameter on composition.

Most of the experimental studies have been carried out on crystalline metals deposited in epitaxial layers. Gradient energy effects have been demonstrated and in most cases can be explained quantitatively. In some systems only nearest-neighbor atomic interactions are important; in others longer-range interactions have been invoked. Very low *bulk* interdiffusivities $\tilde{D}$ have been measured; agreement with conventional measurements of interdiffusivities and tracer diffusivities at higher temperatures is good, indicating Arrhenius behavior over up to 12 orders of magnitude. In some cases $\tilde{D}$ is strongly composition dependent. Strain effects are apparent in the dependence of the interdiffusivity on the degree of coherency. Coherent spinodal temperatures have been estimated. Some sharp deviations in $\tilde{D}$ at small wavelengths have been tentatively attributed to Brillouin zone effects.

In amorphous metals, the high sensitivity of the modulated film method has permitted both the measurement of $\tilde{D}$ in the absence of crystallization and the detection of structural relaxation effects on $\tilde{D}$. Gradient energy effects have been observed. So far, the linear theory has been applied to systems with a large composition variation; in future work the analysis could be improved by using the nonlinear theory or by reducing the modulation amplitude.

In semiconductors, the measurement of low interdiffusivities has again proved useful. The composition dependence of $\tilde{D}$ is large and has been taken into account in the analysis, but gradient energy effects have been ignored. These effects appear to be present and should be included in future analyses.

Artificial compositionally modulated materials are unstable. The layered structure may either disappear gradually (in homogenizing systems) or coarsen (in phase-separating systems). In both cases, a low bulk interdiffusivity is desirable for stability. The effective interdiffusivity is unaffected by extended lattice defects, except moving grain boundaries. Coarsening kinetics are enhanced when there are faults in the composition modulation.

ACKNOWLEDGMENTS

We thank R. C. Cammarata for useful discussions. The authors' research in this area has been supported by the Office of Naval Research, contract No. N00014-77-C-0002, and by the National Science Foundation, contract No. DMR80-20247.

REFERENCES

Askill, J. (1970). "Tracer Diffusion Data for Metals, Alloys and Simple Oxides." Plenum, New York.

Bardeen, J. (1954). *Phys. Rev.* **94,** 554.

Becker, R. (1938). *Ann. Physik* **32,** 128.

Birac, C., and Lesueur, D. (1976). *Phys. Stat. Solidi. A* **36,** 247.

Bloch, F. (1932). *Z. Phys.* **74,** 295.

Born, M., and Huang, K. (1954). "Dynamical Theory of Crystal Lattices." Oxford Univ. Press, Oxford.

Brown, A. M., and Ashby, M. F. (1980). *Acta Metall.* **28,** 1085.

Cahn, J. W. (1961). *Acta Metall.* **9,** 795.

Cahn, J. W. (1962). *Acta Metall.* **10,** 179.

Cahn, J. W. (1966). *Acta Metall.* **14,** 1685.

Cahn, J. W. (1968). *Trans. Metall. Soc. AIME* **242,** 166.

Cahn, J. W., and Hilliard, J. E. (1958). *J. Chem. Phys.* **28,** 258.

Cahn, R. W., Evetts, J. E., Patterson, J., Somekh, R. E., and Kenway-Jackson, C. (1980). *J. Mater. Sci.* **15,** 702.

Cammarata, R. C., and Greer, A. L. (1984). *J. Non-Cryst. Solids* **61** and **62,** 889.

Cantor, B., and Cahn, R. W. (1983). "Amorphous Metallic Alloys" (F. E. Luborsky, ed.), p. 487. Butterworths, London.

Chang, L. L., and Koma, A. (1976). *Appl. Phys. Lett.* **29,** 138.

Chen, H. S., Kimerling, L. C., Poate, J. M., and Brown, W. L. (1978). *Appl. Phys. Lett.* **32,** 461.

Christian, J. W. (1975). "The Theory of Transformations in Metals and Alloys," 2nd ed., Chapter 6. Pergamon, Oxford.

Cook, H. E. (1966). Ph.D. thesis. Northwestern Univ., Evanston, Illinois.

Cook, H. E., and deFontaine, D. (1969). *Acta Metall.* **17,** 915.

Cook, H. E., and deFontaine, D. (1971). *Acta Metall.* **19,** 607.

Cook, H. E., and Hilliard, J. E. (1969). *J. Appl. Phys.* **40,** 2191.

Cook, H. E., deFontaine, D., and Hilliard, J. E. (1969). *Acta Metall.* **17,** 765.
Darken, L. S. (1948). *Trans. AIME* **175,** 184.
deFontaine, D. (1966). *In* "Local Atomic Arrangements Studied by X-Ray Diffraction" (J. B. Cohen and J. E. Hilliard, eds.), pp. 51–94. Gordon and Breach, New York.
deFontaine, D. (1967). Ph.D. thesis. Northwestern Univ., Evanston, Illinois.
Dinklage, J., and Frerichs, R. (1963). *J. Appl. Phys.* **34,** 2633.
DuMond, J., and Youtz, J. P. (1940). *J. Appl. Phys.* **11,** 357.
Edelin, G., and Tete, C. (1981). *Scr. Metall.* **15,** 739.
Eisen, F. H., and Birchenall, C. E. (1957). *Acta Metall.* **5,** 265.
Eltoukhy, A. H., and Greene, J. E. (1978). *Appl. Phys. Lett.* **33,** 343.
Eltoukhy, A. H., and Greene, J. E. (1979). *J. Appl. Phys.* **50,** 505.
Eltoukhy, A. H., and Greene, J. E. (1980). *J. Appl. Phys.* **51,** 4444.
Eltoukhy, A. H., Zilko, J. L., Wickersham, C. E., and Greene, J. E. (1977). *Appl. Phys. Lett.* **31,** 156.
Fleming, R. M., McWhan, D. B., Gossard, A. C., Wiegmann, W., and Logan, R. A. (1980). *J. Appl. Phys.* **51,** 357.
Ginsburg, V. L., and Landau, L. D. (1950). *Zh. Eksp. Teor. Fiz.* **20,** 1064.
Golding, B., Moss, S. C., and Averbach, B. L. (1967). *Phys. Rev.* **158,** 637.
Goldstein, B. (1961). *Phys. Rev.* **121,** 1305.
Graham, L. D., and Kraft, R. W. (1966). *Trans. Met. Soc. AIME* **236,** 94.
Greer, A. L. (1983). *In* "Modulated Structure Materials" (T. Tsakalakos, ed.). NATO ASI Series, Applied Sciences, Martinus Nijhoff, The Hague. In press.
Greer, A. L., Lin, C.-J., and Spaepen, F. (1982). *Proc. Int. Conf. Rapidly Quenched Metals, 4th* (T. Masumoto and K. Suzuki, eds.), pp. 567–572. Japan Institute of Metals, Sendai, Japan.
Guggenheim, E. A. (1952). "Mixtures." Clarendon Press, Oxford.
Guinier, A. (1963a). "X-ray diffraction," pp. 169, 209–212. Freeman, San Francisco, California.
Guinier, A. (1963b). "X-Ray Diffraction," pp. 279–281. Freeman, San Francisco, California.
Gupta, D., and Asai, K. W. (1974). *Thin Solid Films* **22,** 121.
Gupta, D., Tu, K. N., and Asai, K. W. (1975). *Phys. Rev. Lett.* **35,** 796.
Gupta, D., Tu, K. N., and Asai, K. W. (1982). *Thin Solid Films* **90,** 131.
Henein, G. E. (1979). Ph.D. thesis. Northwestern Univ., Evanston, Illinois.
Hillert, M. (1956). Sc.D. thesis, Massachusetts Institute of Technology, Cambridge, Massachusetts.
Hillert, M. (1961). *Acta Metall.* **9,** 525.
Hilliard, J. E. (1957). *Acta Metall.* **5,** 29.
Hilliard, J. E. (1970). *Phase Transform. Pap. Semin. Am. Soc. Met.* p. 497. ASM, Metals Park, Ohio.
Huston, E. L., Cahn, J. W., and Hilliard, J. E. (1966). *Acta Metall.* **14,** 1053.
Johnson, W. A. (1942). *Trans. AIME* **147,** 331.
Kolomets, I. D., and Smirnov, A. A. (1962). *Fiz. Met. Metalloved.* **14,** 3.
Langer, J. S. (1971). *Ann. Phys.* **65,** 53.
Luborsky, F. E., and Bacon, F. (1982). *Proc. Int. Conf. Rapidly Quenched Metals, 4th* (T. Masumoto and K. Suzuki, eds.), p. 561. Japan Institute of Metals, Sendai, Japan.
Ludeke, R., Esaki, L., and Chang, L. L. (1974). *Appl. Phys. Lett.* **24,** 417.
Mathiot, D., and Edelin, G. (1980). *Philos. Mag.* **41,** 447.
Mitsui, T., and Furuichi, J. (1953). *Phys. Rev.* **90,** 193.
Monma, K., Sutor, H., and Oikawa, H. (1964). *Nippon Kinzoku Gakkaishi* **28,** 188.
Moss, S. C., and Clapp, P. C. (1968). *Phys. Rev.* **171,** 764.
Nanao, S., Kuribayaski, K., Tanigawa, S., and Doyama, M. (1972). *Phys. Lett.* **38** *A,* 489.
Norman, N., and Warren, B. E. (1951). *J. Appl. Phys.* **22,** 483.
Palfrey, H. D., Brown, M., and Willoughby, A. F. W. (1981). *J. Electrochem. Soc.* **128,** 2224.
Paulson, W. M. (1972). Ph.D. thesis. Northwestern Univ., Evanston, Illinois.
Paulson, W. M., and Hilliard, J. E. (1977). *J. Appl. Phys.* **48,** 2117.

Petroff, P. M. (1977). *J. Vac. Sci. Technol.* **14,** 973.
Philofsky, E. M. (1968). Ph.D. thesis. Northwestern Univ., Evanston, Illinois.
Philofsky, E. M., and Hilliard, J. E. (1969). *J. Appl. Phys.* **40,** 2198.
Preston, G. D. (1938). *Proc. R. Soc. London, Ser. A,* **167,** 526.
Purdes, A. (1976). Ph.D. thesis. Northwestern Univ., Evanston, Illinois.
Röll, K., and Reill, W. (1982). *Thin Solid Films* **89,** 221.
Rosenblum, M. P. (1979). Ph.D. thesis. Harvard Univ., Cambridge, Massachusetts.
Rosenblum, M. P., Spaepen, F., and Turnbull, D. (1980). *Appl. Phys. Lett.* **37,** 184.
Siebhoff, H. (1979). *Phys. Lett.* **71** *A,* 265.
Spaepen, F. (1981). *In* "Physics of Defects" (R. Balian *et al.,* eds.), p. 133. Les Houches Lectures XXXV, North-Holland Publ., Amsterdam.
Spreng, D. T., Hilliard, J. E., and Kaufman, J. W. (1972). *J. Appl. Phys.* **43,** 2040.
Taub, A. I., and Spaepen, F. (1980). *Acta Metall.* **28,** 1781.
Thomas, D. E., and Birchenall, C. E. (1952). *Trans. AIME* **194,** 867.
Tsakalakos, T. (1977). Ph.D. thesis. Northwestern Univ., Evanston, Illinois.
Tsakalakos, T. (1981a). *Thin Sol. Films* **86,** 79.
Tsakalakos, T. (1981b). *Scr. Metall.* **15,** 225.
van der Waals, J. D. (1890). *Z. Phys. Chem.* **5,** 133.
Vrijen, J., and Radelaar, S. (1978). *Phys. Rev. B.* **17,** 409.
Woodilla, J. E., and Averbach, B. L. (1968). *Acta Metall.* **16,** 225.
Yamauchi, H. (1971). Master's thesis. Northwestern Univ., Evanston, Illinois.
Yamauchi, H., and Hilliard, J. E. (1972). *Scr. Metall.* **6,** 909.
Yang, W.M.-C. (1971). Ph.D. thesis. Northwestern Univ., Evanston, Illinois.

INDEX

N

R

T

V

W

X

Z

MATERIALS SCIENCE AND TECHNOLOGY

EDITORS

A. S. NOWICK
Henry Krumb School of Mines
Columbia University
New York, New York

G. G. LIBOWITZ
Solid State Chemistry Department
Materials Research Center
Allied Corporation
Morristown, New Jersey

A. S. Nowick and B. S. Berry, ANELASTIC RELAXATION IN CRYSTALLINE SOLIDS, 1972
E. A. Nesbitt and J. H. Wernick, RARE EARTH PERMANENT MAGNETS, 1973
W. E. Wallace, RARE EARTH INTERMETALLICS, 1973
J. C. Phillips, BONDS AND BANDS IN SEMICONDUCTORS, 1973
J. H. Richardson and R. V. Peterson (editors), SYSTEMATIC MATERIALS ANALYSIS, VOLUMES I, II, AND III, 1974; IV, 1978
A. J. Freeman and J. B. Darby, Jr. (editors), THE ACTINIDES: ELECTRONIC STRUCTURE AND RELATED PROPERTIES, VOLUMES I AND II, 1974
A. S. Nowick and J. J. Burton (editors), DIFFUSION IN SOLIDS: RECENT DEVELOPMENTS, 1975
J. W. Matthews (editor), EPITAXIAL GROWTH, PARTS A AND B, 1975
J. M. Blakely (editor), SURFACE PHYSICS OF MATERIALS, VOLUMES I AND II, 1975
G. A. Chadwick and D. A. Smith (editors), GRAIN BOUNDARY STRUCTURE AND PROPERTIES, 1975
John W. Hastie, HIGH TEMPERATURE VAPORS: SCIENCE AND TECHNOLOGY, 1975
John K. Tien and George S. Ansell (editors), ALLOY AND MICROSTRUCTURAL DESIGN, 1976
M. T. Sprackling, THE PLASTIC DEFORMATION OF SIMPLE IONIC CRYSTALS, 1976
James J. Burton and Robert L. Garten (editors), ADVANCED MATERIALS IN CATALYSIS, 1977
Gerald Burns, INTRODUCTION TO GROUP THEORY WITH APPLICATIONS, 1977
L. H. Schwartz and J. B. Cohen, DIFFRACTION FROM MATERIALS, 1977
Zenji Nishiyama, MARTENSITIC TRANSFORMATION, 1978
Paul Hagenmuller and W. van Gool (editors), SOLID ELECTROLYTES: GENERAL PRINCIPLES, CHARACTERIZATION, MATERIALS, APPLICATIONS, 1978
G. G. Libowitz and M. S. Whittingham, MATERIALS SCIENCE IN ENERGY TECHNOLOGY, 1978
Otto Buck, John K. Tien, and Harris L. Marcus (editors), ELECTRON AND POSITRON SPECTROSCOPIES IN MATERIALS SCIENCE AND ENGINEERING, 1979
Lawrence L. Kazmerski (editor), POLYCRYSTALLINE AND AMORPHOUS THIN FILMS AND DEVICES, 1980
Manfred von Heimendahl, ELECTRON MICROSCOPY OF MATERIALS: AN INTRODUCTION, 1980
O. Toft Sørensen (editor), NONSTOICHIOMETRIC OXIDES, 1981
M. Stanley Whittingham and Allan J. Jacobson (editors), INTERCALATION CHEMISTRY, 1982
A. Ciferri, W. R. Krigbaum, and Robert B. Meyer (editors), POLYMER LIQUID CRYSTALS, 1982
Graeme E. Murch and Arthur S. Nowick, (editors), DIFFUSION IN CRYSTALLINE SOLIDS, 1984
Leroy L. Chang and B. C. Giessen (editors), SYNTHETIC MODULATED STRUCTURES, 1985

In preparation:
Paul Hagenmuller (editor), INORGANIC SOLID FLUORIDES, 1985